CHROMATOGRAPHIC AND ELECTROPHORETIC TECHNIQUES

Volume II

ZONE ELECTROPHORESIS

CHROMATOGRAPHIC AND ELECTROPHORETIC TECHNIQUES

Edited by

IVOR SMITH

Ph.D., D.Sc., F.R.C.Path., F.R.I.C., M.I.Biol.

Fellow American Association for the Advancement of Science
Reader in Biochemical Education,
Courtauld Institute, Middlesex Hospital, London

VOLUME II

ZONE ELECTROPHORESIS

FOURTH EDITION

LONDON

WILLIAM HEINEMANN MEDICAL BOOKS LTD

First Published March, 1960
Reprinted May, 1961
Reprinted January, 1962
Reprinted April, 1963
Reprinted July, 1965
Second Edition (*reset*) 1967
Third Edition (*see note below*)
Fourth Edition 1976

No Third Edition of this volume was published. With the present substantial revision of the two volumes, the opportunity has been taken to bring them into line by denoting both volumes as Fourth Edition.

ISBN 0 433 30505 5

Text set in 10 pt. Monotype Modern, printed by letterpress, and bound in Great Britain at The Pitman Press, Bath

CONTENTS

THE EDITOR

IVOR SMITH, D.Sc., Ph.D., F.R.I.C., M.I.Biol., Fellow AAAS
Reader in Biochemical Education, Courtauld Institute, Middlesex Hospital Medical School, London, W1P 5PR

THE AUTHORS

D. BEALE, B.Sc., Ph.D.
A.R.C. Institute of Animal Physiology, Babraham, Nr. Cambridge

A. BROWNSTONE
National Institute for Medical Research, The Ridgeway, Mill Hill, London, NW7 1AA

C. F. DORÉ, A.I.S.T.
National Institute for Medical Research, The Ridgeway, Mill Hill, London, NW7 1AA

A. FEINSTEIN, B.Sc., Ph.D.
Institute of Animal Physiology, Babraham, Nr. Cambridge

W. B. GRATZER, M.A., B.A., Ph.D.
M.R.C. Biophysics Research Unit, King's College, London, WC2

J. KOHN, Med. Dip. (LWOW), F.C.Path., D.C.P.
Consultant Pathologist, Queen Mary's Hospital, Roehampton

D. LEABACK, B.Sc., Ph.D.
Biochemistry Dept., The Institute of Orthopaedics, Brockley Hill, Stanmore, Middlesex

J. W. PAYNE, B.Sc., Ph.D.
Botany Dept. University of Durham, Science Laboratories, South Road, Durham, DH1 3LE

E. G. RICHARDS, B.Sc., Ph.D.
M.R.C. Biophysics Research Unit, King's College, London, WC2

W. J. RITSCHARD, Ph.D.
Chemische Forschungsabteilung, Hoffmann-La Roche & Co., CH4002 Basle, Switzerland

C. R. SHAW, M.D.

Chief, Section of Medical Genetics, The University of Texas System Cancer Center, Texas Medical Center, Houston, Texas, 77025 and Associate Professor of Biology, Department of Biology, University of Texas, M.D. Anderson Hospital, Houston, Texas

M. SICILIANO, B.S., M.S., Ph.D.

Assistant Professor of Biology, Dept. of Biology, M.D. Anderson Hospital, Houston, Texas, 77025

MARGARET J. SMITH, M.R.C.S., L.C.R.P.

(*c/o* DR. IVOR SMITH)

A. L. TARNOKY, Ph.D., F.R.I.C., F.R.C.Path.

Biochemist, Clinical Chemistry Dept., Royal Berkshire Hospital, Reading, RG1 5AN

J. M. B. VERSEY, B.S., Ph.D.

Dept. of Chemical Pathology, Westminster Medical School, London, SW1

C. W. WRIGLEY, B.Sc., Ph.D.

Wheat Research Unit, c/o Bread Research Institute, Private Bag P.O., North Ryde, NSW 2113

JACQUELINE B. WEISS, M.Sc., D.C.C.

Senior Lecturer in Medical Biochemistry, Dept. of Rheumatology, Medical School, University of Manchester M13 9PL

PREFACE TO THE FOURTH EDITION

THE newcomer to the subject and literature of electrophoresis is confronted with a great array of commercially available apparatus, often of very similar quality, as well as an almost infinite variety of designs of home-made equipment. However, in practically every case it is very difficult, if not impossible, to find an adequate discussion of the correct techniques or those most suitable for a particular method.

Briefly, therefore, the purpose of this work is threefold: to bring together in as concise a form as possible tried and tested methods of zone electrophoresis; to discuss all those methods which can be applied routinely; and to provide all the essential information to enable the reader to apply these techniques immediately and without recourse to the literature for further details of technique. To this end discussion on the history of electrophoresis and various theories which, though valuable in themselves, provide no further technical information has been omitted as this has been covered elsewhere many times. Similarly, discussion on moving boundary methods is left out as it does not fall within the definition of routine zone methods.

All the work discussed herein has been carried out on commercially available apparatus as this appears to offer many advantages to the majority of laboratories who prefer to buy apparatus rather than to construct it themselves. Nevertheless, details are given as to overall dimensions of apparatus as well as to electrical circuits and requirements so that local construction is possible if desired.

In order that each chapter can be read as a separate unit it has also been necessary to duplicate some of the discussion on the various factors involved in the production of satisfactory results. Although this duplication has been kept to a minimum, it is emphasized that the reader can only gain by a reading of those chapters which may not be of immediate interest to his particular laboratory as the general discussions in each chapter are relevant to the overall technique of zone electrophoresis.

CHAPTER 1

GENERAL PRINCIPLES OF ZONE ELECTROPHORESIS

Ivor Smith

THE principle of electrophoresis is very simple, namely that a charged ion or group will migrate towards one of the electrodes when placed in an electric field. Hence if two electrodes are inserted, one into each end of a U tube containing a dilute solution of proteins, the proteins will each migrate in one direction only. As the proteins are originally present throughout the whole solution there is little chance of achieving any separation. Zone electrophoresis overcomes this as follows. The mixture of substances to be separated is placed as a narrow zone or band at a suitable distance from each electrode such that, as migration occurs, the different components, which move at different rates, slowly draw away from each other to produce a separation in the direction of migration. A further problem is that of fixing the substances permanently at the positions to which they have migrated after the run is terminated because, being in solution, the possibility of diffusion is present. Thus electrophoresis is rarely carried out in a free solution but, instead, the solution is held in a stabilising medium such as a sheet of paper or a gel. In the former case, the paper is rapidly dried in an oven at the termination of the electrophoresis run. In the latter case the gel is either placed in a fixative which precipitates the substances being examined with the result that the separated substances remain in their correct positions of migration or into some enzyme stain which results in the precipitation of an insoluble reactions product. Subsequently the substances are located as a number of separate, discrete zones or bands. All the techniques discussed in this book therefore are examples of zone electrophoresis in different stabilised media.

The simplest and earliest form of zone electrophoresis is that carried out on paper as the stabilising medium and in Fig. 1.1 is shown an early version of apparatus used. A strip of paper is moistened with buffer, blotted to remove excess liquid and placed horizontally on a glass sheet such that each end dips into a separate beaker containing buffer. The sample is applied at a suitable position on the paper which is then covered with a further plate of glass to prevent evaporation. Current is supplied via a 120 volt battery and carbon electrodes dipping into each beaker. After a suitable time the strip is rapidly removed and dried in an oven at 110°C and then treated with a reagent to locate the separated substances. Although the substances being investigated move towards one electrode only, it must be remembered that the positive and negative buffer ions are also migrating towards one or other electrode as well. In the above case the buffer on the paper and in the beakers is the same. This is referred to as a continuous buffer system and is the usual procedure in paper electrophoresis, because it was originally thought necessary to maintain conditions as constant as possible for optimum results. Subsequently it was found that advantage did sometime accrue from the use of a more dilute buffer in the beaker and this is used to effect in the cellulose acetate (CA) electrophoresis technique where there is a tendency for sharper bands to result. More recently it was found that a discontinuous buffer system in which the beaker buffer is chemically different and has a different *p*H to that of the stabilising medium results in a great sharpening of the faster moving zones and this is now the standard procedure in starch gel electrophoresis. The mechanism of this sharpening effect is not too

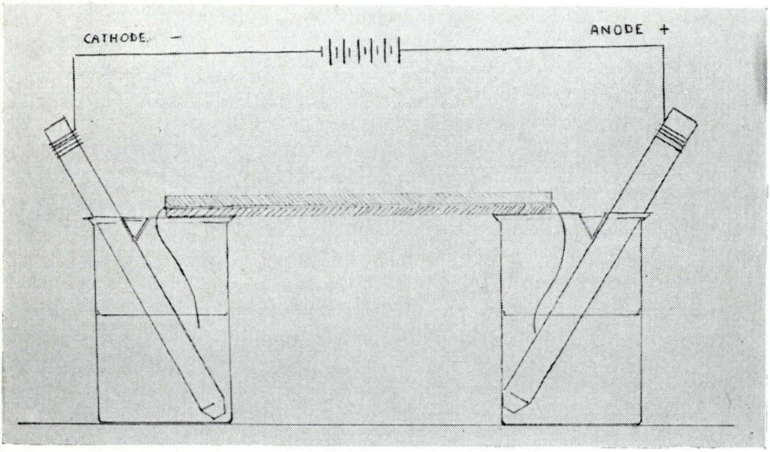

Fig. 1.1. The simplest form of electrophoretic apparatus. Two beakers, containing buffer, support a glass sandwich between the plates of which is laid a length of Whatman 3 MM paper, the ends dipping into the buffer at either end. Carbon pencils form the electrodes and are connected to a 120 V H.T. battery. The sample is applied as a band across the strip.

well understood. Thus it was postulated by the inventors of the acrylamide gel technique, the most recent gel technique in current use, that discontinuous systems were essential to the procedure but it has since been shown that practically identical results can be obtained with a continuous buffer system.

During the passage of current, electrolysis occurs at the electrodes. In order to minimise any effects of this electrolysis the electrode is isolated in a separate compartment from that into which the paper dips and these two compartments are connected by means of a paper wick, glass fibre wick or gel bridge, the former being the most convenient. Hence most apparatus has four electrode chambers, a pair of cathode and a pair of anode chambers, each pair being connected by means of the wick just described.

In the case of paper electrophoresis the ends of the strip dip into the buffer and make direct connection. With CA, agar and starch gel, however, the medium is connected by means of a further paper wick overlapping the medium by about 1 cm. and whose other end dips into the buffer. In all such cases the medium must be perfectly horizontal and above the buffer level to prevent both syphoning and waterlogging.

However, it has been found possible to dispense with the double buffer compartments and the paper wicks in the acrylamide gel techniques without adverse effects. pH changes are overcome by using relatively large buffer volumes, or, by continuously mixing and recirculating the buffer solutions.

Factors Affecting Speed of Migration and Separation

The Stabilising or Supporting Medium. Many supporting media are in current use but they fall into two main classes:

1a. paper, cellulose acetate, glass fibre paper, thin layer materials, agar gel, single cellulose fibres. Analytical and preparative.

1b. pevikon, starch and gypsum blocks, sponge rubber. Preparative only.

2. starch gel, acrylamide gel. Analytical and preparative.

Before discussing the effect of these media on separations, it is worth considering briefly the uses to which the different media are put. Paper, particularly at high voltage but also at low voltage, is used for the separation of small molecules such as the aminoacids, for medium sized molecules such as peptides and nucleotides obtained by enzymic hydrolysis of proteins and nucleic acids, and with low voltage only for large molecules such as proteins, enzymes and nucleic acids. Thin layer materials are used mainly for small and medium sized molecules. Gels were used exclusively for large molecules but by decreasing the pore size of the gel it can also be used for peptides and nucleotides. Blocks are used for preparative purposes.

Supporting media in Group 1 above are relatively inert and have little effect on the compounds being separated (paper has a slight denaturing effect on proteins which results in minor tailing of the bands back towards the origin or point of sample application). Hence separation is dependent on the net effect of charge and mass; thus a large molecule with many charges may move to the same position as a smaller molecule with an equal charge to mass ratio. It is generally thought that separation occurs by migration in and through the liquid held stationary within the stabilising medium.

Supporting media in Group 2 exert a positive influence on the compounds. The gel can be considered as a porous medium in which the pore size is of the same order as the protein molecule with the result that a molecular sieving effect is impressed on the normal protein separation and molecules of similar charge to mass ratio but with different molecule size can be separated. This is well illustrated with serum proteins. Two identical samples are run on paper and one is subsequently stained to locate the proteins when one albumin and four globulin bands are observed. Each globulin band is cut from the unstained wet paper, placed in starch gel, run at the same *p*H and stained as usual when each will be seen to have given rise to a number of bands which must differ only in mass as the electrophoresis conditions were the same in both runs. With starch gel, a standard procedure is commonly used which results in a standard pore size although it is known that different buffer salts do have minor effects on the pore size. However, acrylamide gels can easily be varied in pore size such that molecules which penetrate a gel with larger pores may be excluded from a tighter gel with smaller pores and this can be particularly useful in the study of a single smaller protein both analytically and preparatively.

Supporting media in Groups 1a and 2 were all originally designed for analytical purposes which require that an optimum volume, namely one which results in the maximum number of sharply defined bands, be applied. Subsequently it was found that larger volumes could be applied and although this resulted in less well-defined separations, this was more than compensated for by the fact that preparative amounts could be recovered. Simultaneously other preparative methods were sought whereby millilitre rather than microlitre quantities could be separated and these are now referred to as the block techniques. Most recently the gel techniques have been found to operate in the preparative range intermediate between the analytical and the block ranges. Blocks, being inert also, yield separations similar to the Group 1a materials so that, frequently, a preliminary separation is carried out on a block and the recovered fraction is then further fractioned on a preparative acrylamide gel.

pH. A substance can only migrate if it carries a charge, i.e. it is ionised. A weak acid will remain at the origin at *p*H's equal to or below its isoelectric point and move to the anode at *p*H's above this point; its rate of migration is proportional to its degree of ionisation and hence the higher the *p*H the faster it travels. The converse is true for weak bases. Most interest is centered around ampholytes such as aminoacids and proteins which migrate to the anode at *p*H's above their iso-electric points and in the reverse direction at *p*H's below this. Indeed one method of determining iso-electric points is based on this effect. The best *p*H for any given mixture is a matter for empirical investigation; one *p*H may be best if it is desired to obtain maximum separation of all the components present whereas another *p*H may be much more suitable if it is desired to isolate a single component.

The Buffer, Chemical Composition, Concentration and Ionic Strength. The value of the buffer was originally thought to be that of controlling the *p*H of the medium at a constant value but this is a gross over-simplification. The buffer does indeed maintain a constant *p*H when used in a continuous system and also ensures that each component will maintain a constant charge throughout the experiment because the ionisation of each compound is stabilised and this is particularly important for the ampholytes. Later it was realised that buffer ions, particularly phosphates, often bind to proteins and this may result in different absolute and relative mobilities of a mixture of compounds when investigated with different buffers of the same *p*H and ionic strength. Likewise borate buffer ions bind to glyco-compounds.

The more concentrated the buffer the slower will the other compounds move because, as current is carried by the ions present, the greater the quantity of buffer ions relative to other ions, the greater the proportion of current can they carry. Furthermore, the movement of ions surrounded by ions of opposite charge is retarded by the attraction of these ions so that increased buffer concentration has a double effect in reducing migration rates. However, to compensate for this disadvantage, the zones are found to be sharper.

Buffers are available for almost the whole *p*H range and certainly for the range over which proteins remain soluble and apparently undenatured. Thus a protein or other ampholyte mixture can be separated first with an alkaline buffer in which the components are all present as anions, the partially separated fractions can be recovered and separately rerun in an acid buffer in which they exist as cations; this being similar to ion exchange separations on DEAE cellulose followed by CMC.

It is now necessary to consider the question of ionic strength, u, although it must be said that there is some confusion in the literature as to the definition of the term.

$$u = \tfrac{1}{2} \Sigma\, mc^2$$

where m is defined either as molarity (gram ions/litre) or molality (gram ions/1000 ml. solvent) and c as the charge on the ion. One work defines u on the basis of molality and offers an explanatory calculation based on molarity. In two different works the identical buffer is quoted in one case as 0·05 and in the other as 0·06u. The author has often been tempted to consider what real value the term has in electrophoresis, and no-one has yet had the temerity to calculate ionic strengths for any but the simplest buffers. Two examples of such calculations follow; these being based on molarity: Veronal or barbitone buffer *p*H 8·6: sodium diethylbarbiturate, 10·3 g. (M/20), and barbituric acid, 1·84 g. (M/100), per litre, assuming free acid completely unionised.

$$\mathrm{u} = \frac{1}{2}\left(\sum \frac{1}{20} \times 1^2 + \frac{1}{20} \times 1^2 \right) = 0{\cdot}05$$

Acetate buffer *p*H 4: sodium acetate.$3H_2O$, 13·61 g. (M/10), plus acetic acid 25·59 ml. (0·45M) per litre, assuming acid unionised.

$$\mathrm{u} = \tfrac{1}{2}(\Sigma\, 0{\cdot}1 \times 1^2 + 0{\cdot}1 \times 1^2) = 0{\cdot}1$$

In practice, buffers are made up by weighing solids or measuring volumes of liquid and variations in composition are based on changing these weights or volumes; ionic strengths are calculated subsequently.

It is now worth considering the effect due to the use of different buffer concentrations in the electrode compartment and on the medium. Invariably the electrode buffer is more concentrated than the medium buffer in gel electrophoresis and often of different *p*H. Frequently the buffer used in the electrode reservoir is different from that used to prepare the gel. Briefly the net effect is to produce a voltage discontinuity at the interface of the two buffers and as this interface travels through a protein band, the band compacts to produce a narrower, tighter zone. This effect is discussed in some detail in the chapters on starch and acrylamide electrophoresis.

It has been said that, for electrophoresis to occur, the substances must exist in an ionic form. Hovever, neutral molecules can be separated if they can easily be converted to carry a charge. Thus sugar molecules are neutral but, as they react with borate to form sugar-borate ions, they can be separated in borate buffers.

Voltage, Current and Heat Effects. An electrophoresis medium, such as a strip of paper moistened with buffer, exerts a resistance to current flow which is a function of the medium, the buffer and its concentration. For a given applied voltage, the resistance of the medium will determine the current passed and, consequently, the heat generated during the experiment. Thus applied voltage, V, across a resistance, R ohms, will result in a current flow of I amps, the consumption of W watts, and the generation of heat as calories, C; these terms being related as follows:

$$V = IR \qquad \text{Ohms Law}$$
$$W = IV$$
$$C = Wt/4{\cdot}18 \qquad (t = \text{seconds})$$

It is the ions which carry the current and in order to increase the rate of migration it is obvious that the current must be increased. As the resistance of the medium is fixed it follows that the applied voltage must be raised if it is desired to obtain a higher current.

However, as the experiment proceeds heat is generated and therefore the strip heats up. A number of effects follow from this. First, the resistance of the strip falls and, therefore, the rate of migration increases. Second, distillation from the warmer strip on to the colder walls of the apparatus occurs with the result that the resistance is again altered and is usually decreased. It follows that unless either the current or voltage is stabilized, both will alter with time. It is usual to include either a voltage or current stabilizer in the power pack for these reasons.

Constant Voltage. With a constant applied voltage the current will increase during the experiment because the resistance of the strip is continuously dropping. The heating will also increase with consequent continuous distillation off the strip. Therefore the rate of migration will also increase continuously until an equilibrium is established in the apparatus. With voltages of not more than 100 volts or so (2–3 volts/cm. length of strip) this effect can be ignored and at temperatures around 20°C the current will increase only slightly for a 4 cm. wide strip. With greater voltages, distillation is large and continuous and the rate of migration will vary throughout the experiment.

Constant Current. With a constant applied current the voltage will fall continuously because the resistance of the strip is falling. This results in a lowering of distillation and the rate of migration is more nearly constant for the whole of the experiment.

The above discussion has centered around a single strip but it is now necessary to examine the effect of a number of strips run simultaneously. If two similar strips are run in parallel with the same applied constant voltage as before the resistance of the apparatus is now halved and therefore the current is doubled

$$1/R = 1/r_1 + 1/r_2 \ldots + 1/rn; \quad R = r/n \text{ and } I = nV/R$$

where R is the total resistance of the apparatus, r_1 and r_2 are the resistances of each strip (these will be identical) and n is the number of strips in the apparatus. In order to obtain the same degree and length of separation, therefore, either a constant voltage must be applied irrespective of the number of strips used or the current on the meter must be adjusted by multiplying by n.

The resistance of the strip is proportional to its length and in order to quote figures which can be reproduced elsewhere it is essential to quote not just the voltage or current used for a given separation, which is quite useless, but the voltage drop per centimetre length of strip not dipping into the buffer and the current density per centimetre width of strip as well as the temperature.

Constant Current or Constant Voltage. With paper electrophoresis at low voltage it matters little which choice is made as the heat generated is small and this is easily dissipated. However, with all gels and blocks the thickness may vary from 0·3–1 cm. and heat dissipation becomes a problem particularly when one realises that the heat in the middle of

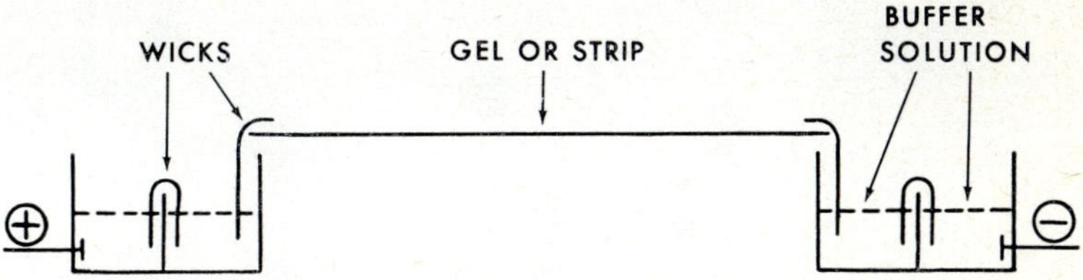

FIG. 1.2. The figure illustrates the fact that current must flow through buffer solutions and various wicks as well as through the electrophoresis medium and that each will, therefore, contribute towards the total resistance of the apparatus. Thus the voltage drop across the medium is only a proportion of that shown on the power pack.

the gel or block must be rapidly removed in order to avoid a temperature gradient within the thickness of the medium. With these latter procedures, constant current tends to be used as this results in less heat.

One final point deserves mention if only because it is so seldom remembered. The resistance of an apparatus, shown diagrammatically in Fig. 1.2, is the sum of the resistance of the individual components. Here the resistances present are due to buffer solution, wicks and supporting medium. The resistance varies with the number of thicknesses of paper wick connecting buffer solutions and the number of thicknesses connecting buffer to medium on both anode and cathode sides. Hence the applied voltage recorded on the power supply is not that across the medium and may, in fact, be up to ten times that which does cross the medium. It is always worthwhile measuring the voltage drop across the medium in quoting experimental details and a correct voltage gradient will refer to this figure, rather than the figure seen on the power pack, divided by the medium length.

Power Supplies. There is no power pack available suitable for all forms of electrophoresis. Indeed this is not even desirable as such an apparatus would be far too costly for most purposes which require a less sophisticated piece of equipment. The important features a satisfactory apparatus should possess are both constant voltage and

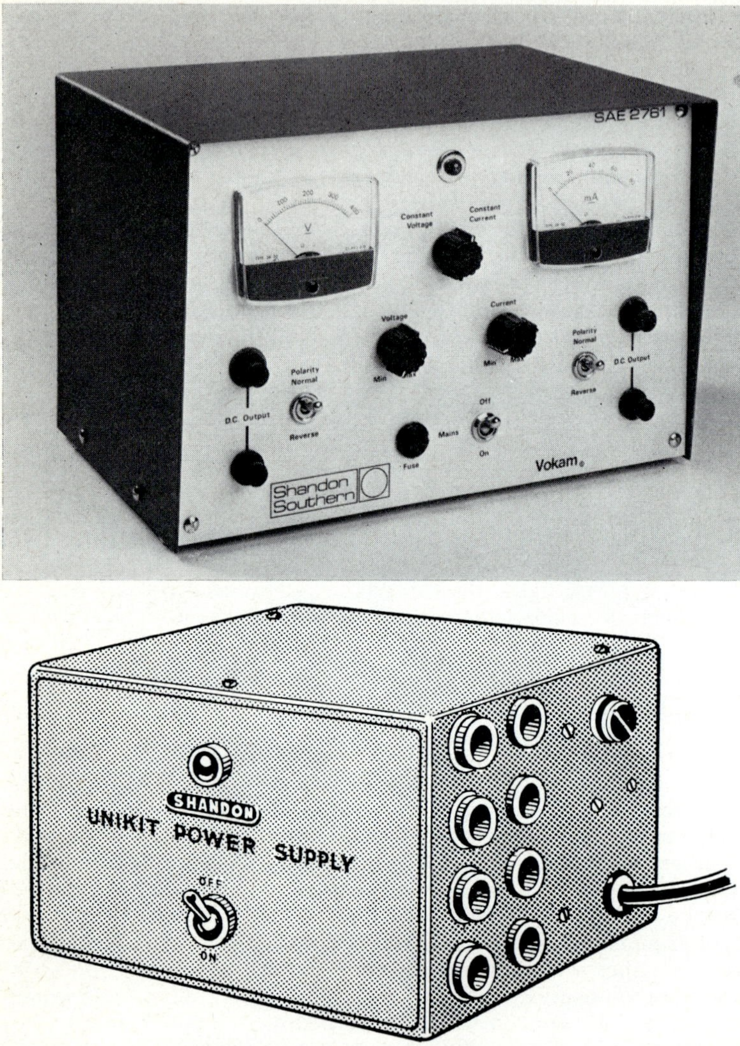

FIG. 1.3. (*a*) The Vokam power pack. (*b*) The Unikit power pack with four independent outlets.

amperage controls (both cannot operate simultaneously) which means that the apparatus should include both a voltmeter and an ammeter covering the desired range.

It has been stressed that the important figures are the voltage and amperage across the strip or gel. However, once the experiment has been standardised, the figures shown on the meters can be used to adjust the conditions for subsequent runs.

One power pack satisfying the above conditions and suitable for electrophoresis on paper, acetate, agar, starch gel and discs is the Vokam (*vo*lts, *k*onstant, *am*ps) which has a range of 50–300 volts at constant voltage with a maximum of 80 m.amps and 3–80 m.amps. at constant current with a maximum of 300 volts. When two power packs are available they can be run in series to produce a much higher voltage. By leading the negative of pack A into the positive of a second B and connecting the positive of A and the negative of B to the electrophoresis tank, approximately double the maximum voltage can be obtained although the exact figure should be monitored with an Avometer or VOM (volt-ohm-milliammeter). The two outlets of the Vokam are wired individually in parallel. Hence, on constant voltage, no adjustment is necessary to run either one or two identical tanks but, on constant current, twice the current must be used when two tanks are in use. Moreover if a single electrophoresis tank, at a given voltage, results in only a small amperage then two or more tanks can be run in series from the same pair of output terminals.

The Unikit power pack is also worthy of mention here. It was originally designed as a simple and cheap smoothed but unregulated apparatus in which all four outlets are wired independently so that four class experiments could be run simultaneously. Each outlet shows approximately 350 volts on open circuit with a short circuit current of 10 mA. through a 35,000 Ω protective resistor. As no voltage or current regulation is possible, the time required per experiment will be dependant on the buffer concentration etc. It has been used successfully for such procedures as two way screening, paper electrophoresis of carbohydrate-protein compounds, cellulose acetate electrophoresis, etc.

Care should always be taken to ensure that colour convention of the lead wires connecting the unit to the mains is understood as serious consequences can occur if the connections are wrongly made. Attention is drawn to this as colour conventions for use with electric wiring vary from country to country. Similarly power packs are constructed to take input voltages of 100–250 V. and this should be checked before connecting a new apparatus to the mains.

Migration Velocity. The migration velocity, M, of a substance is defined as the distance travelled from the origin per second at a field strength of 1 volt/cm. (constant voltage), and is a very small term.

$$M = \text{cm}^2/\text{volt-sec.}$$

To take one example, in paper electrophoresis albumen migrates roughly 12 cm. in 16 hours when a potential of 80 volts is applied across a strip 32 cm. long. Here

$$M = \frac{12}{16 \times 3600} \times \frac{32}{80} = 8 \times 10^{-5}$$

Electro-osmosis. When two different chemical substances such as paper and water come into contact they may become relatively charged and here the water becomes the positive. In electrophoresis the paper is fixed but the water is free to move and so it streams toward the cathode carrying the buffer salts and components of the mixture with it. In the case of proteins this stream is in the opposite direction to that of migration and, for this reason, slow moving gamma globulins are carried back beyond the origin and appear to have travelled in the reverse direction to that expected. Neutral molecules are also carried by the electro-osmotic flow and are incorporated into the mixture solution (or in a parallel lane) if it is desired to determine the true origin or the backward migratory effect. Usually glucose or urea is used for experiments with small molecules and PVP, blue Dextran, etc. for experiments with large molecules.

Diffusion Effects. In the absence of the applied voltage, any substance may diffuse to a more blurred or overlapping position with a nearby compound. This is a relatively minor problem but should not be overlooked.

Two Way Procedures. Electrophoresis on any given medium is normally a one way procedure. However, two way runs are in very general use and the following examples will illustrate the many applications.

1. Paper electrophoresis followed by paper chromatography for the separation of aminoacids and related compounds and for peptide fingerprinting (chapter 3).
2. TLE followed by TLC—see chapter 4. Applications similar to 1 above.
3. PE followed by inserting paper into starch gel for second direction. For proteins including transferrins, ceruloplasmins, comparative species studies, etc. (chapter 7).
4. PE followed by punching out circles and applying to disc gels. For proteins (chapter 9).
5. Acetate electrophoresis followed by starch or disc gels. For proteins as in 3 and 4 above.

In all the above cases, use is made of what is, in effect, two different techniques. In 1 and 2 above, the techniques are based on different principles whereas in 3, 4 and 5 the principle is the same but the extra sieving effect of the gel is used to advantage either with the same or, more usually, a different buffer and *p*H.

Collection of the Sample

Serum. Whole blood should be allowed to clot and retract for a sufficient time as too rapid use of the sample can lead to false results. At least a few hours is required or overnight in the refrigerator between —2 and +4°C. Freezing and thawing alters the lipoprotein pattern and, if freezing is necessary for storage, the sample should be split into many small fractions so that each fraction is thawed once only and then rejected.

Plasma. Blood should be collected and cooled rapidly on ice. Subsequent centrifugation will provide the specimen.

CSF and Urinary Proteins. Both of these fluids must be concentrated first in order to obtain locatable amounts of protein after electrophoresis on all media with the occasional exception of acrylamide.

Concentration of Protein Solutions

The concentration of protein solutions, such as CSF, is often an essential preliminary to electrophoresis. However, just as frequently, fractions from columns, eluates from gels, blocks or preparative disc

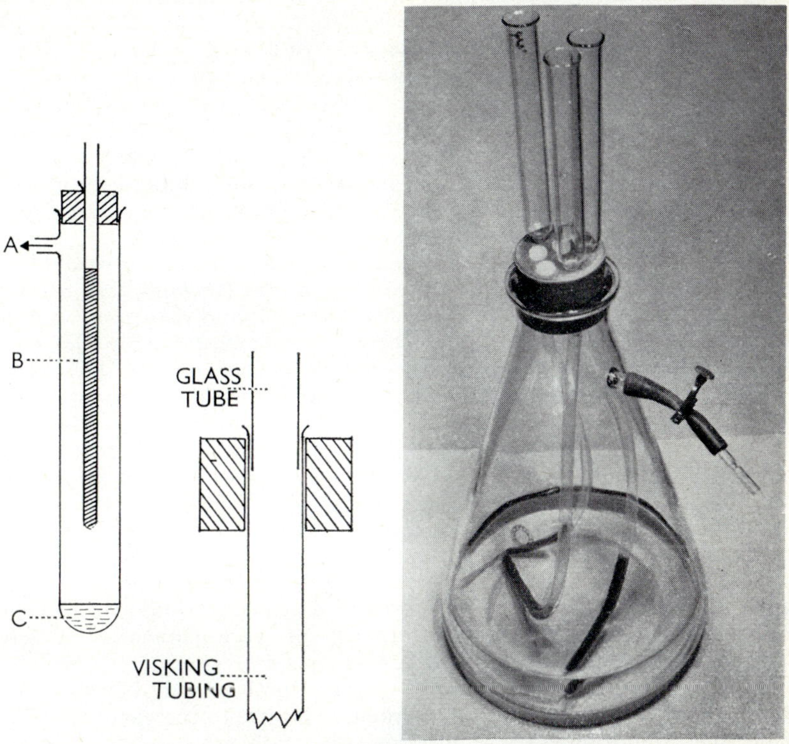

(*a*) (*b*) (*c*)

Fig. 1.4. Apparatus for ultrafiltration: (*a*) the boiling tube design may be used for small volumes or adapted with long glass columns to take 5 ft. lengths of tubing; (*b*) enlargement to show method of holding tubing secure; (*c*) alternative large-scale version of apparatus with part of tubing immersed in water or the ultrafiltrate to prevent final small volume going to dryness.

electrophoresis must be concentrated prior to further investigations. A number of good procedures are currently available, the method of choice depending on the individual specimen and the subsequent procedures.

Ultrafiltration. Ultrafiltration is usual carried out through Visking seamless tubing as used for dialysis. Such tubing contains small amounts of many types of impurity and should always be first cleaned by boiling in 10% sodium carbonate solution for 20 minutes followed by a number of rinses with distilled water. The cleaned material should be kept wet until required and not allowed to age over more than a few weeks. The procedure to be described may be used over a very wide range of volumes. Thus a 6″ length of 8/32″ tubing may be set up in a large boiling tube or a 60″ length may be set up in long glass tubes especially made for this purpose. We have used this to concentrate fractions of 5–10 ml. down to a few tenths of a ml. and also to concentrate litres of original solution down to 10–20 ml. A 10 cm. length of 8/32″ tubing has an approximate capacity of 2·5 ml.

A suitable length of clean moist tubing is taken and inserted through a hole in a rubber bung or stopper. A glass tube with polished ends is carefully inserted into the open tubing and gently forced into the stopper thereby holding the membrane in position. The opposite end of the tubing is then knotted and pulled tight but taking care not to stretch the membrane. The assembly is mounted inside a large test tube with side arm or a second tube can be inserted through the stopper (with no membrane) when an ordinary boiling tube can be used. For larger volumes either a longer tube or a 1–5 litre Buchner flask can be used, cutting membranes of appropriate length, and many scuh tubes can be suspended from a single stopper by increasing the number of holes available. The side arm or spare tube is connected to the water-vacuum pump which is slowly turned on and samples are admitted to the tubes by means of a fine teat pipette in which the glass end is protected by covering with a small piece of plastic tubing to avoid puncturing the membrane. When the membranes are fairly full the vacuum should be raised to about 600 mm. of mercury and, at this stage, the vacuum tube can be clipped off as it is not essential to run the pump further. The tubes can now be completely filled and, if desired, a syphon arrangement set up to continue addition of sample until the whole has passed into the tube. When large volumes are being filtered it is best to arrange the capacity of the apparatus such that only the lowest part of the tube is eventually covered by the filtrate. This serves two purposes namely that the protein sample cannot be concentrated to dryness by an overlong exposure to the vacuum filtration and, simultaneously, a dialysis effect is exerted on the sample remaining in the bag as, frequently, salts and other dissolved solids pass through more slowly than water. A further variant on this procedure is as follows. A 50 cm. length of tubing is set up in a five litre Buchner flask containing two litres of 0·2 M sodium chloride or other dilute buffer. The specimen is added to the tubing and concentration continued to the desired amount. The tubing dips into the buffer and cannot dry out but ultrafiltration continues even after the level of specimen has fallen below that of the surrounding buffer due to the fact that water but not air can pass through the membrane pores.

Concentration Dialysis. This procedure is based on the principle of dialysis against a substance of high molecular weight.

Fluid containing protein is placed in a glass tube or container, preferably with a small protuberance at the bottom end having approximately the smallest volume to which the fluid is to be concentrated. The size and shape of the glass container should be such as to provide the greatest possible dialysing surface. Using small glass tubes and narrow tubing, even a few millilitres of fluid containing protein can be successfully concentrated. The polyethylene glycol ("Carbowax" 20 *M*, made by Union Carbide, also available from G. T. Gurr), a waxy, flaky substance, with a molecular weight of 20,000, is broken up and an appropriate quantity of the dry granulated material is poured into the dialysis tubing. This can easily be achieved by

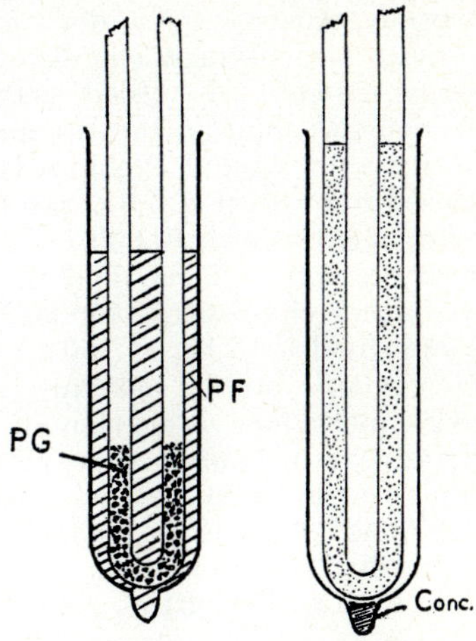

FIG. 1.5. Apparatus for concentrating fluids containing protein. *PF*, fluid containing protein; *PG*, polyethylene glycol; *Conc.*, concentrate.

inflating the tubing and using a funnel. "Visking" dialysis tubing is very satisfactory, but any similar tubing would be suitable. The proportion of polyethylene glycol to fluid containing protein is not critical, but there should not be less than 1 part of polyethylene glycol to 10 parts of fluid. Polyvinylpyrrolidone (PVP) in dry powder form is also satisfactory; it is, however, much more expensive. The dialysis tubing containing the polyethylene glycol is then bent double and is placed in the glass vessel containing the protein solution (Fig. 1.5). The length of the tubing should be such as to allow for the rising column of water inside it. It is advisable to moisten the polyethylene glycol inside the tubing with a small quantity of water just before it is placed in the fluid containing protein. The level to which the tubing should be pushed down depends on the final concentration required. The concentrating process ceases, of course, when the fluid containing protein reaches the level of the bottom of the tubing. The whole procedure is thus automatic. With larger quantities of fluid the tubing can be bent several times, providing a larger dialysing area and accelerating the process. The electrolyte content of the concentrate will be the same as in the fluid containing protein at the start, as there is a free passage of electrolytes across the membrane.

The method is rapid, reliable, and no special equipment is required. It has also the great advantage that the concentrate can be collected from clean glass and need not be scraped out from the inside of a sticky tube. Under suitable conditions a concentration, for example, of 10 ml. of urine to about 0·25 ml. is easily achieved in 3–4 hr. The high efficiency of the method is also due to the fact that the hydrophilic agent, namely, polyethylene glycol, does not contain water and has, therefore, full absorptive capacity. No denaturation effects were observed. Sephadex and other polymers can also be used.

Concentration using acrylamide gel. Dried acrylamide gel regains water when placed in aqueous solution but, because of its high cross-linkage, acts as a molecular sieve and excludes large molecules. The gel described here has a water regain of 5·5 g. per 1 g. gel and excludes substances of molecular weight above 10,000.

Glass tubes, internal diameter 0·5 cm. and up to 20 cm. long are dipped into Drifilm or other chlorosilane for 30 seconds, washed for 10 minutes in water and dried for 2 hr. at 150°C. The silicone film so formed on the tube remains stable indefinitely. Dissolve 15·2 g. acrylamide and 0·8 g. methylenebis-acrylamide (BIS) in 100 ml. water. Add immediately 0·4 g. ammonium persulfate and 0·2 ml. dimethylaminopropionitrile, mix and load into the glass tubes which are closed at the lower end. Care is taken not to trap air bubbles. Allow the gel to form for 30–60 minutes. Remove the tube stopper and gently push out the gel with a close fitting rod. Wash the gel with a number of changes of water over 3–4 days to remove contaminants. The gel is cut in 3–6 cm. lengths and exposed to the draught from the cooling unit of a cold room for about 4 days when the gel rods shrink to about 20% of the original diameter and 90% of the length; drying at higher temperatures fractures the gel. Rods may be stored over P_2O_5 in a desiccator till required. On complete water imbibition, these gels swell to about 0·75 cm. diameter.

For the concentration of aqueous solutions, the quantity of gel rod to be used is calculated from the original volume and the degree of concentration required. Solution and rod are placed in a tube of about 1 cm. diameter and left overnight in the cold. Even small volumes, such as 0·4 ml. can be concentrated ten times by this method with no loss of high molecular weight compounds.

Application of the Sample to the Medium

Perhaps the best and most widely applicable method of applying the sample is by means of a microliter disposable pipet. This consists of a teat with a hole in the top, a length of glass tube and a rubber cap also

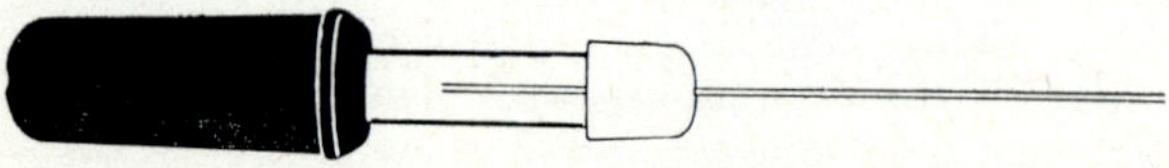

FIG. 1.6. Microcap micro-pipettes; see text.

with a hole in it (Fig. 1.5). Micropipettes are made in a variety of sizes to hold 2, 5, 10, 25, 50 μl. One of a suitable size is selected and gently inserted through the cap hole so that its end is visible through the glass tube. Held almost horizontally, the tube sucks up the complete volume of sample. The finger is placed over the hole in the teat and the sample is applied by means of slight pressure. For intermediate sized samples, the tube need be only partly filled. Thus exactly the correct volume is dispensed and the paper, gel, etc. cannot be overloaded.

CHAPTER 2

PAPER ELECTROPHORESIS AT LOW VOLTAGE

TECHNIQUES OF PAPER ELECTROPHORESIS

Ivor Smith

THIS chapter is concerned chiefly with paper electrophoresis at low voltages, i.e. voltages not exceeding about 400 V. with a consequent potential drop of not more than 15 V/cm. However voltage gradients of up to 30 V/cm. are discussed. Paper electrophoresis was the first of the electrophoretic techniques to come into wide general use. However, in the majority of hospital laboratories it has been displaced by other more suitable techniques. The macromolecules, such as serum proteins and hemoglobins, are now separated routinely using cellulose acetate strips or cellogel. Small molecules can be better examined using one of the high voltage techniques on paper or thin layers of cellulose. In both these cases there is a tremendous saving in time and the resultant separation is much cleaner. Nevertheless paper electrophoresis is still used in some clinical laboratories for lipoproteins and other families and it remains the simplest and cheapest way of teaching the principles of electrophoresis. For this latter purpose, the Unikit teaching kit can be highly recommended.

Apparatus for Paper Electrophoresis

Apparatus in common use for paper electrophoresis is of two types, namely, the vertical type where the paper strips or sheets are suspended over a nylon string or glass rod and hang down, or the horizontal or flat bed type where the strip or sheet is held horizontal. In practice both types give satisfactory and almost identical separation patterns and, as in paper chromatography, it is the pattern which is most important and most useful for identification purposes.

The Vertical Tank. As will be seen from figure 2.1, the tank base is a single moulded unit possessing two pairs of electrode-buffer compartments; the outer compartment of each pair contains the electrode wire whereas the paper strips dip into the inner compartment, connection between the members of each pair being made by paper wicks running the length of the apparatus. Paper strips or sheets are suspended over the tufnol unit which holds an upper Polytetrafluorethylene wire and two lower PTFE wires so that the paper does not contact the tufnol frame; up to seven 5 cm. strips can be placed on the frame or a single sheet of 38 cm. width can be used. After the strips are loaded in position with their ends dipping into the inner buffer compartments, the apparatus is closed with a transparent Perspex (Lucite) cover. At the end of the experiment the tufnol frame with the papers still in position can be transferred directly to an oven for drying.

The Horizontal Tank. A suitable tank is that described in Chapter 5,

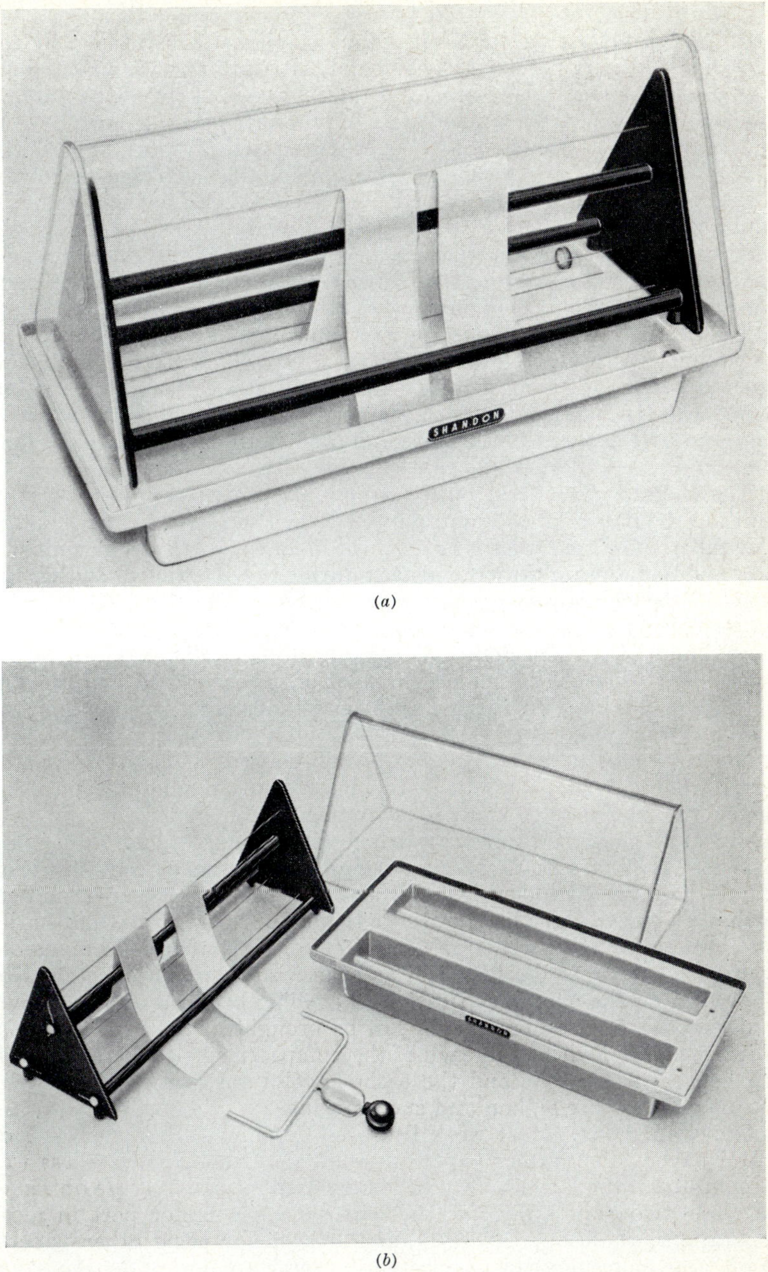

(*a*)

(*b*)

FIG. 2.1. Vertical electrophoresis apparatus: (*a*) assembled; (*b*) illustrating individual parts. *Shandon Southern Instruments.*

and p. 19. The cross bridges are placed at the maximum distance apart, leaving a gap of 18 cm. between them, and the papers are held by the strip holders. A paper length of 30 cm. is satisfactory with widths from 4–20 cm.; the length is so arranged that the ends of the strips dip into the buffer. Samples are applied at an appropriate position depending on the charges on the compounds to be separated. 500 ml. of buffer is placed in the tank and levelled by raising one side such that the buffer forms a continuous level above the partitions; after a few moments the tank is lowered to the horizontal position again. This volume can be used between five and ten times during the course of one week provided that the current is run in opposite directions across the tank for alternate runs to avoid the effects of electrolysis.

The Wicks. Both types of apparatus are available with glass wool wicks but it is preferable to use paper wicks which stretch the whole length of the apparatus. Wicks are prepared as paper strips about 6 cm. wide and are folded over the compartment dividers, care being taken to flatten them to the sides and to smooth out air bubbles along the top of the wicks. These can be replaced as necessary.

The Buffers. The term buffer is now used in a very wide sense to indicate the nature of the solution with which the strip is impregnated. The function of the buffer is to keep the strip wet so that a current will flow and to maintain the *p*H constant during the experiment so that the compounds will migrate under reproducible conditions. The correct buffer for any particular separation is found by experiment but within these limits other considerations come into play. For example the *p*H must not be that of the iso-electric point of the substance concerned or it will not migrate. Furthermore, the substance must be soluble in the buffer.

Distillation occurs from the apex or mid-point of the paper due to the generation of heat. Consequently buffer solution migrates into the paper and buffer salts increase in concentration in this area with a tendency to produce front sharpening of the bands.

Where buffers are used a number of times, the current direction should be reversed for each new run. It is usual to use fresh buffers each week and to add a little water to the buffer reservoir to compensate for evaporation if this is large. The level of the buffer solutions in all compartments must be identical as otherwise a siphon effect will be impressed on the overall separation. Levelling between the two anode and two cathode compartments can be brought about by means of a single, small hole drilled in each of the compartment dividers. Levelling between the anode and the cathode solutions is best attained by means of a Y-piece siphon and suction tube.

Choice of Paper. Most work has been done on Whatman No. 1 or 3 MM paper. The former paper is preferable although it has less wet strength than the latter. Very satisfactory spots or bands are obtained on these two papers and, as the paper plays a minor part in most separations, there appears to be little reason for using other papers. With large molecules, such as proteins, a small amount of denaturation and adsorption occurs and it is rarely possible to obtain a pure white background.

Cutting and Marking the Papers. For the vertical tank, strips 36 cm. long are cut, folded in half and the fold marked with a small pencil line; the origin is thus at the mid-point of the strip. Sheets are treated similarly. For the horizontal tank, strips 31 cm. are cut and the origin is marked 8 cm. from the cathode end leaving a migration distance of 18 cm. before the end of the bridge is reached; sheets are treated similarly. These measurements apply to serum protein separations.

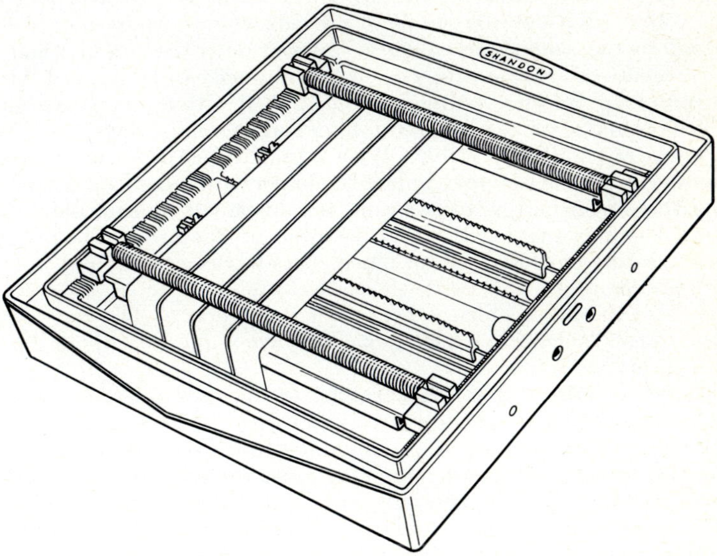

FIG. 2.2, Horizontal paper electrophoresis in the Kohn Tank. Dimensions: 25 cm. long × 24 cm. wide. *Shandon Southern Instruments.*

However, many cases are known where it is better to choose an origin nearer one of the electrode ends in order to obtain a greater length of separation and the final position chosen for the origin is dictated by experience.

Wetting the Papers. A number of procedures are available for applying the buffer to the paper. Where only small quantities of sample are to be applied, e.g. 10–20 μl. serum, the whole strip or sheet is dipped through a fresh sample of the buffer and blotted between clean sheets of paper. The paper is then placed in the apparatus and the sample is applied as described below; this procedure is very satisfactory. An alternative procedure is to place the dry strips in the apparatus and to allow the buffer to soak up or along the strip but this is much more time consuming (up to 45 minutes being required for the vertical strips) and also results in some chromatographic separation of the buffer salts. Where large amounts of sample must be applied, with intermittent drying of the origin, it is necessary to apply the sample to the dry paper and then to dip each side separately, allowing the solvent to come within 1–2 cm. from the origin, and then after placing the strip or sheet in position the solvent rises up to the origin by capillary action. In so doing it compacts the origin into a sharper band. Alternatively, the dry paper with sample applied is placed in position and is moistened by pipetting fresh buffer down the paper, from 1–2 cm. below the origin, until the whole is just wet. The origin itself must not be made wet but the buffer is allowed to ascend the last centimetre by capillary attraction.

Application of the Sample. Whatever method is used it is essential not to scratch the paper as the torn surface will affect the separation. For protein solutions the simplest procedure is to apply a measured amount to the applicator which is lowered onto the origin when the sample soaks in. Single drops or small volumes can be applied with a platinum loop. Larger volumes can be applied with a pipette whose tip has been rounded off to avoid scratching and care must be taken to keep the origin as small as possible. With horizontal papers the applicator technique is not satisfactory unless the papers are hung over a separate nylon thread. A margin of 2–3 mm. should be left on each side of the paper. The micro-cap method can be used satisfactorily. The punched spot procedure (as described for blood spots works well for both electrophoresis techniques but, with the vertical method, it must be placed away from the apex which is bent over the support. The spot can be placed about 1·5 cm. from the apex or anywhere down to about 2·5 cm. from one end.

Placing the Strips in the Apparatus. In the vertical apparatus the strips should be hung over the thread and with the lower ends touching the inner Perspex divider; they should not touch the wicks. Each strip should be adjusted so that it is hanging quite vertical. In the horizontal apparatus the blotted papers are tensioned across Perspex bridges so that they are fully stretched and do not sag in the middle. All the strips should be placed in position when the ends become wetter, and each can then be re-tensioned by drawing the ends down over the bridge a second time. They then remain taut for the experiment.

The Electrophoresis Experiment. As soon as possible after placing the sample on the strip or placing the strip in the apparatus the top plate is placed in position and the current is applied; otherwise diffusion commences and the band widens. The chosen current or voltage is turned up on the power pack and the apparatus left for the required time. With voltages not exceeding 100–150 V there is little evaporation and this often amounts to no more than a clouding of the walls. However, with voltages of up to 300 V there may be excessive evaporation and large droplets will form on the top plate. If there is any danger of these falling back on to the paper then the top plate can be rapidly removed and wiped down although the tank should not be opened if this is not essential. Changes in *p*H in the outer electrode compartments are quite large and, in order to overcome this, the direction of current flow is reversed after each run or the buffer is used once only.

Removal and Drying of the Papers. As soon as the current is turned off the papers should be removed and dried, again to prevent diffusion of the separated substances. A stand, about 5 cm. tall somewhat shorter than the length of the paper and wide enough to take the maximum of strips the tank will hold, is made. Headless nails are tapped into this so that each strip will lie horizontally and be held by one nail at each end. Alternatively any oblong Pyrex dish can be used and the papers held taut by means of a small bulldog clip at each end. As soon as all the papers are out of the tank they are placed in an oven at 105–110°C and dried. A Number 1 paper requires about 10 mins. and a 3 MM about 20 mins.

Location of Substances. Most substances can be located using the chemical reagents previously described in Vol. 1 for chromatography. However, the high salt content from the buffer may adversely affect or react with certain reagents so that a modification of the reagent may be necessary. This does not apply, of course, if a volatile buffer has been used as this evaporates off the paper during the drying process. The various simple and conjugated proteins are mainly located by protein or lipid dyes although substances carried by the proteins (cholesterol, phosphate) still require chemical reactions for their identification.

The Unikit Assembly. A simple but versatile apparatus[1], originally designed for teaching purposes, is also available. It is valuable for the occasional run, for testing new buffer compositions, for one way electrophoresis prior to chromatography, etc. The adjustable height V support allows for sheets or strips from 20 to 40 cm. length and up to 10 cm. width.

Electrode Assembly. The base unit is moulded as a three-compartment vessel of polythene. Threaded sockets are provided so that the two polythene electrode pillars screw in vertically. The pillars are hollow to allow the electrodes to run the full length of the tube; the parts immersed in the buffer solution are platinised. At the lower end of each pillar are four holes and a circular groove; and at the upper end an air vent hole. The threaded sockets have, near their ends, a slit, which permits the circulation of buffer solution through the four pillar holes and past the platinised electrode. The apparatus is assembled as follows: the electrodes are screwed, firmly but carefully, into the base vessel. The Vee-shaped paper holder is placed on one electrode and clamped about 15 cm. above the base. The glass rod with button ends is placed across it, parallel to the partitions in the base. Buffer is added to each compartment, until it reaches the height of the groove on the electrode. As this height is above that of the four holes, buffer passes into contact with the platinised electrodes. About 100 ml. buffer is required in total, equal volumes in each compartment to prevent syphoning from one compartment to the other, via the paper strip.

Single Drops of Sample. The paper is dipped through the buffer and blotted well between clean sheets of filter paper. The blotted paper is picked up by one end, the mid-line is placed over the button-ended glass rod which is then replaced on the Vee-support. Each lower end of the paper is then carefully placed in one of the buffer compartments so that the paper hangs as an inverted Vee. The sample or samples are applied at the origins using a micropipette or a melting point tube whose end has been rounded off in a flame, so that it will not damage the paper surface.

Many Drops of Sample or Samples to be Applied, or Electrophoresis after Chromatography. Drops of sample are applied to the origin of the dry paper strip with drying between each application. A dried one-way chromatogram may be used directly. The paper is now to be wetted at both ends and a dry strip left at the origin. To do this it is held firmly at each end and dipped by inserting it about 2 cm. from the origin line and drawing it through the buffer towards one end of the paper. This is repeated on the dry part of the paper on the other side

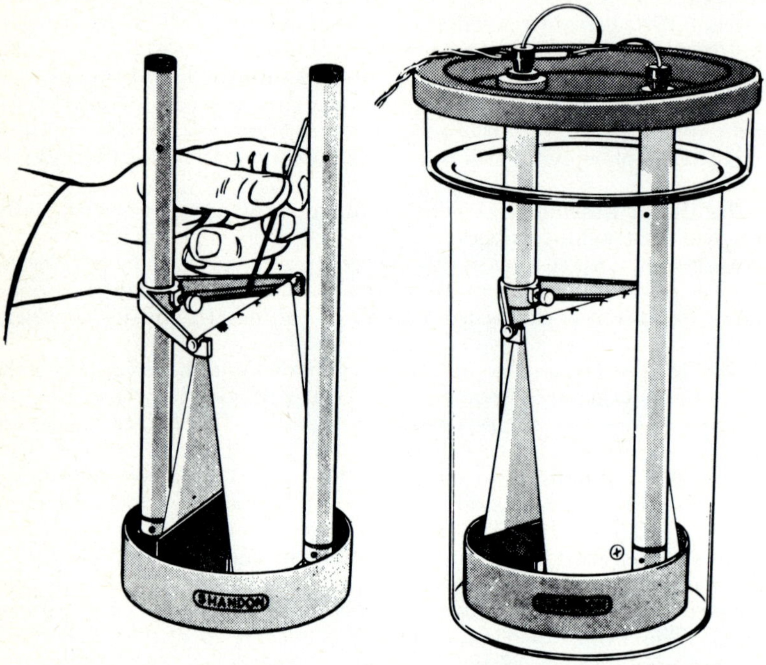

FIG. 2.3. The "Unikit" Tank holding sheets 10 cm. wide and 20–50 cm. long over an adjustable Vee-supported glass rod or nylon wire. Samples can be applied at the apex after assembly or at any point on the dry sheet before buffer application. Shandon Southern Instruments and Ohaus Scale Corp.

of the origin. Care is taken not to wet the origin line and the paper is carefully and rapidly blotted. The blotted paper is picked up by one end, the midline is placed over the button-ended glass rod which is then replaced on the Vee-support. Each lower end of the paper is then carefully placed in one of the buffer compartments so that the paper hangs as an inverted Vee. Using a capillary tube containing buffer, wet the dry origin area so that buffer diffuses into the sample. Do not wet the sample directly.

Operation. Pick the unit up carefully and steadily (to avoid spillage of the buffer) and lower it into the tank. Place the lid on the tank; the electrodes will protrude through the holes in readiness for power pack connections. At the end of the experiment, turn off the current, remove the leads from the electrodes, take off the lid and lift the electrophoresis assembly carefully out of the tank. Remove the paper with the aid of a second glass rod, inserted near the lower end of the side of the sheet and used to hold the paper, and blot the ends which were immersed in the buffer by gently touching on to a clean filter paper.

Dry the paper rapidly, by hanging in an oven at 105°C, or in front of an infra-red heater. About 10 minutes is sufficient. In some cases drying in a draught of cold air is to be preferred. Unscrew the electrode pillars and wash thoroughly. The platinised wire can be adequately washed by dipping into water up to the level of the circular groove. Shake off excess water. Do not heat the polythene.

THE SEPARATION OF SERUM PROTEINS

The separation of serum proteins illustrates both the need for absolute standardization of all variables (voltage, time of run, time of drying, dye used and staining time, etc.) and the complete lack of such standardization pertaining at the present time. Variation in any of the above factors will result in small or large variations in results even when all the other factors are kept constant.

A neat trick for producing identical widths of separation is as follows. A dot is marked at 12 cm. from the origin and the paper is set up as previously described. A paint brush is moistened with buffer, drawn to a fine tip and dipped into solid bromphenol blue so that one small crystal is picked up. The crystal is then deposited at one edge of the origin and migrates with the albumen. The run is continued until the blue spot reaches the 12 cm. mark. The paper is then removed and dried as usual. With experience, the correct amount of dye can be deposited at the origin. Too small amounts fade completely and too large amounts interfere with the subsequent dyeing procedure. The technique can be used whatever dye is finally chosen to stain the proteins.

Buffers. Two buffers are in common use. These are the original barbitone buffer of Flynn and DeMayo[1] which is prepared as required, and the "Oxoid" buffer* which is a premixed pack completely soluble in water but of undivulged composition. In spite of this latter fact, the Oxoid buffer is to be recommended as it yields much narrower bands than the barbitone buffer. In both these buffers the proteins exist as anions and so migrate to the anode. However, due to the electro-endosmotic flow the γ-globulin may appear to remain stationary at the origin or even to move slightly in the reverse direction.

Barbitone Buffer.	Sodium Diethylbarbiturate	. 41·2 g.
	Barbituric Acid . .	. 7·32 g.
	Thiomersal or Sodium Azide	. 0·4 g.

The sodium salt is dissolved in 500–600 ml. cold water, the barbituric acid is dissolved in 100–200 ml. warm or hot water and added to the previous solution. Thiomersal or azide is added as an optional preservative and the whole made to 1000 ml. This stock solution is diluted 1 volume into a total of four volumes for use; $\mu = 0{\cdot}05$; pH 8·6.

* Oxo Ltd., London.

Some workers prefer a slightly more concentrated buffer, $\mu = 0{\cdot}075$, in which case one volume is diluted to 2·67 volumes.

Barbitone "Oxoid" Buffer. 16·5 g. is dissolved in two litres with the addition of preservative; $\mu = 0{\cdot}05$.

With both buffers satisfactory separations can be obtained using approximately 100 V and 0·1 mA/cm. width for an overnight run (16 hr.) at about 20°C and in the vertical apparatus. The Oxoid buffer cannot be used overday but the barbitone is very satisfactory when using 200 V and 0·23 mA/cm. for 7 hrs. Separation widths of 12 cm. are obtained. The figures quoted vary slightly from laboratory to laboratory and also with the season of the year, particularly when the temperature variation is great.

Barbitone-Calcium. Sodium diethylbarbiturate, 12·76 g., barbituric acid, 1·66 g., calcium lactate, 0·384g., to 1 litre, *p*H 8·6. The effect of the calcium ion is to split the β band into two component bands.

Tris Buffer.[2]	Trishydroxymethylaminomethane .	60·5 g.
	Ethylenediaminetetracetic Acid .	6·0 g.
	Boric Acid	4·6 g.

The three salts are dissolved in 1 litre, *p*H 8·9. With a voltage gradient of 6·5 V/cm. for 16 hrs., a significant increase in the number of bands is observed and nine often can be seen.

Acetate buffer.	Glacial Acetic Acid .	25·59 ml.
	Sodium Acetate, $3H_2O$.	13·61 g. to 1 litre. *p*H 4·0

Paper. Most work is now done on Whatman No. 3 MM or No. 1. Strips of width 4 or 5 cm. for individual samples or sheets may be used. The 5 cm. strip is useful for two determinations on the sample, e.g. proteins and lipo—or glyco—proteins when the sample is run as a single band and the strip is subsequently cut lengthwise down the middle. With sheets the bands should be 4 cm. wide and separated by 1 cm. gaps.

Volume of Sample. About 4 μl. serum per cm. width of strip should be used for serum protein analyses and double this quantity for other protein components. This enables minor components to be observed easily without overloading of the major components. The procedure of choice is the following. The strips are dipped in fresh buffer, blotted and placed in the apparatus, the samples applied with the applicator or microcap as quickly as possible and the current turned on. After the chosen time, the current is turned off, the strips removed to the holder and immediately dried. The bromphenol blue marker is frequently used for maintaining identical runs over varying times and temperatures during the year. Urinary proteins must be concentrated 10–20 times before running as described in Chapter 1.

Lipoproteins are often prestained before electrophoresis. Sudan black, 2% in diacetin, is aged for a week or more and then filtered. 0·1 ml. of this is added to 1·0 ml. of serum and shaken well. After standing 30 mins., the mixture is centrifuged at 3000 rpm. for a further 30 mins. and the supernatant used in the usual manner. The following alternative prestaining procedure is also satisfactory provided that short runs of about 2–3 hours with 200 V are used. With the long overnight runs, the dye colour disappears and nothing is observed on the strip the following morning. 500 mg. of acetylated sudan black[3] is added to 50 ml. ethanol and heated to 60–70°C in a water bath for 10 minutes. The solution is then cooled and filtered. One volume of dye is added with stirring to 10 volumes of serum and after standing for one hour the mixture is centrifuged to remove dye particles. Eight to ten μl. of dyed serum/cm. width are used.

Dyes and Staining Procedures

The choice of a dye for staining proteins appears to be a matter of personal preference and it would seem essential that one dye should be chosen and adhered to in order to obtain comparable results. Three main points should be noted. First, most dyes are mixtures and therefore each protein band is being stained simultaneously by a number of different dyes. Second, individual proteins take up different amounts of any one dye. Third, the quantity of dye taken up by a particular protein often depends on the quantity of protein on the paper, larger amounts of protein often taking up proportionately less dye than smaller amounts. The aggregate of these points is summarized in Fig. 2.4. The protein separations were carried out simultaneously in the vertical apparatus, dried simultaneously and stained at approximately the same time in the first four dyes described below. Lipoproteins were located either with sudan black or acetylated sudan black and glycoproteins with Schiff's reagent. Other serum components can be located with reagents previously described for chromatography, e.g. serum cholesterol with phosphotungstic acid and organic phosphate with the phosphate reagent (Vol. I).

Naphthalene Black 12B (Amidoschwartz). A saturated solution of the dye (approximately 1 g./100 ml.) in 10 per cent acetic acid in methanol is used. The strips are dyed for 10 minutes, washed in successive baths of fresh 10 per cent acetic acid in methanol until the background is light blue, finally washed in methanol and air dried.

Bromphenol Blue. One per cent dye in 95 per cent ethanol saturated with mercuric chloride (30 g./100 ml.) is used. The strips are dyed for 5 minutes, washed either in 1 per cent acetic acid or in tap water and finally in pure methanol. The blue bands may turn green in acid atmospheres in which case a further rapid wash in 2 per cent Na_2CO_3 in 50 per cent methanol-water is used to render the colours permanent.

Procedure. The whole strip can be dyed or the strip is trimmed some 5 cm. from the origin in the cathode direction and 15 cm. in the anode direction. The protein-containing portion is wound round in a spiral and completely immersed for the necessary time in the dye contained in a 100 ml. beaker. Alternatively, it is lain flat in a dip tray or Pyrex dish and just covered with dye solution; a cover should be placed over the dish to prevent evaporation. During the dyeing process the buffer salts dissolve in the dye solution and so it is preferable to use minimal amounts of solution and to reject this after a few uses. Washing can be carried out in a similar way using wash solution instead of dye reagent.

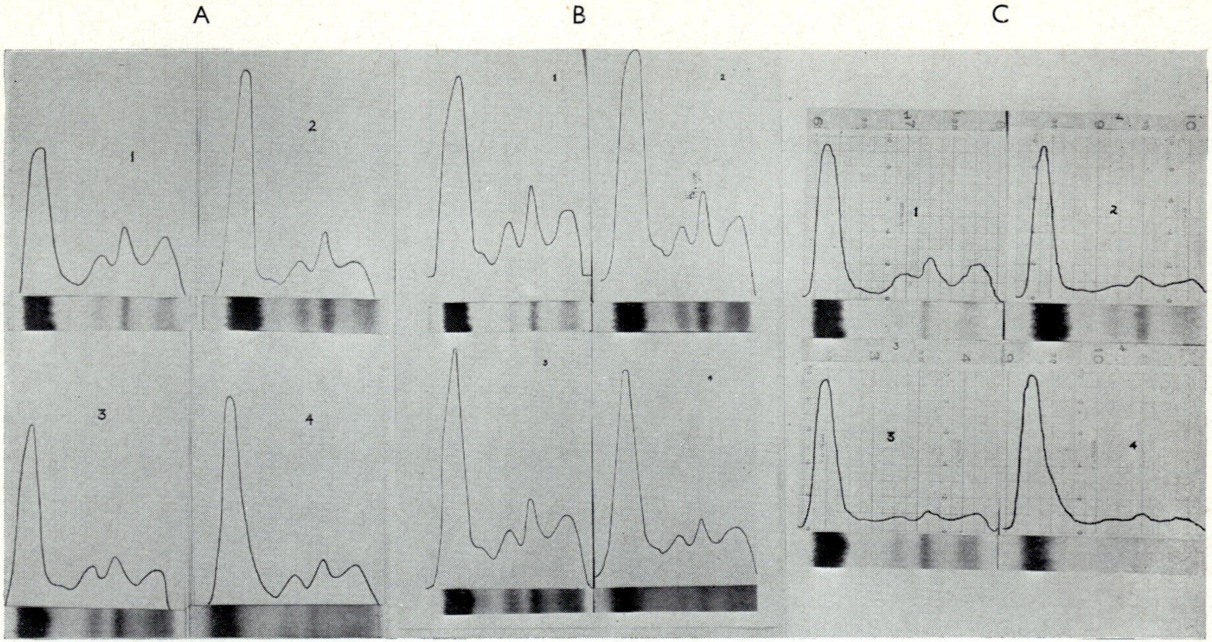

FIG. 2.4. Four strips of the same serum were run overnight and simultaneously in the barbiturate buffer, dried and stained with the four commonly used dyes, namely azocarmine (1), naphthalene black (2), light green (3) and bromphenol blue (4). These four strips were then scanned in three different commercially available densitometers; A = EEL scanner, (Evans Electro Selenium, Ltd., Harlow, Essex), B = Double-Beam Reflectance Densitometer[4] (Joyce, Loebl & Co., Vine Lane, Newcastle-upon-Tyne 1), C = Locarte-Laurence Automatic Logarithmic Recording Densitometer[5] (The Locarte Co., 24 Emperor's Gate, London, S.W.7). It is stressed that the same four strips were used in each machine but the photographs were taken on different days and it is extremely difficult to photograph the four different dyes simultaneously; in these examples the bromphenol blue strip shows up least satisfactorily.

Strip Scanning and Quantitation. From what has been said on the question of dyes and dye uptake it will be obvious that absolute quantitation of paper strips is theoretically unsatisfactory at the present time. In fact, the only value of such quantitation is that it allows of the comparison of strips prepared in the same laboratory when a rigorous and exactly reproducible procedure is used and when the strips compared are not too abnormal or dissimilar in quantity and distribution of protein. Fig. 2.4 shows typical scans obtained using three different commercially available instruments. Densitometers using either transmission or

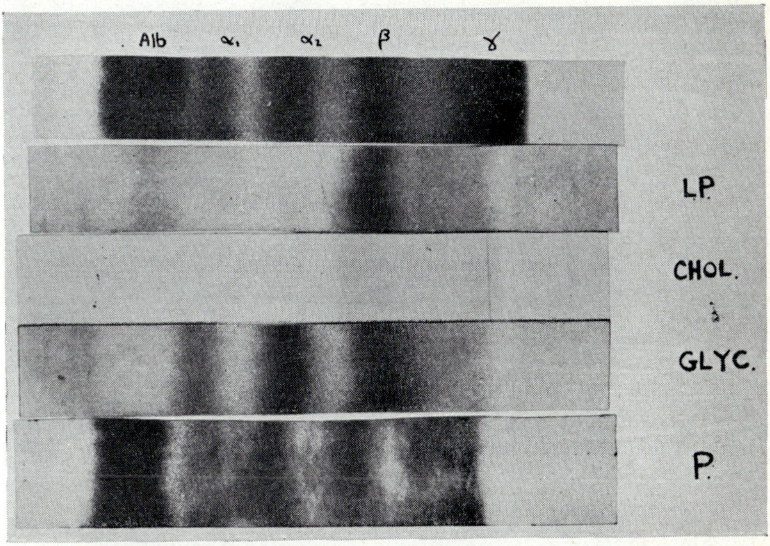

Fig. 2.5. Five strips were run simultaneously, overnight in barbiturate buffer, dried, and stained for proteins (naphthalene black), lipoproteins (sudan black B, post-stained), cholesterol (phosphotungstic acid, this strip did not photograph well), glycoproteins (Schiff's reagent) and phosphate (molybdate reagent).

reflection methods are available and this introduces yet another variable to the overall procedure.

It seems to be generally agreed (certainly in most London Hospitals) that scanning is of little value at present, as abnormal conditions produce gross variations which are quite obvious on visual inspection. Minor variations can usually be ascribed to the normal individual variation or to slight deviations in the drying-staining-washing procedure. Further, it must be appreciated that no scanning device yields protein ratios similar to those found by nitrogen determinations on the separated bands. Many workers have, therefore, determined their own conversion factors for each band.

Sudan Black B. A saturated solution of sudan black in 55 per cent ethanol is prepared by adding an excess of the dye to the warm ethanol solution. The solution is cooled, allowed to stand a few days

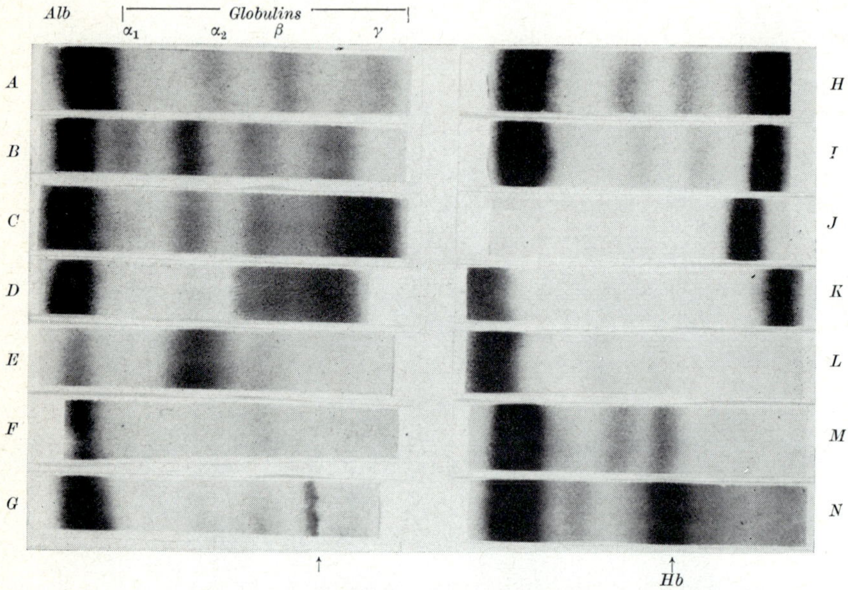

FIG. 2.6. Serum protein patterns in disease. A, normal serum; B, the non-specific disease response; C, disseminated lupus erythematosus; D, cirrhosis of the liver; E, nephrotic syndrome (serum); F, nephrotic syndrome (urine); G, normal plasma (f = fibrinogen); H, benigh purpuric hyperglobulinaemia; I, myelomatosis (serum); J, myelomatosis (urine); K, macroglobulinaemia; L, primary hypogammaglobulinaemia; M, seconday hypogammaglobulinaemia (chronic lymphatic leukaemia); N, serum containing haemoglobin (Hb).

and filtered. The strips are stained for at least 30 minutes and then washed first with a quick rinse in 50 per cent ethanol and then thoroughly in 40 per cent ethanol until most of the background has disappeared; the background never disappears completely.

Schiff's Reagent

(*a*)	Periodic acid, 4 per cent in water . .	2 vol.	mixed fresh
	$CH_3COONa.3H_2O$, 2·72 per cent in water	1 vol.	
	Ethanol	7 vol.	
(*b*)	Potassium iodide, 5 g., plus $Na_2S_2O_3.5H_2O$, 5 g., in 100 ml. water . . .	100 ml.	mixed fresh
	Ethanol	150 ml.	
	HCl, 2N	2·5 ml.	

(*c*) Fuchsin sulphite soln. (Schiffs reagent, Hopkins and Williams)

(*d*) Potassium metabisulphite, 0·4 g., conc. HCl, 1 ml. in 100 ml. water

(*e*) Absolute alcohol and 70 per cent alcohol

The dried strip is immersed in reagent (*a*) for 5 minutes and rinsed twice for two minutes each in 70 per cent alcohol. It is then immersed in (*b*) for 5–8 minutes, again rinsed twice with 70 per cent alcohol blotted on clean filter paper and immersed in (*c*) for 30–45 minutes when the maximum colour intensity is obtained. It is then rinsed three times for three minutes each in (*d*), dehydrated with alcohol and allowed to dry.

Volatile Buffers for aminoacids and related compounds.

*Buffer p*H 1·9. Formic acid/acetic acid. 58 ml. of glacial acetic acid and 26 ml. of 25 per cent (*w*/*w*) formic acid are diluted to 2 l.

*Buffer p*H 3·1. Ammonia/formic acid. 50 ml. 90 per cent (*w*/*w*) formic acid and 20 ml. S.G. 0·880 ammonia to 2·5 l.

*Buffer p*H 5·3. Pyridine/acetic acid. 25 ml. pyridine and 10 ml. glacial acetic acid to 2·5 l.

*Buffer p*H 6·5. Pyridine/acetic acid. 125 ml. pyridine and 5 ml. glacial acetic acid to 2·5 l.

*Buffer p*H 7·0. Collidine/acetic acid. 23·5 ml. collidine and 6·9 ml. glacial acetic acid to 2·5 l.

*Buffer p*H 10·0. Triethylamine/acetic acid. 16·5 ml. triethylamine and 5·6 ml. glacial acetic acid to 2·5 l.

*Buffer p*H 10·4. Triethylamine/acetic acid. 17·5 ml. triethylamine and 4·6 ml. glacial acetic acid to 2·5 l.

*Buffer p*H 11·7. Triethylamine. 1 per cent (*v*/*v*) aqueous triethylamine.

By reducing the concentration of components the voltage can be increased without the need to cool; e.g. the formic-acetic buffer can be diluted eight times and voltage increased to 800 V. (32 V./cm.).

Mobility Measurements. Relative mobilities are calculated conveniently by reference to the migration of a fast component. Ornithine is convenient as a reference standard at *p*H 7 and below this value; aspartate may be used at *p*H values above 7. At low voltages the relative mobilities are found to be remarkably constant for a given buffer if the experimental conditions are carefully standardised. It has not been found necessary to correct for electro-endosmotic flow when calculating relative mobilities.

The reagents described in Volume I for aminoacids are applicable here but may have to be adjusted to compensate for the residual buffer salts on the paper. Salts may modify or even completely inhibit the location reagent. Volatile buffers are free from this disadvantage.

Neutral Ninhydrin. 0·2 g. ninhydrin in 5 ml. water plus 95 ml. acetone. Used for papers run in *p*H 4 or 6 buffer.

Acidified Ninhydrin. 0·2 g. ninhydrin in 5 ml. water plus 5 ml. glacial acetic acid plus 90 ml. acetone. Used for highly alkaline papers, i.e. buffers *p*H 8·6, 10 and 11·5.

Alkaline Ninhydrin. 0·2 g. ninhydrin in 100 ml. ethanol plus 0·5 ml. aqueous N KOH; or, 0·2 g. ninhydrin in 95 ml. acetone plus 5 ml. lutidine, although the colours are slightly weaker and the lutidine gives off a rather pungent odour on heating. Used for highly acidic buffers, i.e. *p*H 1·9.

Papers are dried after electrophoresis and then dipped rapidly through the appropriate reagent. After the solvent has evaporated, the papers are again heated to 110°C until maximum colour intensity is obtained. The ninhydrin colours can be rendered permanent as described in Vol. I, Chap. 5.

Other Reagents and Techniques. Most of the reagents described in Vol. I, Chap. 5 for aminoacids are applicable here. Oxidation of the sulphur-containing aminoacids is useful since products with different mobilities may be obtained.

REFERENCES

1. The Unikit Apparatus in "Paper and Thin Layer Chromatography and Electrophoresis." Ivor Smith and J. G. Feinberg. Publ. Shandon Southern Instruments, Camberley, Surrey and Sewickley, Pa. and Longmans, London.
2. Aronsson, T., and Gronwall, A., *Science Tools*, 1958 **5**(**2**).
3. Wilcox, A. A., Wertlake, P. T., Haley, M. L., and Peterson, J. E. *Proc. Soc. Exptl. Bio . Med.*, 1958, **98,** 718.
4. Block, W. D., Rukavina, J. G., and Curtis, A. C. *J. Lab. Clin. Med.*, 1956, **47,** 357.
5. Hobbs, J. R., Slot, G. M. J., Campbell, C. H., Clein, G. P., Scott, J. T., Crowther, D. and Swan, H. T. *Lancet*, 1966 **2,** 614, No. 7464.

CHAPTER 3

HIGH VOLTAGE PAPER ELECTROPHORESIS

D. Beale and Ivor Smith

HVPE offers a rapid and reproducible method for the separation and identification of small molecules such as aminoacids, peptides, indoles, phenols, purines, pyrimidines, nucleotides and imidazoles. It has certain advantages over paper and thin-layer chromatography, notably speed and the fact that samples need not be desalted, unless the salt concentration is much higher than physiological or a very large sample is to be applied. HVPE is also particularly useful as one of the stages in two dimensional separations. Since 1–10 kV at 50–400 mA is generally used, the main technical problems are ones of heat dissipation and prevention of evaporation of buffer from the paper electrophoretogram. Two main types of apparatus have been developed to overcome these difficulties. In one, the electropherogram is pressed down onto a water-cooled metal plate by means of a sheet of plastic foam to ensure complete and even contact and, hence, cooling of the paper. In the other, the paper is immersed in a cooled, water-immiscible organic solvent. Both techniques produce extremely similar results and the choice of technique is entirely a matter of personal preference. One author uses only the cooled plate method whereas the other, who has both apparatus in the laboratory, uses whichever is free at the time.

Power Supplies

Because of the extreme potential danger involved in the use of high voltages up to 10 kV, power supplies and safety precautions will be discussed first. Fatalities have occurred due to ignorance and lack of adequate earthing of apparatus and so safety must be a cardinal consideration. It should never be assumed that a manufacturer has earthed an apparatus adequately because, even though all reputable manufacturers now do this with all high voltage power packs, errors in manufacture and testing can occur.

A variety of power supplies or power packs is available and the two Shandon power packs have been found very satisfactory over many years (fig. 3.1). In practice it is rare to go above 5000 volts for most separations, i.e. 100 V/cm. at maximum, and frequently 2–3 kV is adequate; 5 kV power packs are also much cheaper than 10 kV packs. However, the very long separation used by Morris[1] for quantitative aminoacid analysis from protein hydrolysates uses the higher voltage machine. These two powerpacks have a number of fail-safe devices. Apart from the usual switch on the mains plug, the first switch on the apparatus is a key which turns on the mains low-voltage supply and the high voltage can only be made available by means of a dial. Should a leak or short circuit exist at the beginning or develop during the experiment, an electronic relay cuts off the high voltage even when the leak is only 50 or 100 volts respectively for the 5 and 10 kV packs. Finally, a new experiment can only be started by returning the output control to zero before high voltage can flow so that the apparatus cannot be turned on accidentally if the whole apparatus has been incorrectly turned off directly at the mains on a previous occasion. Of course, the electrophoresis apparatus should also have fail safe devices and these will be described later.

FIG. 3.1. A high voltage electrophoresis power pack.
Shandon Southern Instruments, Camberley, Surrey.

Safeguards against electrocution. In general, the following points should be observed:
1. before purchase, the number and type of safety devices provided on the particular power pack should be checked to ensure the apparatus will be safe in normal laboratory use.
2. The apparatus should be of the fail-safe type.
3. Equipment should be checked and installed by a competent electrician, all earth leads should be connected and preferably to water pipes and, when the first experiment is set up, a final electrical check should be made.
4. Other laboratory personnel should be warned that a high voltage apparatus has been installed and, frequently, a warning notice is placed close by.
5. No untrained personnel should be allowed to use the equipment.
6. Where more than one apparatus is in use in the laboratory, a special area should be set aside for this work. If scaffolding is used in order to save space, then it too should be earthed.

Cooled plate apparatus is normally purchased, but many laboratories do construct their own water-immiscible organic solvent type of apparatus. It is even more essential to ensure good electrical work is carried out in the laboratory. Heavy duty switches should be placed in such a a way that, if the enclosure is opened, the equipment will de-energize (cut-out or switch off). Connections should be in earthed, metal conduit. Indicator lights should be installed to signal if switches fail and switches should be wired in series to prevent current flow if one fails.

Occasionally an electrical connection inside the apparatus may break or become corroded, or the buffer may dry up. Under these conditions the power supply may store a large amount of electricity in the smoothing condenser which will be discharged only very slowly after the power supply has been switched off. For this reason there should be a bleeder resistor across the high voltage outlet of the power supply to discharge any residual electricity rapidly when the unit is shut off. A shorting bar should also be installed in the enclosure to the apparatus which operates when the cabinet is opened—this is usually a microswitch. Most apparatus now incorporate such bleeder bars and microswitches but their existence should be confirmed before using the apparatus. Finally the operator should ensure that the voltmeter and ammeter both read zero and the mains is off before beginning a new experiment and at the end of an experiment before the apparatus is opened.

The Cooled Metal Plate (Flat Bed) Apparatus

This type of apparatus was first described by Gross[(2)]. Here the paper sheet, resting on a slightly larger sheet of insulating polythene, is pressed down evenly but firmly against a horizontal metal plate by means of a firm, plastics foam pad. The plate is cooled by means of water circulating through an internal labyrinth path or piping sealed to the underside of the plate.

The apparatus. The two apparatus illustrated (fig. 3.2) are exactly the same in principle and differ only in the size of base plate. The larger plate takes paper 60 cm. long by up to 28 cm. wide (24″ × 11″) although narrow strips of only 5 cm. width can be run satisfactorily. Basically it is designed for simple one way electrophoresis separations or two-way separations using chromatography for the second direction as described later. The smaller plate is 28 cm. (11″) square and was originally designed to take 25 cm. (10″) punched papers which could then be loaded onto a suitable frame for chromatography in the second direction—the frame technique is described in volume 1, chapter 2. However, it is equally possible to use electrophoresis for the second direction because the plate and paper are square. The setting up and use of the two apparatus are identical and so the methods below may be taken to apply to both.

Setting up the Apparatus. The apparatus is placed on the bench and levelled by means of the four adjusting screws under each corner. Mains water is connected to the inlet at the back of the plate and a fast flow should be observed at the exit tube which empties into an unimpeded

(*a*)

(*b*)

FIG. 3.2. High voltage electrophoresis apparatus both based on the same principle as described in the text. (*a*) 60 × 28 cm. (24 × 11″) and (*b*) 28 cm. (11″) square plates. *Shandon Southern Instruments, Camberley, Surrey.*

sink or waste area. The buffer troughs are placed in position and connected to the terminals. One insulating melinex or polythene sheet (0·01″ thick) is cut so that it just overlaps the whole plate in width and runs into the buffer trays at each end. The top plate is closed and clamped and the transparent top cover is also brought down and closed when the microswitches should be heard to operate; lifting this lid up and down should always be accompanied by listening for the switch. Connect the apparatus to the powerpack which should be off at the mains or disconnected. Connect the power pack to the mains and switch on, turn the high voltage dial down to zero and switch on the connection to the apparatus by means of the key or other device on the powerpack; many power packs have a green light which goes on at this stage. Finally turn on the high voltage and follow the increase in volts on the voltmeter; many power packs have a red light which goes on at this stage and the green one turns off. The increasing voltage indicates that the apparatus is applying a potential difference to the electrodes. However the ammeter should remain at zero amps thus showing that no current is flowing, i.e. there is no short circuiting and this should be confirmed by the red light which remains on because, as mentioned earlier, a good power pack should cut out if a leak to earth occurs. Finally raise the outermost plastic cover—not the hinged upper plastics foam lid—when the microswitch should actuate the cut out and turn off the high voltage; the high voltage should not come on when the lid is lowered as this should only happen after the high voltage dial has been returned to zero.

When all these checks have shown that the apparatus is electrically safe, it can be used for a first experiment with electrophoresis paper in position. The paper is wetted with buffer, blotted and placed on the melinex sheet already in position in the apparatus. Paper or special encapsulated wicks are connected from the electrophoresis sheet to the buffer trays, a second sheet of melinex is placed over this and the apparatus connected up as before. In this case the increasing high voltage should generate amps on the meter to show that current is now flowing correctly through the apparatus. 3–5 kV should be allowed to pass for about 30 minutes during which time little or no heat should be generated although a slight clouding may be observed. The outflow water should be measured with a thermometer and should not be more than a few degrees above the inflow temperature provided that high water flow is being used.

Polythene or Melinex Insulation Sheets. The role of these sheets is to enclose the electrophoresis paper, which carries the current, such that it does not come in contact with the metal cooling plate or any other conducting part of the apparatus except the buffer solutions which are also insulated in their plastics troughs (fig. 3.3). Hence the insulating sheets should always be of greater dimensions than the paper and should overlap the cooling plate and run down into the buffer troughs. Likewise the top sheet should cover the paper and also pass down into the troughs. Insulating sheets should be free of creases and pinholes as these are a source of leaks to earth. They should be washed after each run as they are a potential source of cross contamination from one paper to the next. The plastics foam pad on the underside of the lid should also be enclosed in plastics sheeting. Good quality polythene sheeting, 0·01″ thick, or melinex supplied with the apparatus should be used.

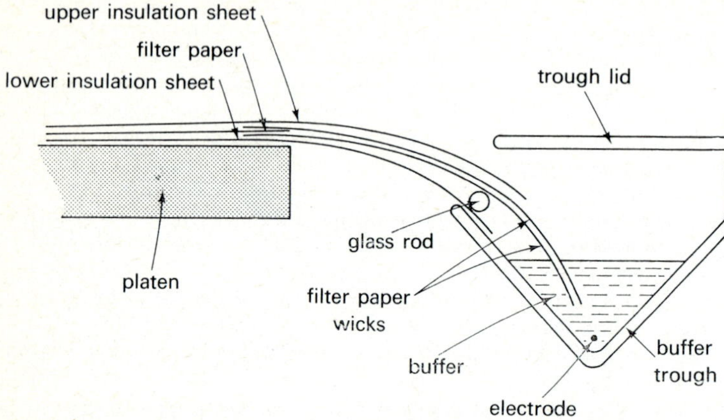

FIG. 3.3. Diagram showing interleaving of insulating sheets, electrophoresis paper and encapsulated wick. Note also buffer trough with electrode.

The Electrodes. Commercial apparatus normally uses platinum for the electrodes but good quality stainless steel is perfectly adequate and much cheaper. The wires are often run through perforated plastics tubing as a protection against rough treatment or accidental breakage. The electrodes must be covered by buffer during the experiment and one end of the wire either passes through the buffer vessel wall to be connected to cable to the power pack or is joined by cable at a point above the buffer level to avoid corrosion.

Buffer Troughs. These troughs are usually two-compartment units in which one compartment holds the electrode and this is connected by means of a paper wick to the second compartment into which dips the main wick (fig. 3.3). However more recent designs use a single compartment trough with a protected electrode. In all cases the trough should sit firmly on the base board of the apparatus so that it cannot be displaced by accidental moving of the apparatus; this is frequently accomplished by means of feet under the trough which fit holes in the base. Troughs usually require a minimum of 200 ml. buffer.

Wicks. Wicks are usually made from ordinary filter paper often using two thicknesses of wick both above and below the paper. Wicks, which should be slightly wider than the electrophoresis sheet but not wide enough to overlap the insulation sheets, are placed to overlap the electrophoresis sheet by about 0·5 cm. where they remain by capillary attraction, and run down into the buffer trough. Their role is to make electric contact between the experimental sheet and the buffer (fig. 3.3).

A number of manufacturers now supply encapsulated wicks in which the paper is enclosed in plastics sheeting containing perforations of varying size to ensure electric contact. These have the advantage that they can be used for very many experiments and are much easier to use. However, when not in use, they must be kept moist in plastics bags as once they dry out they crinkle and will not lie flat again after resoaking. Wicks are wetted with the appropriate buffer by soaking for a few minutes and then heavily blotted before use; they retain sufficient moisture not to impede current flow. Although not essential many troughs incorporate a glass rod to hold the wick away from the side to avoid any syphoning.

Marking out the Paper. The point of application or origin at which the sample is applied depends on the nature of the compounds to be separated, the pH of the buffer, etc., but it must always be located within the area of the plate (fig. 3.4). For example, all aminoacids, except

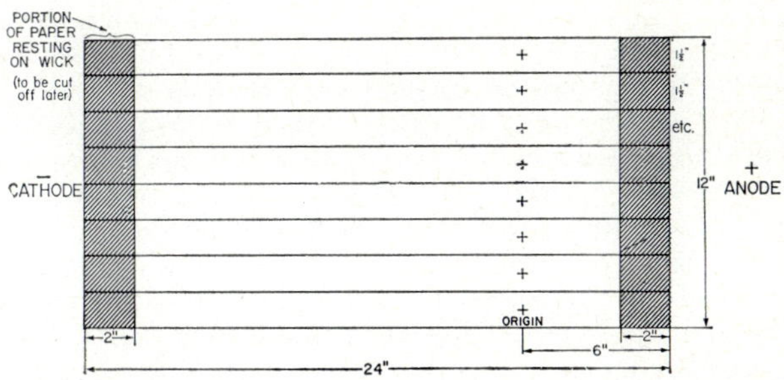

Fig. 3.4. Preparation of paper for HVPE followed by chromatography. In this case, eight samples are to be chromatographed on a sheet of Whatman No. 3 MM paper 24 in. long and 12 in. wide. After electrophoresis the papers are cut into 1½ in. strips, each containing one sample.

cysteic acid, move towards the cathode at pH 2 and so the origin should be placed near the anode end. Conversely, all organic acids move towards the anode in appropriate buffers and so the origin should be near the cathode. Mixtures containing both anions and cations should be placed near the mid point if it desired to examine all the components present. However, having decided on this general position of the origin, a run with the origins in slightly different positions on a single paper will indicate the best position. With the long sheets of paper a suitable origin is 7–8 cm. from the edge but with the 25 cm. sheets, the origin may be placed 3–5 cm. from the edge. The origin line is drawn in pencil across the paper and a number of origins applied if one-way runs are to be made or a single origin is marked for two way runs. Origins should be marked with pencil crosses about 3 cm. apart—2 cm. are adequate for short sheets but up to 4 cm. may be used for a 50 cm. run as the run may not be entirely straight over such a distance.

Wetting the Paper and Applying the Specimen. If sufficient of the compounds of interest is contained in only a few μl. then the specimen can be applied to the wet paper but if a much larger volume must be applied then this must be done to the dry paper. The specimen is applied in 2–5 μl. amounts from a micropipette or platinum loop, allowing each drop to dry before the next is applied. With many origins, the first tends to dry by the time the last is applied but drying can be hastened with a hairdryer if the spots are stable to heat. Desalting is rarely necessary as the salts migrate much more rapidly than the organic ions and so move right out of the areas of interest. For one-way runs the specimen is best applied as a small streak along the origin line but, for two-way runs, a round spot must be applied. The origin spot or line is raised above the filter paper by inserting two glass rods under and on either side of the line. Although many workers apply the buffer by dipping the paper through the buffer whilst leaving the origin line dry this can be difficult with long or thick sheets as they tend not to stay even and flat. A much better procedure is to lay the sheet on a number of sheets of large filter paper and to apply the buffer by means of a shaving brush which is continually wetted and drawn across the paper in sweeps parallel to the origin line. Buffer is applied up to but not touching the line itself. As buffer moves towards the origin line, from both sides, the buffer flow causes the spots to be compacted toward the origin line into a quite thin streak which is advantageous to maximum separation. The paper is heavily blotted along its whole length except for the origin line and placed carefully onto the lower insulating sheet. The two ends are then sandwiched between paper wicks or covered by an encapsulated wick with an overlap of about 1 cm. for short papers and 2–3 cm. for long papers. The upper insulating sheet is placed in position and the top plastics-foam plate is brought down and closed; the differences in thickness caused by the wicks is taken up easily by the compressibility of the foam.

Applying the Voltage. The transparent lid is closed, a check is made that the water is flowing and the voltage is applied carefully as described previously. Electrophoresis is carried out for the required time at an appropriate voltage which is determined experimentally. Voltages up to 100 V/cm. may be used.

Removing the Paper. At the end of the experiment turn off the high voltage by taking the voltage dial back to zero and switching off at the mains. Open the apparatus and expose the electrophoresis paper. If thick paper such as 3 MM has been used it may be picked up and hung directly in a fume cupboard or oven to dry. With thin papers, it is best to prepare pieces of dry paper or polythene sheet the same width as the paper and to place these around the edge so that the paper will not tear as it is gently picked up and taken to dry. Location of the separated compounds is carried out using the dip or spray reagents described in volume 1.

Electrophoresis followed by Electrophoresis. This is best done in the 25 cm. apparatus. The dried paper is returned to the apparatus but placed at right angles to the previous direction so that the line of separation now forms a new origin. Using a different buffer, the whole is set up and run as before.

Electrophoresis followed by Chromatography. For the smaller sheets, a number of papers are run by electrophoresis, one after the other.

When five or other appropriate number are ready, these are transfered to a frame taking five sheets and run as a single chromatography experiment. The papers are dried and examined with appropriate reagents. However when electrophoresis is carried out on long sheets, the dried sheet is cut into strips with one sample in the middle of each. Each strip is then sewn into another sheet using a machine with a zigzag stitch as shown in fig. 3.5. This type of stitch produces a very close contact between the papers and allows the compounds undergoing separation to migrate across the join without distortion. Even those compounds, which finally migrate to a position on the stitching, yield compact spots. Examples of the above different techniques are shown in the figures.

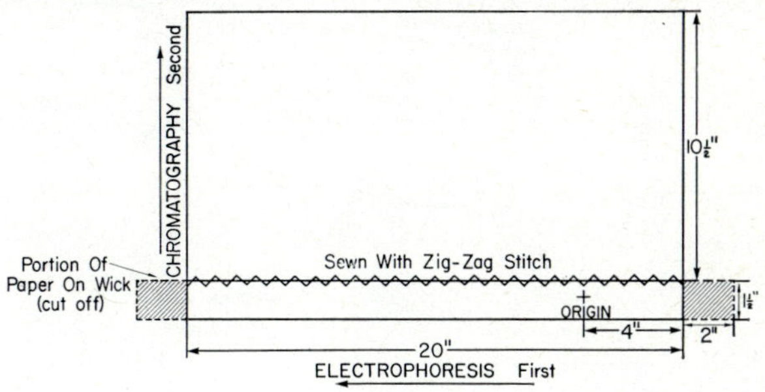

FIG. 3.5. Preparation of $1\frac{1}{2}$ in. strip containing compounds separated by HVPE for ascending chromatography in the second dimension. A clean sheet of Whatman No. 3 MM paper is stitched onto the strip. For descending chromatography, a second paper wick is stitched onto the other side of the strip so that it will fit in a trough for descending chromatography.

Technique Using a Water Immiscible Organic Liquid

The original apparatus of Michl[3,4] as developed by Ryle *et al.*,[5] has received many modifications[6] and several commercial types are available. The paper electrophoretogram is immersed vertically in a tank of water-cooled, organic liquid that is water immiscible. Such liquids are of course non conductors of electricity.

The Organic Liquid. Numerous liquids such as carbon tetrachloride, heptane, chlorobenzene, toluene and Varsol (Esso White Spirit 100) have been used. Carbon tetrachloride, although being non-inflammable, has a higher density than water and is therefore technically inconvenient. Toluene, although inflammable and toxic, has been widely used particularly with buffers containing a high proportion of pyridine. However, Varsol 100 is probably the best choice for use with buffers lacking or low in pyridine. Tanks employing Varsol and low pyridine buffers can be used almost indefinitely as long as the buffer is replenished from time to time. However, when toluene and high pyridine buffers are employed a very noticeable discolouration of buffer and coolant is apparent after a few days routine use. The rate of deterioration will depend on the conditions but with routine operation using 3 kV and 200 mA it is advised to change the buffer each week and the coolant every 2–3 months.

The Tank. Simple glass chromatography tanks such as the Shandon Panglass Chromatank have been widely used. Some workers have employed perspex tanks but it should be remembered that many organic liquids have a long term action on perspex so that only Varsol 100 is recommended. Even then it is advisable to screw any joints in the perspex as well as using cement.

The Cooling Coil. Heat dissipation is of course greatly improved by having a cooling coil preferably fitted to a special lid as shown in Fig. 3.6. However, although less efficient, a coil can be simply suspended on one of the walls inside the tank.

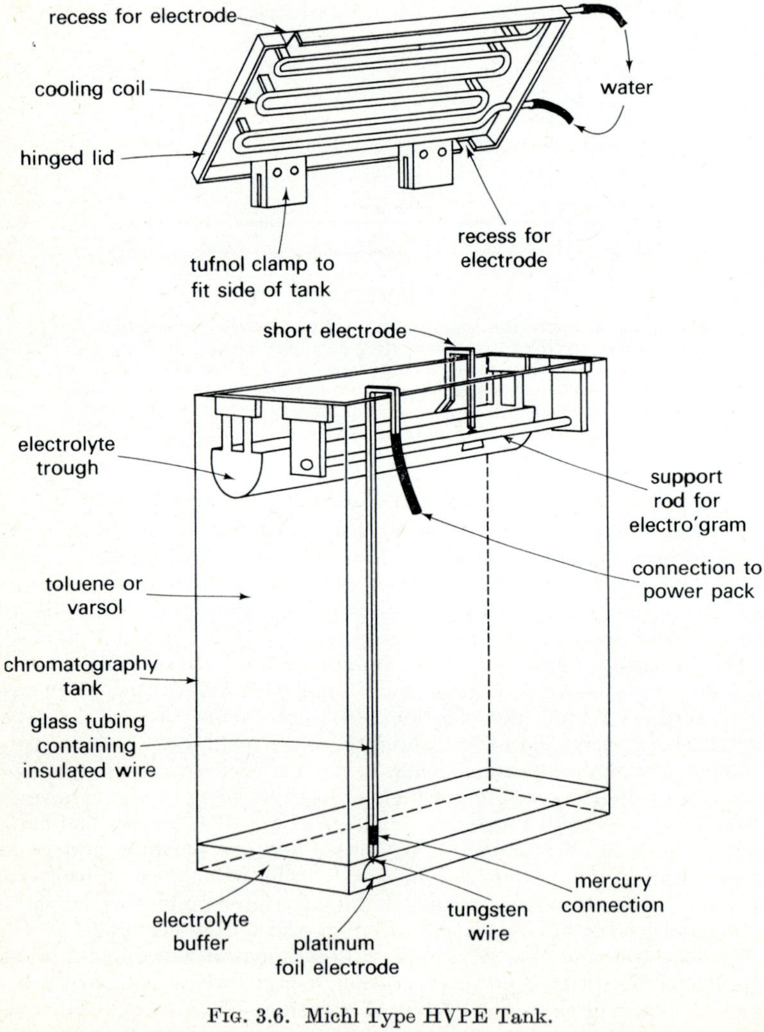

Fig. 3.6. Michl Type HVPE Tank.

The Electrolyte Compartments. Normally one of these is simply the bottom of the tank. The other compartment can be a tall jar standing inside the tank but this can obviously seriously restrict the space available for the electrophoretogram. A more favoured arrangement is to use a somewhat larger than normal chromatography trough fitted at the top of the tank (Fig. 3.6). Obviously the space must be allowed above the trough for any cooling coil fitted to the tank lid. Some workers have used a perspex construction as shown in Fig. 3.7 which can be easily moved in and out of the tank. It consists simply of a divided trough above which is attached a support for the electrophoretogram.

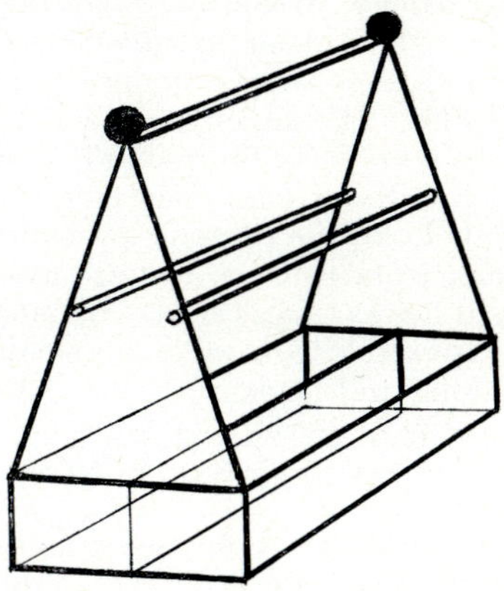

FIG. 3.7. Twin perspex electrolyte compartments and support frame for electrophoretogram.

Such a device places both electrolyte compartments at the bottom of the tank and enables the paper to be suspended in a compact tent like arrangement hence giving considerable economy of space.

The Electrodes. A widely used type of electrode consists of a piece of platinum foil spot welded to a short length of tungsten wire, the other end of which is sealed through into a suitable length of glass tubing (Fig. 3.6). A mercury connection is then made inside the tubing between the tungsten wire and a length of well insulated electrical wire connecting to a power pack. Simpler, although perhaps less safe, electrodes of platinum wire have also been used.

The Power Supply. Heat dissipation in a Michl tank apparatus is considerably less efficient than in the Gross type plate system. Hence voltages above 5 kV are not recommended unless very narrow paper strips are used. In most cases a 5 kV, 500 mA power pack has been used and many separations have been carried out quite satisfactorily at 2–3 kV.

Application of the Sample. The procedure is similar to that described above for the metal plate apparatus. With a tank of the size shown in Fig. 3.6. a whole sheet of Whatman 3 MM paper (57 × 46 cm.) can be used as the electrophoretogram. However, when such a large sheet is wetted with buffer it is not easy to handle so that half sheets 57 × 23 cm. are often more convenient. Obviously the number of samples to be applied may also determine the width of the paper. The point of application of the samples is determined by their nature and by the buffer being used to achieve their separation. If aminoacids are to be separated at pH 2 the sample is usually applied 10 cm. from the anode end of the paper. If tryptic peptides are to be run at pH 6·5 the sample is usually applied 15–20 cm. from the anode end. In nearly all separations the anode is at the bottom of the tank. Typical arrangements are shown in Figs. 3.12–15.

Safeguards Against Toxicity. It is strongly advised that any apparatus using a Michl type HVPE tank or pyridine in the buffers should be installed in a fumecupboard or at least under a vented hood.

Safeguards Against Fire. The use of toluene or Varsol in a Michl type HVPE tank obviously involves a fire risk. If a single HVPE tank is used it should be mounted above an open metal or fibre-glass tank of sufficient capacity to hold all the contents of the HVPE tank. This of course reduces the risk from spillage, leakage and tank fracture. Although such equipment has at times been set up more or less on a normal laboratory bench it is advisable to install it in a small cubicle or cupboard having an asbestos lining and fitted with automatic fire extinguishing equipment. With toluene or pyridine a fumecupboard is advised and in this case an automatic switch off for the fan should be linked to the fire equipment.

If several Michl type tanks are to be used a more elaborate system is necessary. A small room should be fitted with sufficient fumecupboard space to house all the HVPE tanks. The spillage tanks underneath should have a wide lead away pipe to a safe place outside the building or be connected to a mains solvent separation system. Automatic fire equipment should be installed and automatic switches linked to the extraction fan, power pack, water supply and a siren. A fire expert should obviously be consulted on all these matters.

Separation of Aminoacids

These amphoteric substances can easily be separated by electrophoresis particularly when high voltages are used. The very basic compounds can be separated from the more acidic ones throughout a wide range of buffers. It is more difficult to effect a good separation of the monoamino, monocarboxylic aminoacids but, with high voltages, compounds with only a small difference in isolectric point can often be separated without overlap of spots. When large numbers of aminoacids are present in a single specimen there is almost certain to be an overlap of spots so that the electrophoresis should be followed by chromatography in the second dimension.

The most useful buffer for the separation of large numbers of aminoacids present in biological fluids is a pH 1·6, 8% formic acid buffer. Other buffers are occasionally useful for the separation of specific aminoacids. For example, the cyclic anhydrides of arginino-succinic acid separate very well from each other and from the parent substance when a pH 5·3 buffer (100 ml. pyridine, 32 ml. glacial acetic acid, made up to 10 l. with water) is used.[7] Alkaline buffers are more useful for the separation of arginine, lysine and ornithine, but above pH 10 considerable streaking occurs.

In the examples given below a metal plate apparatus has been used but a Michl type tank is just as effective. Paper sizes and points of application would of course be different.

Preparation of Specimens.[8]

Urine. Urine can be applied directly to the paper, provided that it is free from protein. If protein is present by any of the standard tests, the specimen should first be deproteinized. This can be done rapidly, without significant dilution of the specimen, by precipitating the protein with picric acid. To 2 ml. urine in a test tube, 50 mg. crystalline picric acid is added. The mixture is then centrifuged for 3 minutes at 2,500 r.p.m. The yellow supernatant is decanted. If the supernatant is clear, the specimen has been adequately deproteinized. In an occasional specimen from a patient with marked proteinuria, the supernatant may still be cloudy, in which case the addition of a further 50 mg. picric acid should usually produce a clear supernatant. Picric acid does not interfere with electrophoresis at *p*H 1·6 as it moves slightly towards the anode, whereas all amino acids, except cysteic acid move towards the cathode The urine, as has previously been pointed out, needs no desalting. Three volumes of ethanol can be used instead of picric acid.

A volume of urine containing 12·5 μg. of creatinine is a suitable quantity for study in a one-way separation. If overloading is desired, however, 10 or 20 times this amount can be run, but in this case two-way separation would ordinarily be necessary to avoid overlap of large spots. For two-way separations by electrophoresis and chromatography, 25 μl. or a volume of urine containing 15 μg. creatinine is a suitable quantity.

Plasma can also be deproteinized with picric acid and run without desalting. Most 2 ml. plasma samples are adequately deproteinized by the addition of 150 mg. picric acid. An occasional sample, particularly one containing hæmolysed red cells may require the addition of a further 50 mg. picric acid to produce a clear supernatant. Approximately 1·3 ml. of protein-free plasma is recovered from the initial 2 ml. sample. Of this volume, 625 μl. is a suitable quantity for study.

Spinal Fluid is deproteinized by the addition of 25 mg. picric acid to 1 ml. of the fluid. As with plasma, desalting is unnecessary. 625 μl. is a suitable quantity for study.

Sweat is treated in the same way as cerebrospinal fluid. The supernatant, in this case, though adequately deproteinized, never becomes completely clear. 125 μl. is a suitable quantity for study.

Cellular Material is weighed, put into a solution of 75 per cent ethanol, homogenized and centrifuged. The supernatant is decanted, evaporated to dryness and taken up in a convenient volume of ethanol. This procedure gives satisfactory deproteinization for electrophoresis.

Alternatively, picric acid can be used for deproteinization; the homogenate is treated in the same way as plasma. A volume of solution equivalent to 30 mg. of tissue in the original specimen is a suitable quantity for study of aminoacids.

One-way Separation by Electrophoresis. In routine separations of aminoacids in biological fluids, the *p*H 1·6 buffer is used. In this case the origin should be marked 3 in. from the anode (right hand) end of the paper strip. Three urine specimens, each containing about 125 μg. of nitrogen or 12·5 μg. creatinine, can be analysed on a single strip on the Shandon apparatus. Whatman 3MM paper can be used, unless two strips are to be run simultaneously, in which case Whatman No. 1 paper is preferable. The specimens are applied to the paper in 10 μl. portions, drying in a hot air stream between applications. 6 kV is applied between the electrodes for 20 minutes to effect separation along 18 in. of paper not held by wicks. Following electrophoresis the paper strip is dried for a few minutes in an oven at 100°C, or by hanging at room temperature for 30 minutes. It is then ready for the application of the location reagent, usually ninhydrin. A few drops of pyridine should always be added to the ninhydrin solution immediately before dipping to overcome the acidity of the paper. When alkaline buffers are used, a few drops of glacial acetic acid should be added, as well as pyridine.

Electrophoresis followed by Chromatography. When a two-way separation is desired, electrophoresis should always be used for the first separation for three reasons. First, it automatically desalts the specimen. Second, it separates the aminoacids from picric acid if the specimen has been deproteinized by this method. Third, it avoids leaving chromatographic solvents on the paper as these may interfere with electrophoretic separation.

Following electrophoresis, butanol-acetic acid-water (12:3:5) produces excellent separation of amino acids, superior to that produced with lutidine. Any other chromatographic solvent can, of course, be used in the second dimension, depending on the type of separation desired (see Figs. 3.4 and 3.5).

Results. The positions assumed by aminoacids during a one-way electrophoretic separation in the *p*H 1·6 buffer can be determined from Fig. 3.8 which is a map of the positions taken by aminoacids in a two-way separation, using butanol-acetic acid-water (12:3:5) as the solvent for the chromatographic separation. Figs. 3.9–11 show photographs of the aminoacids present in normal plasma, normal cerebro-spinal fluid, and normal and abnormal urines.

From the above discussion, it will be apparent that there are several distinct advantages of two-way separation, using electrophoresis as the first separation as contrasted with a two-way chromatographic separation. First, the method is more rapid. Time is saved by (*a*) replacing one chromatographic run by a 20 minute electrophoresis; and (*b*) eliminating the need for desalting. Second, the spots are rounder and more compact than those produced with two-way chromatography using phenol as the first solvent without desalting. Third, very small quantities of aminoacids are often seen in two way "electrochromatograms,"

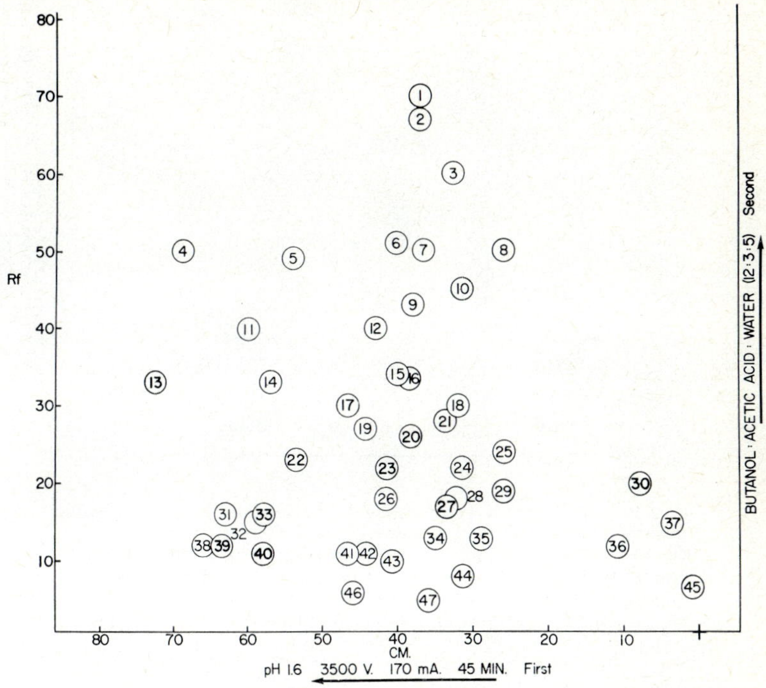

FIG. 3.8. Map of the positions taken by ninhydrin-positive substances in a two-way separation, using electrophoresis for the first separation and butanol-acetic acid-water for the second (chromatographic) separation. The aminoacids in standard solution were applied at the origin (+). Electrophoresis was carried out in the *p*H 1·6, 8 per cent formic acid buffer, using 3500 V for 45 minutes, developing a current of 170 mA. All aminoacids move towards the cathode. The chromatography was ascending. The spot identifications are: (1) leucine; (2) isoleucine; (3) phenylalanine; (4) delta-aminovaleric acid; (5) beta-aminoisobutyric acid; (6) valine; (7) methionine; (8) tryptophan; (9) kynurenine; (10) tyrosine; (11) gamma-aminobutyric acid; (12) alpha-aminobutyric acid; (13) ethanolamine; (14) beta-alanine; (15) proline; (16) 3-hydroxykynurenine; (17) alanine; (18) homocitrulline; (19) sarcosine; (20) threonine; (21) glutamic acid; (22) glycine; (23) serine; (24) hydroxyproline; (25) aspartic acid; (26) homocystine; (27) glutamine; (28) citrulline; (29) gamma-hydroxyglutamic acid; (30) taurine; (31) carnosine; (32) arginine; (33) methyhistidines; (34) asparagine; (35) penicillamine-penicillamine; (36) phosphoethanolamine; (37) cysteine sulfinic acid; (38) ornithine; (39) lysine; (40) histidine; (41) argininosuccinic acid C, cyclic anhydride, 5-membered ring; (42) argininosuccinic acid A; (43) cystathionine; (44) penicillamine-cysteine; (45) cysteic acid; (46) argininosuccinic acid B, cyclic anhydride, 6-membered ring; (47) cystine.

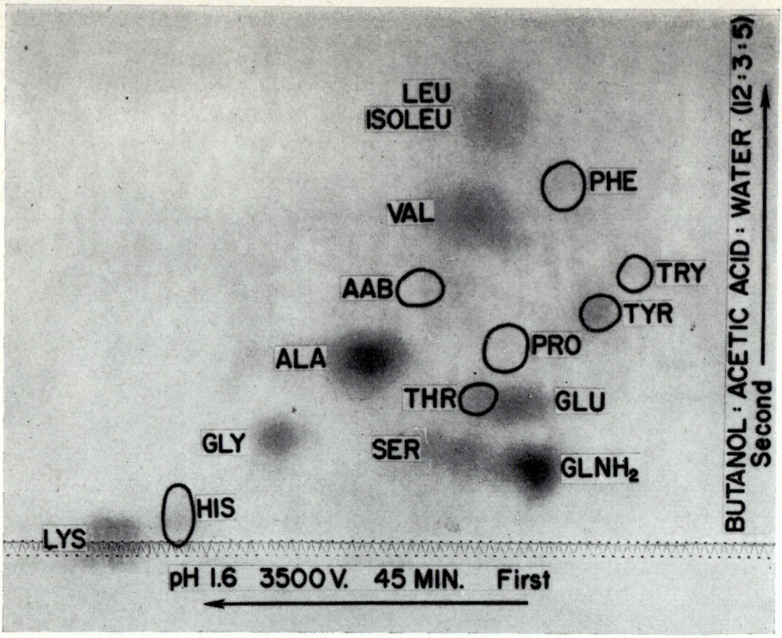

FIG. 3.9. Photograph of aminoacids present in 1 ml. of normal plasma. Electrophoretic separation was from right to left, using the *p*H 1·6 formic acid buffer. Following HVPE the paper strip was stitched onto a clean sheet of paper for chromatography. Butanol-acetic acid-water was then used as the solvent for ascending chromatography, using Whatman No. 3 MM paper.

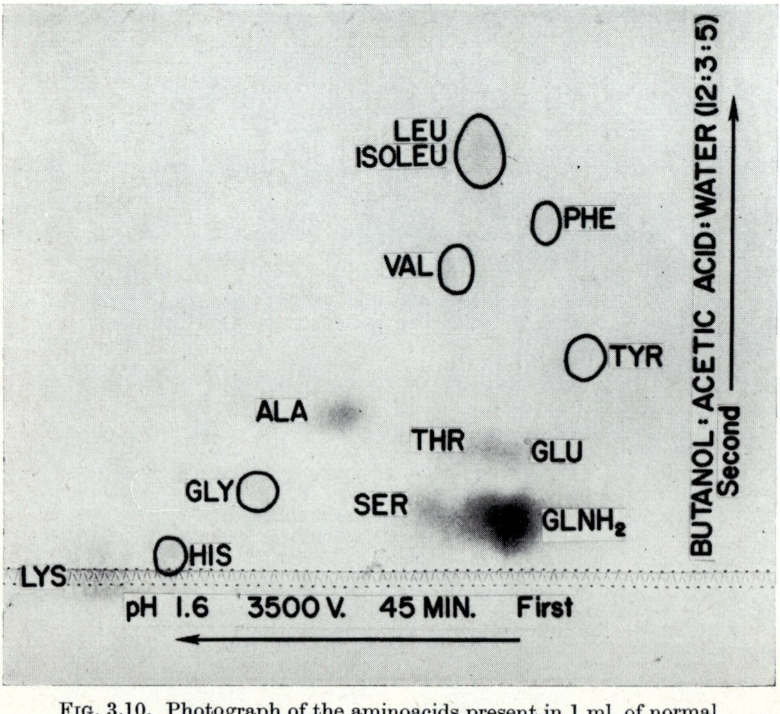

FIG. 3.10. Photograph of the aminoacids present in 1 ml. of normal cerebro-spinal fluid. Conditions as in Fig. 3.6.

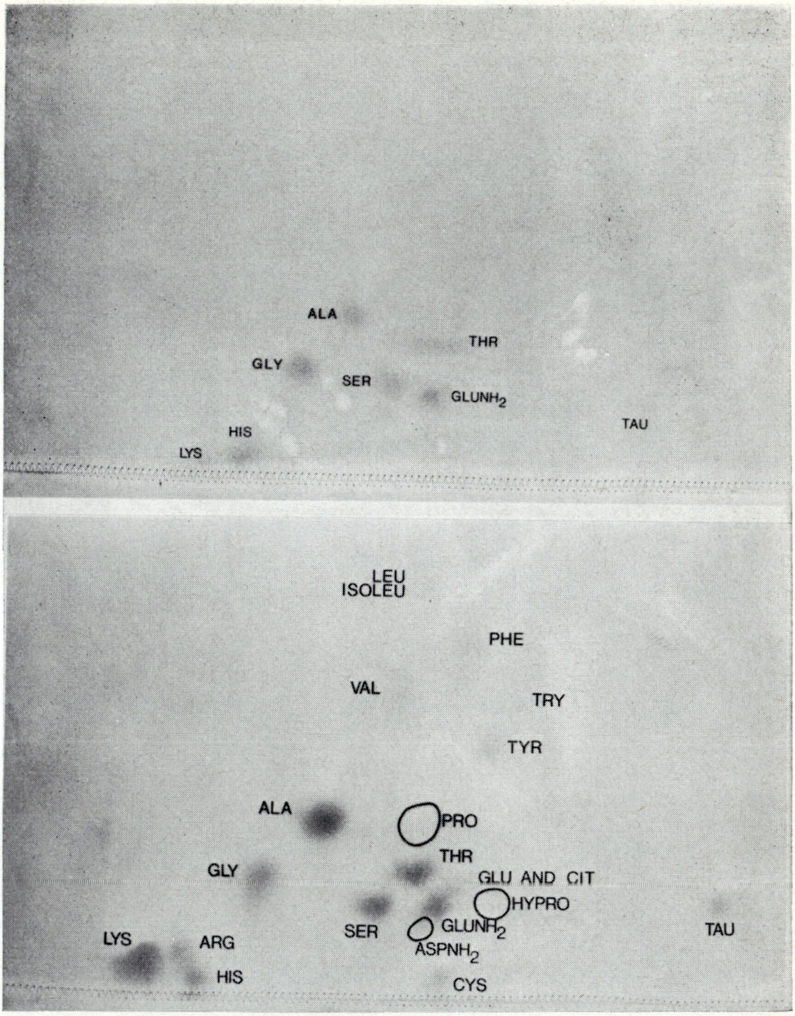

FIG. 3.11a & b. Photograph of the aminoacids present in normal urine and various abnormal conditions. An amount of urine equivalent to 15 micrograms of creatinine was applied at the origin. (*a*) Normal urine; (*b*) generalized aminoaciduria; (*c*) urine of neuroblastoma patient excreting cystathionine; (*d*) Hartnup disease; (*e*) argininosuccinic aciduria; (*f*) citrullinemia.

when they are not detected in the same specimen by two-way chromatography, even after desalting. Fourth, there is better separation of certain aminoacids. For example, methionine separates from valine so that oxidation is not necessary to detect this substance on electrochromatograms. Asparagine produces a distinct spot. Cystathionine, argininosuccinic acid and phosphoethanolamine are more distinctly separated from each other. In the phenolic solvents used in chromatography, these three important substances occupy nearly the same position.

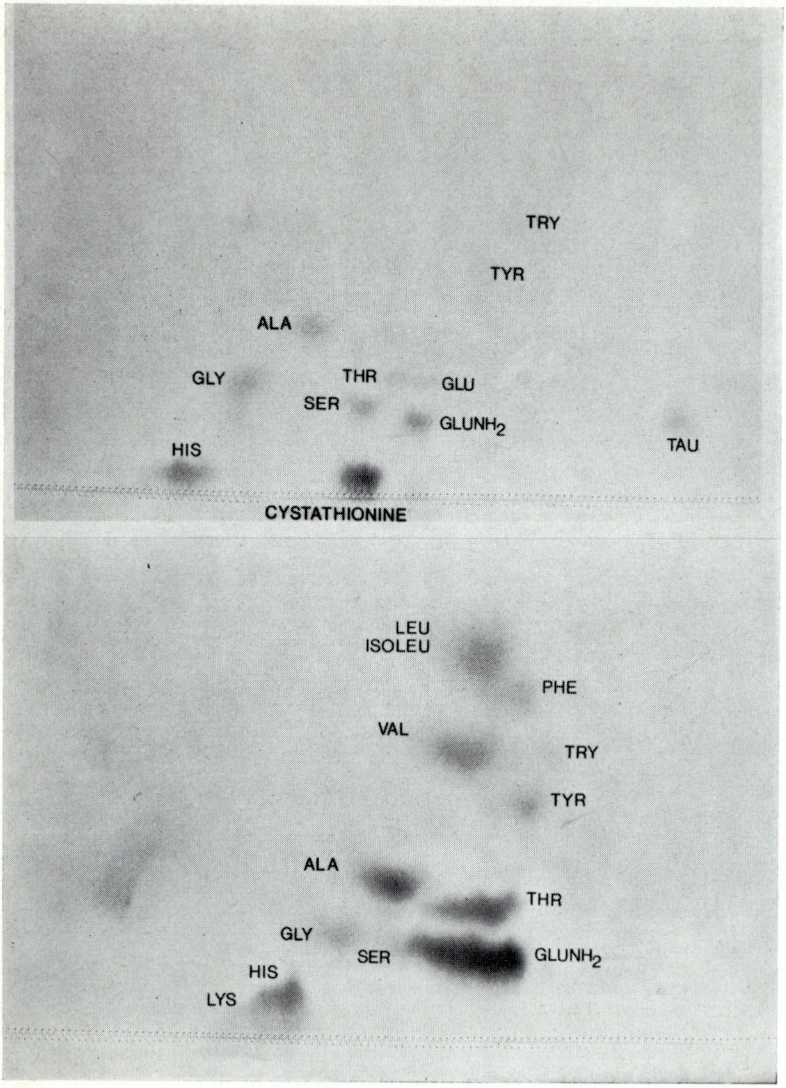

FIG. 3.11c & d.

Separation of Peptides

In most cases peptides requiring separation have been obtained from proteins by proteolytic digestion. Indeed this forms the basis of the "fingerprint" technique that has been extensively employed in the study

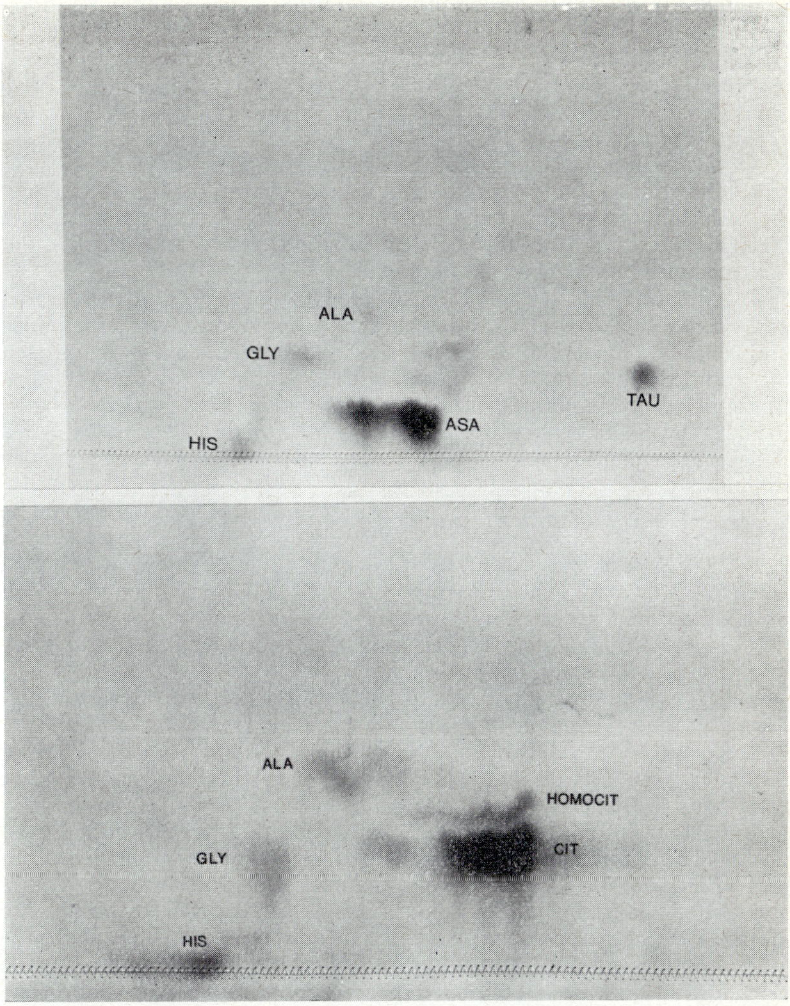

FIG. 3.11e & f.

of abnormal haemoglobin. In many cases the proteolytic enzyme is trypsin because of its high specificity for lysine and arginine residues. However, chymotrypsin and pepsin have also been used although they have wider specificities. In the examples given below a Michl type tank apparatus has been used for the electrophoretic separations but of course the cooled metal plate system would be just as efficient.

Proteolytic Digestion. A tryptic digestion of haemoglobin can be obtained as follows:

1. 0·2 ml. of a 5 per cent aqueous solution of hæmoglobin is added by means of a dropper to 4 ml. of a 1·5 per cent (v/v) solution of conc. HCl in acetone in a 6 ml. centrifuge tube. The acetone is cooled to 0°C just before use and is vigorously stirred during the addition.

2. The precipitated globin is centrifuged off and washed with 3 × 2 ml. acetone and finally with 2 ml. of ether. Residual solvent is removed by evaporation in a vacuum desiccator.

3. The globin (10 mg.) is dissolved in 1 ml. water, 0·1 ml. of an aqueous solution of trypsin (2 mg./ml.) is added followed by 0·1 ml. 1 M NH_4HCO_3 to give a *p*H of 8·5. The tube is then sealed with parafilm and incubated at 37°C for 2 hours.

4. After incubation the *p*H is lowered to 6·5 with 1 M acetic acid and the digest placed in a boiling water bath for 2 minutes to precipitate and coagulate insoluble "core" peptides. The precipitate is centrifuged off and the supernatant containing the soluble tryptic peptides is freeze-dried or evaporated in a beaker placed in a vacuum desiccator over NaOH pellets and conc. H_2SO_4.

Separated peptide chains can be digested with trypsin in a similar manner.

Tryptic digests of the light and heavy chains of γ-globulin can be obtained as follows:

1. 5·0 ml. of a 1 per cent solution of γ-globulin in 0·5 M tris/HCl (*p*H 8·2) is treated with mercaptoethanol to give a final concentration of 0·2 M and the reaction mixture left at room temperature for 1 hour.

2. The reaction mixture is then cooled in ice and treated with 1·1 ml. of 1 M iodoacetate and left for 1 hour.

3. Dialysis against 1 per cent NaCl solution is then carried out followed by dialysis against 1 M propionic acid.

4. The S-carboxymethyl chains are then separated using a column of sephadex G75 (4 × 55 cm.) eluted with 1 M propionic acid. 5 ml. fractions are collected.

5. Protein material is detected by UV absorption at 280 mμ and the heavy chain fractions and light chain fractions are combined in each case and freeze dried after dialysis against saline and water.

6. Intra-chain disulphide bridges are then oxidized by dissolving 10 mg. light chain or 20 mg. heavy chain in 0·5 ml. 98 per cent formic acid and adding 0·5 ml. of performic acid (made from 9·5 ml. of 98 per cent formic acid and 0·5 ml. 30 per cent hydrogen peroxide). The reaction mixture is left in an ice bath for 3 hours.

7. 50 vols. of water are then added and the solution freeze-dried. The residue is dissolved in 5 ml. water and again freeze-dried.

8. The residue of oxidized chain is dissolved in 1 ml. of 0·1 M NH_4HCO_3 and digested with 0·1 ml. trypsin solution (2 mg./ml.) at 37°C for 4 hours.

9. The *p*H is then lowered to 6·5 with 1 M acetic acid and any slight precipitate is centrifuged off. The supernatant is freeze-dried or evaporated in a vacuum desiccator.

Paper Electrophoresis and Chromatography. This is normally carried out on Whatmann 3 MM paper using volatile electrolytes and solvents; The most widely used electrolytes are:

pyridine:acetic acid:water (1:10:89), *p*H 3·5;
pyridine:acetic acid:water (25:1:225), *p*H 6·5;
pyridine:acetic acid:water (1:2:250), *p*H 4·4; and
ammonium carbonate 1 per cent (w/v), *p*H 9·0.

A potential gradient of 50–60 v/cm. is the most satisfactory, and a sample load of 1–2 mg./cm. width is generally applied.

The usual chromatography systems for separating peptides are:

butanol:acetic acid:water (4:1:5, top phase) (12:3:5, monophasic);
butanol:acetic acid:pyridine:water (150:30:100:120); and
iso-amyl alcohol:pyridine:water (30:30:35);

used in ascent or descent at room temperature for 12–24 hours.

The soluble tryptic peptides from the light or heavy chains of γ-globulin can be separated as follows:

1. The mixture of tryptic peptides, obtained from the digestion of 50 mg. polypeptide chain as described above, is dissolved in 0·1 ml. of *p*H 3·5 buffer and applied to a sheet of 3 MM paper (33 × 57 cm.) along a 25 cm. line, 10 cm. from a short edge, Fig. 3.12a using a Pasteur pipette and warm air drier to prevent spreading of the sample.

2. The paper is saturated with *p*H 3·5 buffer using a pipette or wash bottle to within 2 cm. of the application line. The buffer is then allowed to diffuse slowly into the sample so as to avoid spreading. Excess electrolyte is blotted off.

3. The paper is hung in a Michl type high voltage electrophoresis tank filled with "white spirit" with the origin nearest to the anode and a voltage of 3 kV is applied for 1 hour.

4. The electrophoretogram is dried at 50°C and 6 cm. wide strips are cut from each side (see Fig. 3.12b) and developed with ninhydrin. Using these strips as markers the areas of the electrophoretogram that contain peptides are cut out and sewn to new sheets of 3MM paper (29 × 46 cm.) (Fig. 3.12c) and the paper at the back of each sewn on strip is cut away.

5. Descending chromatography in butanol:acetic acid:pyridine: water is then carried out for 12 hours and the peptides are detected as before. Some peptides may need further purification using electrophoresis at *p*H 6·5 or chromatography in *iso*-amyl alcohol:pyridine: water.

6. Purified peptides are eluted from the paper with 1 M acetic acid and freeze-dried or evaporated in a vacuum desiccator over NaOH pellets and conc. H_2SO_4.

The tryptic peptides from 50 mg. digested globin, α- or β-chain can be separated in a similar manner.

1. The peptide mixture is dissolved in *p*H 6·5 buffer and applied to 3 MM paper on a line 15 cm. from a short edge. Electrophoresis is carried out at *p*H 6·5 in a Michl tank filled with toluene and peptides are detected as described above.

2. The areas of the electrophoretogram containing the peptides are sewn onto 3 MM paper and further separated by ascending chromatography in *iso*-amyl alcohol:pyridine:water. Some peptides may

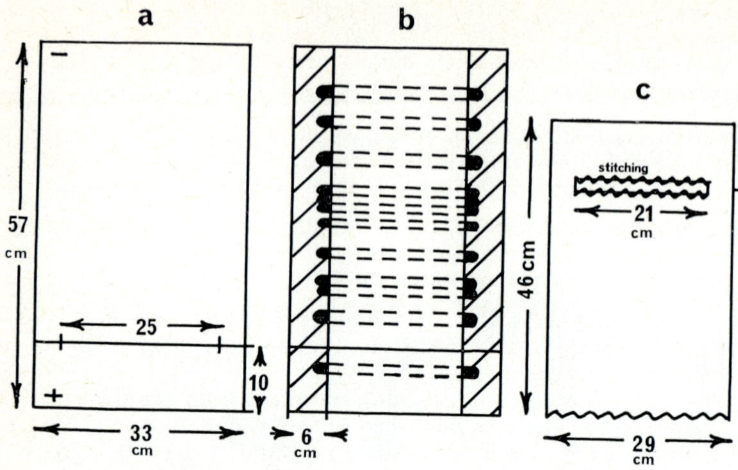

FIG. 3.12. The separation of peptides by paper electrophoresis and chromatography.

(*a*) The peptide mixture is applied to a 25 cm. line leaving 4 cm. margins and subjected to high voltage electrophoresis in a Michl type tank. The dimensions shown are for the separation at *p*H 3·5 of the tryptic peptides from 50 mg. digested protein. For separations at *p*H 6·5 the application line should be moved to 15–20 cm. from the end of the paper.

(*b*) 6 cm. wide strips (shaded areas) are cut from the edges of the electrophoretogram and developed in ninhydrin. Using the strips as markers the regions of the electrophoretogram containing peptides (dotted in areas) are cut out.

(*c*) Each cut out area is stitched to a new sheet of 3 MM paper and the paper beneath the sewn on strip is cut away. (It is important to use zig-zag stitching which overlaps the areas of the sewn on strips.) Chromatography is then carried out. The dimensions shown are for descending chromatography, and the bottom of the paper is serrated. For ascending chromatography the strip is stitched 4 cm. from the bottom of the paper and of course no serration is necessary.

The size of paper illustrated is convenient when a Shandon Panglass Chromatank is used, both for electrophoresis and chromatography. Other types of tank may of course necessitate an alteration in the dimensions shown.

require a third stage using electrophoresis at *p*H 3·5 or descending chromatography in butanol: acetic acid: water.

Fingerprinting. This is a very useful technique for the rapid comparison of two proteins or peptide chains. Globin can be fingerprinted as follows:

1. A mixture of soluble tryptic peptides from 2–3 mg. digested globin (see above) is applied to a sheet of 3 MM paper (23 × 57 cm.) as a spot 15 cm. from a short edge and 4 cm. from a long edge (Fig. 3.13a).

2. Electrophoresis is carried out at *p*H 6·5 as described above and the electrophoretogram dried at 50°C.

3. Ascending chromatography is then carried out in the second dimension using *iso*-amyl: alcohol: pyridine: water after a 3 MM sheet of equal area has been sewn to the electrophoretogram (Fig. 3.13b).

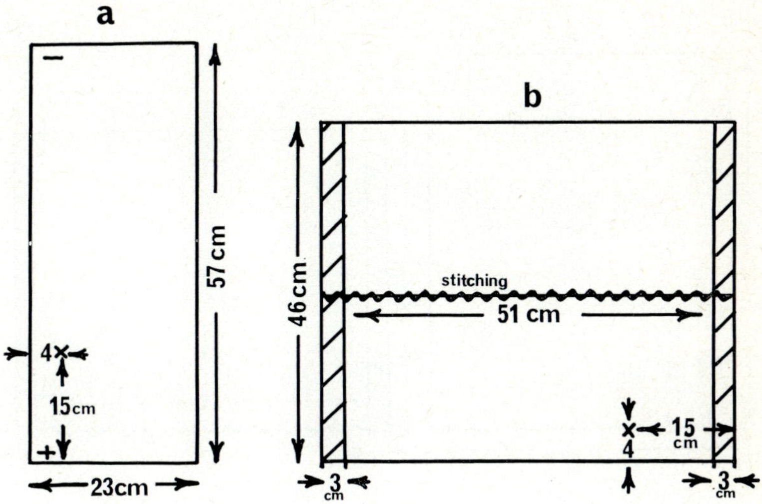

FIG. 3.13. Preparation of fingerprints of hæmoglobin.

(*a*) The tryptic digest of 2–3 mg. protein is applied to the origin (*x*) and subjected to high voltage electrophoresis at *p*H 6·5 in a Michl type tank.

(*b*) A sheet of paper of equivalent area is stitched to the electrophoretogram, a 3 cm. strip (shaded area) is cut off from each side and ascending chromatography is carried out overnight.

The dimensions illustrated are for use with a Shandon Panglass Chromatank.

4. The paper is dried at 50°C and developed in 0·2 per cent ninhydrin in acetone at 60–70°C for 5 minutes. Some peptide spots may give a very pale colour at first but this will deepen after a few hours at room temperature. Peptides with N-terminal glycine often give an initial yellow colour with ninhydrin and serine peptides a brownish-grey. Histidine, arginine, tyrosine, methionine, and tryptophan residues can be detected on the paper using the methods described in Vol. 1 chap. 5 Ninhydrin developed papers will gradually fade unless kept in the dark and in the cold. A permanent colour can be obtained using the cadmiumninhydrin reagent described in Vol. 1; see also chap. 4.

The light and heavy chains of γ-globulin can be fingerprinted as follows:

1. The digest from 2–3 mg. of peptide chain is applied to a sheet of 3 MM paper (46 × 57 cm.) as a spot 10 cm. from a short edge and 10 cm. from a long edge (Fig. 3.14a).

2. Electrophoresis is then carried out at *p*H 3·5 in a Michl tank as described above.

3. The electrophoretogram is dried at 50°C after which descending chromatography in the second dimension is carried out in butanol: acetic acid:pyridine:water (Fig. 3.14b)

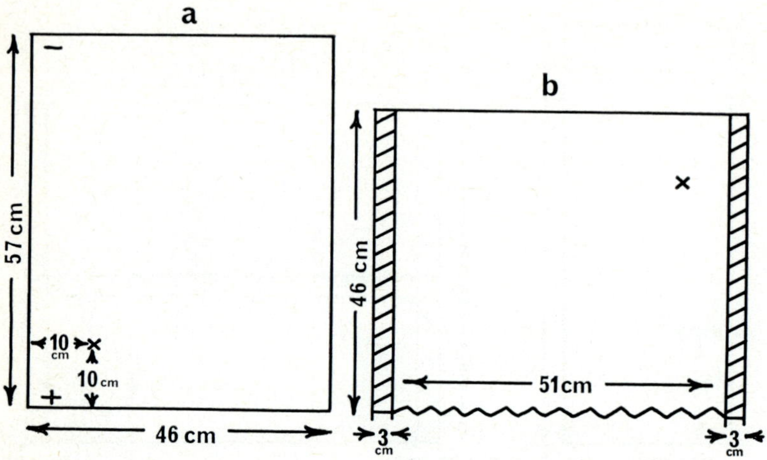

FIG. 3.14. Preparation of fingerprints of γ-globulin chains.

(*a*) The tryptic digest from 2–3 mg. protein is applied to the origin (*x*) and electrophoresis at *p*H 3·5 is carried out in a Michl type tank.

(*b*) A 3 cm. strip (shaded area) is then cut from each end of the electrophoretogram and descending chromatography is carried out in the second dimension after serrating the bottom of the paper.

The dimensions illustrated are for use with a Shandon Panglass Chromatank.

If a number of samples are to be compared it is sometimes advantageous to run several of them on the same electrophoretogram. The electrophoretogram can then be cut into strips and each strip containing the peptides of a sample can be sewn to a sheet of 3 MM paper ready for chromatography (Fig. 3.15).

With experience larger samples of digest (5–8 mg.) digested protein or peptide chain) can be fingerprinted and the time of the electrophoretic and chromatographic runs extended. These fingerprints can be developed with 0·02 per cent ninhydrin in acetone which enables an excess of peptide over ninhydrin to be present. The peptide spots can be eluted directly with constant boiling glass distilled HCl into 0·1 ml. capillary tubes (Fig. 3.16) sealed and hydrolysed. Alternatively the peptide spots can be eluted with 1 M acetic acid and the eluates used for dansyl-Edman degradation or further digestion with proteolytic enzymes. If digestions are to be carried out at *p*H 8·5 the peptide can often be eluted directly with 0·1 M NH_4HCO_3.

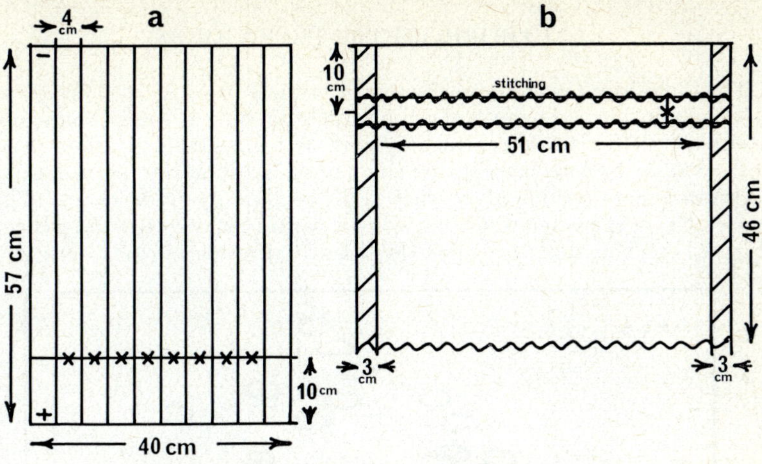

FIG. 3.15. The simultaneous preparation of eight fingerprints.

(*a*) A sheet of 3 MM paper (40 cm. wide) is marked into ten 4 cm. wide strips and samples of tryptic digests are applied (at *x*) to the eight inner strips. Electrophoresis at *p*H 3·5 is then carried out and the electrophoretogram cut into strips. For electrophoresis at *p*H 6·5 the origin should be positioned 15–20 cm. from the end.

(*b*) Each strip containing peptides is sewn to a sheet of 3 MM paper and the area of paper behind the strip is cut out. 3 cm. wide strips (shaded areas) are then cut away from the sides of the chromatogram and descending chromatography is carried out after serrating the bottom of the paper. For ascending chromatography each electrophoretic strip is sewn to a sheet of paper 4 cm. from the bottom.

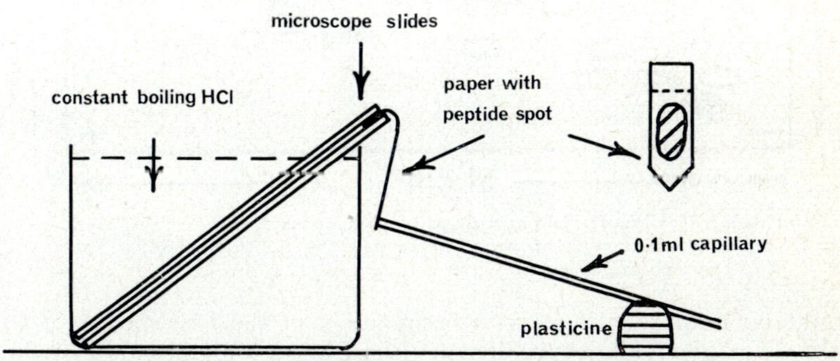

FIG. 3.16. The elution of peptide spots from fingerprints. A pointed strip of paper containing the peptide spot is cut out and the flat end is placed between two microscope slides lying diagonally in a dish and protruding a little way over the rim. The dish is filled with constant boiling HCl (for aminoacid analysis), water, or 1 M acetic acid (for dansyl-Edman degradation and further proteolytic digestion) and the end of a 0·1 ml. capillary is placed against the tip of the pointed end of the paper strip. Elution of peptide spots with HCl or acetic acid takes about 30 minutes. If water is used the elution is slower and it is generally necessary to collect the eluate into two capillaries to ensure complete recovery of the peptide. The capillary tube can be conveniently held in position by means of a small piece of plasticene. Peptides which are to be digested at *p*H 8·4 can often be eluted directly with 0·1 M NH_4HCO_3.

The fingerprint technique has been widely used in the study of hæmoglobin variants most of which differ by only a single aminoacid residue.[9] Thus sickle-cell hæmoglobin or hæmoglobin S differs from normal adult human hæmoglobin by having a valine residue instead of a glutamic acid residue at position 6 of the β-polypeptide chain. This causes the displacement of one of the peptide spots on a fingerprint (Fig. 3.17). The fingerprint has also found extensive use in the study of

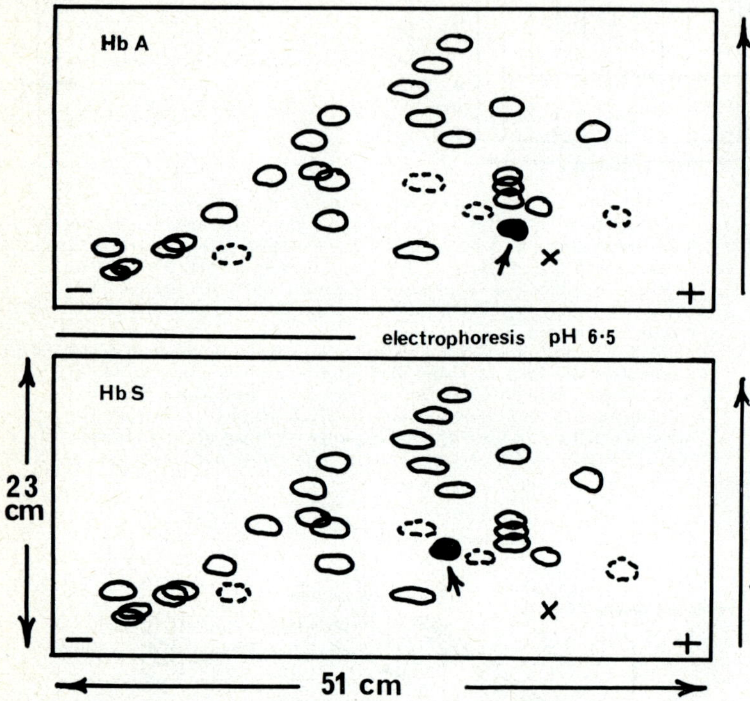

FIG. 3.17. Fingerprints of human hæmoglobins A and S. The black spot marked with an arrow indicates the peptide which differs in the two fingerprints.

immunoglobulins.[10,11] Thus a comparison of the fingerprints of the light (k) chains of allotype Inv(a+) and Inv(a−) human γ-globulins showed one tryptic peptide to be different (Fig. 3.18) due to the substitution of a leucine residue by valine.

The Separation of Indoles

A rapid and reproducible separation of indoles and other Ehrlich-reactors can be effected by HVPE using the following buffers.

*Pyridine-Acetic Acid Buffer p*H 6·1. Pyridine 400 ml. plus glacial acetic acid 32 ml. Make up to 10 l. with water and adjust *p*H if necessary with acetic acid or pyridine. This buffer is useful for the separation of indoles.

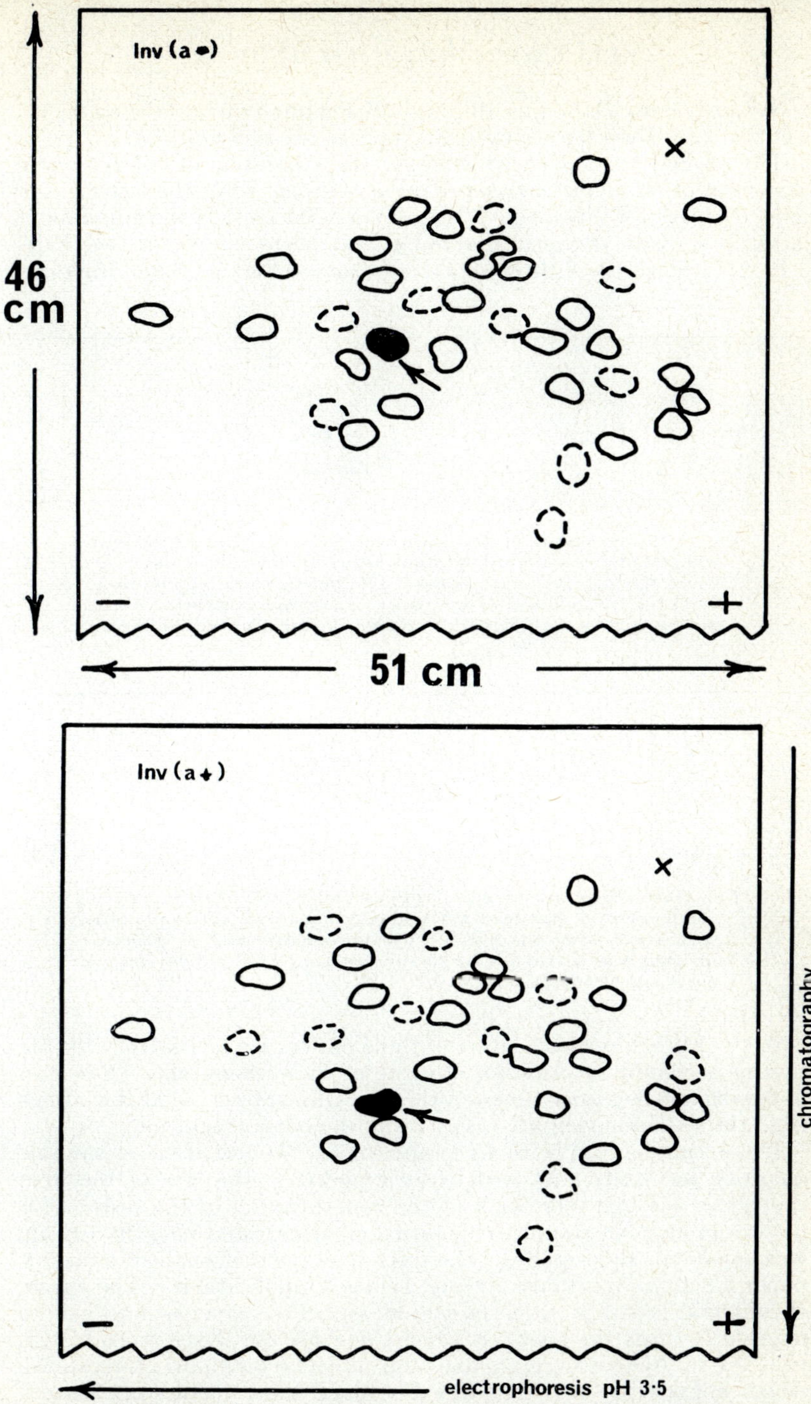

FIG. 3.18. Fingerprints of human γ-globulin light (K) chains Inv (a+) and Inv (a−). The black spot marked with an arrow indicates the peptide which differs in the two fingerprints.

Sodium Borate Buffer pH 10. Add 2M sodium hydroxide to an 0·05M solution of sodium borate until the desired pH is obtained.

The former is the most effective for the separation of indolyl acetic acid from other indoles likely to appear in urine, while the latter is the most effective for separating 5-hydroxy indolyl acetic acid from indolyl lactic acid, indolyl acetyl glutamine and porphobilinogen (see Figs. 3.19–21). Since these three substances have different colour reactions

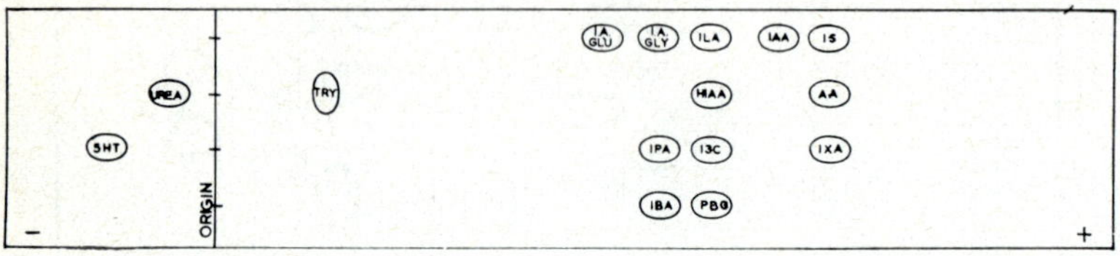

FIG. 3.19. Diagram of the positions taken by Ehrlich-positive compounds in standard solution after one-way electrophoresis, using the pH 10 borate buffer. The indoles were applied 3 in. from the cathode end of the paper. Electrophoresis was carried out for 30 minutes, using 6 kV and developing a current of 60 mA.

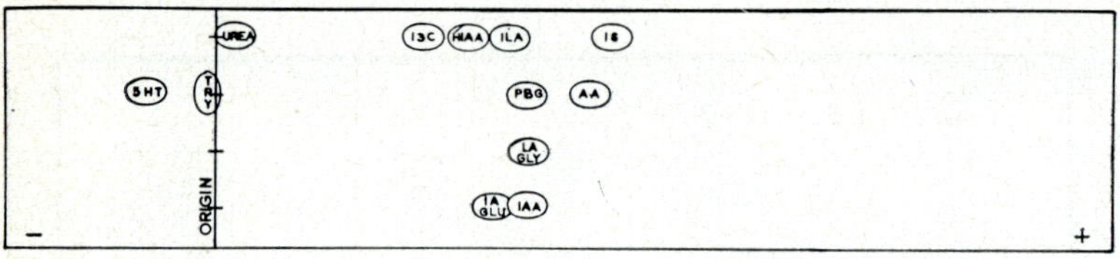

FIG. 3.20. Diagram of the positions taken by Ehrlich-positive compounds in standard solution after one-way electrophoretic separation using the pH 6·1 pyridine-acetate buffer. Electrophoresis was carried out for 30 minutes using 6 kV and developing a current of 60 mA.

with the Ehrlich reagent, either of these buffers are satisfactory for the routine screening of urine for any one of these chemicals. These two buffers should be used at precisely the pH values stated; small deviations from the recommended pH values result in poorer separation.

Urine can be run without preparation. 50–100 μl. is a suitable quantity for study. As with chromatography, the size of the urea spot gives a rough indication of the concentration of the urine. For more reproducible results, a quantity of urine containing 50–100 μg. creatinine should be applied. The specimen should be applied to the dry paper in 10 μl. portions, drying between applications. The origin should be at least 3 in. from the cathode end of the paper as most indoles migrate towards the anode at pH 6·1 and pH 10. Exceptions to this rule are 5 hydroxy-tryptamine and other tryptamine derivatives, which migrate towards the cathode. Since many indoles are decomposed by light, better results are obtained if the specimen is applied to the paper and dried in subdued light. After drying at room temperature the spots are located wtih Ehrlich reagent, although another location reagent may occasionally be useful.

Applications and Results. There are few Ehrlich-reactors in normal urine. Because of this paucity of indoles and because the substances can be identified by their colour reactions as well as by their positions on the paper, a one-way separation is suitable for the rapid screening of large numbers of urine samples. This is particularly true when screening for a particular substance such as porphobilinogen in acute porphyria or hydroxyindolyl acetic acid in a suspected case of argentaffinoma (see Fig. 3.19). In specimens containing large quantities of different indoles

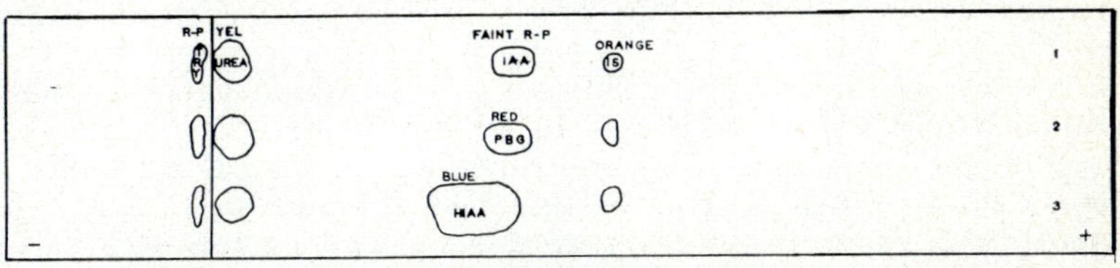

FIG. 3.21. Diagram of the positions taken by indoles in urine. The upper strip (1) shows the Ehrlich-positive compounds present in normal urine. The middle strip (2) shows the spot of porphobilinogen present in 100 μl. of urine from a patient with acute porphyria. The lower strip (3) shows the large spot produced by HIAA in 100 μl. of urine from a patient with an argentaffinoma. Electrophoresis was carried out in the *p*H 6·1 pyridine-acetic acid buffer, using 6 kV for 30 minutes.

(e.g. in urine after a tryptophan load) one-way separation is not adequate for final definition of the indole pattern, because of overlapping of the huge spots. If overlapping spots are present after one-way electrophoresis, the specimen should be re-examined using two-way chromatography or by electrochromatography, using *iso*-propanol ammonia as the solvent during the chromatographic run following electrophoresis.

The Separation of Purines, Pyrimidines and Related Derivatives

These substances are amphoteric, therefore the direction of migration varies with changes in the *p*H of the buffer. Their separation by HVPE is rapid and reproducible.

Preparation of Specimens. Urine can be run without preparation, although the presence of protein will cause slight streaking of the purine spots. This does not ordinarily interfere with interpretation but a more perfect result can be obtained after deproteinization with three volumes of 95 per cent alcohol.

Buffers. In the *p*H 1·6, 8% formic buffer, all substances tested moved, if they moved from the origin at all, towards the cathode. On the other hand, in the 0·05 M borate buffer at *p*H 9·2, 8% all except cytosine moved, if they moved at all, towards the anode. (See Figs. 3.22 and 3.23.)

The origin should be drawn at least 3 in. from the cathode end when the *p*H 9·2 buffer is to be used (to ensure that cytosine does not move off the end of the paper). Electrophoresis for 30 minutes at 6 kV gives a good separation. The paper strip can be dried in an oven for a few minutes at about 100°C.

Location Reagents.

Ultra-violet Light. Most of these substances absorb ultra-violet light strongly and this is a sensitive location reagent. The paper must be well-dried before viewing under the U.V. lamp. In addition, after the *p*H 1·6 buffer is used, the absorption is much more marked if the paper has been hung for a few minutes in a glass jar in an atmosphere of ammonia before viewing. This overcomes the acidity of the buffer (see Vol. I).

Bromophenol Blue Reagent. This is a good location reagent for these compounds see Vol. 1, chapter 8. In addition to the blue staining of many of the compounds, after electrophoresis in the *p*H 9·2 buffer a distinct brown colour is given by uric acid, thymine and uracil. Apparently this is due to the reaction of silver with purines on a paper impregnated with alkali. The same effect can be achieved, almost as successfully, after the *p*H 1·6 buffer is used if a few drops of ammonia (Sp. Gr. 0·880) are added to the reagent immediately before dipping.

Clinical Applications. These have been few so far, but since the description of xanthinuria[12] and of orotic acid,[13] many clinics have become interested in the screening of urine from children with non-opaque renal stones for the presence of xanthine and patients with megaloblastic anæmia for orotic acid. Electrophoresis at *p*H 9·2 in the borate buffer is an excellent, rapid and accurate method for the diagnosis of these conditions. 50 μl. of a random urine specimen is a suitable quantity for study. The specimen is applied 3 in. from the cathode end of the paper and, after evaporating to a small volume, in 10 μl. portions, drying in a hot air stream between applications. Electrophoresis is continued for 30 minutes. The paper is dried in an oven at 100°C for 5 minutes and is then ready for location with the bromophenol blue reagent. Xanthine moves farther towards the anode than does uric acid; xanthine stains blue whereas the uric acid spot is a distinct brown colour. Hence the identification can be made without difficulty (see Fig. 3.23). The entire process can be completed within an hour and three 50 μl. specimens can be analysed on a single strip. Orotic aciduria is easily diagnosed by HVPE since orotic acid migrates rapidly towards the anode in the *p*H 9·2 buffer; this provides a rapid screening for the disorder.

Imidazoles

Imidazoles are basic substances; hence they migrate rapidly toward the cathode on HVPE. Fig. 3.24a is a map of a standard mixture of imidazoles separated by HVPE at *p*H 1·6 followed by chromatography in butanol-acetic acid-water (12:3:5). In histidinemic urine, imidazole lactic and imidazole acetic acids can be seen as distinct spots after electrophoresis in this system (Fig. 3.24b). These compounds can be

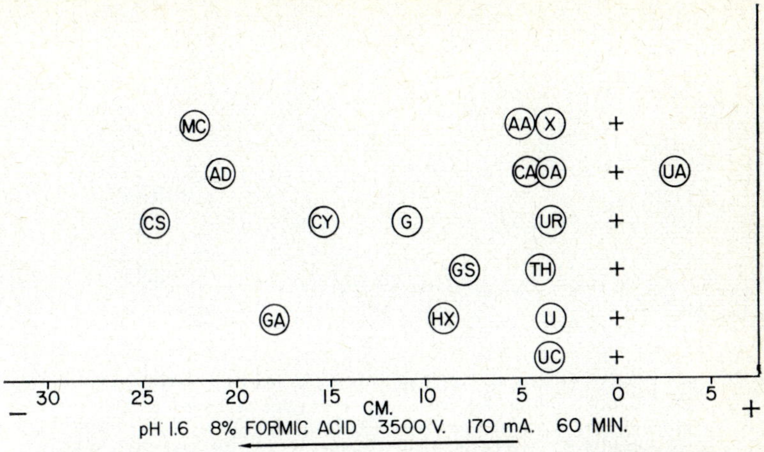

FIG. 3.22. One-way separation of purines, pyrimidines and related substances by HVPE at *p*H 1·6. The spots were located with UV light and traced. In the *p*H 1·6 formic acid buffer, all substances tested moved, if they moved from the origin at all, towards the cathode. On the other hand, in the 0·05 M borate buffer at *p*H 9·2, all except cytosine moved, if they moved at all, toward the anode (Fig. 3.23). UA = uridylic acid; UC = uracil; UR = uridine; U = uric acid; TH = thymine; CA = cytidylic acid; CS = cytosine; CY = cytidine; AA = adenylic acid; AD = adenine; GA = guanylic acid; GS = guanosine; G = guanine; X = xanthine; HX = hypoxanthine; OA = orotic acid; MC = 5-methyl cytosine.

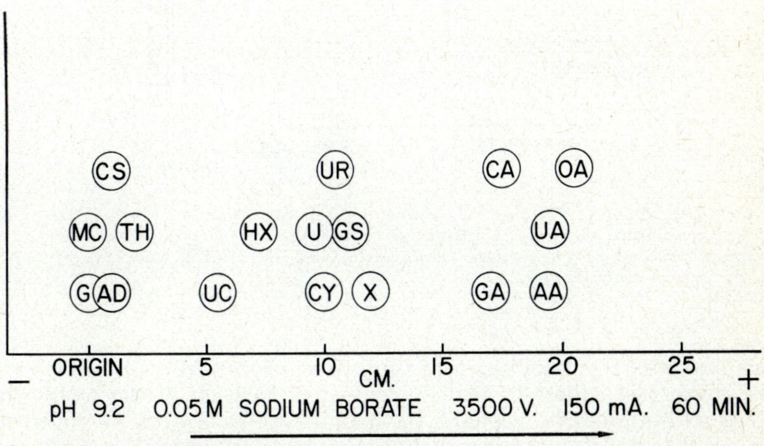

FIG. 3.23. One-way separation of purines, etc. at *p*H 9·2. For key see Fig. 3.22.

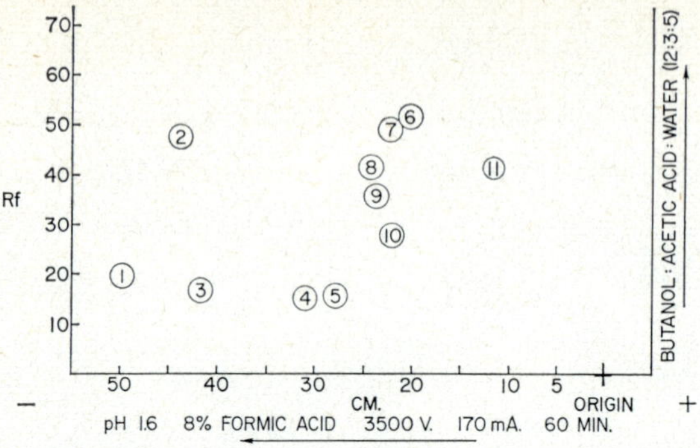

FIG. 3.24a. Separation of imidazoles by HVPE in the pH 1·6 formic acid buffer followed by ascending chromatography in butanol-acetic acid-water (12:3:5). (1) Histamine; (2) imidazole; (3) histidinol; (4) carnosine; (5) histidine; (6) urocanic acid; (7) dihydrourocanic acid; (8) 4-amino-5 carboxamide imidazole; (9) imidazole acetic acid; (10) imidazole lactic acid; (11) tyrosine.

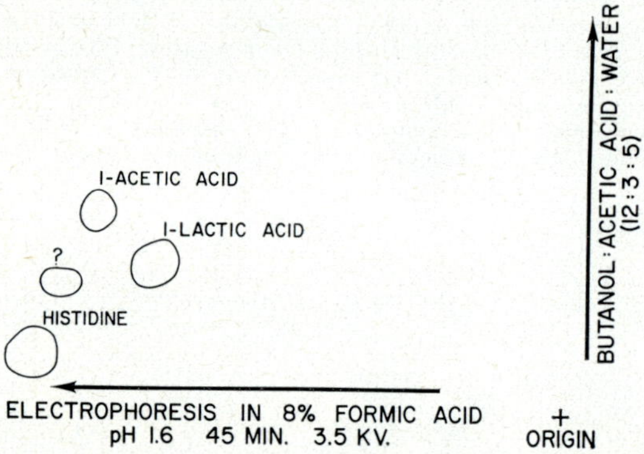

FIG. 3.24b. Tracing of imidazoles in histidinemic urine. An amount of urine containing 15 micrograms of creatinine was applied at the origin. Separation was by HVPE followed by chromatography.

demonstrated by staining the aminoacid electrochromatogram with the diazotized sulfanilic acid reagent after location of the aminoacid spots with ninhydrin reagent. In the author's laboratory we routinely dip in diazotized sulfanilic acid reagent all ninhydrinstained electrochromatograms which show any increase in the size of the spot in the position of histidine and the methyl-histidines. If the large spot is due mainly to histidine it stains red; if the patient has histidinemia the spots of imidazole lactic and imidazole acetic acids are clearly demonstrable. If the spot is due mainly to the methylhistidines there will be no red colour in the histidine position.

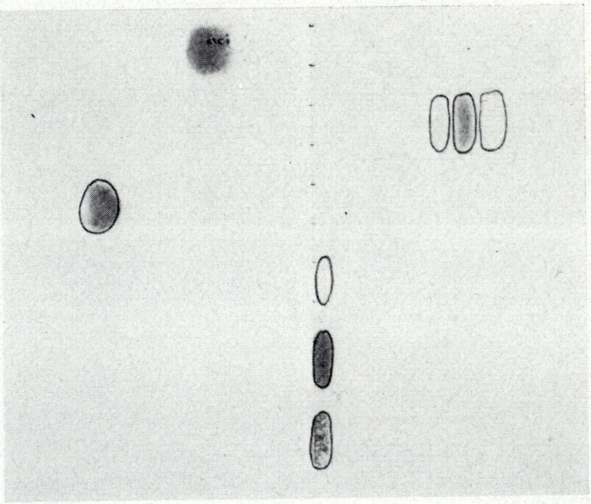

FIG. 3.25. One way separation of aromatic standards and mixtures on 10″ papers using 2 kV for 10 mins. at *p*H 5.3. Reading from the top down: 1. mixture of aromatic acids derived from 2. mixture of amines (phenylethylamine, tyramine and dopamine), 3. sulphate esters (p-hydroxyphenylacetic acid sulphate), 4. tyramine-O-sulphate, 5. tyrosol, 6. N-acetyl tyramine.

A family with imidazole aciduria has been described.[7] A sample of urine from one of these patients showed a large spot just to the cathode side of histidine, in the position of carnosine. The spot stained with ninhydrin and with diazotized sulfanilic acid reagent. Since carnosine separates well from histidine by HVPE, this technique offers a simple screening method for detection of carnosinuria.

Aromatic Amines and Metabolites

Aromatic amines, acids and their metabolites are readily separated into groups at *p*H 5·3.[16] Fig. 3.25 shows a one way electrophoretic separation of synthetic compounds and mixtures. The acids travel as a group, amines show some separation and neutral compounds or zwitterions remain at the origin. Each origin is a one cm. streak and these are separated by 1 cm. gaps. Two way separations of a single mixture found in urine after ingestion or injection of 1-^{14}C-tyramine in man and rabbit is shown in fig. 3.26. These separations were carried out on the smaller 10″ squares of paper. The method is a valuable, rapid way of examining such metabolites.

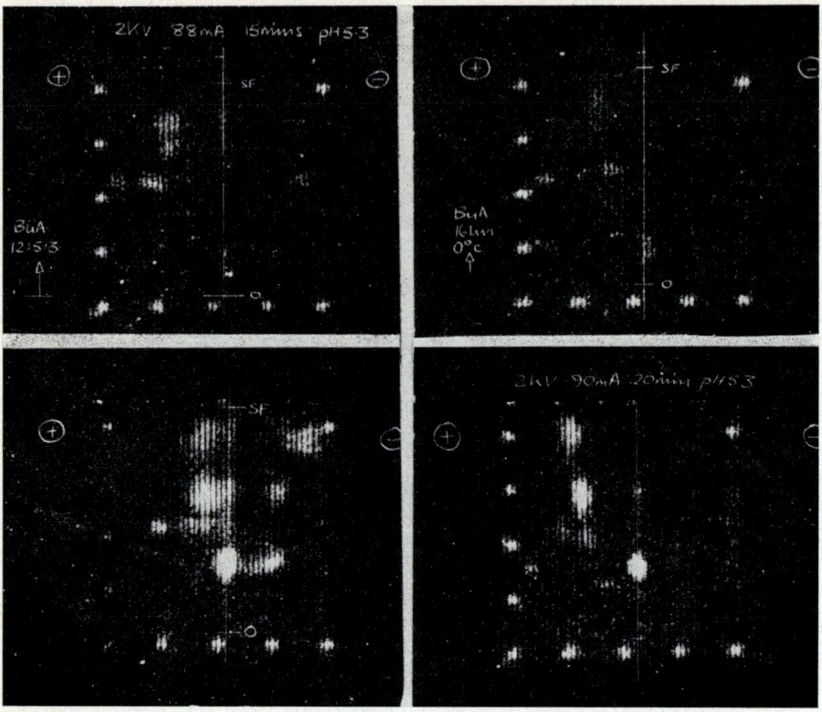

FIG. 3.26. Two-way separation by chromatography (butanol-acetic acid—water) and electrophoresis (2 kV, *p*H 5.3, 15 mins.) of 1-C^{14} tyramine metabolites following chloroform-methanol extraction of dried urine. Upper pair from rabbit and lower pair from man; left pair following intravenous injection and right pair after oral administration. Complete experiment on 10″ square paper with origin 12.5 cm. from side and 5 cm. up from bottom. The spots were located using a radiochromatogram spark chamber scanner with a 10 minute exposure as described in Volume 1 Chapter 15. Even spots at edges represent radioactive grid for aligning papers.

REFERENCES

1. Gross, D., *Nature*, 1953, **172**, 908. *Nature*, 1954, **173**, 487.
2. Atwood, G., and Morris, C. J. O. R. *Biochem. J.*, 1961, **81**, 606.
3. Michl, H. Monatsch. *Chem.*, 1951, **82**, 489.
4. Michl, H., *Chromatog.*, 1959, **1**, 11.
5. Ryle, A. P., Sanger, F., Smith, L. F., and Kitai, R., *Biochem. J.*, 1955, **60**, 541.
6. Katz, A. M. Dreyer, W. J., and Anfinsen, C. B. *J. Biol. Chem.*, 1959, **234**, 1897.
7. Bessman, S. P. and Baldwin, R., *Science*, 1962, **135**, 789.
8. Efron, M. L. *Bibchem. L.*, 1959, **72**, 691.
9. Ingram, V. M. *The haemoglobins in Genetics and Evolution.* Columbia University Press. London and New York, 1963.
10. Frangione, B. and Franklin, E. C., *J. Exp. Med.*, 1965, **122**, 1.
11. Baglioni, C., Alescio Zonta, C. and Cioli, D. *Science*, 1966, **152**, 1517.

12. Dent, C. E. and Philpot, G. R. *Lancet*, 1954, (**I**), 182.
13. Huguley, C. M., Jr., Bain, J. A., Rivers, S. and Scoggins, R. *Blood*, 1959, **14**, 615.
14. Studnitz v., and Hanson, A. *Scand. J. Clin. and Lab. Invest.*, 1959, **11**, 101.
15. Armstrong, M. D., McMillan, A. and Shaw, K. N. F. *Biochim. Biophys. Acta*, 1957, **25**, 422.
16. Smith, I, March, S. E., and Gordon, A. J. *Clin. Chim. Acta.*, 1972, **40**, 415.

Acknowledgements. figs. 3.4, 3.5, 3.8–11, and 3.19–24 are taken from the work of Dr. Mary Efron (see previous edition volume 2 Chapter 5).

Equipment Suppliers

Shandon Southern Instruments Ltd., Camberley, Surrey.
Shandon Inc. Sewickley, Pa.
Savant Instruments Inc., Hicksville, New York.
Gilson Medical Electronics, Madison, Wisconsin.

CHAPTER 4

THIN-LAYER ELECTROPHORESIS

W. J. Ritschard

ELECTROPHORESIS is a very powerful method for separating macromolecules having charged groups. Especially in protein chemistry, the several forms of electrophoresis are of great value both for analysing protein mixtures, and for preparative work. In order to avoid denaturation, these electrophoresis runs are mainly carried out at low voltages and often in the cold.

Chromatography is the method of choice for analytical or preparative separation of low molecular weight substances. Based on an alternative principle, high voltage electrophoresis offers a complementary method for their separation. By combining both methods an excellent separation of complex low molecular mixtures can be obtained. Ingram(1) first described this technique, producing fingerprints, or peptide maps in the case of proteolytic digests. As a support he used paper sheets on which tryptic or chymotryptic digests of proteins yielded a characteristic pattern of peptide spots. Later, this method was improved by Katz, Dreyer and Anfinsen.(2)

The equipment for preparing fingerprints on paper sheets is rather bulky and complex. Large power supplies for generating 5,000–10,000 V are used and special systems are needed to cool the paper sheets. It was therefore logical to adapt thin-layer techniques for the combined procedure. Thin-layer chromatography and electrophoresis require much simpler equipment, and the combined procedure allows a considerable saving of time and smaller amounts of material are needed. The procedures described here are very useful for analysing enzymatic protein digests and fractions from column chromatography as well as for metabolic studies. Thin-layer electrophoresis also has been used for the separation of amines, amino acids,(3,4,5,6,7) peptides,(8,9) proteins,(10,11,12) phenols, naphthols, phenol carboxylic acids,(4,13) dyes(4,14) and also some inorganic substances.(15)

General Techniques

Apparatus. The apparatus* described here (Fig. 4.1) combines several advantages resulting from our own experience with high voltage paper electrophoresis and thin-layer chromatography.(8) It consists of two electrode vessels (inside dimensions 200 mm. × 50 mm. × 40 mm., each holding 200 ml. of buffer), a cooling platform of aluminum (190 mm. × 190 mm.) through which water can be circulated, the thin-layer plate and the cover plate. A power supply with an output of

* This apparatus and power supply are available from Shandon Southern Instruments, Camberley Surrey. Desaga G.m.b.H, Hendellaux, Heidelberg Germany, and Carnag CH-4132 Muttenz, Switzerland.

0–1,000 V and 0–200 mA is not expensive and shortens the time of the electrophoretic runs considerably. It is also suitable for making micropreparative runs. Tap water is sufficient for cooling the thin-layer plate; a refrigerated brine is not necessary and immersion in a water immiscible organic solvent can be avoided. The short distance between the thin-layer plate and the cover plate acts like a moist

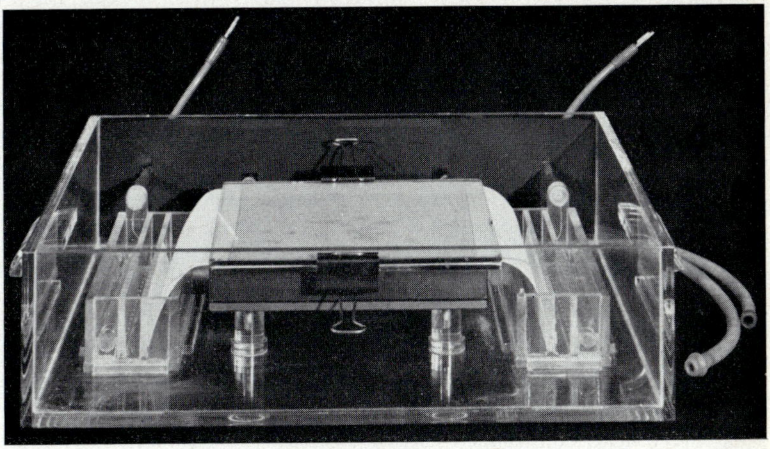

A

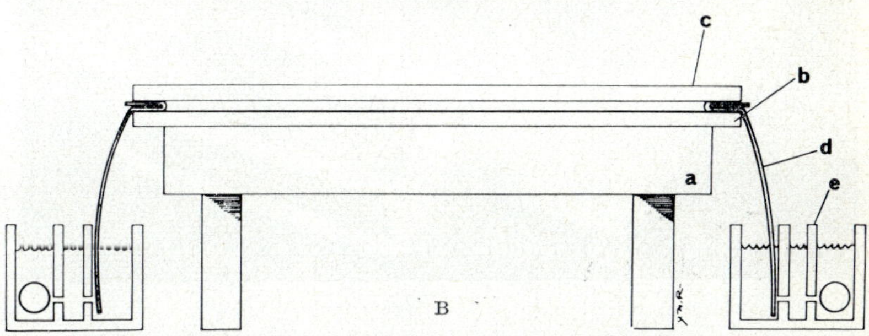

FIG. 4.1. Electrophoresis chamber. (A), Assembled chamber with electrical connections and rubber tubing leading to the cooling coil. (B), Schematic drawing. (*a*) Aluminum platform (190 mm. × 190 mm.). (*b*) Coated thin-layer plate. (*c*) Cover plate. (*d*) Refolded paper connection. (*e*) Electrode vessel. (C), Shandon apparatus based on this design.

chamber and prevents the thin-layer plate from drying out. Standard glass plates of 200 mm. × 200 mm. are used and the known methods for spreading the thin-layer supports can be employed. The electrode vessels are linked to the thin-layer plate by connections of Whatman No. 3 paper.

Thin-layer Supports. Silica gel, kieselgur, alumina and cellulose are commonly used as thin-layer supports. Inorganic supports have the advantage over cellulose, that the plates can be sprayed with aggressive reagents for detection of the substances. For preparing peptide maps silica gel and cellulose have been used. The best resolutions and the least tailing for the electrophoretic separation of dyes were obtained with kieselgur and silica gel, alumina giving rather poor results.[14]

Other materials include polymer sheets for the separation of alkali metal mixtures which is an interesting new application.[16] Sephadex G 100 and G 200 combines gel filtration and electrophoresis for the bidimensional separation of proteins[10] (see Chap. 3 and Vol. 1, Chap. 33).

Buffers. For the separation of amino acids and peptides volatile buffer systems are exclusively recommended. By drying the plates after electrophoresis, the buffer is practically eliminated and therefore

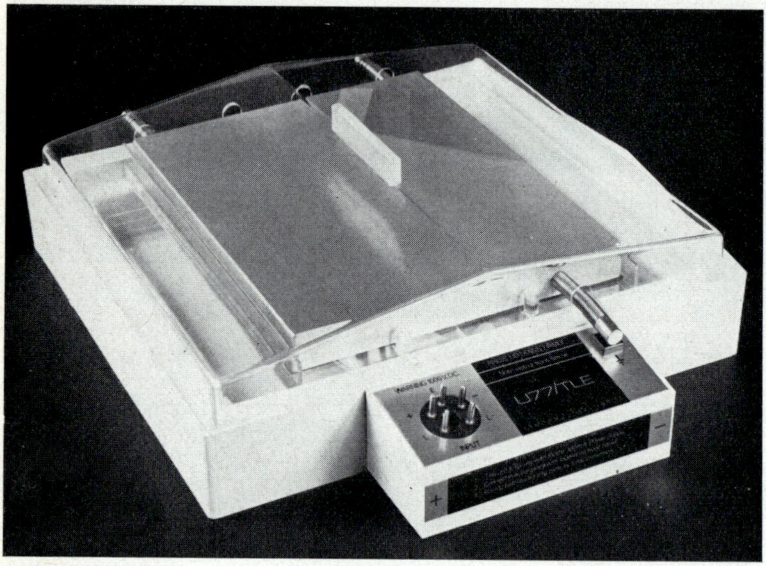

FIG. 4.1. C

interference with the spraying reagent is excluded. The removal of salt is also important when micropreparative electrophoresis is performed, the peptides being scratched off the plate and then eluted from the support.

The following buffers are useful for separating amino acids and peptides:

1. Buffer *p*H 3·5: 1 ml. pyridine plus 10 ml. of acetic acid made up to 500 ml. with distilled water.

2. Buffer *p*H 6: 100 ml. pyridine plus 10 ml. of acetic acid made up to 1,000 ml.

3. Buffer *p*H 8: 0·1M or 0·05M ammonium carbonate adjusted to *p*H 8 with carbon dioxide. [$(NH_4)_2\,CO_3.H_2O = 114{\cdot}1$].

Other buffer systems:

Buffer *p*H 2·4: equal volumes of 0·4M acetic acid and 0·12M formic acid.

Buffer *p*H 5·5: 30 ml. of pyridine plus 10 ml. of acetic acid made up to 3,000 ml.

For the separation of coal-tar food colours the following buffers are recommended[17]:

1. 1M acetic acid.
2. Buffer solution *p*H 4·0: potassium hydrogen phthalate 750 ml. of 0·1M or 15·32 g., plus 6 ml. 0·1M sodium hydroxide diluted to 1·5 litres.
3. Buffer *p*H 6·0: potassium dihydrogen orthophosphate 750 ml. of 0·1M or 10·21 g., plus 85·5 ml. 0·1M sodium hydroxide diluted to 1·5 litres.
4. Buffer *p*H 8·0: potassium dihydrogen orthophosphate 750 ml. of 0·1M or 10·21 g., plus 702 ml. 0·1M sodium hydroxide diluted to 1·5 litres.
5. Buffer *p*H 9·2: 0·05M borax. ($Na_2B_4O_7.10H_2O = 19{\cdot}07$ g/l.).
6. 0·1M ammonium hydroxide.

For thin-layer electrophoresis of inorganic ions 0·005M lactic acid and 0·1M NaOH have been employed.[15]

Preparing the Plates. The thin-layer plates are prepared according to Brenner *et al.*[18] without activation. However, a more diluted slurry is used: 25 g. silica gel G in 100 ml. of water. This proportion gives better results especially when preparing fingerprints, since the resulting thin-layer is thinner and more uniform than usual. After drying at room temperature overnight these plates have good mechanical properties and are suited for chromatography and electrophoresis. They can be stored over long periods giving reproducible results.

Applying the Sample, Spraying the Plates, Electrophoresis. The samples are applied to the dry plate with a micropipette in the form of small spots. The plate is then evenly sprayed with the selected buffer, avoiding any excess of buffer; 20–40 ml. buffer per plate is usual. After spraying, the plate should have a dull appearance. If the surface shines, too much buffer has been applied. Paper wicks (Whatman No. 3, 190 mm. × 100 mm.) are prepared by wetting them in the buffer. After blotting between two filter paper sheets they are refolded once. The electrode vessels are then joined by these paper bridges to the thin-layer plate, which is laid on the cooling platform, the cover plate is brought into position, the two plates are clamped and the current switched on (Fig. 4.1).

Preliminary electrophoretic runs are made as follows: various amounts of the sample are spotted on a line in the middle of the thin-layer plate, allowing at least 10 mm. between the spots. After a run of 15 or 30 minutes at 950–1,000 V the plate is removed, dried for 10 minutes at 100°C and the separated substances developed with a suitable reagent. These results decide the optimal sample concentration and time of the run and whether the sample should be applied near to the electrode or in the middle of the plate.

Freeze-drying of Thin-layer Plates. In order to overcome the difficulties of zone-migration of dyes which occurs on drying thin-layer plates

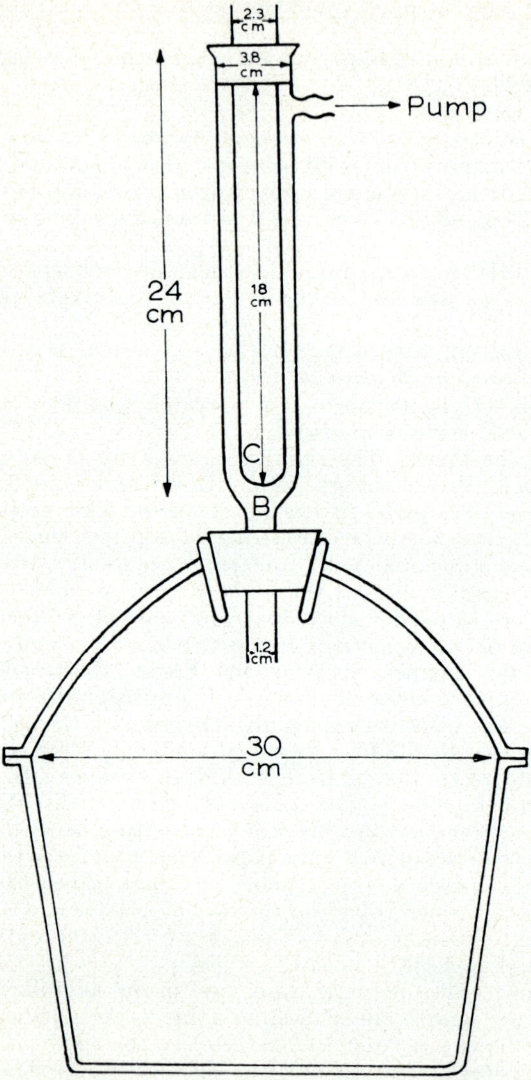

FIG. 4.2. Apparatus for freeze-drying thin-layer plates 200 mm. × 200 mm.[19]

at elevated temperatures, freeze-drying is recommended by Criddle, Moody and Thomas.[19] This method is also useful when substances sensitive to heat are separated on thin-layer plates.

Immediately after electrophoresis the freezing is carried out by placing the thin-layer plate on a 2·5 cm. thick platform of brass or aluminum which is surrounded by powdered carbon dioxide. The adsorbed water freezes in about 20 seconds. The plate is then dried in a desiccator to which a condenser with a cold finger is fitted (Fig. 4.2). Vacuum is applied and after about 30 minutes the plate is free of water. In most cases it is sufficient to connect the desiccator to an ordinary cooling trap by a large joint and vacuum tubing, thus avoiding the cold finger.

Peptide Maps

Peptide maps, or fingerprints have proved very useful in characterizing proteins by the pattern of peptides resulting from the application of chromatography and electrophoresis to a proteolytic digest. In our experience the most reproducible results for making peptide maps on thin-layer plates are obtained when the conventional procedure is reversed by carrying out the chromatography in the first dimension, followed by electrophoresis in the second dimension. If electrophoresis is employed first, the plates have to be heated at 100°C before chromatography. Heat, however, will change the silica gel layer in an uncontrollable way and thereafter chromatography leads to unreproducible peptide separations.

Digestion of Proteins. Many enzymes may be used for the digestion of proteins. Trypsin and chymotrypsin are the most specific ones, but subtilisin, papain and pepsin are also frequently employed. The most convenient procedure for preparing a digest with trypsin and chymotrypsin is that of Katz *et al.*,[2] since the final peptide mixture is free of salts. Ammonium bicarbonate 0·2M or 0·1M at *p*H 8·5 is used as buffer. At the end of the reaction the digest is directly applied to the thin-layer plate and the buffer salt volatilized in a stream of air.

Example: Preparing a tryptic digest of protamine sulfate. 5 mg. of protamine sulfate are dissolved in 0·1 ml. of distilled water in a small centrifuge tube. 0·01 ml. of 1M ammonium carbonate *p*H 8·5 is added (buffer prepared by bubbling CO_2 through a 1M solution of ammonium carbonate to reach a final *p*H of 8·5) and then 0·01 ml. of a trypsin solution (5 mg. of trypsin in 1 ml. of 0·001M HCl). Bacterial contamination is prevented by the addition of a small grain of thymol. The tube is then covered and the reaction mixture left at room temperature for 15 to 24 hours. After centrifugation the enzymic digest is ready for preparing the peptide map (Fig. 4.4). (See Vol. 1, Chap. 4.)

Chromatography-Electrophoresis. In order to find the optimal solvent mixture, preliminary chromatographic runs are made with 200 mm. × 50 mm. plates. The best solvent systems for peptide separation are listed in Table 4.1. At the end of the run, which takes between 1 and 2 hours, the plates are dried at 100°C for 10 minutes and developed with ninhydrin spray. Chromatography, with the chosen solvent for preparing the peptide maps is then carried out on 200 mm. × 200 mm.

plates. Electrophoresis in the second dimension is carried out as previously described (see p. 69).

When longer runs are necessary for adequate chromatographic separations, a modified B-N chamber is useful[8,20] (Fig. 4.3). It allows a continuous flow of the solvent over the thin-layer plate. The solvent evaporates at the open end of the chamber to which an electric heating element is clamped to accelerate evaporation. A cooling platform keeps the chamber at an even temperature during the chromatographic run.

TABLE 4.1

Solvent systems for chromatography of enzymatic protein digests

A. Neutral systems	
1. 96 per cent Ethanol-Water	70:30
2. *n*-Propanol-Water	70:30
B. Basic systems	
1. 96 per cent Ethanol-34 per cent Ammonium hydroxide	70:30
2. *n*-Propanol-34 per cent Ammonium hydroxide	70:30
3. Chloroform-Methanol-34 per cent Ammonium hydroxide	40:40:20
C. Acidic systems	
1. 96 per cent Ethanol-Water-Acetic acid	70:20:10
2. *n*-Propanol-Water-Acetic acid	70:20:10
3. *n*-Butanol-Water-Acetic acid	80:20:20

Sample Application, General Procedure. The solvent system for chromatography and the buffer for electrophoresis is chosen according to the results of preliminary chromatographic and electrophoretic runs. The direction of peptide migration during preliminary electrophoresis determines whether the spots for preparing the peptide maps should be applied at a corner or in the middle of an edge of the plate. Usually, migration occurs towards the cathode, permitting corner application for chromatography. The amount of peptide mixture applied varies from 0·05 to 0·5 mg. To obtain highly concentrated deposits, repeated application can be made and the spots, not exceeding 4 mm. in diameter, are dried with a hair dryer (excessive heat will damage the coating on the plate). Following chromatography, the plate is heated for 10 minutes at 100°C, cooled, sprayed with the selected buffer, and subjected to electrophoresis in the direction perpendicular to that in which the chromatography was carried out. With the pyridine-acetic acid buffer electrophoresis at 950–1,000 V and 30 mA is completed in 1 hour. The plate is then dried at 100°C and the peptide spots are developed with ninhydrin, or by using the chlorine *o*-tolidine test.[21,22] Use of fluorescamine is described on p. 86. The locations of the spots should be marked with a pencil immediately after applying the reagent because the colour diffuses and fades rapidly.

Examples:

1. The peptide map of a tryptic digest of protamine sulfate[8] is illustrated in Fig. 4.4. Experimental details are given on p. 74.

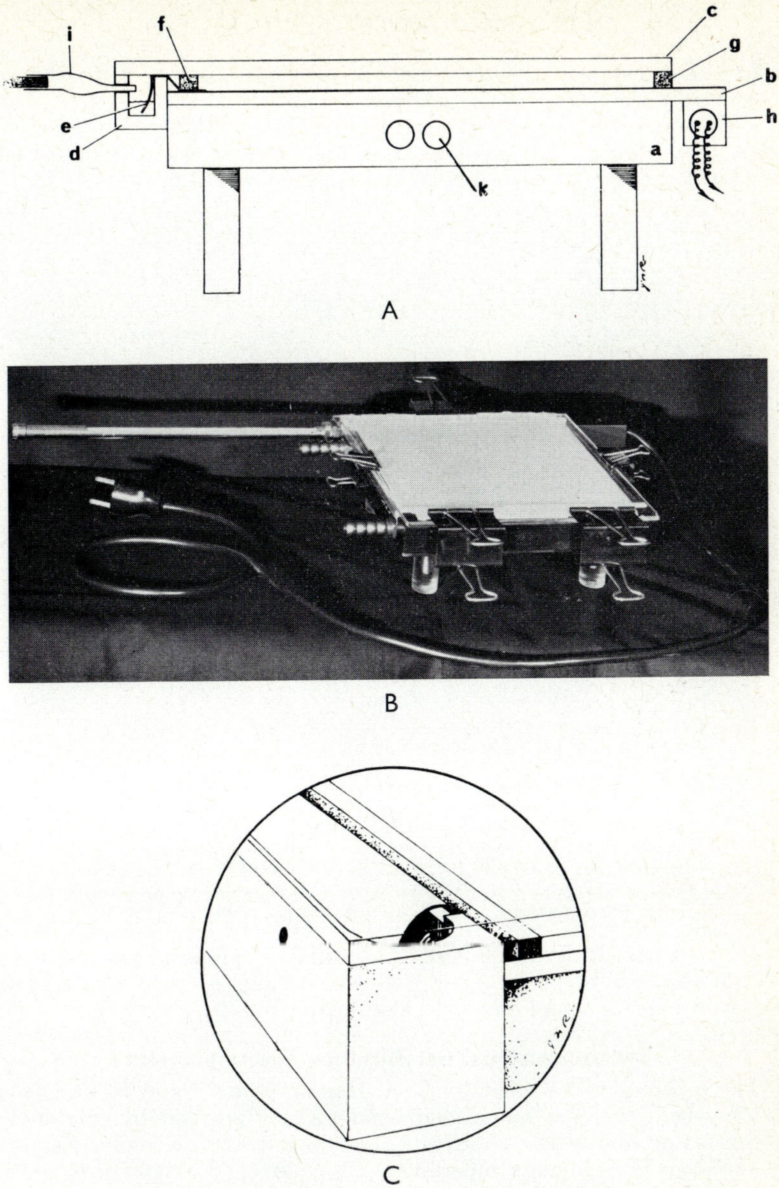

FIG. 4.3. Modified B-N chamber. (A), Schematic drawing. (*a*) Aluminum platform (190 mm. × 175 mm.) with cooling coil (k) inside. (*b*) Coated thin-layer plate. (*c*) Cover glass plate. (*d*) Solvent trough. (*e*) Paper connection (175 mm. × 30 mm.). (*f*) Teflon strip (200 mm. × 50 mm. × 30 mm.). (*g*) Small teflon spacers (5 mm. × 5 mm. × 3 mm.). (*h*) Heating element. (*i*) Pipette with polyethylene catheter for introducing the solvent. (B), Assembled chamber. (C), Detailed view of the solvent trough.

2. Fig. 4.5 shows a peptide map of a tryptic digest of myosin of which more than 60 peptides were separated.[8]

3. Wieland *et al.*, prepared peptide maps on thin-layer plates coated with silica gel S.[9] The separation of a tryptic digest of ribonuclease (oxidized with performic acid) is illustrated by Fig. 4.6.

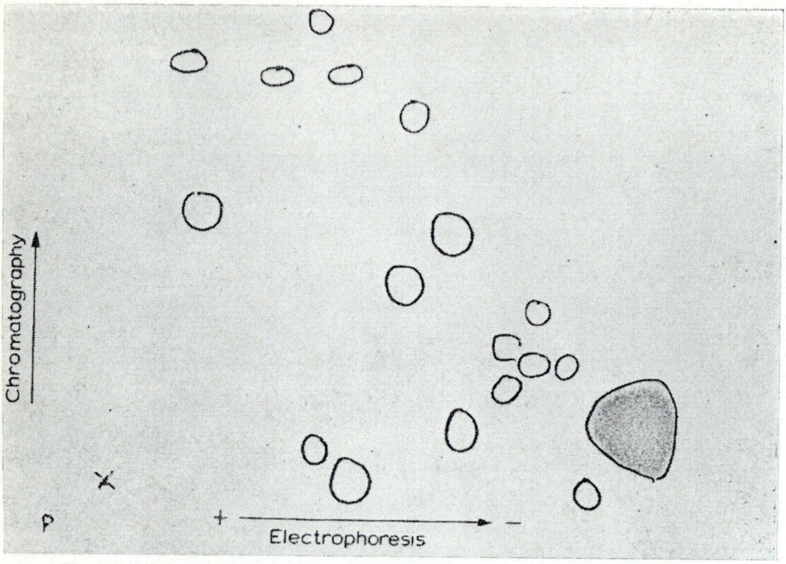

FIG. 4.4. Peptide map of the tryptic digest of protamine sulfate.[8] Enzyme to substrate ratio = 1:100. Buffer: 0·1M ammonium carbonate, *p*H 8·5. Incubation time: 24 hours. First dimension: chromatography. Solvent system: chloroform-methanol-ammonium hydroxide (40:40:20). Time: 60 minutes. Second dimension: electrophoresis. Buffer: pyridine-glacial acetic acid-water (1:10:489). Current: 980 V, 30 mA. Time: 1 hour.

4. Silica gel H and cellulose as thin-layer support was used by Stegemann and Lerch.[6] Fig. 4.7 shows the peptide map of a tryptic digest of horse hæmoglobin; see also Burns and Turner[24].

Two-dimensional Gel Filtration-Electrophoresis

The techniques described by L. A. Hanson *et al.*[11] and P. Fasella *et al.*[12] permit the two-dimensional separation of proteins by combining gel filtration and electrophoresis. According to the molecular weights of the samples Sephadex superfine G 25, G 50, G 75, G 100 or G 200 is used. By modifying these techniques slightly, standard glass plates of 200 mm. × 200 mm. can be used and electrophoresis can be carried out with the electrophoresis apparatus described earlier.

Preparing the Plates. The Sephadex gels are allowed to swell for several hours in the buffer according to Table 4.2. The suspension is homogenized by shaking the slurry in a rubber-stoppered Erlenmeyer flask and transferred to a graduated 100 ml. cylinder. After sedimentation the Sephadex layer without the buffer above should measure

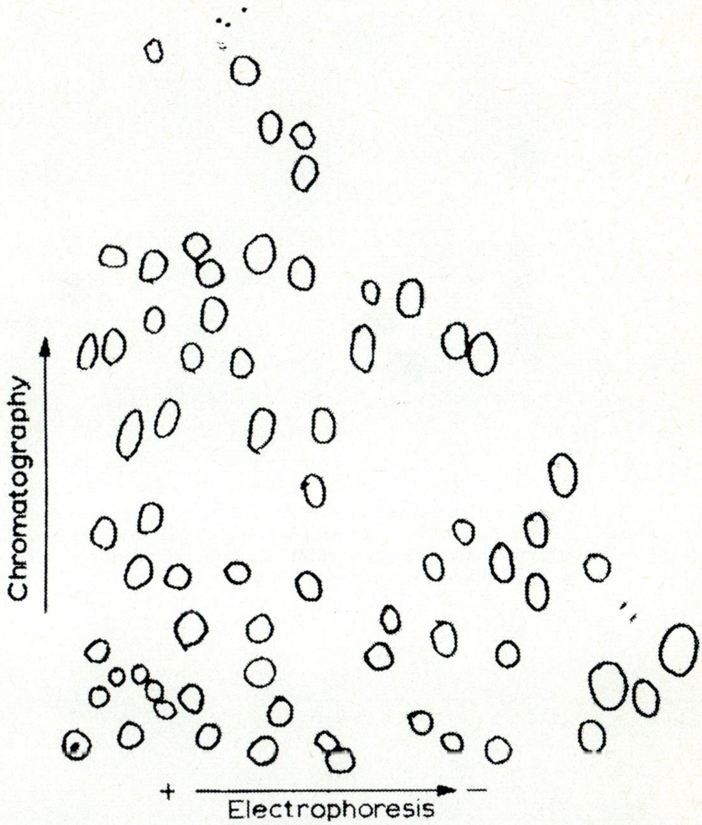

FIG. 4.5. Peptide map of the tryptic digest of myosin.[8] Enzyme to substrate ratio = 1:100. Buffer: 0·1M ammonium carbonate, *p*H 8·5. Incubation time: 24 hours. First dimension: chromatography. Solvent system: chloroform-methanol-ammonium hydroxide (40:40:20). Time 60 minutes. Second dimension: electrophoresis. Buffer: pyridine-acetic acid-water (1:10:489). Current: 980 V, 30 mA. Time: 1 hour. Silica gel G plate 200 mm. × 200 mm.

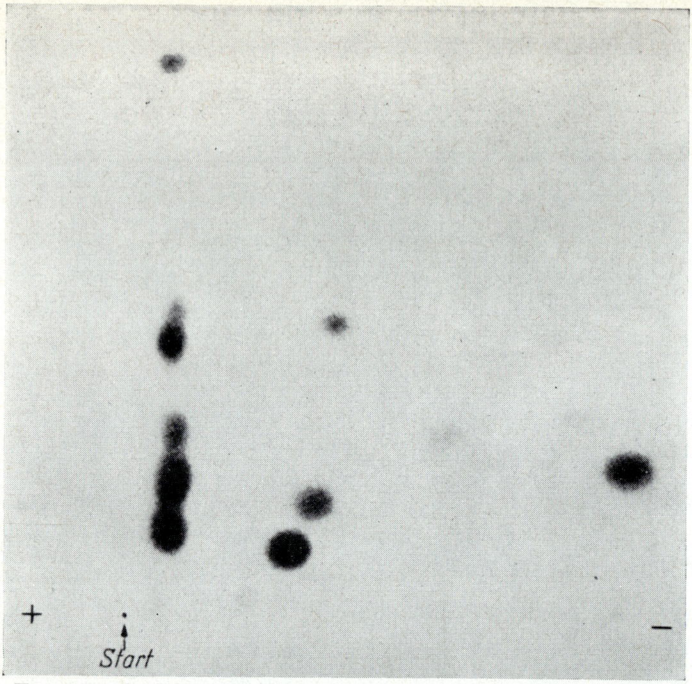

FIG. 4.6. Peptide map of the tryptic digest of oxidized ribonuclease.[9] First dimension: electrophoresis at *p*H 6·5. Second dimension: chromatography (pyridine-acetic acid-butanol-water [40:14:68:25]). Silica gel S plate 200 mm. × 200 mm.

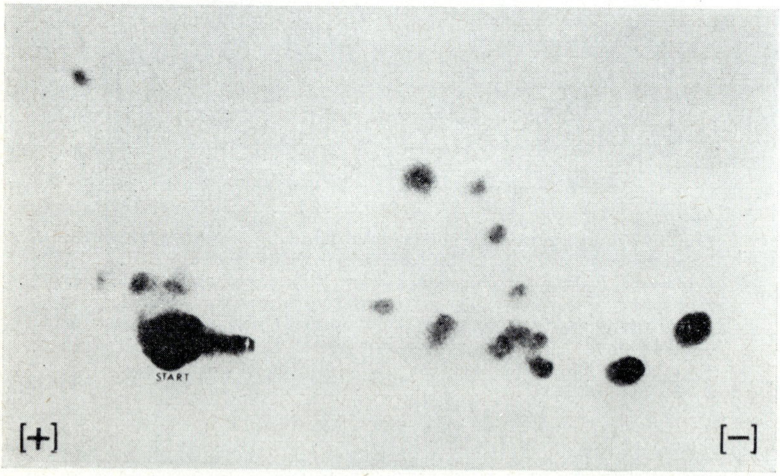

FIG. 4.7. Fingerprint of a tryptic digest of horse hæmoglobin on silica gel H.[6] Silica gel H plate 200 mm. × 200 mm. Electrophoresis: 40 min. in pyridine-acetic acid-water (3:12:485), *p*H 3.9. Current: 950 V, 32 mA. Chromatography: 2½ hours in butanol-acetic acid-water (4:1:1). Sprayed with 0·2 per cent ninhydrin in acetone.

40 ml. The slurry is then homogenized again by shaking and poured onto a very clean glass plate 200 mm. × 200 mm., which has been carefully levelled. With a spatula the gel is spread evenly over the plate and left for about ten minutes. In order to soak up the excess of buffer, four strips of Whatman No. 1 paper (200 mm. × 500 mm.) are placed on the layer along the edges of the plate, reaching about 3 to

TABLE 4.2.

(*recommendations of Pharmacia, Uppsala, Sweden*)

Type of Sephadex Superfine	*Amount of gel g/100 ml*	*Swelling time on water bath*	*Fractionation range (globular proteins)*
G–50	10·5	1 hr	1,500– 30,000
G–75	7·5	2 hrs	3,000– 70,000
G–100	6·5	3 hrs	4,000–100,000
G–150	4·8	3 hrs	5,000–150,000
G–200	4·5	3 hrs	5,000–250,000

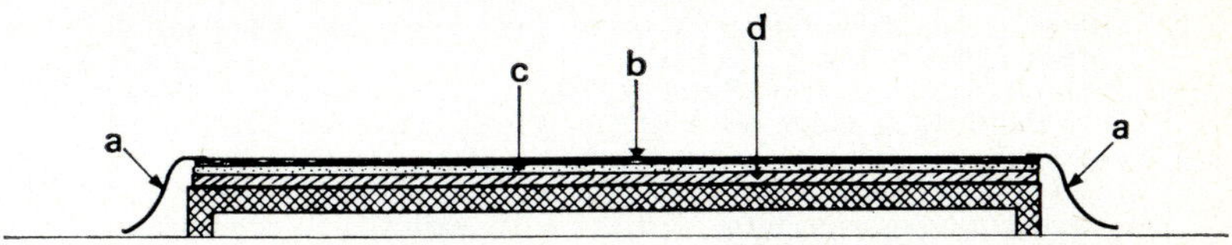

FIG. 4.8. Support used for preparation of Sephadex thin-layer plates. (According to P. Fasella *et al.*,[12] modified for 200 mm. × 200 mm. glass plates). (*a*) Filter paper, (*b*) Supernatant fluid, (*c*) Sedimented Sephadex, (*d*) Glass plate.

5 mm. into the Sephadex layer (Fig. 4.8). As soon as the layer becomes rigid the strips are removed and the plate may be stored for a few days in a moist chamber. (See Vol. 1, Chap. 33.) The Sephadex suspension, free of supernatant, can also be spread in the usual way.

Gel Filtration. The thin-layer plate is placed on the supporting system shown in Fig. 4.9 and equilibrated by allowing the buffer to flow through the gel layer for about two hours. 0·01 to 0·05 ml. of a 0·5 to 5 per cent protein solution are then applied at the upper end of the layer. In uni-dimensional separations it is possible to analyse several samples simultaneously. In order to explore the optimal time of the run it is recommended to apply the same sample at different time intervals. It takes 1 to 8 hours to complete a run. When gel filtration is to be followed by electrophoresis, the sample is applied near a corner or in the middle near the upper edge of the plate.

Electrophoresis. The plate is arranged as usual on the cooling platform of the electrophoresis apparatus (see Fig. 4.1). However, since the Sephadex gel layer on the plate is thicker than a normal silica gel layer, spacer strips (200 mm. × 5 mm. × 3 mm.) of Teflon are placed between the cover plate and the paper wicks which connect the buffer vessels to the thin-layer plate. At 300 V an electrophoretic run requires 3 to 15 hours and tap water provides sufficient cooling. For separating serum proteins a 0·05M barbital buffer *p*H 8·6 gives good results.

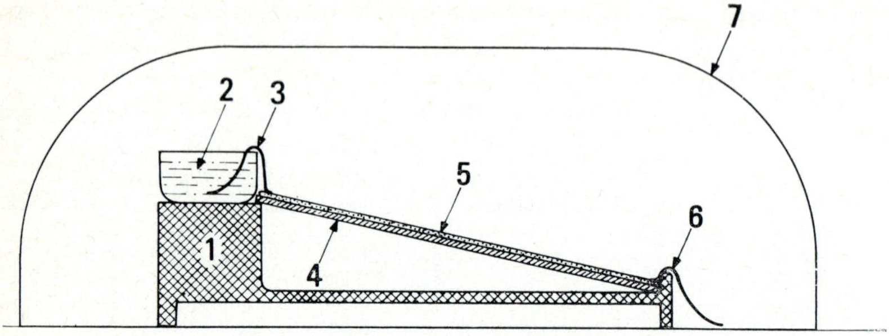

FIG. 4.9(a). Gel filtration on Sephadex thin-layers. (According to P. Fasella *et al.*,[12] modified for 200 mm. × 200 mm. glass plates). (1) Supporting system. (2) Trough containing the buffer or solvent used for the development of the chromatogram: the level of the fluid in the trough must be about 1 cm. above the upper end of the Sephadex layer. (3) Sheet of Whatman No. 1 paper for the transport of fluid from the trough to the Sephadex layer. One end of the sheet is immersed (2 to 3 cm.) in the fluid contained in the trough, the opposite end lies on the upper end of the Sephadex layer. (4) Glass plate with Sephadex layer; the plate should be inclined so as to form a 15–20° angle with the horizontal. (5) Sephadex layer. (6) One sheet of Whatman No. 3 MM filter paper (for the removal of the fluid at the lower end of the Sephadex layer). (7) Glass or plastic bell for protecting the system and for preventing the excessive evaporation of fluid from the Sephadex layer.

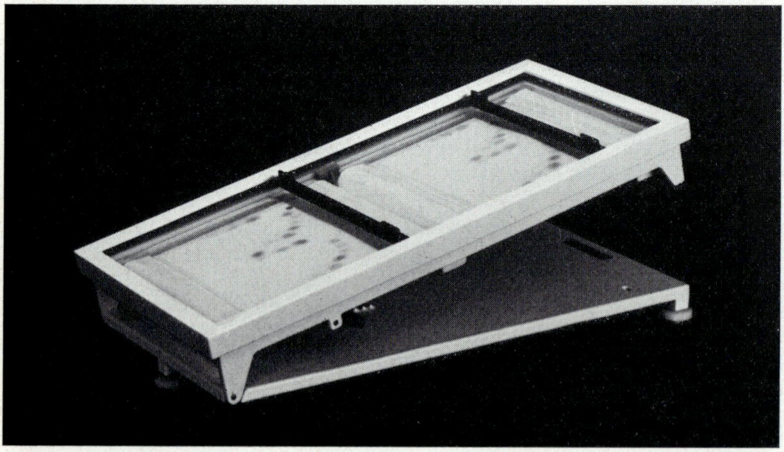

FIG. 4.9(b). Pharmacia TLG-apparatus (*Pharmacia Fine Chemicals AB, S–751 04 Uppsala 1, Sweden*)

Detection of the Separated Protein Fractions. After completion of the chromatographic and electrophoretic run the plate is covered carefully with a damp Whatman No. 1 paper in order to obtain a print of the protein fractions. After five minutes of contact the sheet is removed from the gel layer and dried immediately in an electric oven at 110°C. The protein fractions are stained with amido black or with another protein dye.

Fig. 4.10 illustrates the separation of normal human serum by gel filtration electrophoresis on Sephadex G 200.[11]

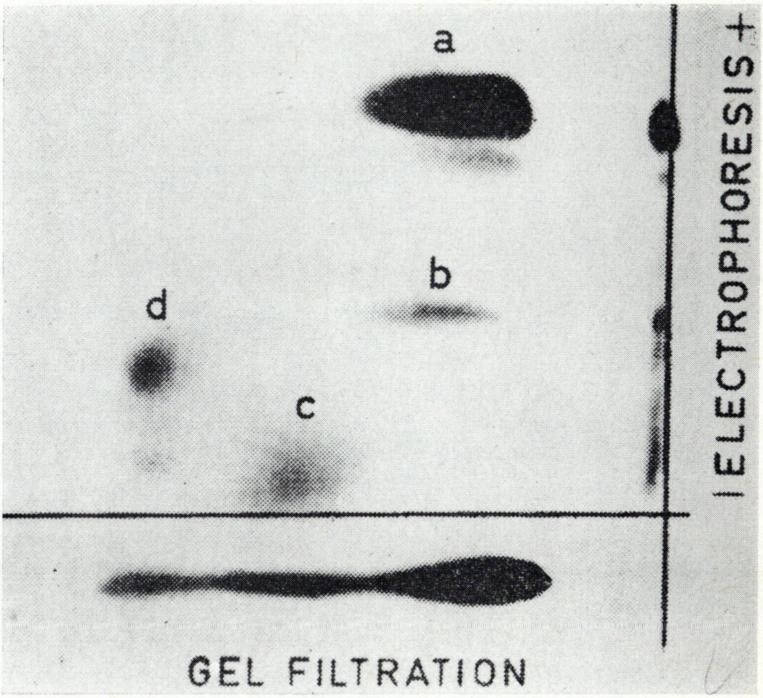

FIG. 4.10. A

Other Applications

Metabolic Studies. The separation of metabolic products of drugs is a very difficult task. Fingerprints on thin-layer plates are especially useful in this field because of the small amount of material needed and the high resolution obtained by this method. The fingerprint of the metabolites of Natulan®[= 1-methyl-2-p(*iso*propylcarbamoyl)benzylhydrazine hydrochloride], a new cytostatic compound, was prepared by Schwartz[23] (Fig. 4.11).

Experimental Conditions. Two male Wistar rats of 150 g. each received by intraperitoneal injection 8 mg. 1-methyl-2-p(isopropylcarbamoyl^{14}C)benzylhydrazine hydrochloride with a specific radioactivity of 11·1 μC/mg.

The animals were placed in metabolic cages. The urine of both animals was collected in containers cooled with ice over a period of 24 hours (= 8·5 ml.). 1 ml. pooled urine and 9 ml. chilled ethanol were mixed, left for 15 minutes in an ice bath and centrifuged. The supernatant was concentrated to dryness under vacuum at 40°C and the

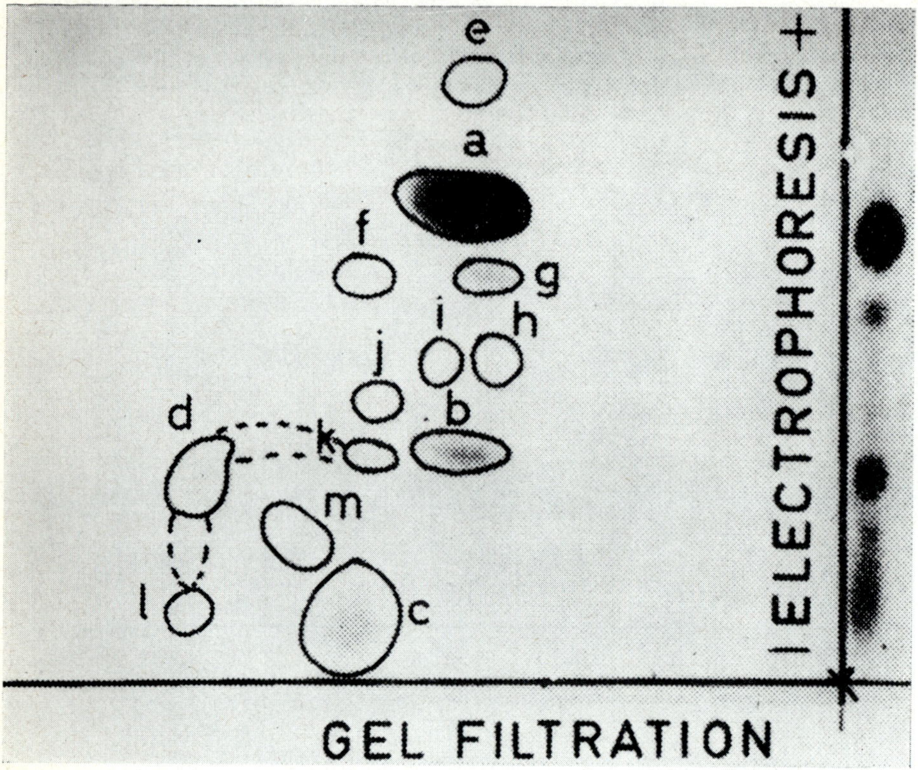

B

FIG. 4.10. (A) Separation of normal human serum by gel filtration-electrophoresis on Sephadex G 200.[11] The same serum was used as a reference in the gel filtration as well as in the electrophoresis. The designations are: (*a*) albumin, (*b*) transferrin, (*c*) γ-globulin, (*d*) α_2-macroglobulin. (B) Separation of normal human serum by gel filtration-electrophoresis with all visible components encircled.[11] The same designations are used as in Fig. 4.10A, and in addition: (l) contains γ_M-globulin, (*m*) contains γ_A-globulin, (*f-k*) are unidentified components.

residue dissolved in 50 μl methanol. 20 μl. of this solution (= 8 μC) were used for preparing the fingerprint.

Separation. A silica gel G (Merck) plate 200 mm. × 200 mm. was used as support. The thin-layer plate was first washed by a preliminary run with methanol-benzene (1:1) in order to remove impurities and left two days to dry in the air.

First dimension: high voltage electrophoresis, 40 minutes, 850 V, 40–50 mA. Buffer *p*H 5·5 (30 ml. pyridine, 10 ml. acetic acid, diluted to 3 litres with distilled water).

Second dimension: Chromatography. System: *n*-butanol-dichloromethane-acetic acid (30:70:0·5).

Autoradiography: X-ray film (Gevaert). Exposure: 2 days.

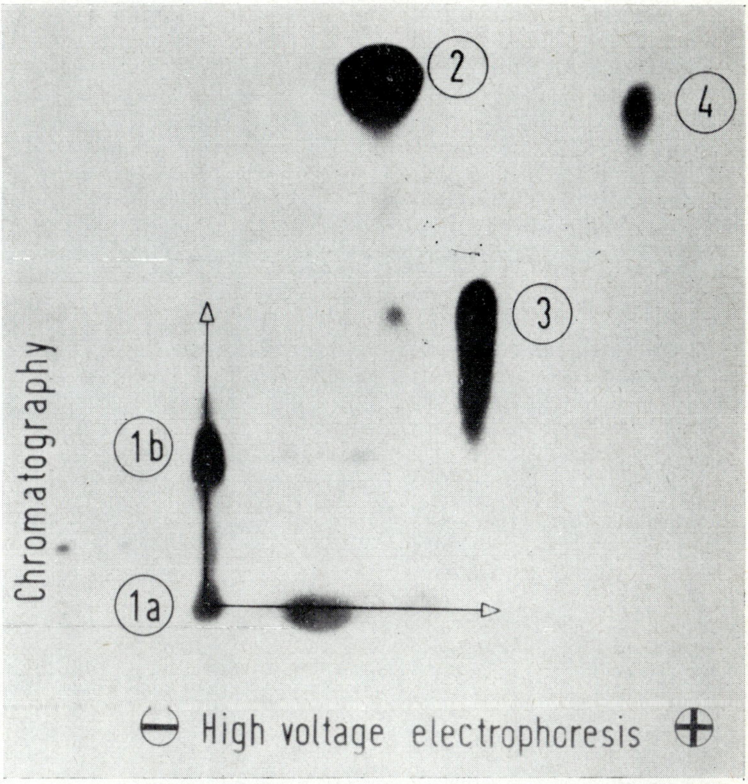

FIG. 4.11. Metabolites of Natulan® in rat urine. Fingerprint on silica gel G.[23] First dimension: electrophoresis 850 V, 40–50 mA. Time: 40 minutes. Buffer: *p*H 5·5 (30 ml. pyridine, 10 ml. acetic acid diluted to 3 litres). Second dimension: chromatography (butanol-dichloromethane-acetic acid [30:70:0·5]). Autoradiography on X-ray film. Exposure: 2 days. Silica gel G plate 200 mm. × 200 mm.

Bidimensional Separation of Aminoacids. Several authors have described the bidimensional separation of aminoacids by electrophoresis and chromatography on thin-layer plates. Nybom[5] used plates coated with cellulose to separate the aminoacids of commercial casein hydrolysates (Fig. 4.12). Electrophoresis with formic acid (0·7 per cent [v/v]) as buffer was carried out in the first dimension followed by chromatography in the second dimension. Electrophoretic mobility and R_f values for the more common aminoacids in three different solvents are listed in Table 4.3.

Thin-layer Electrophoresis of Coal-tar Food Dyes. Electrophoresis of coal-tar food colours was studied by Criddle, Moody and Thomas.[14] The following dyes were separated in the buffer systems listed on p. 150.

Ponceau 4R	Orange RN	Violet BNP
Amaranth	Red 2G	Green S
Fast Red E	Sunset Yellow FCF	Red FB
Carmoisine	Red 6B	Yellow 2G
Black PN	Red 10B	Tartrazine
Ponceau SX	Chocolate Brown HT	Indigo Carmine
Ponceau MX	Yellow RFS	Chocolate Brown FB
Orange G	Yellow RY	Erythrosine BS
	Brown FK	
	Blue VRS	

TABLE 4.3.

Electrophoretic mobility and R_f values for the more common amino acids in three different solvents.[5] *BuOH = n-butyl alcohol. HAc = conc. acetic acid. PrOH = n-propyl alcohol. Emk = ethylmethylketone. HFo = conc. formic acid.*

No.	Substance	Electrophoretic mobility U_{pr}	Chromatographic mobility in solvents					
			4 BuOH 1 HAc 2 H_2O		5 PrOH 1 Pyridine 2 H_2O		24 Emk 1 HFo 6 H_2O	
			R_f	R_{pr}	R_f	R_{pr}	R_f	R_{pr}
1	Tryptophan	0·84	0·62	1·22	0·48	1·44	0·56	2·19
2	Hydroxyproline	0·86	0·40	0·80	0·24	0·71	0·14	0·56
3	Aspartic acid	0·93	0·37	0·72	0·18	0·53	0·19	0·73
4	Tyrosine	0·94	0·57	1·12	0·46	1·38	0·51	2·00
5	Proline	1·00	0·51	1·00	0·33	1·00	0·26	1·00
6	Phenylalanine	1·00	0·66	1·30	0·58	1·73	0·55	2·13
7	Glutamine	1·03	0·34	0·68	0·14	0·42	0·12	0·47
8	Asparagine	1·05	0·27	0·54	0·11	0·33	0·09	0·36
9	Glutamic acid	1·05	0·43	0·84	0·23	0·70	0·25	0·99
10	Methionine	1·07	0·60	1·19	0·51	1·53	0·50	1·94
11	Threonine	1·13	0·42	0·83	0·26	0·78	0·22	0·86
12	Serine	1·15	0·35	0·68	0·18	0·54	0·13	0·51
13	Leucine	1·15	0·70	1·38	0·64	1·90	0·57	2·22
14	Valine	1·20	0·60	1·19	0·48	1·44	0·46	1·79
15	Alanine	1·32	0·47	0·92	0·28	0·83	0·27	1·04
16	Glycine	1·45	0·37	0·73	0·16	0·49	0·16	0·63
17	Arginine	1·97	0·33	0·65	0	0	0·07	0·28
18	Histidine	2·03	0·27	0·52	0·06	0·17	0·04	0·15
19	Lysine	2·09	0·29	0·58	0	0	0·04	0·15
20	γ-amino-butyric acid	2·16	0·54	1·07	0·22	0·65	0·37	1·44
21	β-alanine	2·22	0·48	0·95	0·20	0·59	0·30	1·18

U_{pr}: Relative electrophoretic rates of migration in relation to proline. (Buffer: 0·7 per cent [v/v] formic acid).

R_{pr}: Relative chromatographic rates of migration in relation to proline.

Concluding Remarks. Thin-layer electrophoresis offers a very useful complementary method to thin-layer chromatography. The combination of both techniques is of special interest and the many supports available will open new applications in chemistry and biology, not only

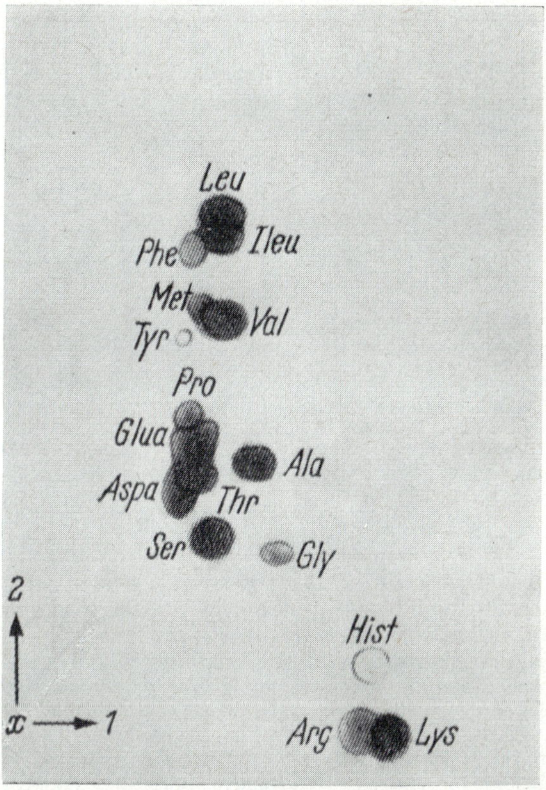

FIG. 4.12. Bidimensional separation of aminoacids of a casein hydrolysate on cellulose.[5] First dimension: electrophoresis. Buffer: 0·7 per cent formic acid. Second dimension: chromatography. Solvent system: *n*-propanol-pyridine-water (5:1:2).

for the separation of substances of low molecular weight, but also for macromolecules. The advantage of this method lies in its rapidity and the small amounts of material needed. In the future, thin-layer electrophoresis will certainly also be useful for micropreparative work. The equipment is relatively simple and inexpensive.

APPENDIX

Fluorescamine ("FLURAM"*) a new fluorogenic reagent for the assay of primary amines. Its use as a reagent for localization of primary amines on thin-layer plates.

W. J. Ritschard and U. Handschin

FLURAM (Fluorescamine; 4-phenylspiro [furan-2(3H), l'-phthalan]-3, 3'-dione) is a new sensitive fluorogenic reagent.(1) It reacts with primary amines, amino acids, peptides and proteins to yield fluorescent compounds as shown in figure 4.13.

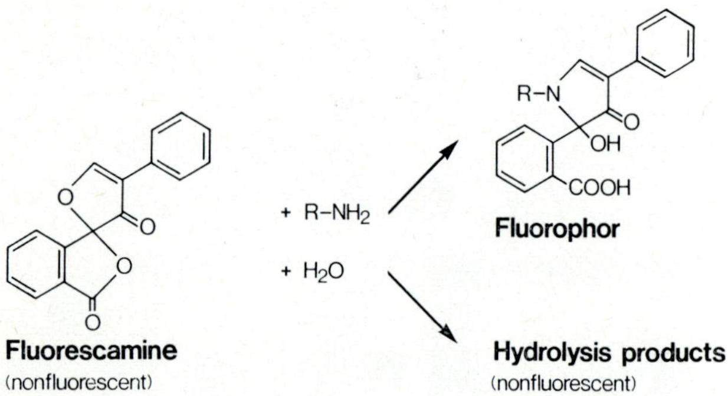

FIG. 4.13.

At *p*H 9 and room temperature reaction with primary amines proceeds almost instantaneously. Fluorescamine, as well as its hydrolysis products, are not fluorescent. Fluorescence of the reaction products is measured in a spectrofluorometer at 390 nm excitation and 475 nm emission, or in an appropriate filterfluorometer(2).

These properties make it a very useful analytical reagent with interesting applications, as for example in protein chemistry.

Applications

Purification of Proteins and Peptides: Assay of the eluates of column chromatography(3–5).

Amino Acid Analysis: The sensitivity of this quantitative method lies in the picomole range. Since only primary amines react with fluorescamine, proline and hydroxyproline do not give fluorescence. With N-chlorosuccinimide, however, they can be converted to detectable primary amines(6–9).

* FLURAM = Trade Mark. Available, also in a spray container,

in Europe from	in USA from
F. Hoffmann-La Roche & Co. Limited Company Diagnostica CH-4002 Basel Switzerland	Roche Diagnostics Division of Hoffmann-La Roche Inc. Nutley N. J. 07110 USA

Electrophoretic Separation of Proteins: The separation of fluorescamine-prelabelled proteins can be monitored[10].

Solid Phase Peptide Synthesis: Rapid detection of trace amounts of uncoupled products is possible[11].

Reagent for Detecting Primary Amines on Thin-layer Plates: Detection of biologically important amines, amino acids and peptides[12–14].

Fluorescamine for Detection of Primary Amines on Thin-layer Plates. Some recent publications describe the use of fluorescamine for the localization of amino compounds on thin-layer plates.

Amino acids, peptides, catecholamines and other primary amines were first reacted with fluorescamine, followed by thin-layer chromatographic separation of the fluorescent derivatives. While the sensitivity of this method for detecting amines is very high (detectable range between 10 and 100 picomoles), the chromatographic separation of complex mixtures of fluorescamine labelled amines is not always satisfactory[12].

An alternative method uses a solution of fluorescamine as a spray reagent after chromatographic separation of the primary amines on thin-layer plates. The sensitivity of this procedure is 250–500 picomoles[13].

In our experience the detection of peptides by spraying thin-layer plates with a fluorescamine solution frequently gives unreproducible results, even though the plates were sprayed first with a buffer (in order to establish optimal reaction conditions) and then with fluorescamine. It seems that this inconsistency is due to variations of the adsorbent. The quality, including purity of the silica gel and the binding material, thickness of the gel layer on the plate, and type of gel (silica gel, cellulose, aluminiumoxide) play an important role in obtaining reproducible results. Commercial plates, and plates freshly prepared in the laboratory, yield different degrees of fluorescence. In our laboratory a procedure has been developed which includes additional treatment of the thin-layer chromatograms with a solution of sodium hydroxide before spraying with fluorescamine. In this way the sensitivity and reproducibility of the fluorescence can often be improved, while quenching effects and secondary fluorescence appear to be minimal. By this method better penetration of the adsorbed peptides by the fluorescamine spray solution is probably achieved, thus giving a more intense fluorescence[14].

EXPERIMENTAL

1. Method for Detection of Fluorescamine-labelled Primary Amines on Thin-layer Plates[12].

Amino acids, peptides, catecholamines, polyamines and other primary amines may be prepared and identified according to this method.

1.1. Preparation of Fluorophors

Method A. A solution containing 4–80 nanomoles of each amine in 0·1 M sodium borate, *p*H 9·0, is added to a 13 × 100 mm. tube. With the tube held on a vortex mixer, 200 μl. of fluorescamine in acetone

(20 mg/100 ml.) is added at room temperature. Amounts of 10 to 200 picomoles of each amine are spotted.

For catecholamines, 100 μl. of each amine in water is first mixed with 100 μl. of fluorescamine in acetone (20 mg/100 ml.) and then with 200 μl. of 0·1 M sodium bicarbonate, pH 8·3. With catecholamines the reaction is performed at 4°C.

Method B. An aliquot of solution containing 20–200 picomoles of the amine is evaporated to dryness under nitrogen at 40°C in a 6 × 50 mm. tube. Five microliters of 0·1 M sodium-borate or sodium bicarbonate is added to the tube. With the tube held on a vortex mixer, 5 μl. of fluorescamine in acetone (20 mg/100 ml.) is added from a Hamilton syringe. Five microlitres of each solution is transferred with a Hamilton syringe to the thin-layer plates*. Reaction tubes (6 × 50 mm.) were siliconized with 1% siliclad prior to use**.

1.2. Thin-layer Chromatography

Chromatography is performed on 10 × 10 cm. or 10 × 20 cm. glass plates of silica gel. The following solvent systems are used: A. isopropanol, chloroform, 28% ammonia, water (70:30:18:7); B. n-butanol, acetic acid, water (100:1:100); C. n-butanol, acetic acid, water (4:1:1); D. benzene, dioxane, acetic acid (90:25:5). Chromatograms are developed in the dark for 30–90 min. After development, the plates are air-dried for a few minutes and sprayed with 10% triethanolamine in chloroform. The plates are then dried in a stream of cold air for a few minutes and sprayed again with 10% triethanolamine for 20 sec. The plates are observed under a long-wave (366 nm.) ultraviolet light.

The sensitivity of this method is below 100 picomoles. With many compounds as little as 10 picomoles can be detected. When sprayed with triethanolamine the spots on the plates are stable for several days when the latter are stored in the dark.

1.3. Rf-values of Various Fluorescamine Derivatives of Primary Amines in Two Solvent Systems (see table 4.4 over page).

Solvent system 1: isobutanol, acetic acid, water (100:1:100)
Solvent system 2: benzene, dioxan, acetic acid (90:25:5).

2. Method for Using Fluorescamine as a Spray Reagent for Thin-layer Chromatography[13].

Amino acids and peptides can be separated in a variety of solvent systems on thin-layer plates (using silica gel G without indicator) in the usual way. The plates are dried at 110°C for 10 minutes, cooled to room temperature and then treated according to the procedure in table 4.5 in order to detect the individual spots.

* Precoated silica gel plates (Brinkman, Westburg, N.Y., U.S.A.).
** Siliclad (Clay Adams; Parsippany, N.J., U.S.A.).

Table 4.4

	Solvent system 1	2
Alanylalanine	0·22	—
Carnosine	0·38	—
Glycylproline	0·19	—
Phenylalanylglycine	0·09	—
Tyrosylleucine	0·87	—
Glycylglycylalanine	0·45	—
Bradykinin	0·07	—
3,4-Dihydroxyphenylalanine	—	0·07
Dopamine	—	0·28
Norepinephrine	—	0·15
3-Methoxytyramine	—	0·52
Normetanephrine	—	0·35
Putrescine[a]	0·73	—
Spermidine[a]	0·34	—
Spermine[a]	0·07	—

[a]A second minor spot appeared on the plate with each sample. This may be due to the presence of a small percentage of monosubstituted polyamine in addition to the disubstituted polyamine.

Table 4.5

Fluorescamine Spray Procedure

Step	Treatment
1.	Spray with a solution of 10% triethylamine in methylene chloride; air dry for several seconds.
2.	Spray with a solution of 0·05% fluorescamine in acetone; air dry for several seconds.
3.	Re-spray with a solution of 10% triethylamine in methylene chloride.

The plates are then viewed using a long wave (366 nm.) ultra-violet light source. Fluorescence fades only slightly after 24 hours. The sensitivity of this method ranges from 250 to 500 picomoles.

Other Applications

Fluorescamine may also find interesting applications in other fields, such as toxicological analysis. The detection of amphetamine in urine extracts by fluorescamine, analysed by thin-layer chromatography, has been described recently[15].

3. Method for Detecting Peptides on Peptide Maps using Fluorescamine as a Spray Reagent[14]

The technique for preparing the peptide maps is described on page 71. It is recommended to prepare the plates according to this method, which gives a thinner and better coating than normally obtained. The reproducibility of the peptide pattern, the speed of chromatography and of electrophoresis will depend on it. The intensity of the fluorescence is also improved considerably.

As the fluorescence intensities of peptides can vary considerably, several peptide maps should be prepared by applying various volumes (usually 0·1; 0·5; 1 μl.) of the peptide digest (see Chapter 3 and p. 69). Otherwise some of the peptides might escape detection.

After completing the peptide maps, the thin-layer plates are dried at 110°C. The plate is then evenly sprayed first with a solution of 1 M NaOH (important: safety goggles and plastic gloves must be worn!) and then with a solution of 1 M sodium acetate. This procedure is followed by spraying the plate with a 0·05% solution of fluorescamine in acetone, and drying at 110°C for 10 minutes.

Fluorescence is then observed under a UV-lamp at about 360 nm. Although the fluorescence is stable for at least 24 hours when the plates are stored in the dark, it is recommended to mark the spots with a pencil immediately after development with fluorescamine.

REFERENCES

1. Ingram, V. M. *Nature*, 1956, **178,** 792.
2. Katz, A. M., Dreyer, W. J., and Anfinsen, C. B. *J. Biol. Chem.*, 1959, **234,** 2897.
3. Honegger, C. G. *Helv. Chim. Acta*, 1961, **44,** 173.
4. Pastuska, G., and Trinks, H. *Chemiker-Ztg.*, 1962, **86,** 135.
5. Nybom, N. *Physiologia Plantarum*, 1964, **17,** 434.
6. Stegemann, H., and Lerch, B. *Anal. Biochemistry*, 1964, **9,** 417.
7. Munier, R. L., and Sarrazin, G. *Bull. Soc. chim. France*, 1966; 1363, 1365, 1367.
8. Ritschard, W. J. *J. Chromatog.*, 1964, **16,** 327.
9. Wieland Th., and Georgopoulos, G. *Biochem. Zeitschr.*, 1964, **340,** 476.
10. Johansson, B. G., and Rymo, L. *Biochem. J.*, 1964, **92,** 5P.
11. Hanson, L. A., Johansson, B. G., and Rymo, L. *Clin. Chim. Acta*, 1966, **14,** 391.
12. Fasella, P., Giartosio, A., and Turano, C. "Thin-Layer Chromatography." Edited by G. B. Marini Bettòlo, Elsevier Publishing Company Amsterdam/London/New York, 1964.
13. Pastuska, G., and Trinks, H. *Chemiker-Ztg.*, 1961, **85,** 535.
14. Criddle, W. J., Moody, G. J., and Thomas, J. D. R. *J. Chromatog.*, 1964, **16,** 350.
15. Moghissi, A. *Anal. Chim. Acta*, 1964, **30,** 91.
16. Pierce, T. B., and Peck, P. F. *Analyst*, 1964, **89,** 662.
17. Crossley, J., and Thomas, J. D. R. *Analyst*, 1958, **83,** 462.
18. Brenner, M., and Niederwieser, A. *Experientia*, 1960, **16,** 378.
19. Criddle, W. J., Moody, G. J., and Thomas, J. D. R. *J. Chromatog.*, 1965, **18,** 530.
20. Brenner, M., and Niederwieser, A. *Experientia*, 1961, **17,** 237.
21. Reindel, F., and Hoppe, W. *Chem. Ber.*, 1954, **87,** 1103.
22. Stahl, E. "Dünnschicht-Chromatographie"; Springer-Verlag, Berlin-Heidelberg and New York, 1967, p. 822.
23. Schwartz, D. E. Personal Comm.
24. Burns, D. J. W., and Turner, N. A. *J. Chromatog.*, 1967, **30,** 469.

REFERENCES TO APPENDIX

1. Weigele, M., de Bernardo, S. L., Tengi, J. P., and Leimgruber, W. *J. Amer. Chem. Soc.*, 1972, **94,** 5927.
2. Udenfried, S., Stein, S., Böhlen, P., Dairman, W., Leimgruber, W., and Weigele, M. *Science*, 1972, **178,** 871.
3. Böhlen, P., Stein, S., Dairman, W., and Udenfried, S. *Arch. Biochem. Biophys.*, 1973, **155,** 213.

4. Böhlen, P., Stein, S., Imai, K., and Udenfriend, S. *Anal. Biochem.*, 1974, **58,** 559.
5. Nakai, N., Lai, C. Y., and Horecker, B. L. *Anal. Biochem.*, 1974, **58,** 563.
6. Stein, S., Böhlen, P., Stone, J., Dairman, W., and Udenfriend, S. *Arch. Biochem. Biophys.*, 1973, **155,** 203.
7. Felix, A. M. and Terkelsen, G. *Arch. Biochem. Biophys.*, 1973, **157,** 177.
8. Weigele, M., de Bernardo, S. L., and Leimgruber, W. *Biochem. Biophys. Res. Commun.*, 1973, **50,** 352.
9. Georgiadis, A. G., and Coffey, J. W. *Anal. Biochem.*, 1973, **56,** 121.
10. Pace, J. L., Kemper, D. L., and Ragland, W. L. *Biochem. Biophys. Res. Commun.*, 1974, **57,** 482.
11. Felix, A. M., and Jimenez, M. H., *Anal. Biochem.*, 1973, **52,** 377.
12. Imai, J., Böhlen, P., Stein, S., and Udenfriend, S. *Arch. Biochem. Biophys.*, 1974, **161,** 161.
13. Felix, A. M., and Jimenez, M. H. *J. Chromatog.*, 1974, **89,** 361.
14. Ritschard, W. J., and Handschin, U. unpublished results.
15. Klein, B., Sheehan, J. E., and Grunberg, E. *Clin. Chem.*, 1974, **20,** 272.

CHAPTER 5

CELLULOSE ACETATE ELECTROPHORESIS AND IMMUNO-DIFFUSION TECHNIQUES

J. Kohn

THE principles of cellulose acetate membrane (CAM) filter electrophoresis are essentially the same as those of the well-known filter paper electrophoresis and will, therefore, not be discussed here.

The differences in the techniques and apparatus employed are a result of the type and texture of the supporting medium which is fine, thin and microporous. The advantages of cellulose acetate membrane filter paper as a supporting medium for zone electrophoresis are:

1. Adsorption is minimal. This is particularly valuable as it eliminates the very troublesome "tailing" and renders the background completely colourless, which, in turn, results in a sharp separation of well defined bands, greatly improving the accuracy of quantitative determinations.

2. The material is homogenous, microporous, and is chemically relatively pure. There are no hemicelluloses and lignins present and the heavy metal traces are very low indeed. This may be of particular importance for analytical purposes.

3. Separation of α_1 from the albumin fraction is invariably excellent.

4. Rapid separation, with a great saving of time and material. Using the small-scale technique, perfect separation is achieved in $\frac{1}{2}$–2 hours. This, combined with the rapid staining, drying and washing, enables the whole electrophoretic procedure to be performed in one hour if required.

5. Very small quantities of protein can be successfully separated using the same apparatus, and without resorting to complicated techniques. Using Nigrosin staining, samples as small as 0·1 μl. (0·0001 ml.) with a protein content down to 5 μg., still produce clearly recognizable patterns.

6. The strips can be cleared by using a suitable swelling agent or by immersion in an appropriate clearing fluid to glass-like transparency, thus reducing the background error even further for scanning procedures. This is also of great help for photographic purposes particularly in immuno-diffusion techniques.

7. It can be used on its own or in combination with agar for immuno-diffusion techniques.

8. It is eminently suitable for electrophoretic and immuno-diffusion studies of isotope labelled protein.

9. It can be easily cut into small sections after the proteins have been separated. This is particularly valuable for enzyme and immunological studies.

10. Electrophoretic separation of small quantities of protein reveal

the presence of some normal fractions which may be masked by the diffusion area of the adjacent large fractions. This applies also to the detection of small paraprotein bands (minor peaks) which may be missed on filter paper.

11. Glyco-protein staining is invariably successful and the globulin fractions containing glyco-protein are discrete and well separated.

12. Certain proteins which do not separate well on filter paper can be successfully separated on cellulose acetate, e.g. insulin, lysozyme, histones. Also the separation of some animal sera (rat, mouse) is more satisfactory on cellulose acetate.

Large- or Small-scale Electrophoresis. For most purposes, and particularly for routine clinical investigations, small-scale cellulose acetate membrane filter electrophoresis is the method of choice (6–8 cm.). It is very rapid, reliable and economical. The large-scale method (20 cm.) should be reserved for research purposes and it may be used for micro preparative electrophoresis. In the latter case, fractions obtained from several strips can be pooled in order to obtain sufficient quantities.

Apparatus for Cellulose Acetate Electrophoresis

An apparatus for a thin, microporous supporting medium such as cellulose acetate membrane filter strips must have some special features which will prevent excessive and uneven evaporation.

Most horizontal electrophoretic tanks have the cathode and anode buffer compartments separated by a glass or Perspex plane over which the strip is stretched. As the strips heat up under the current, considerable temperature differences develop between the area round the warm centre of the strip and the comparatively cool areas over the buffer compartments. This in turn leads to the occurrence of convection currents inside the tank with their distorting effect on the separation patterns, particularly when a thin, microporous supporting medium is used. With filter paper these effects are less noticeable because of the comparatively large fluid content of the saturated strips which tends to counteract the effects of evaporation and convection currents.

A number of electrophoresis tanks are commercially available—and can be used for cellulose acetate electrophoresis. The electrophoretic chamber* described in this chapter has been developed and used for electrophoresis in my Department but this does not imply that other instruments and apparatus of a similar type are not capable of producing equally satisfactory results.

It has been designed in two versions, both are of the horizontal type

(1) Large universal type, Fig. 5.1*a*, which can also be used for filter paper, starch and agar gel. The details of this standard universal model are shown on the diagram Fig. 5.1*b*. It is made of Perspex and has four compartments, two central electrode compartments and two buffer compartments on the outside. The central partition separates the anode from the cathode compartments, the two lateral partitions

* Shandon Southern Instruments, Camberley, Surrey.

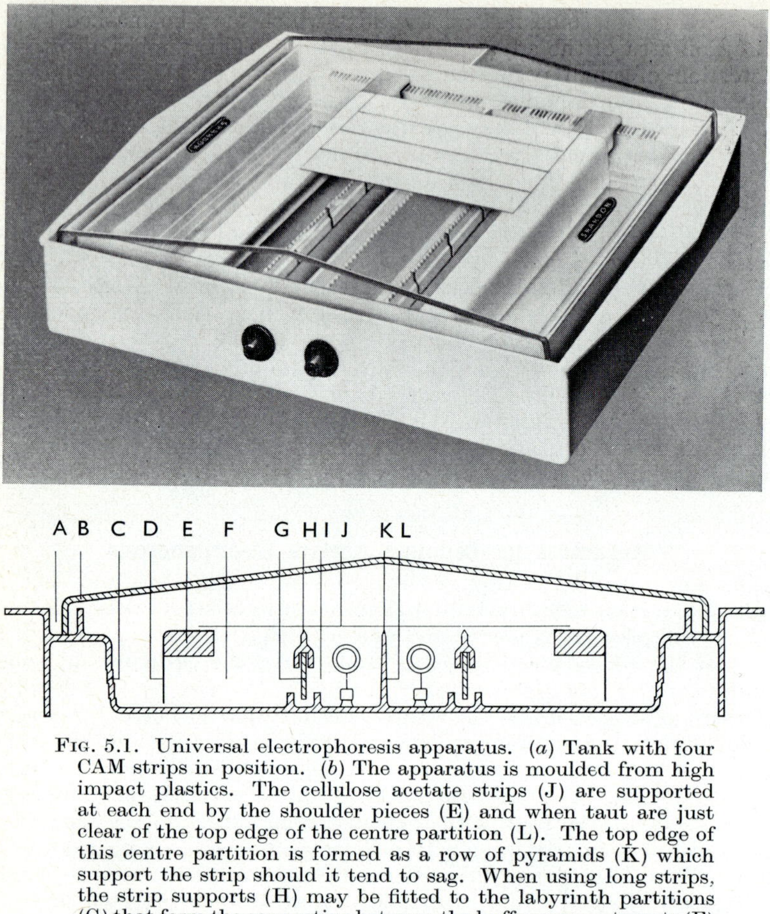

FIG. 5.1. Universal electrophoresis apparatus. (*a*) Tank with four CAM strips in position. (*b*) The apparatus is moulded from high impact plastics. The cellulose acetate strips (J) are supported at each end by the shoulder pieces (E) and when taut are just clear of the top edge of the centre partition (L). The top edge of this centre partition is formed as a row of pyramids (K) which support the strip should it tend to sag. When using long strips, the strip supports (H) may be fitted to the labyrinth partitions (G) that form the connection between the buffer compartments (F) and electrode compartments (I). Filter paper wicks (D) connect the cellulose acetate strips to the buffer compartments. The internal sides of the tank are stepped all round (C) as an aid to checking buffer level. The lid (B) fits in a recess (A) moulded all round the tank.

separate the electrode from the buffer compartments. The buffer solution fills all the compartments, and the connections between the electrode and buffer compartment are provided by means of a labyrinth in the latest models.

The top is made in the commercial type, by a flanged Perspex lid. These flanges fit into grooves in the walls of the tank, a perfect air-tightness is achieved. Platinum electrodes run through the whole length of the compartments and are fixed to the opposite wall by means of a stainless steel or nylon screw.

Fitments. Shoulder or bridge. The gap between the shoulder pieces, across which the strip is stretched, can vary between 4–16 cm. depending on the length of strip required. For small-scale and micro-electrophoresis a gap of 4–8 cm. and for long strips a gap of 8–14 cm. is recommended. Buffer impregnated pads of filter paper, e.g. Whatman No. 3 MM, are placed on the shoulders. It is strongly recommended to use sheets of cellulose acetate instead of individual strips.

(2) Small scale model, specially designed for routine serum electrophoresis performed on CAM sheets. Fig. 5.2*a* and *b*. The shoulder pieces are flat and the CAM sheet is held taut in position by adhesion. The spring loaded nylon rods are not necessary.

Assembling and Setting up of the Apparatus:

1. Fill the compartment with buffer to the marked level.
2. Level the buffer by tilting the tank so that the buffer covers the partitions, wait a while for levelling to be completed and then lower the tank back into position.
3. Place shoulders or frame in position.
4. Place buffer-moistened filter paper pads on shoulder pieces (see Fig. 5.1); these should be replaced daily.
5. Connect power supply, switch on current and check for leaks; an applied voltage, say 400 volts, should produce no current flow. The polarity of the tank should be reversed after each run if the same buffer is used.

Cellulose Acetate Membrane Filter Strips and Sheets. These were developed from the existing bacteriological membrane filters. The colour is pure white, opaque. The surface is very smooth and uniform.

Various types and makes of cellulose acetate membrane are now available and some are mentioned here:

British: Celagram (cellulose acetate) Shandon Southern Instruments. Camberley, Surrey.
Oxoid, Oxo, Ltd., London, E.C.4. (Courtaulds, Ltd.)
German: Sartorius Membranfilter Gesellschaft GmbH, 34 Gottingen.
Italian: Cellogel, Soc. Chemetron Chimica, Milan.
U.S.A. Sepraphor, Gelman Instrument Company, Ann Arbor, Michigan.
Titan, Hypergel, Cam, Mylar backed.

These membranes differ from each other by the degree of acetylation, thickness, pore size, texture, etc., but the separation patterns obtained are essentially the same. Physical and chemical characteristics are supplied by the manufacturers and will not be discussed here in detail.

Buffer Solutions. It is generally best to use more dilute buffer solutions for CAM electrophoresis than are employed for paper. For routine serum electrophoresis a barbitone buffer of 0·05–0·07 molar at *p*H 8·6 is recommended. This can be readily prepared as follows:

Diethylbarbituric acid 7·36 g.
Sodium diethylbarbiturate 41·2 g.
Make up to 4 litres
Add 20 ml. 5 per cent thymol in isopropyl alcohol

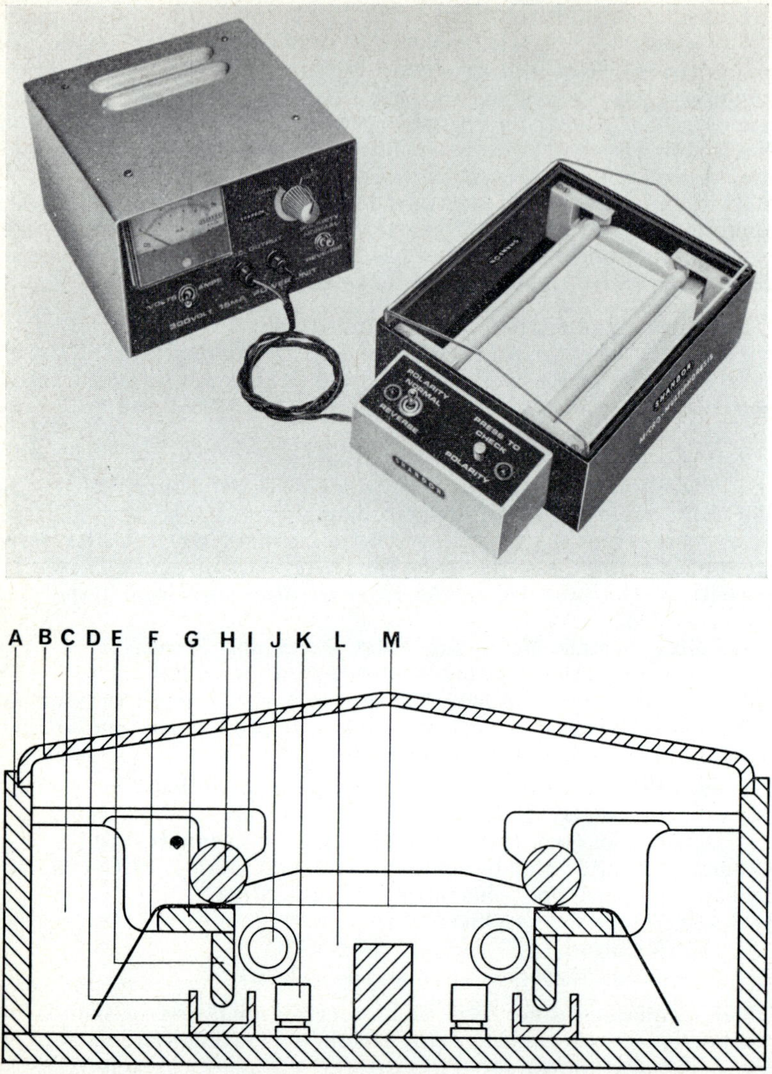

Fig. 5.2. "Microcel" electrophoresis apparatus. The apparatus is made of high impact plastics. The cellulose acetate strips (M) are supported by the shoulder pieces (G), lined with filter paper wicks (F) and tensioned by the grooved retaining rods (H) which are clamped in position by the plastic spring clips (I). The shoulder pieces are formed with the partition walls (E) which, with their associated channel sections (D) on the bottom of the tank, form labyrinth connections between the buffer compartments (C) and the electrode compartments (L). The inclined transparent lid (B) fits into the rebated top edge of the tank walls (A). The terminals (J) are connected to the full length electrodes (K).

The purpose of the thymol is to prevent the growth of micro-organisms. "Merthiolate" 1:10,000 may also be used for this purpose.

The "Oxoid" buffer described previously for paper electrophoresis is a very suitable alternative at the manufacturers recommended concentration. Other buffers are described subsequently.

The buffer concentration should be watched very closely. This can be done by using the voltmeter and ammeter on the power pack as conductivity meters. In routine use this buffer should give the required 25 V/cm. strip length when using a shoulder gap of 8–10 cm. and a current of 0·4–0·5 mA/cm. strip. If the current reading is higher the buffer should be diluted, while if it is lower a buffer of higher concentration should be used. A low buffer concentration has the effect of increasing the migration rate and increasing the width of the bands. A high buffer concentration produces the reverse effect, with crowding of the bands. Due to condensation buffer concentration may increase rapidly, particularly in high ambient temperature.

It is therefore desirable that buffer solutions should be changed at frequent intervals particularly in hot weather. This is most important, as too concentrated buffer solutions can seriously affect or even ruin the separation pattern. Perhaps the best way to ensure constant, reproducible results is the use of fresh buffers for each run. This can be easily and economically achieved by the use of only very small buffer volumes, just sufficient to cover the bottom of the compartments. This procedure has obvious advantages and has been claimed to give more rapid separation and better resolution. Some electrophoretic chambers have a system of baffles connecting the buffer with the electrode compartment, and naturally require more buffer solution. These chambers can still be used with small buffer volumes if filter paper wicks connecting the two compartments and reaching to the bottom are provided.

Strip Size. Cellulose acetate is at present available in various sizes of strip and sheet. For routine electrophoresis of serum proteins, CAM sheets rather than individual strips should be regarded as a method of choice. CAM sheets, e.g. Celagram (150 × 78 mm.) are a most convenient format, particularly if multiple runs have to be carried out. If every third sample on a sheet of CAM is a normal control, a pattern is created in which each test serum is always beside a normal control. This arrangement greatly facilitates accurate comparison. In a great majority of cases a confident assessment can be made without having to perform a quantitative estimation.

Marking the Strip. This is carried out before impregnation with any marker which does not diffuse in either acid or alkaline solutions, and which adheres to cellulose and does not migrate in the electrical field. A grease pencil is quite satisfactory, but some of the cheaper ball-point pens have been found to be particularly suitable for the purpose.

Impregnation of the Strip. Before impregnation, the strip should be thoroughly inspected under a good light for the presence of imperfections such as parallel ridges or spots. Strips showing those faults may produce irregular bands and imperfect separation, and are, therefore, better discarded. Impregnation is performed by floating the strip on the surface of the buffer in a flat container, e.g. plastic sandwich box, letting it soak up from underneath, and only then dipping it into the buffer solution. This is essential—quick submersion traps air and creates opaque spots which take a long time to soak up. The placing of the strip into any liquid, be it staining solution or oil, should be performed in a similar manner.

The impregnated strip is now removed from the buffer and lightly blotted first on one side and then on the other between sheets of filter paper, so that no excess moisture is seen. It should not, however, show any white opaque areas which indicate drying out. If this happens the strip should be once more impregnated and blotted.

Placing the Strip in Position. Dental forceps with curved tips are a most convenient instrument for handling the strips at all stages. CAM should be touched as little as possible with the fingers, particularly if they are subsequently to be stained with ninhydrin, nigrosine or lipoprotein stain. In the routine use of Ponceau S for the staining of serum proteins, this factor is not critical and the strips may be handled carefully with clean fingers. The CAM strips, impregnated with the buffer as previously described, are placed on the two shoulder pads with a $\frac{1}{2}$–1 cm. overlap at each end (Fig. 5.3). The ends of the strip are pressed firmly against the pads to ensure proper contact.

Any contact between the strip and the top edge of the partition must be avoided as the latter will condense moisture and thus interfere with the separation pattern.

With a bridge gap not exceeding 8 cm., a support is not necessary. For longer runs a support may be required, in the current model of the electrophoretic chamber (see Fig. 5.1) a serrated edge mounted on the central partition provides this facility.

Adjacent strips should be placed as close to each other as possible without actually touching. At either side of each set of strips it is advisable to use a 1 cm. wide buffer-impregnated blank strip of filter paper or cellulose acetate since the two outermost margins usually show increased evaporation; this results in uneven separation of substances on these side strips and particularly of the faster migrating fractions. Account must be taken of the blank strips in calculating the current required.

Applying the Sample. After the strip has been positioned on the bridge, the sample—e.g. human serum—is applied to the buffer-impregnated and blotted strip in the form of a fine, uniform streak along the previously marked (very thin) line. A capillary tube, micropipette, or a stamp may be used as applicator. Evenness of application is achieved by moving the capillary along a straight edge placed across the tank (a special ruler is available—see Fig. 5.3). As the capillary moves along the edge of the ruler, the flow rate of the sample is controlled with the tip of the finger across the top of the capillary. The capillary is moved steadily backwards and forwards over the strip until the required amount is applied. It is essential that the sample streak finishes at least 5 mm. from each edge of the strip to avoid irregularity in the running of the components. A trace of saturated bromophenol blue solution can conveniently be added to serum samples to complex with the albumin, which then acts as a useful marker, as it is the fastest migrating major component.

A clear serum sample without particulate matter and impurities is absolutely essential for a satisfactory electrophoretic separation. This can be achieved by adequate centrifugation or passing the serum through an appropriate membrane filter.

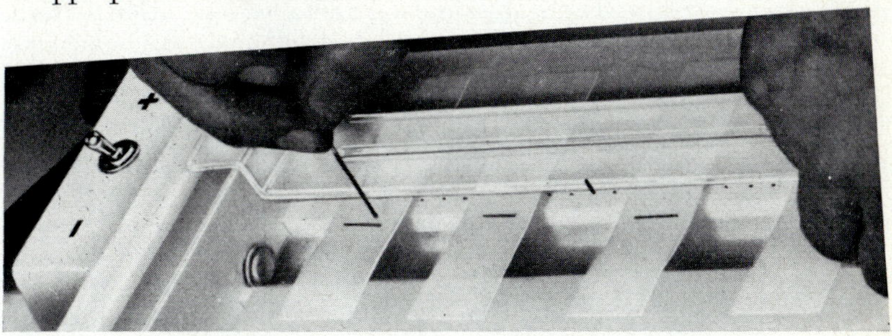

FIG. 5.3. Application of sample to CAM strip in position on the bridge. Note that the capillary is guided along the edge of a special ruler facilitating this procedure. Note also the way strips rest on the filter paper pads, and that the applied sample is easily visible, thanks to the addition of bromophenol blue.

Sample size: The volume of sample applied may vary between relatively wide limits, depending on such factors as purpose of the electrophoretic run, stain to be used, method of evaluation, concentration of the solution, etc. Quantities between 0·1–3·0 μl/cm. can be applied. For routine serum electrophoresis and photometric evaluation, 1μl/cm should not be exceeded, corresponding to ca. 60–80μg. protein. For the micro-method, when only very small volumes have

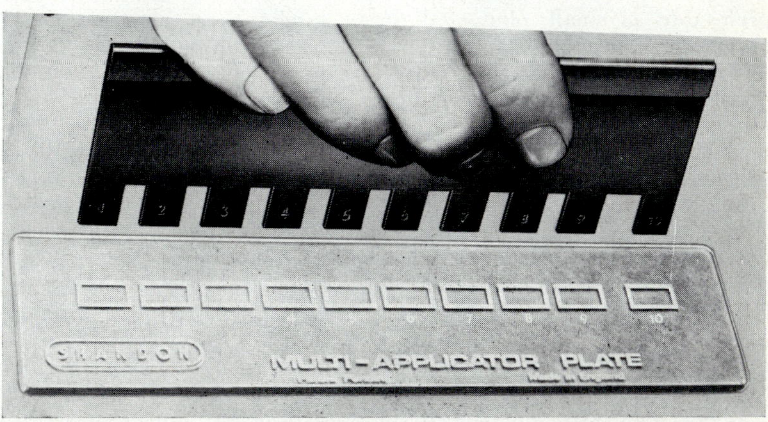

FIG. 5.4. Multisample applicator—the serum samples are applied to the square areas on the sample plate. By touching the samples, the grooves at the end of the projections are filled with the required volume.

to be applied, a finely drawn capillary is inserted into a perforated bung fitted into the end of a blood aspirating tube. Fractions of a μl. can be easily applied by means of this very simple device. For routine purposes it is often advantageous to dilute the serum approximately 2 times with a bromophenol blue barbitone buffer solution. This enables the application of larger samples which is an easier and more convenient procedure. The neatest and most detailed separation patterns, comparable to those obtained by agarose electrophoresis can be achieved by electrophoresis of serum samples, diluted 10–20 times and followed by Nigrosin staining.

The optimal amount to be applied will vary with the method of evaluation to be employed. Elution techniques require larger quantities than photometric evaluation; while the reflectance method, using a cleared strip against a white background, requires the smallest samples. Sample size will also depend on the nature of the substances being evaluated—e.g. lipoproteins and glycoproteins require larger samples. As a general guide, it can be taken that smaller sample volumes favour separation and evaluation, the limiting factor being that the staining of the separate bands is sufficient for optical reading. It should be kept in mind that the light absorption curve on photometric scanning only follows the Lambert-Beer law within certain limits, which should not be exceeded. For best results, the optimal size of sample is best determined by experiment for each particular system employed.

A semi-automatic multiple sample applicator now available on the market greatly facilitates the application procedure. It is a time and labour saving device and to a considerable extent improves the uniformity of the separation patterns. The sample applicator consists of a castellated thin metal plate with grooved projections and a glass or plastic sample plate with separated areas corresponding to the castellations. Serum samples are applied to the numbered areas on the sample plate, preferably as small, elongated drops spread over the surface. The applicator is brought into contact with the serum on the sample plate in such a manner that the grooves become evenly filled. Excess fluid on the outer surfaces of the applicator teeth leads to beady and uneven application lines and should be avoided. The charged applicator is then brought down to a buffer-impregnated and blotted CAM sheet, still resting on a slightly wetted blotting paper. The applicator should be brought down gently and left in contact with the cellulose acetate for a couple of seconds. It should not be slammed down as this leaves marks on the surface. A specially designed device facilitates the application procedure very considerably (Fig. 5.5). The CAM sheet with samples applied is then placed in position in the electrophoretic chamber.

The sixteen samples applicator is particularly convenient for routine serum electrophoresis: the separation pattern obtained is crisp and easy to interpret.

It is absolutely essential that the grooves of the applicator are kept meticulously clean. This can be achieved by using a brush to remove any protein traces and by storing the applicator in a detergent solution when not in use.

FIG. 5.5. Multi sample applicator, with guiding arm, sample plate and CAM sheet in position.

Starting Position. The position of the starting line will influence the separation pattern obtained. A number of factors are concerned in determining the best position of the starting line and this should be established on the basis of the type of material being separated, the length of the strip used, the brand of CAM, the buffer used, and the pattern desired. Initially, it is advisable to apply four samples in different positions on a single CAM sheet. On the basis of the separation patterns, the best starting position can be chosen.

Generally, in the electrophoresis of serum, as the starting line approaches the centre of the strip the globulins migrate further towards the cathode and the overall separation is more rapid. This is due to the effect of the endosmotic flow through the strip, which is more pronounced with CAM than with filter paper. Properly utilized, the endosmotic flow can improve the separation of the serum proteins. With most CAM's the serum sample is best applied between $\frac{1}{3}$ and $\frac{1}{2}$ way along the strip from the edge of the filter paper pad at the cathode end.

Current Supply. A constant voltage, constant current supply is preferred. It should be capable of delivering up to about 500 V and up to about 20 mA. There are many of these units available on the market. A simple direct current supply with an ammeter without stabilizer will also be adequate provided the current output is watched and the current, if necessary, manually adjusted to the desired amperage. If a power pack is not available, dry batteries can also be used.

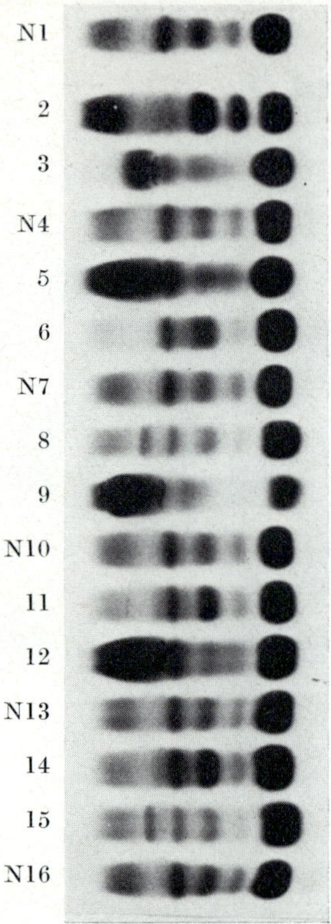

Fig. 5.6a.

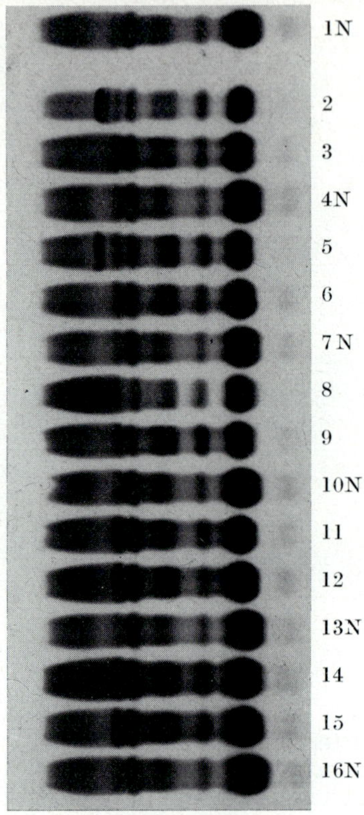

Fig. 5.6b.

Fig. 5.6a. Separation pattern obtained with multisample (16×) applicator on CAM sheet. Normal sera for comparison. Note sample 1 distorted, too close to edge of sheet. N = normal.

Fig. 5.6b. Demonstration of pre-albumin. Pattern obtained with the multi-sample applicator using the standard procedure. The protein separation pattern is very heavy so as to be able to show the faint pre-albumin bands. Note faint pre-albumin band is absent in samples 5 and 8.

Either constant current or constant voltage can be applied. In varying ambient temperatures, however, constant current is recommended. It prevents excessive evaporation from the strip which may result in "burning." Burned up areas are indicated by the appearance of translucent spots or bands.

Due to the heat production the buffer concentration in the strip rises and in consequence the current increases. With a constant current supply the voltage will drop during the electrophoretic run. These resistance changes tend to stabilise after a certain time and the current will steady itself. A rise of the current through the strip is, therefore, to be expected and in the absence of constant power stabiliser the current must be manually adjusted. As a rule 0·4–0·5 mA/cm. width of the strip should not be exceeded; this should result in 15–25 V/cm. measured across the terminals.

It should be emphasized that the voltage and amperage recommended here are measured at the terminals, i.e. on the instruments on the power pack and not across the strip itself. Due to resistance of the filter paper pads, etc., there is a considerable potential drop present and the actual voltage may be lower by 50 per cent or even more.

Electrophoresis. Routine electrophoresis on CAM strips is usually performed at room temperature. The pattern of the electrophoretic separation is affected by the following factors:

Type of CAM.
Concentration and composition of buffer solution.
Current and effective voltage.
Working distances.
Position of starting line.
Time of run.
Temperature.
Type of electrophoresis tank: effectiveness of tank sealing, degree of evaporation from the buffer and electrophoresis strip.
Characteristics of filter paper pads.

By suitable combinations of these variables it is possible to evolve optimum conditions for achieving the desired electrophoretic separation.

For serum electrophoresis, using a barbitone buffer *p*H 8·6, 0·05 M, 0·4 mA/cm. width and. a strip of 16 cm. length, a separation of 12–14 cm. will be produced in 5–7 hours.

For routine purposes using the buffer recommended and a bridge gap of 6 cm., a perfectly satisfactory separation can be obtained in approximately 45–60 minutes.

The optimum time required for the separation depends on the purpose of the electrophoresis and the evaluation method to be employed. It is valuable to add bromphenol blue as a marker for the albumen as previously described. The run should be stopped when the albumen is still more than 1 cm. from the bridge edge as buffer flow may otherwise distort the band and the prealbumin may migrate into the wick.

Crowding of separation patterns should be avoided at all times. A satisfactory electrophoretic run should be at least 3½–4 cm. long. Unfortunately, some instruments available on the market are designed for short runs. The disadvantages should be obvious, i.e. the danger of overlapping and neighbouring fractions running into each other. With a short gamma fraction, it may be very difficult to distinguish minor changes within the pattern, small monoclonal bands can be missed and vice versa, a tight gamma fraction may give the impression of a monoclonal band. Another serious problem is the merging of the small prealbumin fraction with the large albumin band which makes the demonstration and estimation of the prealbumin impossible. On a well-separated serum protein pattern inspection against a strong source of light can easily detect the presence or absence of the prealbumin band and this is of considerable clinical (diagnostic) value. In a series of 6,500 sera from blood donors, it was absent on inspection in only 13 cases. By comparing the prealbumin fraction of the patient's serum, with that of a normal control, a reasonable estimation of the amount present can be made.

Fixing. Preliminary fixation or drying is not essential, the strips can be placed straight in the staining solution provided it contains a strong protein precipitant. This procedure may, however, lead to a rise of *p*H of the staining solution. Hence some authors recommend preparation of the Ponceau S in a solution containing 7·5 per cent trichloracetic acid plus 7·5 per cent sulphosalicylic acid.

Staining

General Rules. On the whole any protein stain can be used provided, of course, the solution does not contain a cellulose solvent. As a rule, aqueous staining solutions are preferred to alcoholic ones. If alcoholic solutions are used it is essential that the strip be passed eventually through an aqueous bath, otherwise the cellulose strips tend to curl and shrink.

All staining solutions recommended for cellulose acetate electrophoresis are less concentrated than those commonly used for filter paper electrophoresis. They can all be used repeatedly without appreciable loss of staining capacity. Plastic boxes of the food or sandwich box type are particularly useful and convenient both for the buffer, and the staining and washing solutions.

It is of utmost importance that only reliable dyes, specially manufactured for electrophoresis and of proven quality, are used. Some stains in spite of the same name contain admixtures and impurities which may completely spoil the results, for instance, by staining the background.

The protein density of the bands per area of supporting medium should not be too great otherwise both the dye uptake and the stain penetration will be adversely affected. It should also be borne in mind that above a certain maximal intensity of colour the dye uptake curve and the light absorption curve of most stains lose their linearity thus making quantitative estimations in the higher values quite inaccurate. This is very important and in consequence, within limits, the smaller the sample volume the more reliable the accuracy of quantitative determinations.

Washing. A very suitable universal washing solution for aqueous stains is 5 per cent aqueous acetic acid. It is cheap, simple to prepare and does not affect the cellulose acetate material.

After the staining is completed the strips are transferred to the washing bath. The washing is continued until the background is white and and washing bath completely colourless if Ponceau S is used. This takes usually only a few minutes. The washing solution may be discarded and a fresh clean one poured into the same container or else the strip is transferred into successive washing solutions. Crowding and "stacking" of stained strips in the washing solution should be avoided as transfer of dye from one strip to another may easily occur, particularly if the strips are left in the bath "sticking" together. Agitation or the use of some shaking device practically eliminates this danger.

Drying of Stained CAM Strips. This should be carried out carefully to avoid curling. For the same reason it is best done at room temperature. The strips are blotted free of excess moisture between sheets of filter paper and left for a few minutes on the bench and then placed, while still pliable, between fresh sheets under a heavy book or similar flat weight. Time permitting, they are best left overnight to dry out completely in the flat state. If time is short, the process may be hastened by resort to a drying oven, taking due care to maintain flat pressure on both surfaces of the strips.

Clearing. There is no need for rendering the CAM transparent unless there is a special reason for doing so, e.g. for scanning or for projection purposes in which case cleared CAM is used as a transparency. Satisfactory photographs can be obtained by transillumination of uncleared CAM. CAM may be rendered glass clear by two different treatments: impregnation with oil, or spraying with a swelling agent followed by heat treatment.

(*a*) *Oil impregnation.* For proper penetration of the oil, the CAM must be completely dry. As with filter paper, it is essential that the oil used for impregnation has the same refractive index as the support. Two oils have proved very successful for all types of CAM namely Whitmore Oil 120 and Ondina Oil 17 Shell. These oils are cheap, odourless, colourless and render the CAM completely transparent. The strip is first floated on the surface of the clearing fluid and left there until it is completely transparent. After the strip has been cleared it is placed between glass plates of suitable dimensions and is ready for scanning. The cleared strips can also be returned to their original dry state by washing them in a fat solvent, e.g. ether. Since dry CAM is brittle, it is advisable to sandwich the strips between transparent self-adhesive plastic film, e.g. "Filmomatt" or "Filmolux."

(*b*) *Swelling agents.* German, American and Italian CAM may all be cleared with oil. However, they can be also permanently cleared with swelling agents, such as glacial acetic acid, or dioxan in accordance with directions supplied by the manufacturers, and will then produce a dry, transparent membrane.

Stains

Protein staining. Of the various stains Ponceau S and Nigrosin have proved to be particularly well suited for cellulose acetate electrophoresis, and will be here described in detail.

Ponceau S Method. Ponceau S 0·2 per cent solution in 3 per cent aqueous trichloracetic acid.

Technique. Place the dried strip into the staining solution as described above. Immediately this has been done a negative of the separation pattern can be discerned. This is the result of the stain soaking first into the protein-free backgound and only then gradually penetrating the protein bands.

The staining time is not critical and depends partly on the density of the bands; usually ten minutes is quite adequate. The strip may be left even overnight without any appreciable difference.

After the staining has been completed, the strip is transferred to the washing solution, i.e. 5 per cent aqueous acetic acid, and is washed and dried as described under General Rules. Ponceau S is a very rapid protein stain. The whole procedure of staining and washing can be accomplished in five minutes if necessary. It is suitable for quantitative determinations and can also be used with excellent results as a general protein stain for the demonstration of precipitation lines in immuno-diffusion methods. For photometric evaluation a 525 mμ filter is used.

If the bands are found to be too faint after staining and washing, the strip can be stained with Nigrosin stain which is much more sensitive.

Nigrosin. This is a most valuable stain for micro methods. It is particularly suitable for the demonstration of fractions, which are usually present in only small proportions, and require a sensitive stain to make them evident. Nigrosin should not be used for sample containing more than about 1g per cent of protein, as its penetration capacity and dye uptake are rather limited so that the centre of dense bands remains understained. The dye uptake curve is not linear and flattens out rapidly with increasing amounts of protein. Nigrosin is, therefore, not suitable for quantitative determination for routine purposes, nor can it be eluted. It can, however, be successfully used for overstaining strips previously stained by some other methods. This may be particularly valuable in cases in which the quantity of protein applied turned out to be too small and the bands too faint. The same applies to the staining of precipitation lines in immuno-diffusion methods.

Staining Solution. Nigrosin (G. T. Gurr) 0·001–0·002 per cent in 2 per cent acetic acid. For rapid staining an alcoholic nigrosin solution is recommended (Nigrosin 1·0 g., glacial acetic acid 150 ml., sodium acetate 0·1 M 400 ml., glycerol 50 ml., Methanol 100 ml., H_2O 1,000 ml.). Using this staining solution results can be obtained in a matter of minutes. Care should be taken that the CAM is well rinsed under a strong jet of water to remove the stain from the background.

Nigrosin solution must be kept clean by filtering or changing at frequent intervals, otherwise heavy deposits tend to accumulate. This in turn interferes with the neatness of results and may cause heavy background staining.

Using the aqueous staining solution (see above) the staining time is not critical and the CAM strips or sheets can be left conveniently overnight. Washing is not required and a rinse under a tap is perfectly satisfactory. Great care should be taken to avoid fingerprints at all stages.

Evaluation of the Electrophoresis Strips. The evaluation of the protein bands on CAM strips does not differ in principle from that on filter paper. However, as has been previously noted, CAM offers a number of distinct advantages; clear separation, sharply defined bands, complete transparency, low blanks due to minimal background staining and possibility of complete solution.

Visual inspection of the stained strip is a most valuable procedure in the hands of an experienced worker who is acquainted with the results obtained by his own particular method. *Inspection should be carried out against a strong source of light.* This is essential and will greatly improve the assessment of the separation pattern, the relative intensity of the separation pattern and the relative intensity of the individual fractions. This is usually quite sufficient for most clinical purposes and easily reveals significant abnormalities, e.g., the presence of monoclonal bands, agammaglobulinaemia, genetic protein abnormalities, etc. It also, to a considerable extent, reveals significant alterations of the main protein fractions.

The only absolute indication for electrophoretic evaluation of serum protein fractions is for quantitation of paraproteins. There is no other reliable routine laboratory method of estimating the amount of paraprotein present. Immunological techniques unfortunately cannot be relied upon due to the variations in the configuration and the antigenic determinants of the "warped" paraprotein molecule. Scanning or elution of the monoclonal band which relies on a general protein stain is not dependent on the antigenic structure of the immunoglobulin molecule.[23]

Other protein stains such as Light Green, Amidoblack, Azocarmine, New Crocein, Procion Brilliant Blue, Pontamine Sky Blue, Chromotrope, Coomassie Blue, Thiazine Red and Bromophenol Blue can all be used for cellulose acetate electrophoresis if preferred. The technique should follow that described under General Rules.

Quantitative determinations can be carried out either by elution or by scanning of the stained strip. For routine clinical purposes, scanning appears to be a simpler and more convenient procedure (Fig. 5.7). Provided the normal values for a given technique and staining method are known, the well-known pitfalls of the dye uptake problem can, to a large extent, be obviated. Automatic scanners are, of course, preferable. Except for some very irregular patterns, "dropping" of perpendiculars is quite adequate. The same rules as for filter paper electrophoresis apply, with the advantage of having no background values to contend with and no errors due to "tailing". An appropriate filter is used, the optimal wavelength depending on the stain. Quantitative results obtained by scanning will vary greatly depending on the method employed and in particular whether cleared or uncleared strips are used. With uncleared strips relatively low albumin values are found, whereas, by scanning cleared strips, due to differential dye up-take, too high values are obtained. In our hands as well as some other workers best results were obtained using reflectance scanning of a cleared strip placed against a white background.

Elution of stained bands from cellulose acetate strips is very satisfactory and rapid; more so than from filter paper.* Ponceau S stained bands can be eluted with Tris or barbitone buffer, or 0·2–0·5 per cent

* For obvious reasons, larger samples must be applied for elution procedures.

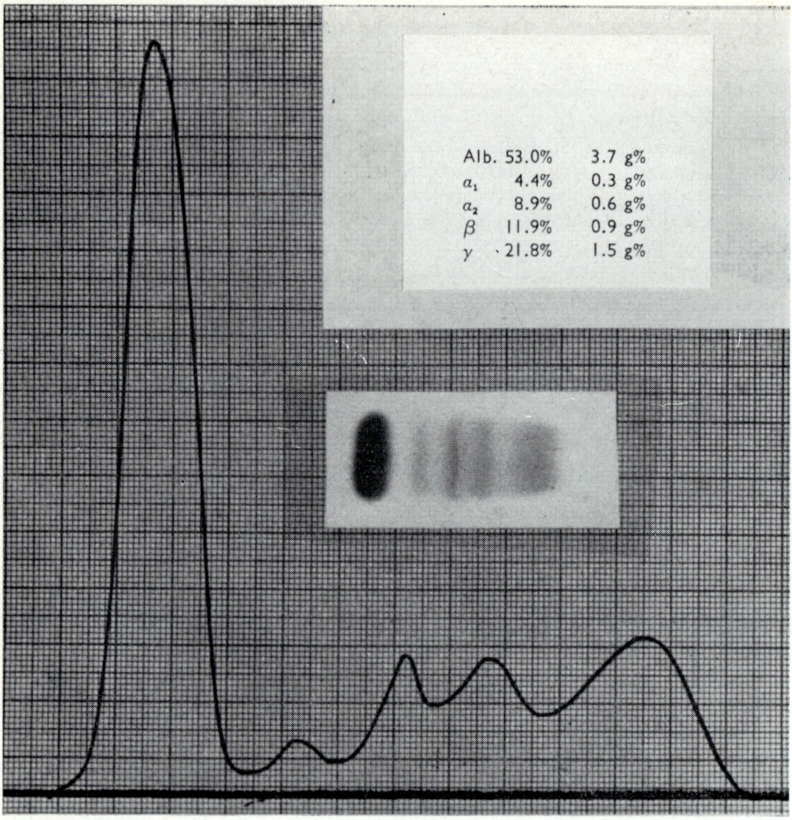

FIG. 5.7. Graph of values obtained by scanning, Ponceau S stained and cleared CAM strip. Scanning by reflection against a white background.

Hemosol. Similar eluting techniques have been described for amido-black and lissamine green.

Alternatively, the CAM may be dissolved as follows: cut up stained fractions and place in correspondingly labelled test tubes, to which appropriate volumes of solvent solution are added, (e.g. chloroform 9 vol.—ethanol 1 vol.). The CAM pieces should be pushed to the bottom. Use a shaker or vibrator to speed up the dissolution. The albumin fraction is dissolved in a larger volume, e.g. 4 mls. as compared with 2 mls. for the other fractions. Read in a suitable cuvette at 525 if Ponceau S is used. Care should be taken that the cuvettes used are well washed, otherwise they will soon be covered with a film of cellulose acetate.

A quantitative procedure which is claimed to eliminate the problem of differential dye uptake makes use of Procion Brilliant Blue R.S.

(ICI) which is bound covalently by protein molecules and therefore is bound almost equally to all the serum protein fractions.

The cellulose acetate strips, after being run in the conventional barbitone buffer, are directly transferred to a staining bath consisting of 0·5 per cent of the dye in absolute methanol to which a small amount of concentrated HCl has been added (1 ml. to 50 ml. dye). Staining proceeds for three to five minutes and the background is then cleared with several (3–5) changes of methanol (the methanol can be reused after decolorizing with charcoal). The blue bands stand out against a white background.

The strips can be cut up into their five main individual fractions and from each of these the bound dye can be eluted with a 1·0 per cent solution of Hæmo-sol in 5 per cent iso-propanol. The elution is carried out in a water bath which has just been taken off the boil (80°C), for 10 minutes. The eluates can then be read for their optical density in a suitable colorimeter using a red filter or the solutions can be decanted into Auto Analyzer cups for automatic evaluation. The quantitative results have shown good agreement with micro-Kjeldahl protein determinations.

In order to be able to judge the quantitative variations of the protein fractions a thorough knowledge of the normal values and patterns is absolutely essential. As yet there is no standardization of electrophoretic procedures and normal values and patterns must be worked out for each technique and staining method before any conclusions can be drawn from quantitative determinations, no matter how accurate they are.

TABLE 5.1

Values obtained by reflectance scanning of CAM using (a) uncleared strips and (b) the same strips cleared and placed on a white background.

CAM	Total Protein g.%	Alb.		α_1		α_2		β		γ	
		%	g.%	%	g.%	%	g.%	%	g.%	%	g.%
Uncleared	4·7	38·9	1·8	3·1	0·1	18·3	0·9	12·8	0·6	26·8	1·3
Cleared		59·0	2·8	5·1	0·2	10·7	0·5	8·4	0·4	16·9	0·8
Uncleared	7·0	45·5	3·2	3·6	0·3	13·0	0·9	11·7	0·8	26·2	1·8
Cleared		53·7	3·8	4·0	0·3	10·9	0·8	11·8	0·8	19·5	1·4
Uncleared	7·4	37·5	2·8	3·4	0·3	15·1	1·1	11·2	0·8	32·8	2·4
Cleared		55·5	4·1	3·5	0·3	11·2	0·8	7·7	0·6	22·1	1·6
Uncleared	6·5	40·0	2·6	4·4	0·3	16·7	1·1	13·5	0·9	25·5	1·7
Cleared		56·6	3·7	4·1	0·3	11·2	0·7	9·2	0·6	18·9	1·2
Uncleared	6·8	38·9	2·6	4·4	0·3	14·0	1·0	11·1	0·8	31·6	2·1
Cleared		57·2	3·9	3·6	0·2	11·6	0·8	9·2	0·6	18·4	1·3

It should be emphatically stressed that even in the best hands, with the best apparatus available and under optimal conditions quantitation of electrophoretically separated fractions is, unfortunately, still far from perfect both as to accuracy and reproducibility and leaves much to be desired. This applies much more to the various quantitation

TABLE 5.2

Normal values obtained on sera from 56 blood donors, using the small-scale cellulose acetate electrophoresis. Ponceau S staining. Evaluation by scanning without correction using **transmitted** *light.*

NORMAL VALUES

56 CASES (BLOOD DONORS)

Ponceau S, cellulose acetate

	Alb.	α_1	α_2	β	γ
% RANGE	57 - 68	1·0 - 5·7	4·9 - 11·2	7 - 13	9·8 - 18·2
AVERAGE %	*64*	*2·9*	*8·9*	*10·6*	*13·6*
ABSOLUTE VALUES in g/100ml					
RANGE	3·5-5·2	0·1-0·4	0·4-0·8	0·5-1·0	0·6-1·3
AVERAGE	*4·3*	*0·2*	*0·6*	*0·7*	*0·9*

TABLE 5.3

Evaluation by **reflectance** *scanning against a white background of Ponceau S stained and cleared CAM strips.*

Main protein fractions	Alb.	α_1	α_2	β	γ
Range of absolute values in g./100 ml.	3·4–5·4	0·1–0·4	0·5–1·0	0·6–1·2	0·7–1·4
Range of relative values per cent	50–67	2–5	6–13	9–14	8–21

It was thought useful to obtain data from a hospital population as well as from so-called healthy normals. After all, probably the great majority of investigations is carried out on in-patients and it is well known that a number of factors connected with the stay in hospital may influence serum protein pattern (e.g. posture, rest, diet, etc.).

Serum specimens were collected from 1,320 patients in a general hospital, comprising a large plastic surgery, orthopædic and rehabilitation units. Normal values were obtained using an adaptation of Hoffman's statistical method and the results are given here.

procedures than the actual separation process. There are many variables involved which are difficult to control and standardize, particularly under ordinary routine conditions. This must be borne in mind and conclusions drawn from relatively small variations can easily lead to misinterpretation; for the same reason, reporting to two decimal points is completely unjustified. Using electrophoresis of serum protein as a diagnostic tool, one should not become obsessed with numbers.

Normal values for serum proteins obtained from separations of sera from 56 healthy blood donors, using the small-scale cellulose acetate method and the Ponceau S staining technique are shown in Table 5.2.

Urine electrophoresis is a most valuable procedure for the detection of Bence Jones protein in the urine, an important feature for the diagnosis of diseases of the myeloma, macroglobulinaemia group. It can and should be performed both on unconcentrated and concentrated urine. The urine sample used for electrophoresis must be clean and should not contain any particulate matter which on concentration may tend to accumulate. For routine purposes it is recommended to apply the urine sample side by side with a $\frac{1}{20}$ diluted serum sample from the same patient. This permits a comparison of the exact location of any protein fractions or abnormal bands present and is very helpful in the interpretation of results. On occasions when a possible significant and relevant band is found close to or on the application line, it may be advisable to apply the sample nearer the anode end where any traces of impurities are not likely to interfere with the interpretation. The presence of haemoglobin in the urine may produce a discrete band on electrophoresis and imitate a monoclonal band, e.g., a Bence Jones protein.

CSF Proteins. CSF electrophoresis can be performed on CAM, even without concentration if a larger sample, e.g. 2 μl. per cm. is applied and Nigrosin used for staining. Concentrated CSF will produce a fair separation pattern but quantitation cannot be regarded as reliable or satisfactory.

With the availablity of immunological techniques for the detection and assay of specific proteins, electrophoresis of CSF protein by itself is a much less satisfactory procedure from the diagnostic and prognostic point of view.

Glycoprotein Staining. This is particularly satisfactory on cellulose acetate (see Fig. 5.8). Electrophoresis is as described for serum proteins but a sample volume 2–3 times greater is used. A technique which has given consistently good and reliable results runs as follows.

Wash the dried strip in industrial alcohol for about ten minutes. Transfer into:

Periodic acid 50 per cent	2·5 ml.
M/5 Sodium acetate	25·0 ml.
Distilled water	250·0 ml.

for about ten minutes. Rinse in N/1,000 HCl. Transfer into:

Potassium iodide	40·0 g.
Distilled water	400·0 ml.
Add before use N/10 HCl	10 ml.

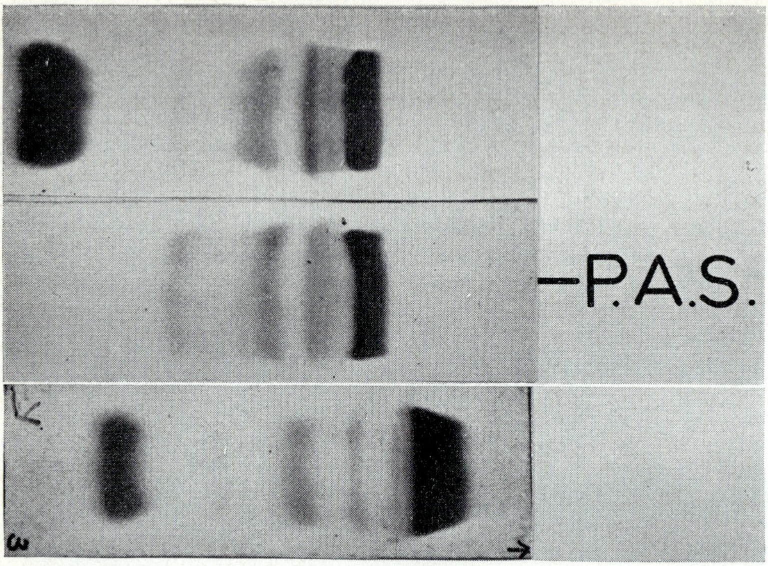

FIG. 5.8. Small-scale strips showing patterns obtained in macroglobulinæmia. P.A.S. shows glyco-protein staining.

The solution and the strips will show a brown coloration after the strip is placed into it. Add from capillary pipette a few drops of Amfix, May & Baker (saturated ammonium thiosulphate) until the brown colour disappears and allow the strip to remain until the stain has just cleared from the iodine stained bands.

Rinse in N/1,000 HCl.

Transfer the strip into Schiff's Stain (G. T. Gurr) and leave it there until the bands have acquired an intense purple colour. This may take 15–30 minutes. Place in three changes of

0·1N A.R. Nitric acid, each for 10–15 minutes.

Rinse again in N/1,000 HCl.

Remove from last washing solution, blot and dry at room temperature, then place between blotting paper sheets under pressure. The times of the various procedures are not strictly critical.

Lipoprotein Staining. The staining methods recommended for the demonstration of lipoproteins by paper electrophoresis are unsuitable for CAM strips. A new staining method combines pretreatment with ozone and subsequent staining with Schiff's reagent (as used for glycoproteins). The method has given very good results and can be used with equal success in CAM and paper electrophoresis.

Procedure. Electrophoresis is performed as described above but appreciably more serum must be applied than for ordinary protein staining.

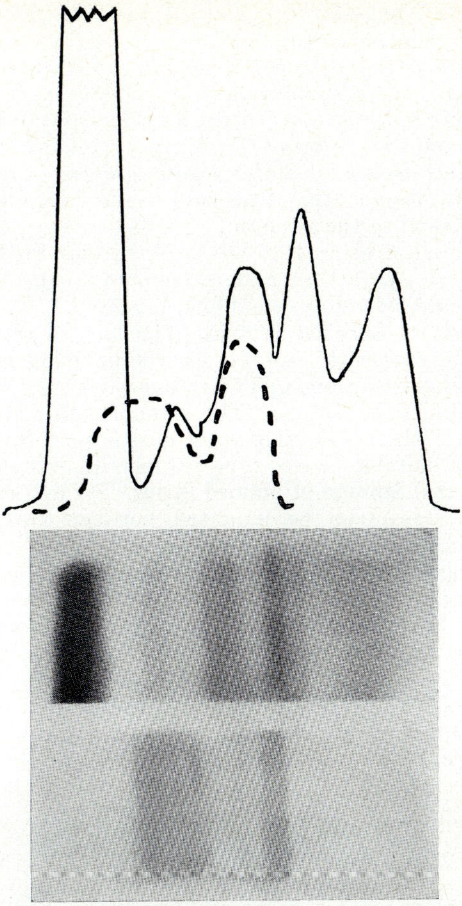

FIG. 5.9. Lipoprotein staining, Kohn's Ozone/Schiff method. Note narrow and sharp α_2 lipoprotein band (corresponding to β lipoprotein on filter paper) and wide and fuzzy α_1 band.

Reagents:

(*a*) *Ozone:* Relatively cheap and simple ozonisers specifically designed for the purpose are now commercially available from German and Japanese sources.

Alternatively a petri dish containing barium peroxide is placed in a large container and ozone is generated by adding conc. H_2SO_4; the beaker or tank is kept well closed.

(*b*) Schiffs reagent—see glycoproteins.

After electrophoresis has been completed, the CAM strips are dried and exposed to the action of ozone for about 15–20 minutes in a suitable container. After the ozone treatment, the strips are rinsed briefly in N/1,000 HCl. They are then stained with the Schiff's reagent as employed for glycoproteins (q.v.).

Normal serum usually shows two bands: a sharply defined band which represents the β-lipoprotein and exhibits α_2 mobility on the CAM strip; and a less clearly defined wide band lying behind the albumin band and representing the α_1 lipoprotein (Fig. 5.9). Usually the albumin band itself will exhibit a faint staining outlined by a more strongly stained margin; this is probably caused by fatty acids which are loosely attached to the albumin.

Staining technique using Red Oil O in various modifications were reported to be a satisfactory and convenient lipoprotein stain. For details the relevant references should be consulted.

Electrophoresis of Labelled Protein. The lack of absorption makes cellulose acetate eminently suitable as a supporting medium for the separation of labelled proteins. The radioactivity is found only in the fractions containing the isotope. The demonstration and quantitative determination of the radioactive substance is performed by scanning the strip with a suitable Geiger-type scanner or by auto-radiography.

Preservation and Storage of Stained Strips. Stained cellulose acetate electrophoretic strips can be mounted between two sheets of self-adhesive plastic. This provides a convenient, tough, transparent and resistant coating which makes the strips eminently suitable for preservation and storage. Oil-cleared strips are kept between glass plates.

Hæmoglobin Separations. The hæmoglobin solutions are prepared as described below under 'estimation of haptoglobins in serum'.

A discontinuous buffer system recommended by Graham and Greenbaum was found to be particularly useful for the separation of abnormal haemoglobins on cellulose acetate in the Shandon tank.

The buffer composition is as follows:

At the Anode:	
Tris	25·2 g.
EDTA	2·5 g.
Boric acid	1·9 g.
Distilled water to	1,000 ml.
At the Cathode:	
Sodium diethylbarbiturate	5·15 g.
Diethylbarbituric acid	0·92 g.
Distilled water to	1,000 ml.

For the impregnation of the cellulose acetate strip itself, both buffers are mixed together in equal volumes.

Sample Application. With the buffer system recommended, the samples are best applied midway between cathode and anode. Known haemoglobin solutions (e.g. HbAS, HbAC, and cord blood for HbF) are applied at suitable intervals for comparison. The application line should be thin and straight and the sample volume small, not more than 0·1 to 0·3 μl./cm. (of approximately 10 per cent haemoglobin solution). The secret of satisfactory separation of abnormal haemoglobins lies in applying the correct sample volume. The optimal size of sample is best established by experience using known haemoglobin marker solution. The sample can be applied by means of a suitable micropipette with a very fine point or very conveniently by means of a fine brush. The applying device, guided along a straight edge, should be kept moving when approaching and coming in contact with the CAM in order to prevent "beading". Between applications the brush is rinsed rapidly and thoroughly in distilled water and blotted against filter paper. This last procedure, at the same time, reveals any contamination from a previous sample. Ample space should be left between the samples and the margins of the CAM. Special haemoglobin applicators similar to those used for serum proteins are now available on the market and are simple and convenient in operation. For HbA_2 determination, particularly if unstained bands are evaluated, relatively large sample volumes are necessary: 10 to 30 μl. of a 10 per cent haemoglobin solution should be applied, preferably as one long streak or a series of shorter streaks, not exceeding 2 to 3 μl./cm. application line. The sample volume must be sufficient to produce enough colour to give a substantial reading on the absorption meter. The Shandon Multi-Applicator can be used, repeating the application several times. Overloading leads to indifferent separation and incomplete elution, which in turn results in relatively too low HbA values and, in consequence, too high HbA_2 values. If stained bands are used for elution, a sample volume of approximately 4 to 5 μl. (1 μl./cm.) is quite adequate.

Electrophoresis. Trial runs with known (preferably heterozygous) haemoglobin markers are performed in order to establish optimal conditions for the type of chamber used and the purpose of the separation. Most satisfactory results were obtained with a bridge gap of 5 to 6 cm. and a potential of 200 to 240 volts obtained by applying constant current of approximately 0·2 to 0·3 ml./cm. width. Under these conditions a perfectly satisfactory separation of abnormal haemoglobins as well as of HbF, if present in appreciable quantities, can be achieved in 30 to 90 minutes. Haemoglobin A_2 will be very clearly separated. The timing of the run depends on the length of the strip, the voltage applied, the temperature and the purpose of the separation; HbF, for instance, may require approximately 80 to 90 minutes. The separation pattern is easily monitored by watching the migration of the Hb bands. Rapid separation of abnormal haemoglobin, including HbA_2 which migrates with the HbC mobility, can be achieved by applying relatively high voltage (i.e. 300 to 350 volts across a narrow bridge gap, for example, 3 cm.). Perfectly satisfactory separation can be achieved in 8 to 15 minutes.

Staining. Following completed separation, the CAM strips or sheets can be fixed and stained, as for routine protein electrophoresis. Care should be taken not to confuse haemoglobin bands with those due to stromal proteins.

Evaluation. Significantly elevated HbA_2 and HbF can be assessed by visual inspection and roughly estimated if the haemoglobin concentration and volume of the sample are similar to that of a control specimen. Quantitative estimation applies essentially to HbA_2 (and possibly HbF). Estimation of HbA_2 is best carried out by elution of the separated bands rather than by scanning, primarily because of the great disproportion of the optical density of the HbA and the HbA_2 bands. Elution can be performed on (1) unstained bands or (2) stained bands (e.g. Ponceau S as for protein electrophoresis). The choice is largely a matter of personal preference and experience. If properly carried out, elution of unstained bands is probably more accurate, as some interfering factors (e.g. uneven separation, irregular dye uptake, stromal protein bands, etc.) can be eliminated. On the other hand, elution of stained bands requires much smaller sample volumes. The separated bands are cut out close to the

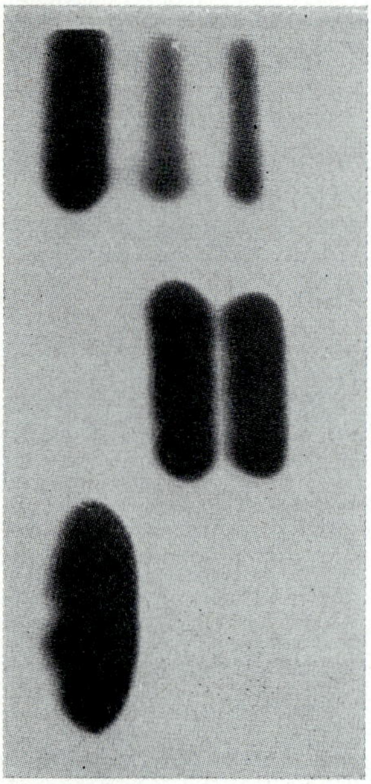

FIG. 5.10. Haemoglobin separation. CAM electrophoresis discontinuous buffer system, bridge gap 6 cm. Sample applied in the middle of the strip; 40 minutes run. ↓ indicates origin which was placed at the mid-point of the strip.

margins and eluted in barbitone or Tris buffer (preferably with agitation) for about 15 to 20 minutes. (The area between the HbA and HbA_2 band is part of the HbA and not the HbA_2.) Subsequent centrifugation helps in obtaining clear eluates. The volume of the eluent used depends on the size of the cuvette and the type of colorimeter available. The HbA_2 band is eluted into the smallest possible volume (e.g. 0.3 to 3 ml. and the HbA into 20- to 30-ml. volumes, for instance, bringing the optical densities (OD) of the eluted fractions within comparable ranges. The optical density is read at 413 μ (Soret band).

Calculations. Use this formula:

$$\frac{\text{OD HbA}_2}{\text{OD HbA}_2 + \text{OD HbA } (\times \text{ dilution factor}) \times 100} = \%\ \text{HbA}_2$$

Stained bands are eluted in a similar manner, and the optical density is read at a suitable wavelength; for Ponceau S at 525 μ. With the method described, the normal range of HbA_2 in the U.K. population was 1·5 to 3·3 per cent and did not exceed 3·4 per cent.

Estimation of Haptoglobins in Serum

Principle. On Electrophoresis free hæmoglobin in serum has a β mobility whereas the hæmoglobin-haptoglobin complex migrates with the α_2. If increasing amounts of hæmoglobin are added to serum until the haptoglobin is saturated a free hæmoglobin band will be seen. The lowest concentration at which a free hæmoglobin band can be demonstrated is taken as the hæmoglobin binding capacity of the serum.

Preparation of Hæmoglobin Solution. Red cells are washed three times in saline. The packed washed cells are lysed with an equal volume of distilled water and the hæmolysate shaken with about 1 ml. of chloroform for a few minutes and then spun. The hæmoglobin concentration of the supernate is estimated and then adjusted to about 800 mg./100 ml. with the electrophoresis barbitone buffer.

Five doubling dilutions (i.e. 400, 200, 100, 50 and 25 mg./100 ml.) of the hæmoglobin solution are then prepared and equal volumes of the hæmoglobin solutions and serum are mixed and left for 30 minutes at room temperature.

Electrophoresis. The hæmoglobin-serum mixtures as well as a sample of serum and a Hb control are submitted to electrophoresis in the usual way preferably on a wide sheet of cellulose acetate. When adequate separation has been achieved (usually about ½–1 hour) electrophoresis is stopped and the lowest concentration at which two hæmoglobin bands are seen is taken as the approximate hæmoglobin binding capacity, expressed in mg. of Hb that just saturates the haptoglobin. If greater accuracy is required further dilutions of hæmoglobin near the approximate value can be prepared and the procedure repeated.

Staining. The hæmoglobin bands can usually be seen without staining but can be made more obvious by staining for a few seconds in freshly prepared 0·1 per cent leucomalachite green in 5 per cent acetic acid to which a few drops of hydrogen peroxide have been added just before use.

Enzymes. Enzymes can be very advantageously separated on CAM strips. After the usual electrophoretic separation the enzymes are generally transferred to an agar gel or a second CAM strip, each containing an appropriate reaction system for identification of the enzyme.

Dehydrogenase Iso-Enzymes. Numerous electrophoretic techniques for the separation of LDH iso-enzymes have been reported using starch gel, agar gel and cellulose acetate. Barnett's LDH method described below is relatively simple, gives consistent results, and all the equipment used is available in most laboratories and does not differ from that employed for routine serum electrophoresis.

Reagents:

Buffer: 0·05 molar barbiturate buffer *p*H 8·6 (the same as for routine serum separations).

Dye substrate reagent: An approximate 0·1M solution of sodium lactate *p*H 7·4 is prepared by mixing 1 ml. 1M lactic acid with approximately 11 ml. of a 0·1M solution of sodium hydroxide; this solution is heated to 90°C for five minutes, cooled, re-heated and cooled again; the *p*H is then adjusted to 7·4 by the addition of more sodium hydroxide.

The complete working reagent is prepared as follows:

The approximate 0·1M solution of sodium lactate	1 vol.
N.A.D., 1 per cent solution, in distilled water	1 vol.
0·1 per cent in distilled water of M.T.T.-3 (4, 5-dimethylthiazolyl-2)-2, 5-diphenyltetrazolium bromide	3 vol.
Phenazine methosulphate, 0·1 per cent solution, in distilled water (prepared freshly before use)	0·3 vol.

The final working reagent is best prepared just before completion of the electrophoretic run. For the demonstration of malic dehydrogenase substitute malic acid for lactic acid and for G6PD substitute G6P.

Fixative.

Formol saline (4 per cent formaldehyde in 0·85 per cent saline)	1 vol.
Distilled water	5 vol.

or

10 per cent acetic acid in water

or

Glacial acetic acid	5 ml.
Ethanol	5 ml.
Distilled water	200 ml

rinse in distilled water

Electrophoresis Apparatus. As described before. A bridge gap of 5–8 cm. is quite adequate.

Cellulose Acetate Strips. One can either prepare separate strips for each determination or use a sheet on which a number of samples can be applied. This enables a control to be incorporated. The strips are prepared in duplicate, one for the separation and one for impregnation with the dye substrate reagent. In order to save the precious reagent the strip used for impregnation can be much shorter, just to cover the separation pattern.

Sample Application. The volume can vary between 2–5 μl. depending on the enzyme activity. It is applied in the centre of the strip as a streak. A drop of bromophenol blue can be added to the serum to act as a marker.

Procedure: The conditions of the electrophoretic run are similar to those described earlier for protein separation. Depending on the lengths of the strip a 40–90 minutes run would be quite adequate.

Preparation of substrate strip: For a strip about $7 \times 2\frac{1}{2}$ cm., about 0·25 ml. of the dye substrate reagent is placed on the bottom of a petri dish and the strip is floated on this solution. In this way the strip is fully impregnated without being overloaded. This should be done rather carefully as excess of the reagent may cause diffusion of the iso-enzyme bands, whereas insufficiency may cause uneven staining. Alternatively substrate reagent-impregnated filter paper can be used. Following impregnation the strip is removed, the excess liquid drained

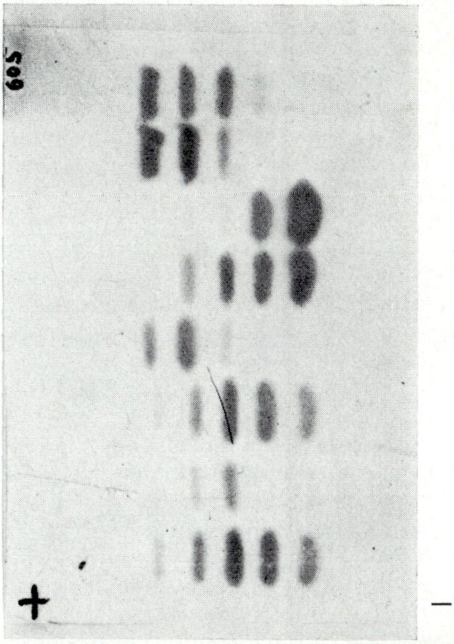

Fig. 5.11. LDH patterns obtained by the CAM electrophoretic technique, using INT Tetrazolium salt—samples obtained from various sources (tissue and serum). (*Courtesy Dr. Saito.*)

off and it is then placed on a clean glass plate in a suitable moist chamber (plastic box, large petri dish, etc.). The electrophoretic strip after separation is removed from the tank, trimmed so as to match the reagent strip in size and is now carefully superimposed on to the reagent strip, top surface facing the reagent strip. Care should be taken to ensure that the contact is perfect, and that no air bubbles are trapped. The moist chamber or the petri dish is now closed and incubated at 37°C for about 15–30 minutes; with weak enzyme activity the incubation period may be prolonged. Once the bands have developed sufficiently the strips are separated from each other and placed in the fixative for 5–15 minutes and then blotted and dried at room temperature, between filter paper sheets under pressure. Using acetic acid ethanol fixative 3–5 minutes is enough. Several variants of the technique using various Tetrazolium salts have been reported to give excellent results. The separation pattern obtained by the use of INT Tetrazolium salts is shown in Fig. 5.11.

Separation of Nucleotides. Electrophoretic procedure is exactly the same as previously described.

Buffer. Ammonium formate-formic acid *p*H 2·9–3·1 and ionic strength of 0·01. The buffer solution should contain 33 per cent by volume of propylene glycol.

Sample. It is recommended that the nucleotide mixture be dissolved in a minimum volume of 0·01 formic acid. The authors recommend the sample volume of a 3 μl. as a 4 cm. streak; larger volumes can also be used.

Current. 0·3 mA/cm. width with a potential drop of 60 V/cm. length. Under these conditions complete separation of adenosine, cytidine, uridine and guanosine monophosphates can be achieved in 15 minutes. The current, size of strips, sample volume can, of course, be changed to suit the particular requirements and the optimal conditions established by experience.

Demonstration of Separated Nucleotides. The strips are removed from the tank, placed on a glass plate and dried in a stream of air. The nucleotides are visible under ultra violet light as dark areas which can be cut out and placed in small test tubes for elution. Similar areas are cut out from sites immediately adjacent to the separated substance and eluted to provide background blanks.

Elution. This is performed with a 0·02N KOH solution in 6 hours at room temperature. About 1·5 μg. of each substance is eluted with 200 μl. Under these conditions a background absorption of ultra violet should not exceed 10 per cent of total reading. This absorption will depend on the type of cellulose acetate membrane. The German (Sartorius) cellulose acetate membranes seem to contain the lowest amount of ultra violet absorbing contaminating material. According to the authors the lower limit of analysis of each nucleotide was about 0·15 μg. and the reproducibility about 3 per cent.

Aminoacid Separation. Good resolution of many aminoacids can be achieved using the standard CAM technique and appropriate buffer solutions, without having to use high voltage. Electrophoresis of urinary amino-acids can be employed as an excellent screening test. Urine is concentrated for this purpose, to a standard level, e.g. by evaporation under vacuum. Electrophoresis is performed in a formic acid-acetic acid buffer *p*H 2 (formic acid 31·2 ml.; glacial acetic acid 59·2 ml.; distilled water to 1,000 ml.) for 20 minutes at 200–300 V. Samples are applied near the anode end. Reference aminoacid mixtures are run for comparison on the same CAM sheet.

Development. The spots are located with a ninhydrin solution which does not affect or dissolve the cellulose acetate:

Ninhydrin, 0·2 g., dissolved in 6 ml. alcohol and then made up to 100 ml. with di-ethyl ether.

Method. After electrophoresis the strips are dried either in an oven or with a hair-dryer. They are then drawn through the ninhydrin solution and dried by waving them in the air for a few seconds. They are then placed between filter paper sheets on a hard support, under pressure, and dried for about five minutes at 80–100°C.

Biologically Active Substances. Since electrophoretic separation is accomplished much more rapidly on CAM strips than on paper, the danger of loss of activity is very much reduced. After separation has been completed a narrow strip is cut lengthwise from the CAM and rapidly stained to locate the fractions. The unstained zones on the main strip are then cut out and used for the appropriate biological evaluation—e.g. organ bath. Using a barbitone buffer, *p*H 8·6 and a terminal voltage of 200 V it was possible to separate bradykinin and histamine within 30 minutes without any appreciable loss of activity.

Some Common Errors. Unsatisfactory results when using CAM can normally be traced to one or more of the following faults in applying the techniques described in this paper.

(a) Buffer solutions

- (i) Usually too high concentration.
- (ii) Less often, too low concentration.
- (iii) Leaving the buffer for long periods in the tank, leading to evaporation and hence increase in buffer concentration.

(b) Setting up tank and positioning of CAM.

- (i) Buffer levels uneven.
- (ii) Insufficient volume of buffer so that the level is below the connection between the electrode and buffer compartment (cotton wool wick or labyrinth) preventing current from reaching the strip.
- (iii) Electrical leak between anode and cathode compartment.
- (iv) Filter paper wicks dirty due to not being changed frequently. If left standing for long periods of time, the buffer concentration at the wicks increases very considerably and unevenly.
- (v) Filter paper wick not reaching into the buffer compartments.
- (vi) Inadequate contact between the CAM and the filter paper wick.
- (vii) Leaving the tank uncovered.

(c) Cellulose acetate membrane (CAM).

- (i) Faulty impregnation, i.e., dipping the CAM instead of initially floating it on the surface of the buffer.
- (ii) Excessive blotting or drying. This will be indicated by white spots or areas.
- (iii) Insufficient blotting—excess buffer left on surfaces.
- (iv) Finger prints.
- (v) Curling or bowing of the CAM—it should be flat or slightly sagging in the middle.

(d) Application of sample.

- (i) Volume too large.
- (ii) Old or dirty serum.
- (iii) Uneven application.
- (iv) Applying sample from edge to edge without leaving adequate margins.

(v) Crowding, if more than one sample is applied.
(vi) Delay between application of samples.
(vii) Too much Bromophenol Blue added to serum.
(viii) Incorrect starting position.
(ix) Splashes or droplets of sample on the CAM: even tiny droplets can spoil the run.

(e) Electrophoretic run.

(i) CAM touching the central partition at any point and at any stage during the run.
(ii) Too long or *too short* a separation.
(iii) Albumin band coming too close to the bridge edge.
(iv) Leaving the strip too long on the bridge with the lid off before applying the sample.
(v) Too high current. This leads to overheating and hence crowding and distortion of the pattern.
(vi) Inadvertently changing the polarity.

(f) Fixing, staining and clearing.

(i) Too high drying temperature. It should not exceed 80°–100°C.
(ii) Staining solution used for too long. It must be born in mind that buffer is continually being added to the staining solution and the *p*H change must therefore be carefully watched.
(iii) Poor quality stain (Ponceau S in particular).
(iv) Too many strips in the same staining dish—preferably, they should not rub against each other.
(v) Inadequate washing.
(vi) Using alcoholic stains without final aqueous wash.
(vii) Excessive heat—too rapid drying of stained CAM.
(viii) Trying to clear the CAM when not completely dry.

IMMUNO-ELECTROPHORESIS

Transfer Method. The technique is a variant of the classical Grabar and Williams method and is based on the same principle. Most of the details are essentially similar, apart from the electrophoretic run which is performed on cellulose acetate strips.

The method consists in applying the cellulose acetate strip on which the protein fractions have been separated (electrophoretic strip) and a narrow filter paper strip impregnated with the antiserum (antiserum strip) on to the surface of an agar gel. The separated fractions and the antiserum diffuse into the agar forming precipitation lines where antigen and antibody meet at equivalent proportions (Fig. 5.12). Thus conditions are created similar to those achieved by the classical methods.

Preparation of Agar Gel Plate. Lantern slides, 8 × 8 cm. or 8 × 10 cm. (American standard size) are particularly convenient, as they are easily available, suitable for multiple patterns and also lend themselves for projection. Clean glass plates are placed on some sort of support on a horizontal table (the level of these plates being checked by a spirit level) and the required volume of molten agar is poured onto the plates, taking care that it reaches the edges at all points. It is advisable to have the glass plates warm so as to avoid premature setting of the agar. The simplest way of pouring out the agar is to use a test tube or a small cylinder marked to the volume required and pour the agar out in one movement and then, with the lip of the cylinder, the agar can be spread out as required.

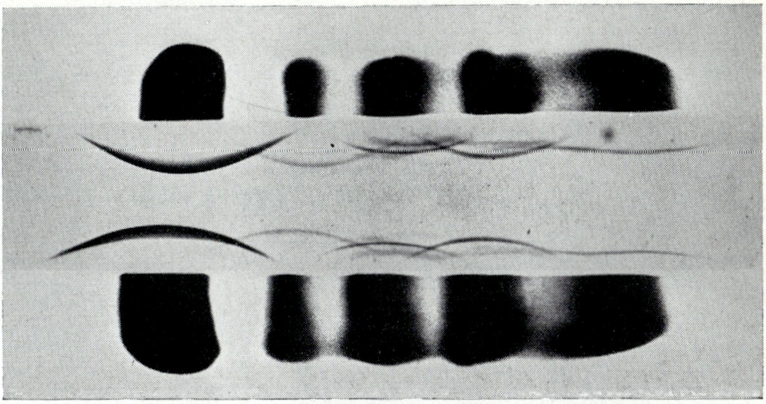

FIG. 5.12. Transfer method. Large-scale. Horse antiserum. Ponceau S stain. Precipitation pattern with corresponding electrophoretic pattern cut from the same strip.

Any *clear* agar can be used. A 1 per cent agar solution containing 1 per cent sodium azide is prepared. The sodium azide acts both as an electrolyte and also as a bacteriostatic agent. Alternatively, saline or barbitone buffer agar (a composition similar to that used for electrophoresis) can also be used. A layer of about 0·5–1 mm. thickness is recommended for most purposes. This, on a 8 × 8 cm. lantern slide, corresponds to approximately 5–8 ml. of the molten agar. The agar plates are left for a short time to set under a suitable cover to prevent dust settling and are then transferred to moist chambers in which they can be stored for several days. Plates prepared the previous day are best.

A pattern indicating the desired relative positions of the electrophoretic and antiserum strips is drawn on white paper. This pattern will be clearly visible through the agar plate when it is placed underneath. With experience, this step can be omitted.

The electrophoretic run is performed on cellulose acetate strips of desired size as described previously. For small-scale immuno-electrophoresis, 8 to 10 × 0·5–1·0 cm. strips are used. Mark the starting line. The sample is applied as a short streak reaching almost to the edges. The volume of the sample may be greater than that applied for ordinary electrophoretic separation. The migration time necessary for an adequate separation is determined by watching the position of the bromophenol blue-albumin band. The length of separation patterns for small-scale immuno-electrophoresis should, as a rule, not exceed 5 cm. leaving about 1·5 cm. clear on both ends of the working length of the strip, i.e. 8 cm. After the electrophoretic run is completed, switch off the current, remove the strip from the apparatus and place it on the surface of the previously prepared agar plate as indicated by the position pattern. The strip will immediately adhere to the wet surface of the agar. Care should be taken to avoid air bubbles under the strip. If these occur, the strip can be partially lifted and replaced in its former position, in such a manner that the strip is in complete contact with the agar.

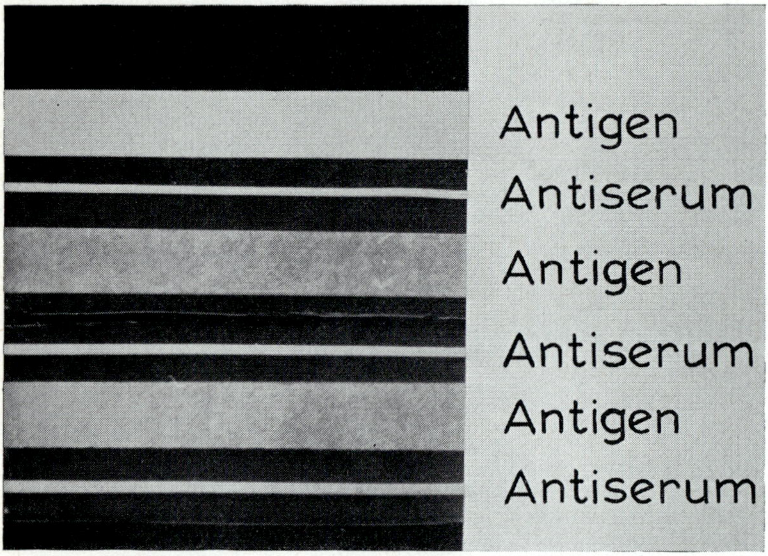

FIG. 5.13. Transfer method. Position diagram. Electrophoretic and antiserum strips in position on agar plate. (*a*) Two electrophoretic strips on both sides of antiserum strip; (*b*) two antiserum strips on both sides of one electrophoretic strip.

Application of Antiserum. Filter paper strips (Whatman No. 3 or 3 MM) approximately 1–2 mm. wide, are impregnated with the antiserum and placed on the gel at a distance of 3–5 mm. according to the required pattern (e.g. one serum sample against two antisera or vice versa). (Figs. 5.13 and 5.15). The impregnation is performed by means of a charged Pasteur pipette, running it down the strip. The appropriate amount of antiserum best for each system is found by experience. With the commonly available antisera a serum-antiserum ratio of about 1:25–30, corresponding to approximately 40–50 μl. of antiserum per strip is usually satisfactory. As the filter paper strips can only hold a limited amount of antiserum, the remaining volume required is added to the impregnated antiserum strip already placed on the gel, care being taken that it does not overrun.

Cut off overlapping ends of electrophoretic and antiserum strips. Place the agar plate in a moist chamber at room temperature so that the diffusion and precipitation process may take place. The precipitation lines usually appear within the first 24 hours and are clearly visible against a suitable background under favourable lighting conditions. The agar plate is left in the moist chamber until the full precipitation pattern is developed. This must be judged by experience. In most cases, e.g. with serum proteins, using the small-scale immuno-electrophoretic technique, 1 to 2 days will usually be quite sufficient to obtain the optimal pattern. When the diffusion and precipitation process is considered to be completed, the agar plates are removed from the moist chamber and all the strips gently lifted from the agar surface. The

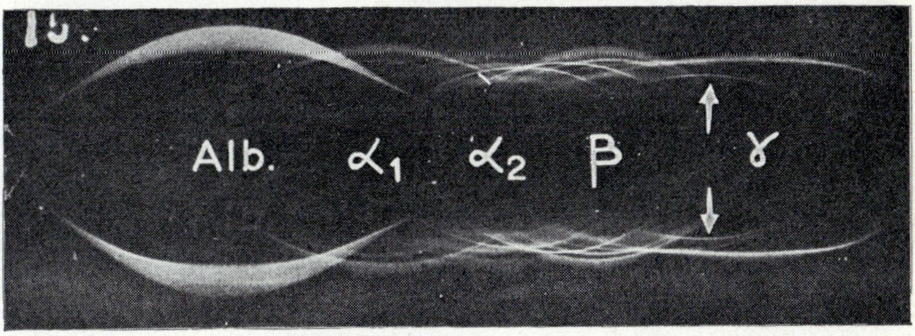

Fig. 5.14. Transfer method precipitation pattern. Horse antiserum. Ponceau S stain. In this case IgM clearly visible. Note separation of α_1 fraction.

agar plates can, at this stage, be photographed either by contact printing, or some enlarging procedure. In order to obtain permanent preparations, however, the agar plate has to be dried and stained. Dried and stained immuno-electrophoretic plates are very convenient both for storage and reference purposes. The results obtained with this method are very gratifying, the precipitation lines show up very clearly (Fig. 5.14) and quite often lines can be seen, which would otherwise be missed on unstained preparations. Special staining techniques, e.g. for lipoproteins and glycoproteins, also give very useful information. The combined Lipid Crimson-, or Oil Red O- Light Green staining method described in Chap. 8 is particularly satisfactory.

A most useful modification consists in carrying out the electrophoresis on a wider strip and applying the sample as a 2 cm. streak across the width of the strip. After separation of the fraction is completed, the strip is cut into three parallel lengths. The central part, about 6–10 mm. wide, is placed in a moist chamber on a suitable support, whilst one or both of the marginal lengths are immediately stained and washed. The central part of the strip is then removed from the moist chamber and is placed in juxtaposition to the stained control, so that the exact location of the fractions can be established and appropriately marked. The sections corresponding to individual fractions, or any desired segments of the pattern, e.g. abnormal protein bands, are cut out and transferred onto an agar gel plate,

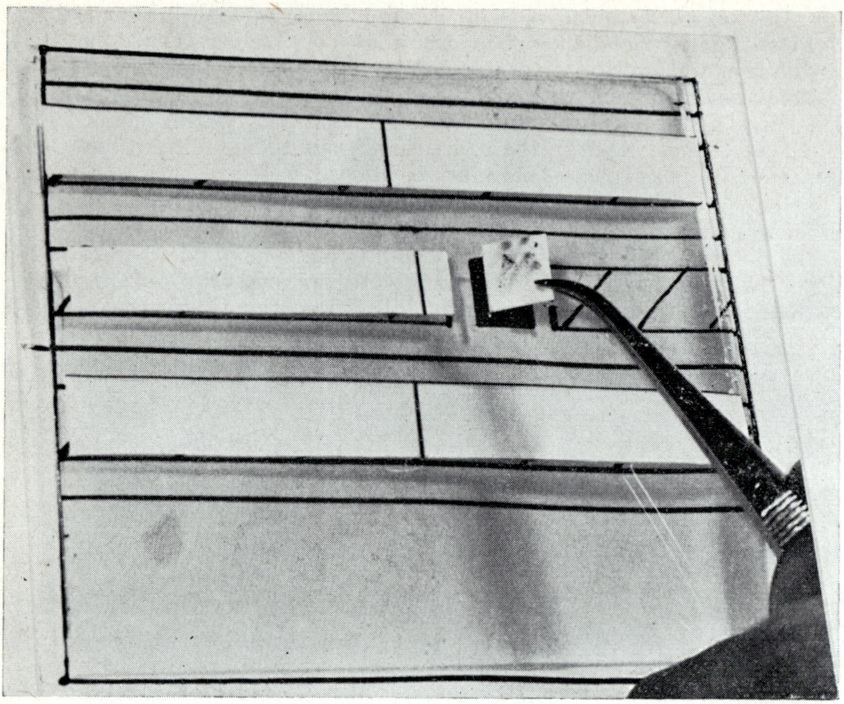

Fig. 5.15. Transfer Technique. Application of CAM eletrophoretic strip to agar plate. Relevant band (in this case a myeloma band M) cut out and applied separately.

separated from each other by 3–4 mm. The rest of the procedure is as described above.

This procedure results in a much clearer pattern, the precipitation lines being less crowded. It also shows clearly the correlation of the antigenic structure with the electrophoretic pattern. Thus the presence of common antigens in two adjacent fractions can be easily demonstrated by the looping of the corresponding precipitation lines (Fig. 5.16).

Drying and Staining. Put a sheet of smooth water-wetted and blotted filter paper on the surface of the agar plate, from which all the strips have been previously removed. The size of the filter paper sheet should be such that it overlaps the agar plate by approximately 2–3 cm. The soluble proteins and the buffer or salt pass into the filter paper, so that washing of the plate is not necessary. This is a major advantage. Place the filter paper covered agar plate into a 37°C incubator, preferably overnight. The agar must be completely dry for staining, otherwise it tends to detach itself from the glass plate. The covering filter paper detaches itself on its own account or, if it is still partially adherent, it can be easily lifted using slight upward pressure *only* if the agar is dry.

It is advisable to rub the agar surface gently with the fingertips under a gentle stream of running water so as to remove any residual filter paper fibres.

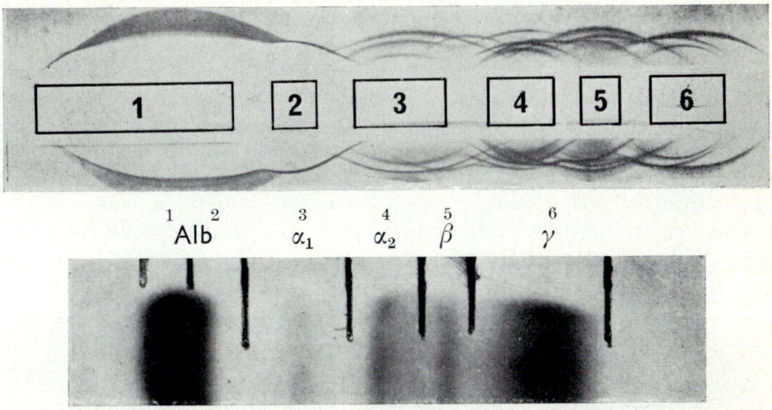

FIG. 5.16. Transfer technique with individual fractions cut out as indicated and placed on agar plate at intervals. Serum sample from patient with raised γ globulin. 1 and 2, albumin; 3, α_1; 4, α_2; 5, β; 6, γ. Note clear pattern of immunoglobulins corresponding to γ fraction. Correlation of electrophoretic to antigenic pattern is easily established. Anti-human serum applied as impregnated filter paper strips.

Staining of Precipitation Lines. Any protein stain can be used. 0·2 per cent Ponceau S in 3 per cent trichloracetic acid as previously described has been found to be particularly convenient and satisfactory. The agar plate is transferred into the staining solution for a few minutes until all the lines are satisfactorily stained. After this has been achieved the agar plate is placed in 5 per cent aqueous acetic acid in order to wash off the excess stain and clear the background. The washing solution is changed a few times until the background is clear and the bath colourless. This is usually achieved in a few minutes. The drying of the plate can be carried out at 37°C or room temperature or under a fan. When drying lean the plate against a support in such a position that the agar surface faces downwards, thus preventing any surface contamination from falling dust.

The alcoholic Nigrosin stain (see p. 104) may be used for the staining of precipitation lines and is particularly convenient as a counterstain with a lipoprotein or glycoprotein staining method.

The transfer technique is quite simple and offers some obvious advantages, particularly in the investigation of paraproteinemias Clear separation provided by the cellulose acetate electrophoresis gives a strongly defined precipitation pattern. There is very little background staining and the volume of antiserum used is relatively small. It is particularly useful for antigen studies of separated fractions and for protein mixtures which do not separate well on agar electrophoresis, but do so on cellulose acetate. The α_1 fraction of human serum, for instance, is invariably very well separated from the albumin (see Figs. 5.14 and 5.16). By cutting off and rapidly staining a parallel length from the electrophoretic strip, the exact positioning of the separated fractions can be easily ascertained. Knowing the location of the individual fractions, these can be cut out, or portions of them punched out with a cork-borer, from the unstained part of the electrophoretic strip, and applied to an agar plate. The above procedure has also proved valuable in comparing unknown antigen mixtures with known fractions or antigens for proof of immunological identity. In this case the separated fractions and the proteins under investigation are arranged in a bidimensional pattern. The protein solutions to be compared are applied as impregnated filter paper discs or placed in wells cut in the agar (Fig. 5.17).

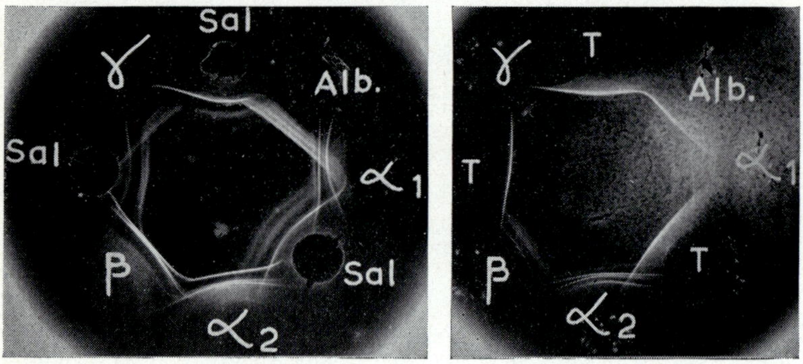

FIG. 5.17. Combined Transfer Immuno-electrophoresis with bidimensional pattern. Horse antiserum. Ponceau S stain. Tears (T) and saliva (Sal) respectively are compared with fractions of human serum. Note looping of identical antigen.

Common Errors: Insufficient volume of anti-serum; careless application of the electrophoretic and anti-serum strips; spilling of anti-serum; letting the agar plate or the electrophoretic strip dry out. This is particularly important when cut-out fractions are transferred.

Cellulose Acetate Immuno-diffusion Methods

Cellulose acetate membrane filter materials fulfil most of the requirements of an ideal support for immuno-diffusion. They are wettable, chemically almost completely inert, do not interfere with the diffusion of large molecules and, as mentioned in the previous section on electrophoresis, do not adsorb proteins, or as far as it is known, any other large molecules (except, perhaps, under certain conditions, macroglobulins).

Advantages. CAM is immediately available for use without preliminary preparations, and there is no need for any supporting materials. No wells or troughs are necessary, hence no cutting devices are needed.

As the continuity of the supporting medium is not interrupted, there is no distortion or masking of the precipitation lines. CAM is easily cut, and the marking is very legible. Economy of reactants results from high sensitivity conferred by the high ratio of sample volume to cross section of supporting medium. This is further enhanced by Nigrosin staining. Whole blood may be used as the reactant, the RBC's and the haemoglobin are removed during the washing procedure. Elution of nonprecipitated reactants is rapid and complete, and there is hardly any background staining left. Storage problems are negligible. CAM can be rendered completely transparent, both for projection and enlarging purposes.

Disadvantages. The formation of precipitation lines cannot be observed without staining. The volumes of samples to be applied are limited. Some care and skill in handling is required. Some reactants may leave traces on the application sites.

Cam Characteristics

CAM for immunodiffusion techniques is the same as that used for CAM electrophoresis. The thickness, the average pore size, the water absorption ratio, and the degree of acetylation vary according to the commercial source. But these factors do not materially affect the suitability of CAM as a supporting medium for immuno-diffusion. The procedure described here was carried out on Sartorius and Shandon CAM, but other brands should be equally suitable.

Procedure for Double Diffusion—Ouchterlony-Elek Technique

Diffusion Chamber. For two-dimensional double diffusion, round or square holes (about 3 cm. in diameter) are cut out from filter paper pads, 4 to 6 mm. thick (for example a few layers of Whatman No. 3 MM). This forms a support which is placed on a piece of filter paper in a suitable flat container (e.g. plastic box about 2 to 3 cm. high) with a tight fitting lid. The lid should be lines with foam which when wetted will maintain a high humidity in the chamber and also absorb condensation. The pad is soaked with an appropriate electrolyte solution, the excess fluid being drained off; it should not be too wet.

Preparation of CAM. The application sites and the appropriate legends are marked on suitable size and shape CAM, e.g. 4 × 4 cm. squares or disks, equal distances from a central point are obtained by means of a pair of dividers. The distance between the application sites will vary, usually about 5 to 10 mm. The CAM is then impregnated with a saline buffer or a suitable electrolyte solution, blotted (as for electrophoresis), and placed on the pad in the diffusion chamber in such a way that the CAM spans the gap, overlapping it by about 5 to 10 mm. Good contact with the underlying filter paper is essential.

Application of Reactants. The samples containing antigen and antibody are applied to the surface of the CAM resting on the supporting pad as discrete drops by means of a micropipette, capillary or suitable automatic dispensing pipette, taking care to avoid "blow out" bubbles.

A much more convenient technique consists in pressing hollows or troughs (Fig. 5.23) to a required pattern in wet impregnated CAM by means of a simple set of templates easily made from readily available materials. This technique applies to double-diffusion, immunoelectrophoresis and cross electrophoresis. Buffer impregnated not blotted CAM sheets or strips are placed between the corresponding parts of the template and adequate pressure is applied to produce the required pattern. The CAM is then blotted, placed in position in the diffusion or electrophoretic chamber. The antigen and antiserum samples are then applied to these depressions in the pattern required. The volumes will

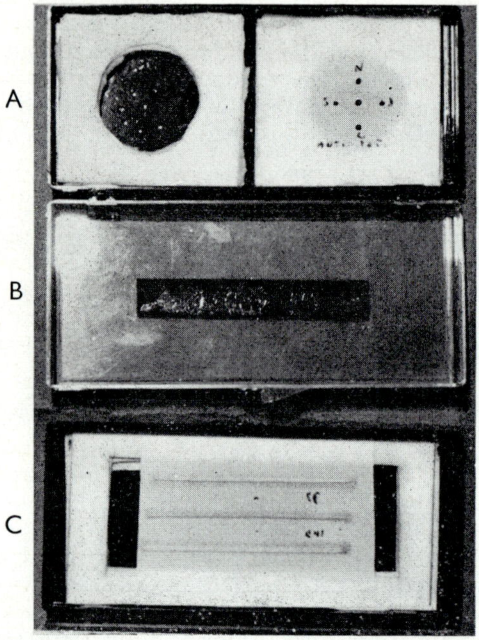

FIG. 5.18. A. Bidimensional immunodiffusion. The figure shows, on the left, a thick paper pad with a circular well and, on the right, one CAM square with samples in position on the pad. B. Lid of box. C. Thick pad with rectangular well for use in immunoelectrophoresis. Electrophoresis strip with two specimens and three anti-serum strips in position on the pad.

depend on the type and concentrations, and the distances between antigen and antibody solutions, 1–3 μl. can easily be applied. The addition of dyes, e.g. bromophenol blue to the reactants is recommended. The stained samples show up against a white background and also indicate whether any spilling or irregular spreading has occurred. The lid is then closed and the diffusion chamber placed on an even surface.

The diffusion time depends on the nature and concentration of the antigen and antibody, the distance of application sites, and the temperature. With 1 cm. distance, overnight diffusion at room temperature is

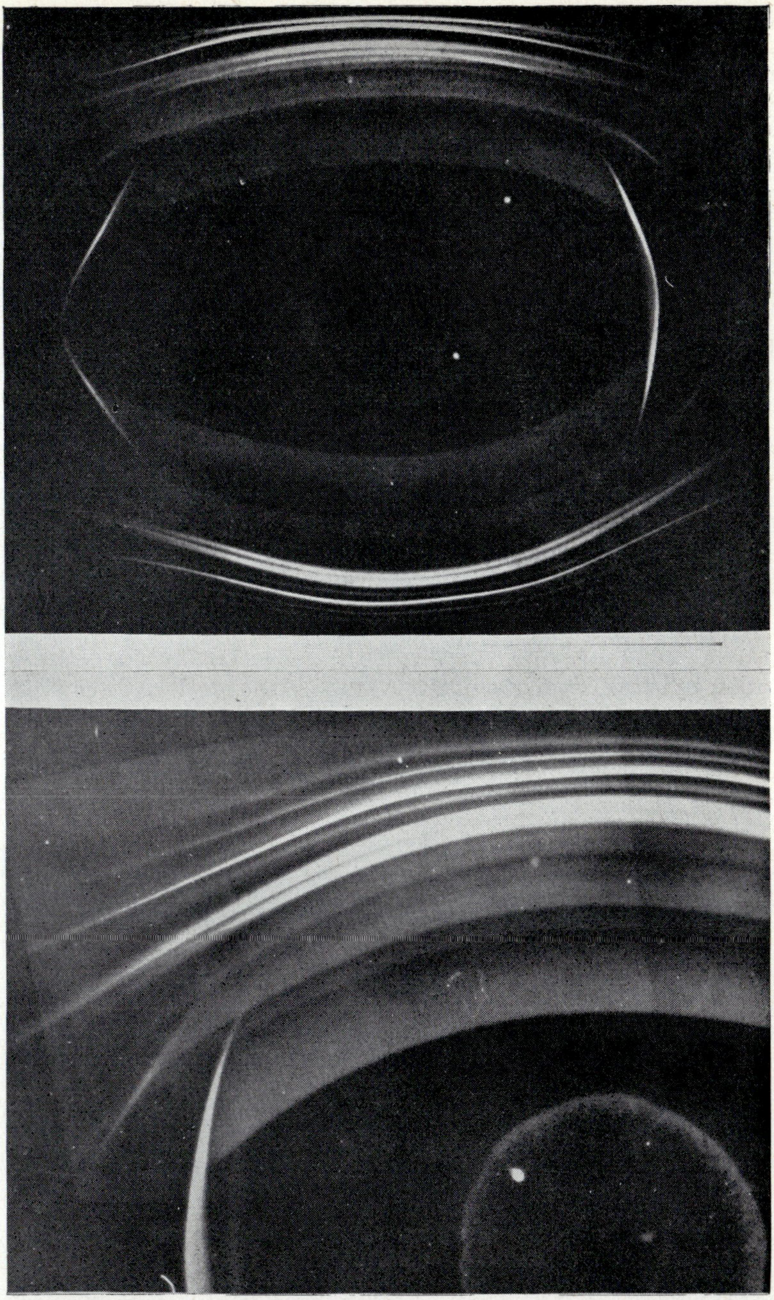

Fig. 5.19. Bidimensional Ouchterlony–Elek method on cellulose acetate. Direct enlargements from cleared cellulose acetate. Horse anti-human serum in centre. Human serum and urine on periphery. Note that precipitation pattern is much easier to analyse on enlargements. Two magnifications from the same specimen, both enlarged to full plate size.

usually adequate. After diffusion the CAM is removed and placed in an eluting fluid, such as 0·2 per cent Haemo-Sol (Meinecke and Co. Inc., U.S.A., Alfred Cox Ltd., U.K.) buffer or saline for about 10 to 20 minutes, preferably with agitation.

An alternative, simple technique of application consists of using small filter paper disks (e.g. 2–3 mm. diam. Whatman No. 1) impregnated with the sample fluid. The impregnated disks are easily placed in exactly the desired position and at the exact distance chosen. More fluid can be added after the first application has partly soaked in, until the desired volume has been applied. Using the technique larger volumes an be applied because of the additional absorptive capacity provided.

After elution the CAM is stained, washed, dried and (if required) cleared as recommended for CAM electrophoresis. For weak precipitates 0·002 per cent Nigrosin stain in 2 per cent acetic acid followed by a rinse under the tap is a most useful stain. For rapid staining an alcoholic Nigrosin solution is recommended. The stained CAM should be inspected against a strong source of light—this is essential. Photographic enlargement of stained and cleared CAM greatly facilitates the analysis of complex, crowded precipitation patterns. The most common errors are careless and irregular application of sample, too large sample volume, letting the CAM dry up.

Titrations. By applying increasing dilutions of one of the reactants placed around the other reactants in the centre, the last line of precipitation to be still seen constitutes a crude "titre". This technique was found to be useful for the estimation of C-reactive protein.

Microspot Technique. This is an immunodiffusion method and is based on the impregnation of the CAM with one of the reactants and the subsequent application of discrete microspots of the other reactant. A strip of CAM is marked into squares, numbered, and impregnated with anti-serum or antigen appropriately diluted in a suitable fluid, e.g. saline-phosphate buffer. Usually the antiserum is diluted 1:10 and 1:20, but the optimal dilutions are best determined for each system. The impregnated CAM is blotted and placed on a suitable support—as for double diffusion. Small drops of the antigenic substances and suitable controls are then placed in the centre of the marked squares. The CAM is then processed as described above. Nigrosin should be used for staining. Precipitates appear as spots or concentric circles against a white background. The microspot test has been found very convenient for the grouping of B-hemolytic streptococci, for instance, and was also applied with success to the detection of autoimmune antibodies. By applying serial dilutions and recording the end point or the diameter of the rings compared with that obtained with a known standard, semi-quantitative estimations can be obtained.

Immunoelectrophoresis

Procedure

Preparation of CAM. The CAM strips should be somewhat longer than for simple electrophoresis to allow for diffusion of the most rapidly and most slowly migrating fractions. For small-scale IEP 5 to 7 cm. by 3 to 5 cm. is a suitable size. (A microtechnique can also be used, it gives satisfactory and more rapid results with consequent economy of materials.) For multiple patterns, wider strips are employed. Buffer impregnation, blotting, and placing in position are as described for CAM electrophoresis. A preservative should be added to the buffer solutions to inhibit growth during the diffusion period.

Application of Sample and Electrophoresis. The antigen is applied in the form of a drop or a short streak according to the desired pattern. The position of the application is as for CAM electrophoresis. The optimal volume depends on the concentration and nature of the sample and the width of the strip. For serum electrophoresis 0·5 to 2 μl. is adequate. The addition of bromophenol blue to serum samples is most useful as this indicates the migration and diffusion of the applied sample during the electrophoresis period enabling termination at the most suitable distance and time. As for CAM electrophoresis, care should be taken that at least 10–15 mm. are left clear between the ends of the separation pattern and the edges of the strip supports. This is best judged by watching the migration of the coloured albumin spot.

Application of Antiserum. After the electrophoretic separation has been completed the CAM strip is removed and placed on a moist pad inside a diffusion chamber similar to that described for double diffusion, with a rectangular well of the required size, e.g. 3 to 4 $\times$ 6 to 8 cm. cut out to accommodate the CAM strip. The CAM strip is thus supported only on its edges, which should overlap the support by about 5 mm. (Fig. 5.18C).

The serum and antiserum samples can be very conveniently applied to hollows or troughs formed in CAM by means of a suitable template. See p. 134.

The appropriate amount of antiserum and the optimal diffusion distance has to be determined by experience for each antigen-antibody system. Essentially the same principles apply as for IEP in gels. For IEP of serum proteins by the commonly available antisera, an antigen-antiserum ratio of 1:25, which corresponds to about 30 to 50 μl. of antiserum per strip, is usually satisfactory. If specific antisera are used the length of the antiserum applicator need only be slightly longer than the expected electrophoretic position of the antigen. This results in economy of antiserum.

The diffusion distance between the margins to the applied sample and the antiserum strip may vary from 3 to 5 mm. A special antiserum applicator for IEP on CAM depositing the correct volume directly to CAM has been designed and is available on the market.

Diffusion. The principle and technique are the same as for double diffusion. The optimal diffusion time is best established by preliminary experiment, as the precipitation lines only become visible after staining. For most IEP of serum proteins, an overnight period at room temperature is adequate.

Elution, Washing, Staining, Drying and Clearing. After the diffusion is considered to be completed, it is then processed as described for double diffusion.

Staining, drying and clearing of the eluted CAM strip follows the rules recommended for simple CAM electrophoresis. Ponceau S is recommended if strong precipitation lines are expected, and Nigrosin for weak

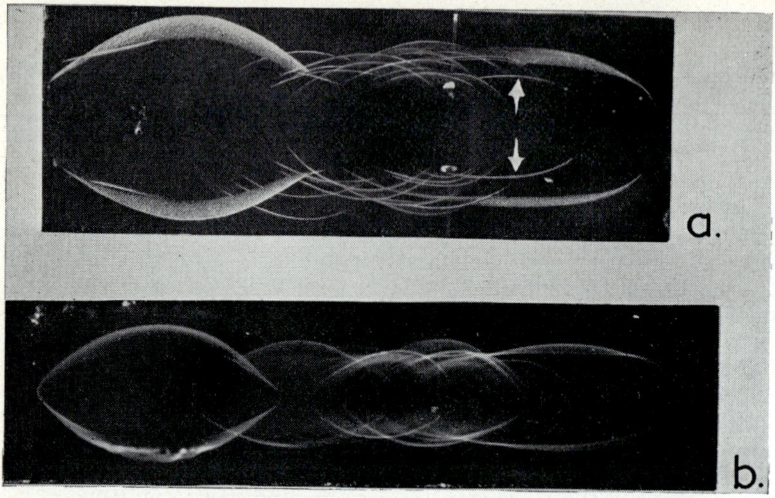

FIG. 5.20 Small-scale Immuno-electrophoresis on cellulose acetate. Horse human antiserum. Ponceau S stain. (*a*) Serum applied as a short streak (arrow indicates IgM line). (*b*) Serum applied as a drop.

precipitates and low concentration of reactants. For these and other staining techniques see previous discussion on stains.

Meticulous cleanliness is essential during the handling of the cellulose acetate. Fingerprints show clearly. All buffer solutions, washing bath and oil used should be clear and clean and must contain some preservative to prevent bacterial growth. For buffers, those used are:

1/10,000 merthiolate

or 5 ml. of 5 per cent thymol in isopropyl alcohol in 1,000 ml. of buffer solution.

CAM-Sephadex Transfer Immunodiffusion. CAM lends itself very well to a combination of immunodiffusion with Sephadex thin layer chromatography of proteins and other antigenic substances. After the termination of the chromatographic separation, two saline impregnated and blotted CAM strips or sheets are placed on the surface of the Sephadex layer. After about 15 to 30 minutes the top sheet is gently lifted, fixed and stained with an appropriate dye, e.g. Ponceau S. This will reveal the separation pattern. Filter paper strips (about 2 mm. wide) impregnated with the appropriate volume of antiserum can now be placed on the first CAM sheet (in contact with Sephadex) in the desired position and at a suitable distance from the separation pattern. The Sephadex plate with the adhering CAM sheet and antiserum strips are then left in a moist chamber for diffusion for a suitable period of time, e.g. 24–48 hours. The rest of the procedure is as described for CAM immuno-electrophoresis. After staining and clearing, the two CAM strips can be superimposed, thus revealing the exact correlation of the chromatographic separation pattern and the immuno-precipitation pattern.

Cross-over Electrophoresis. Cross-over electrophoresis (counter current electrophoresis, electrosyneresis, electroprecipitation, electro immunophoresis, etc.) is a very ingenious and elegant technique for the demonstration of precipitation reactions. The method depends on the fact that due to endosmotic flow antibodies being gamma globulins move towards the cathode, whereas many antigenic substances have a different electrophoretic mobility and migrate towards the anode. By placing the antibody containing specimen on the anode side and the antigen on the cathode side, both reactants can be made to migrate towards each other forming a precipitation line between the application sites. The technique can be performed on most substrates such as agar gel but is very easily and conveniently carried out on cellulose acetate. The impregnated and marked strips or sheets are placed in the chamber as for routine electrorphoresis and the samples are applied to the CAM in position in the form of discrete drops or, much more conveniently, into depressions in the form of round hollows or troughs produced by the use of two templates, one with hollows and the other one with corresponding projections positioned in a desired pattern, see Figs. 5.21, 5.22. Sensitivity can be

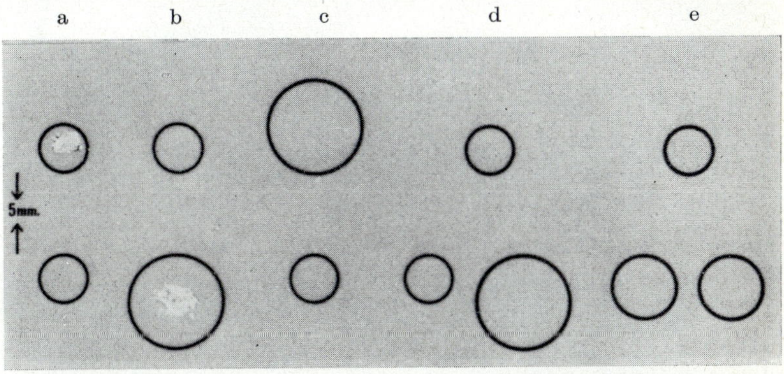

FIG. 5.21. Sample application to hollows on CAM sheet in position in chamber. Note unequal size of cathodic, antigen samples.

greatly increased by using larger antigen samples, thereby making greater amounts of antigen available and avoiding the need for concentration of the sample of re-filling of the wells.

In the conventional pattern the samples are applied on the same electrophoretic axis, about 5 mm. apart, usually midway between cathode and anode.

The optimal proportions for each system are best worked out by experience. A serious disadvantage of the conventional counter current technique is that it does not include a positive control which would provide a reaction of identity. A further disadvantage is the fixed antigen-antibody ratio. This is particularly important when screening for Australia antigen for instance, as the antisera are not yet standardised and their titres vary within a wide range. In order to avoid the prozone

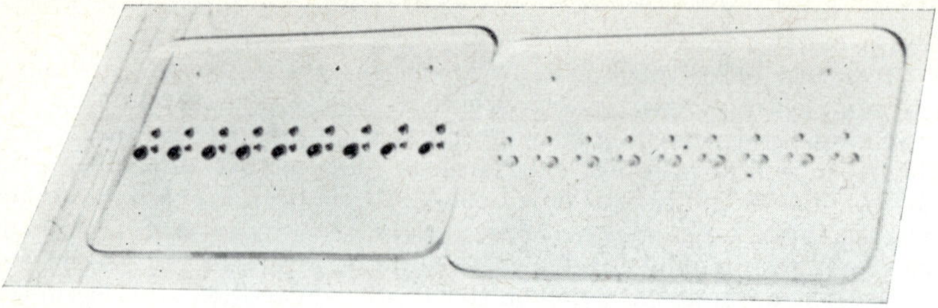

FIG. 5.22. Template for CAM counter-current immunoelectrophoresis.

phenomenon due to antigen or antibody excess, the reactants can be placed in such a manner that a variable antigen-antibody ratio is achieved. Two samples of unequal volume are applied on the cathode side about 2 mm. apart and the antiserum is applied on the anode side (about 3–4 μl.) placed equidistant from both cathode application sites at 5 mm. distance measured along the electrophoretic axis (Fig. 5.21.)

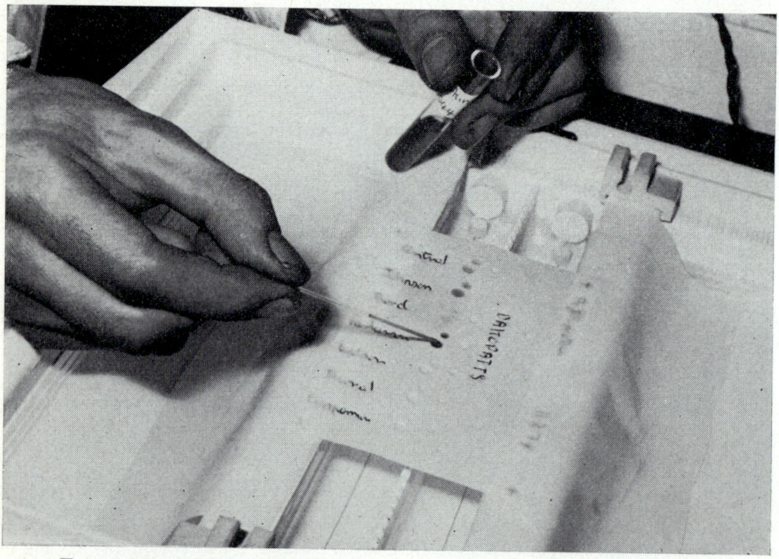

FIG. 5.23. Application patterns. (a) Conventional; (b) For weak antigen; (c) For weak antibody; (d) Screening pattern; (e) Specificity check.

The volumes applied can vary from 1–7 μl., e.g. 1–2 and 5–7 μl. respectively on the cathode and about 3–4 μl. of antiserum (depending on the antiserum titre) on the anode side. Another pattern similar in principle covers an even wider range of sample to antiserum ratios (Fig. 5.21). A positive control is used invariably with each run and it should be used at the highest dilution which still produces a clearly visible, neat precipitation line, providing a check on the specificity and the sensitivity

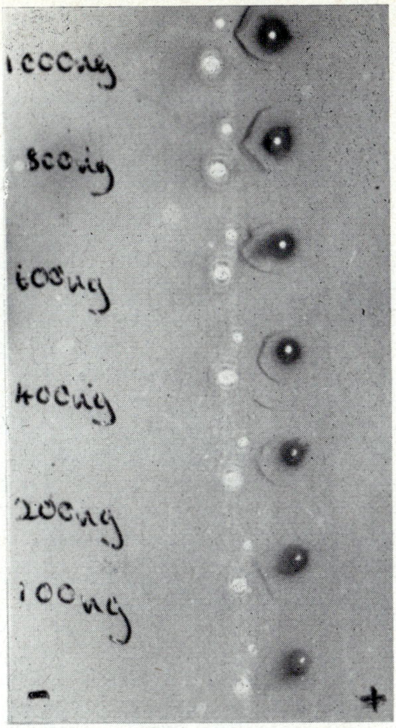

FIG. 5.24a.

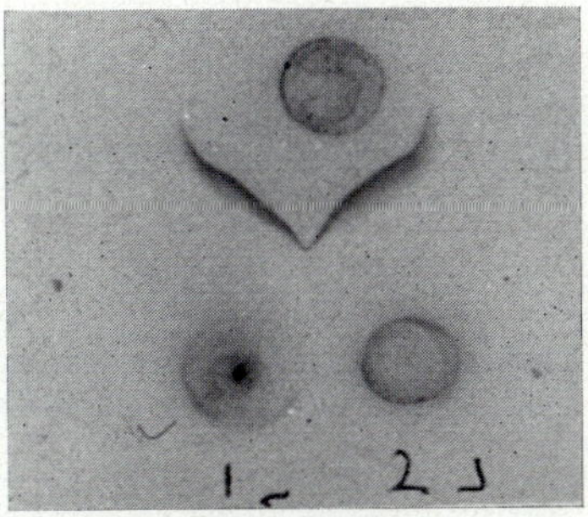

FIG. 5.24b.

FIG. 5.24. Precipitation patterns obtained with counter current immunoelectrophoresis on cellulose acetate.

(a) Screening pattern. In this case increasing dilutions of alpha-1 fetoprotein, note in sample this precipitation line only clearly visible against large size of sample and not against small one due to the very low concentration 100 ng/ml. of AFP.

(b) Specificity check
  (i) patient's serum;
  (ii) positive control.

Note reaction of identity confirming the specificity.

of the technique. The specificity of the positive results is confirmed by establishing a reaction of identity with a known positive control which is applied on the cathode next to the test sample. It is best to use identical volumes and concentrations both for the test sample and for the positive control, using the amount which gave a neat precipitation line (Fig. 5.24b). Electrophoresis is carried out as described for serum proteins using a constant current not exceeding 0·5 milli amps per cm. resulting in approximately 150 volts using the conventional 0·5 M. barbitone. Precipitation lines are not visible without staining. Following electrophoresis, the CAM is placed in an eluting fluid, e.g.: buffer or saline with a trace of detergent, and is gently agitated for 20–30 minutes to wash out unprecipitated proteins. The CAM is then stained with Nigrosin followed by a short rinse under the tap. Staining time is not critical, from 15 minutes (in alcoholic solution) to an overnight period if this is convenient (in aqueous solution). *It is essential that the precipitation lines are viewed against a strong light source*. There is no need for rendering the CAM transparent. It should be mentioned that the procedure described above can be most helpful for the demonstration and identification of any precipitating antigen-antibody system in which the antigen has an electrophoretic mobility other than gamma, e.g. bacterial antigens, thyroid antibody, species identification in forensic medicine, etc. The electrophoretic migration of the antigen can be changed by carbamylation, formalisation and by other procedures modifying the charge and flow of the antigen molecule. This will make any antigen suitable for cross electrophoresis. The method has been found to be particularly valuable in screening for the presence of α_1 Fetoprotein and a sensitivity of 100–200 ng/μl. can be achieved. The double, unequal sample size pattern is particularly useful in systems in which prozone phenomena are known to occur.

REFERENCES

1. Arai, K. *et al.*, Electrophoretic determination of glycoproteins. An improved method with cleared cellulose acetate membranes. *Anal. Biochem.*, 1969, **31,** 71–76.
2. Beckering, R. E. Jr. *et al.*, A rapid method for lipoprotein electrophoresis using cellulose acetate as a support medium. *Am. J. Clin. Path.*, 1970, **53,** 84–8.
3. Beckering, R. E. Jr. *et al.*, Variations in cellulose acetate membranes for lipoprotein electrophoresis. *Am. J. Clin. Path.*, 1971, **56,** 765–66.
4. Box, T. C. *et al.*, Cellulose Acetate Electrophoresis of heparin. *J. Chromatogr.*, 1972, **64,** 193–5.
5. Charman, R. C. *et al.*, Human alpha$_2$ lipoprotein bands visualised by cellulose acetate electrophoresis. *Lipids*, 1969, **4,** 397–8.
6. Fletcher, M. J. *et al.*, A simple method for separating serum lipoproteins or cellulose acetate. *Clin. Chem.*, 1970, **16,** 362–65.
7. Fritsche, H. A. Jr. *et al.*, Cellulose acetate electrophoresis of alkaline phosphatase isoenzymes in human serum and tissue. *Clin. Chem.*, 1972, **18,** 417–21.
8. Grabner, W. *et al.*, Lipoprotein microzone electrophoresis on cellulose acetate strips. *Clin. Chim. Acta* 1970, **28,** 299–310.
9. Harris, R. I. and Kohn, J., The pre-albumin fraction: A useful parameter in the interpretation of routine protein electrophoresis. *J. clin. Path.*, 1974, 986–989.

10. Klibansky, C., Separation of N-acetyl-D-hexosaminadase isoenzymes by cellulose acetate paper electrophoresis—a simple procedure for the diagnosis of Tay-Sachs disease. *Jsr. J. Med. Sci.*, 1971, **7**, 1086–9.
11. Kohn, J., Electrophoresis and immunodiffusion techniques on cellulose acetate membrane. *Meth. Med. Res.*, 1970, **12**, 243–60.
12. Levitt, N. H. *et al.*, Rapid detection of viral antibody by cellulose acetate electrophoresis. *Appl. Microbiol.*, 1971, **22**, 143–4.
13. Magnani, H. N. *et al.*, A quantitative method for blood lipoproteins using cellulose acetate electrophoresis. *J. Clin. Path.*, 1971, **24**, 837–45.
14. Millner, J. N. *et al.*, Detection of proteolytic enzymes after cellulose acetate electrophoresis. *Anal. Biochem.*, 1969, **32**, 154–7.
15. Millner, J. N. *et al.*, Cellulose acetate electrophoresis: the use of discontinuous buffer systems for protein separations. *Clin. Chim. Acta*, 1971, **31**, 427–33.
16. Nerenberg, S. T. *et al.*, Rapid large scale screening of blood donors by use of cellulose acetate electrophoresis on solid support. *Clin. Chem.*, 1972, **18**, 145–9.
17. Nerenberg, S. T. *et al.*, Multi-antisera applicator for use in solid support immunoelectrophoresis on cellulose acetate. *Clin. Chem.*, 1972, **18**, 1021–3.
18. Robin, B. S. *et al.*, A rapid technique for the electrophoresis of unconcentrated urine. *Am. J. Clin. Path.*, 1973, **60**, 359–63.
19. Rhone, D. P. and Mizuno, F. M., Profiles of alkaline phosphatase isoenzymes in serum using cellulose acetate electrophoresis and organ specific inhibitors. *Am. J. Clin. Path.*, 1973, **59**, 531–41.
20. Schwartz, H. G., Immunoelectrophoresis in cellulose acetate membranes. *Am. J. Clin. Path.*, 1972, **57**, 326–31.
21. Sherwin, R. M. *et al.*, Microzone electrophoresis of unconcentrated cerebrospinal fluid using cellulose acetate strips and nigrosin dye. *Am. J. Clin. Path.*, 1971, **55**, 205–12.
22. Wintershoff, D. *et al.*, Method of electrophoresis of 10 mg. liver tissue on cellulose acetate strips. *Z. Klin. Chem. Klin. Biochem.*, 1971, **9**, 31–32.
23. Kohn, J., Laboratory investigations of paraproteinaemia. *Recent Advances in Clin. Path.* Churchill Livingstone, 1973, pp. 363–401.

CHAPTER 6

AGAR GEL ELECTROPHORESIS AND IMMUNO-ELECTROPHORESIS

A. Feinstein

When an animal is injected under suitable conditions ("immunized") with macromolecules ("antigens") foreign to that animal, an immune response may occur. Immune globular proteins ("antibodies") appear in the circulating plasma, with the property of specifically combining with the antigen which generated them, and the serum of this animal can be separated and preserved as an antiserum. The immune globulins are a family of proteins which carry antibody activity. Since immunological methods have been particularly useful in establishing their structural relationships to each other, they are used to illustrate many of the points in this chapter. It will be less confusing if the immune globulins are considered here simply as a family of related antigens, and if we forget for the moment that they also function themselves as antibodies. Since antibodies recognize a relatively small region of the macromolecular surface, it is possible to use cross reactions to establish that some regions in two molecular species are identical and that others differ. In this way it has been possible to confirm the following basic structure for all the immune globulins.[10]

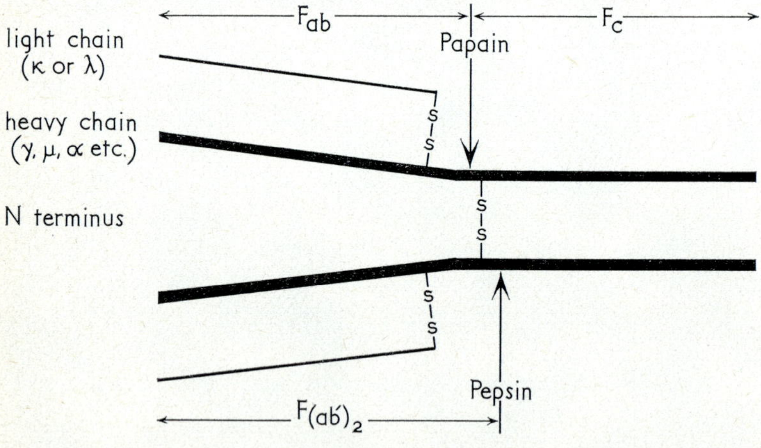

Each consists of two heavy polypeptide chains and two light chains, linked as shown through cystine bridges. Two types of light chain, κ and λ, are found to occur in all immune globulins but each type of immune globulin (IgG, IgM, IgA, etc.) has a characteristic heavy chain (γ, μ, α, etc.). On reduction with a thiol reagent, such as 2-mercaptoethanol, the cystine bridges are split and the resulting heavy and light chains can be separated. When IgG is treated with cysteine-activated papain, the molecule is split to form one Fc fragment and two identical Fab fragments. Free light chains occur in the urine of myeloma cases, when they are known as Bence-Jones proteins.

In one method of analysing antigen-antibody systems, double diffusion in agar gel, two neighbouring wells are first cut in the gel. A mixture of antigens to be examined is placed in one well, and an antiserum containing a mixture of the corresponding antibodies is placed in the other. After diffusion has occurred into the gel from each well,

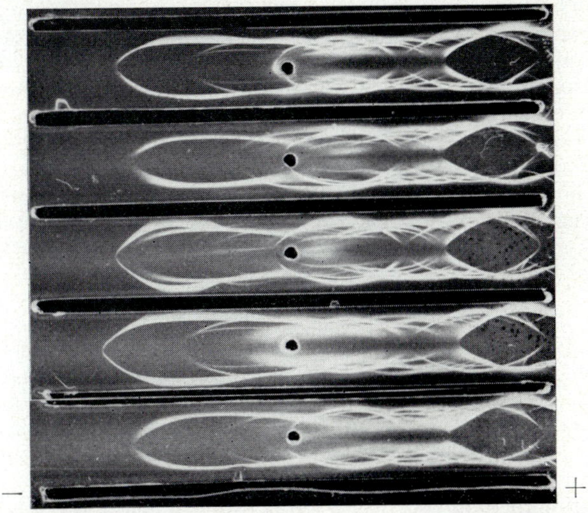

FIG. 6.1. Immuno-electrophoretic patterns of human sera. Upper four wells were filled with pathological sera, lowest well with normal serum. Slots filled with rabbit anti-human serum. The polarity in all subsequent figures is as shown above.

each antigen combines with its specific antibody to form a line of precipitate in the gel where the proportions are optimal. In the technique of immuno-electrophoresis which was developed largely by P. Grabar and his collaborators,(1) a similar diffusional analysis is applied to components which have first been separated by zone electrophoresis. The essential steps may be summarized as follows:

1. Zone electrophoresis of the macromolecular mixture to be examined is carried out, usually in buffered agar gel. At this stage the individual components are partially or completely separated into a linear pattern of spots.

2. A lateral slot is cut alongside the separated components, and filled with an antiserum capable of specifically precipitating components in the mixture.

3. The plate is allowed to develop, during which time each component diffuses to form a precipitate arc where it meets the antibody.

Fig. 6.1 shows the pattern when human serum is examined using a rabbit anti-human serum in the slots.

It may be helpful in following the more detailed description of the procedure to bear in mind the following general points affecting electrophoresis. During electrophoresis buffer salts are necessarily present to maintain a constant *p*H and to reduce molecular interactions. The resulting conductivity leads to heat production and evaporation, and the ionic strength is therefore kept as low as possible. To protect the electrophoresis plate from *p*H changes which occur in the electrode vessels, the latter are connected by wicks to a second buffer compartment, which is in turn connected to the electrophoresis plate. Charged components in the agar result in a considerable electro-osmotic flow of buffer towards the cathode during electrophoresis. Fig. 6.1 shows the resulting final distribution of serum components. It can be seen that in the case of the electrophoretically slowest components of serum, the immune globulins, electro-osmotic flow actually outweighs electrophoretic movement towards the anode, so that the net movement is towards the cathode. For this reason it is usually convenient to make the starting position in agar gel half-way between the anodic and cathodic ends of the plate.

Immuno-electrophoresis

The method described here is based on the micro-technique proposed by Scheidegger.[(2)] Electrophoresis is conducted in a closed tank to reduce evaporation. The Shandon universal electrophoresis tank is suitable; two $3\frac{1}{4} \times 3\frac{1}{4}$ in. gel plates, each carrying 6–7 samples, or eight 3×1 in. microscope slides each carrying up to 3 samples, are used.

Preparation of Agar Gel. Since agars differ in character, a large batch of agar should be secured if consistent results are required over a long period. Using the best commercially available agars, no further purification is necessary. Sufficient agar to give a 1–1·5 per cent gel is dissolved in the nearly boiling buffer, avoiding prolonged heating. It is best to clarify the resulting hot solution by filtration through thick rapid filter paper. The gel is kept in small (50 ml.) bottles, so as to avoid frequent remelting. For most purposes, barbital buffer, *p*H 8·2, I = 0·025 is suitable; a suitable composition is:

Sodium Barbital	3·96 g.
HCl, 0·1N	57·5 ml.
Distilled water to	1,000 ml.

As preservative, either 1:10,000 merthiolate or 1:1,000 sodium azide w/v are added to the buffer. The same buffer or buffer of twice the concentration is used in the electrode vessels.

The glass plates to be coated are carefully cleaned, and placed on a horizontal surface. The agar gel is melted in a boiling water bath, checking that all lumps have dissolved, and pipetted on to the glass plates. If necessary the pipette tip can be used to help the agar run to the edges of the slide. Two mls. on a 3 × 1 in. slide give a gel 1 mm. thick; eight to ten mls. are used on the $3\frac{1}{4} \times 3\frac{1}{4}$ in. plates. In the Shandon system a scraper coats 8 microscope slides with 1 mm. thick agar gel. Improved adherence is obtained if the gel is poured on to slides which have been previously coated with dilute agar gel and allowed to dry down. When the gel has set, circular wells 1 mm. in diameter are cut about one cm. apart. Suitable metal circular cutters providing patterns with up to three wells per microscope slide are commercially available. Where dilute antigens are being examined, these starting wells may be cut much larger. Wicks of thick filter paper (2 sheets of 3 MM), lint or folded gauze are used to connect the plates to the electrode vessels, which may be filled with buffer of twice the concentration in the agar gel. If two rows of plates are to be run simultaneously, a centre wick moistened with buffer connects the two rows. With makes of tank other than the universal tank previously mentioned, care should be taken that the plates do not rest along their full length on the bottom of the electrophoresis tank in order to avoid "shorting" through buffer beneath the plate; if necessary they can rest on thin perspex strips.

Electrophoresis. The samples to be examined are placed in the starting wells, and a field of 5–6 V/cm. is applied, measured directly on the plate. It will be found useful to include a dyed albumin marker in one sample well. For this purpose amidoschwarz 10B in water is added to serum, and becomes attached to the albumin. An excess of dye is to be avoided as it travels as a second faster spot. The electrophoresis is stopped when the albumin has travelled almost to the wick.

Immune Precipitation. After electrophoresis, the plates are removed and slots 1–2 mm. in width are cut parallel to the direction of electrophoretic migration, midway between the starting wells, i.e. about 5 mm. from each well. The slots are most easily cut using a double bladed cutter, which is available commercially. The strips of agar are then easily removed; after cutting across the short ends of the slot with the tip of a Pasteur pipette, the strip can be lifted out with the pipette by applying slight suction. The plates are placed into horizontal wet chambers in readiness for the slots to be filled with the required antisera. This is done using a Pasteur pipette and the patterns are allowed to develop. The intensity of the precipitin arcs which form depends on many factors other than the antigen concentrations, and may mislead one in assessing the relative amounts of antigens present. For example, when testing the purity of a component one may be unduly alarmed (or complacent) at the appearance of arcs indicating contamination. For this reason it is usually advisable to conduct the electrophoresis in duplicate. One plate is then placed in a fixative, dried down, and stained to reveal the electrophoretic pattern. The procedure is as follows.

Fixation is achieved by immersing the plate for thirty minutes in 50 per cent ethanol containing 1 per cent acetic acid. The plate is then covered with filter paper, and allowed to dry in an oven at 37°C. The filter paper prevents cracking of the gel and absorbs buffer salts. A variety of stains for proteins, lipids and polysaccharides are listed by

Grabar and Burtin.[1] The most commonly used protein stains are amidoschwarz 10B, or nigrosine.

Dye	1 gm.
12 per cent acetic acid	500 mls.
1·6 per cent sodium acetate	500 mls.

The fixed plate is stained for at least 15 minutes, and finally differentiated in 2 per cent acetic acid.

Choice of Antiserum

The information to be derived from the immuno-electrophoretic patterns depends on a careful choice of antiserum.

Multispecific Antisera. The most easily obtained and commonly used type of antiserum is one reacting with a large number of different components in a biological fluid; for example, the antiserum may be made in a rabbit by injecting human serum. Fig. 6.1 shows the use of such an antiserum (anti-human serum) to compare a number of abnormal human sera with a normal serum. Following widespread and extensive investigations of human serum components it has been possible to identify many of the serum arcs, as is shown in the diagram,

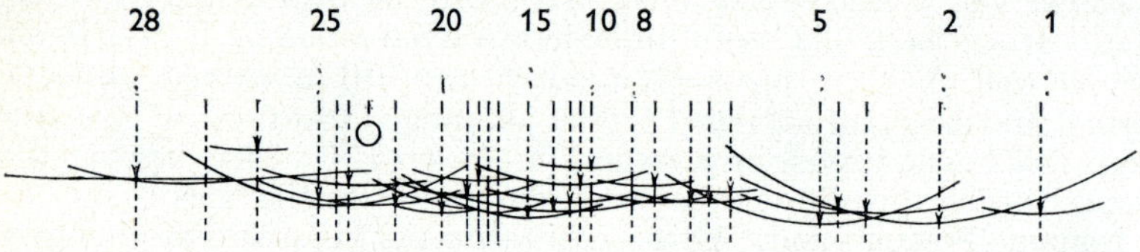

FIG. 6.2. Schematic diagram of immuno-electrophoretic pattern of human serum.
1. Prealbumin. 2. Albumin. 3. α_1-Lipoprotein. 4. α_1-Seromucoid. 5. α_1-Glycoprotein. 6. α_1-Globulin I. 7. α_2-Globulin I. 8. α_2-Globulin II. 9. α_2-Globulin III. 10. Group-specific component I. 11. α_2-Lipoprotein. 12. α_2-Macroglobulin. 13. α_2-Globulin IV (ceruloplasmin). 14. Haptoglobin I-I. 15. Group-specific component 2. 16. Haptoglobin 2-2. 17. α_2-Globulin V. 18. α_2-Globulin VI. 19. α_2-Globulin VII. 20. α_2-Globulin VIII. 21. Hæmoglobin-binding β_1-globulin. 22. Starting point. 23. β_1-Globulin I. 24. Transferrin. 25. β_1-Globulin II. 26. IgM. 27. IgA. 28. IgG.

(Fig. 6.2). Nevertheless, although it is possible to detect in Fig. 6.1 abnormalities in serum patterns, it will be evident that it is not always simple to identify the particular components involved. In such cases, as will be seen later, it may be necessary to prepare less complex antisera. However, if a pure antigen is available it is possible to identify which of the arcs in the pattern of the mixture corresponds to that antigen, using the method of Osserman.[3] Here the slot on one side is filled as before with the antiserum. The slot on the other side of the well is filled with the purified antigen. When the antigen diffuses to meet the antiserum a single line forms across the immunoelectrophoretic pattern parallel to the slots and this line fuses only with that arc corresponding to the antigen.

Multispecific sera are essential when monitoring at various stages during the attempted purification of a single component from a mixture. As an example, Fig. 6.3 shows the patterns of a series of fractions of rabbit serum eluted from a diethyl-aminoethyl-cellulose column in order to prepare pure IgG. Only the first two fractions show a single arc of IgG, in the later fractions arcs of contaminating components are detected. It is clear that for this purpose the ideal antiserum will be one capable of reacting with all the potential contaminants in the mixture. Several courses of immunisation may be needed and a mixture of several antisera may be required.

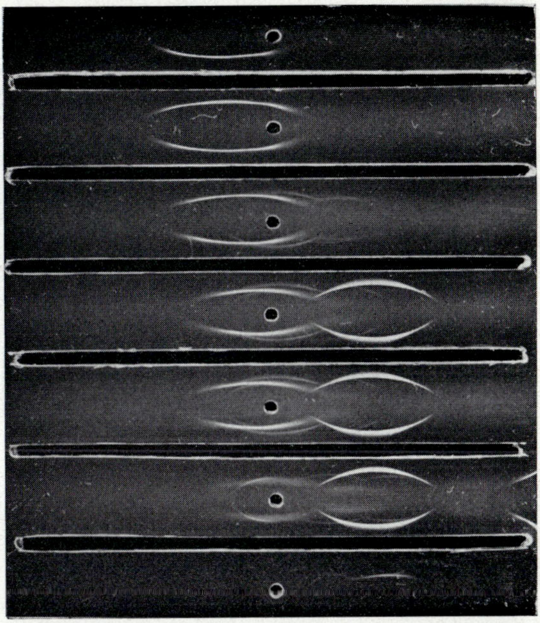

FIG. 6.3. Patterns of rabbit serum fractions progressively eluted during ion-exchange chromatography. Slots filled with sheep anti-rabbit serum.

It is often useful on the other hand to limit the complexity of the patterns. One way of accomplishing this is to precipitate out in advance antibodies to some of the components from the antiserum, by adding selected antigens. For example, we wished to examine the pattern of immune globulins in adult bovine serum. Advantage was taken of the fact that new-born calves lack these immunoglobulins in their plasma but have the full complement of the other plasma proteins. The serum of a new-born calf was freeze-dried and a total of 60 mgm. was added to 1 ml. of a rabbit anti-bovine serum in three lots, spinning off after the first two additions the precipitate which formed. In Fig. 6.4 are compared the patterns developed with adult bovine serum and colostrum using as antiserum either the original rabbit serum (*a*) or the same serum after the selective precipitation (*b*). It is easy in the latter case to see the pattern of the immunoglobulins, unconfused by the arcs of other components. In Fig. 6.4b the (upper) serum pattern shows two IgG arcs partially fusing with spur formation. The IgM arc is visible nearer the well. The (lower) colostrum pattern shows selective secretion of the more anodic IgG component, though all the serum immune globulins are present to some extent.

Unispecific Antisera. Where they are available, antisera capable of precipitating a single component from a mixture are extremely valuable. Such reagents are not easily made, since undetectable traces of contaminants in the purified antigen when injected may elicit disproportionate amounts of unwanted antibody.

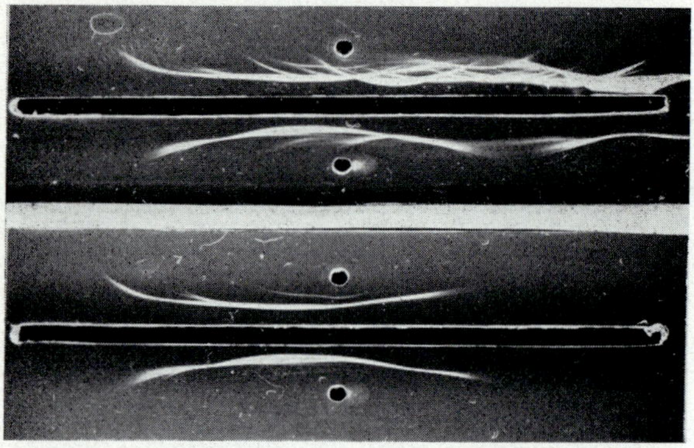

FIG. 6.4. Patterns of bovine serum (upper wells in (*a*) and (*b*)) and colostrum (lower wells). Slots filled with, (*a*) rabbit anti-bovine serum. (*b*) the same serum after precipitation with presuckling calf serum proteins.

The three antisera used in Fig. 6.5 were prepared by immunization with purified human serum IgG, with IgA isolated from human colostrum, or with a mixture of preparations of pathological IgM. The anti-IgG was made specific for the γ-chain by precipitation with an Fab preparation. The other two antisera were made specific for α and μ chains respectively by precipitation with freeze-dried serum from human cord blood, since of the immune globulins only IgG is present in significant amounts in cord blood. Using these antisera it is possible to characterize sera showing a raised immunoglobulin level, by examining the level of each immunoglobulin type independently.

In a similar manner Bence-Jones proteins can be detected in urine and they can be distinguished as type K (κ chains) or type L (λ chains) using specific antisera. An antiserum reacting with both κ and λ chains was made by immunizing a sheep with light chains, isolated from normal human IgG. After precipitating this antiserum with a single Bence-Jones protein of type K, it reacted specifically with λ chains; similarly, after precipitating with a Bence-Jones protein of type L, the antiserum reacted with κ chains only. In Fig. 6.5 five Bence-Jones proteins and a preparation of normal human light chains have been examined using either (*a*) anti-κ or (*b*) anti-λ chains. It can be seen that whereas the normal light chains contain both κ and λ chains and so

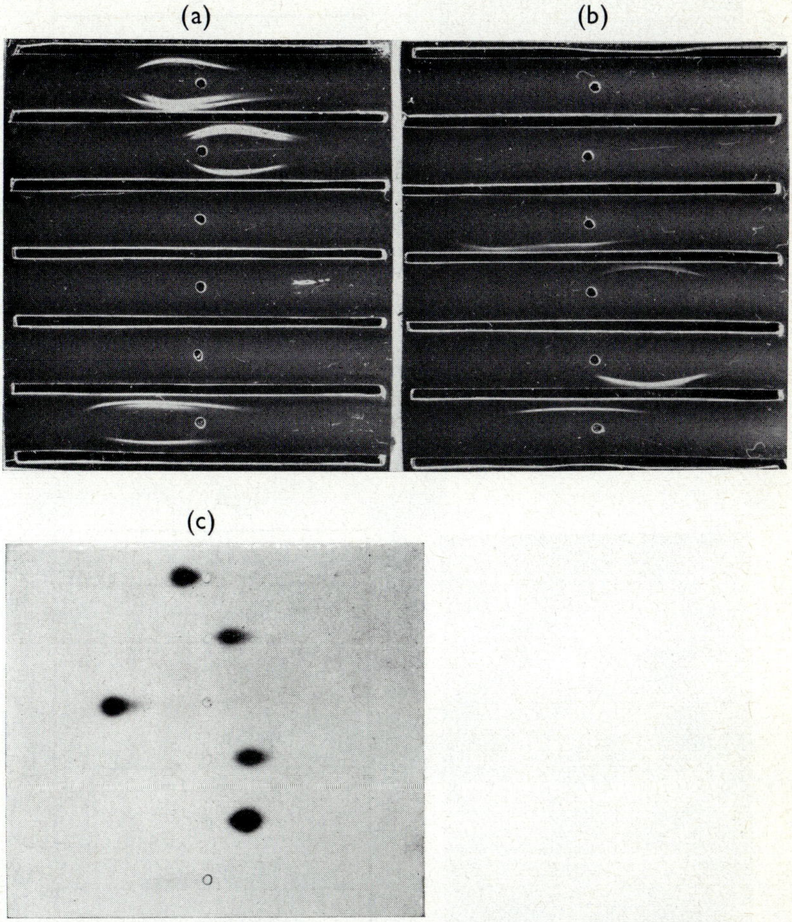

FIG. 6.5. Detection and classification of Bence-Jones proteins. The upper five wells on plates (*a*) and (*b*) were filled with five different urine samples, and the lowest well with normal light chains isolated from IgG. Slots in (*a*) filled with antiserum to type K light chains. Alternate slots in (*b*) were filled with an antiserum to type L light chains. Plate (*c*) shows corresponding electro-phoretic patterns of the Bence-Jones proteins and light chains.

react with both antisera, Bence-Jones proteins which react with one of the antisera never react with the other; two are seen to be of type K, and three of type L.

It is possible to use immune precipitation itself as a purification step

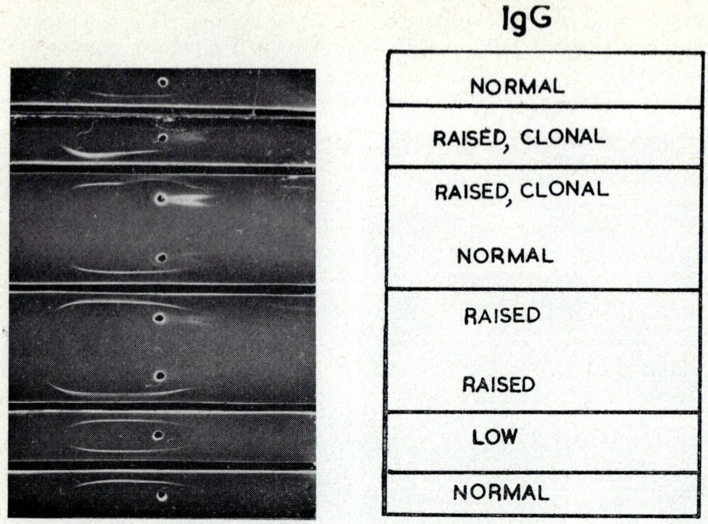

(a)

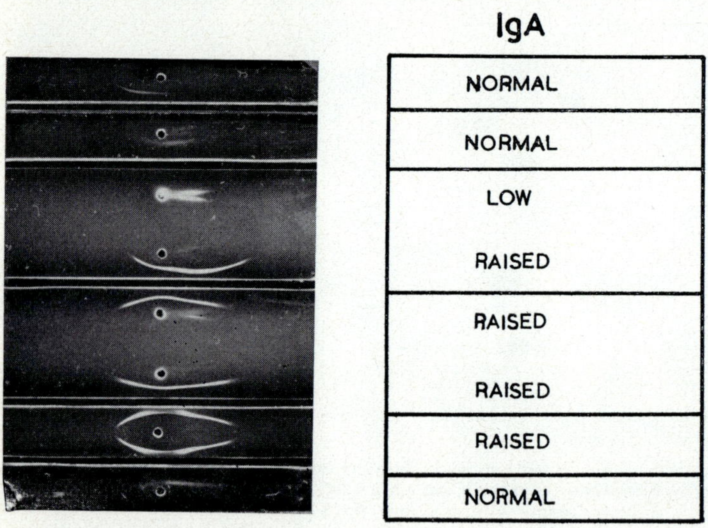

(b)

FIG. 6.6. (i) Comparison of patterns of seven pathological human sera (upper wells) with a normal serum (lowest well). The same eight sera are examined on each plate, but slots are filled with specific antisera to (*a*) IgG, (*b*) IgA, (*c*) IgM. The corresponding stained electrophoretic patterns are shown at (*d*).
(ii) An interpretation of the above patterns.

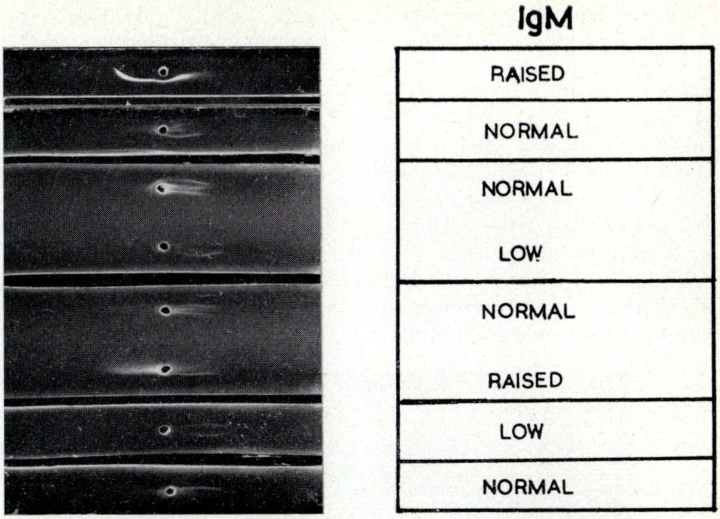

(c)

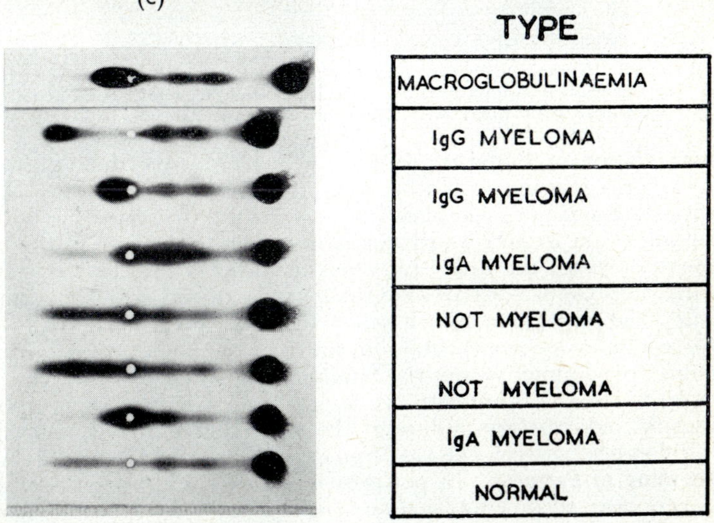

(d)

by immunizing an animal with a single precipitin line cut from the washed agar plate.[4,5] Selective precipitation of the antiserum in the manner described above may also prove useful. Fig. 6.6 shows how a panel of unispecific antisera may be used. Once again pathological human sera are being examined and compared with a normal serum (c.f. Fig. 6.1). Here, however, three unispecific sera are employed, each reacting with one of the three major immune globulins, IgG, IgA, or IgM. As was mentioned in the introduction, an antiserum made to any of these immune globulin types will usually precipitate other types of immunoglobulin by reaction with their light chains. To prepare specific reagents reacting with the heavy chains only, it was necessary to precipitate out in advance, from the antisera, antibodies to light chains.

Degradation Studies. It has been seen that when using either a single antigen, or antiserum specific for a single antigen, a single arc is formed. Such systems have proved useful in degradation studies. For example, Fig. 6.7 shows the pattern of human IgG before and after papain treatment. The molecule is split into two types of fragment by papain[6] and each forms an arc independently of the other. One of the two antisera used has reacted predominantly with the Fc fragment, whilst the other antiserum is directed equally to both fragments.

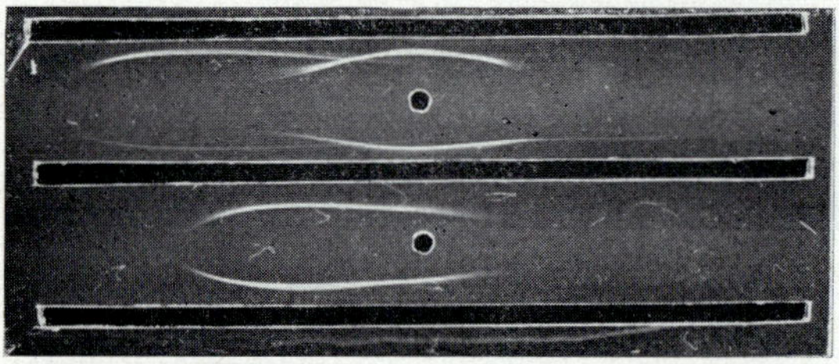

FIG. 6.7. Effect of papain digestion on patterns of human IgG. Lower well contained purified IgG, upper well papain digested IgG. Alternate slots filled with two different anti-human IgG sera.

Interpretation of Patterns. Where, as in the above case, patterns show two arcs crossing each other with no appearance of fusion, this is known as a reaction of complete non-identity. It indicates the failure of the antiserum to detect any common structural feature in the two fragments forming the arcs. In other instances, where two arcs fuse partially, but one or both arcs continue across the other in the form of a "spur", the reaction is known as one of partial identity, and demonstrates common antigenic sites on the two antigenic species. Such reactions are common within the family of immune globulins, (see, for example, Fig. 6.4b). Complete fusion of two arcs, known as a reaction of identity, indicates the failure of the antiserum to detect any antigenic difference between the two species.

Recording of Patterns. In patterns formed by multispecific sera, it is unlikely that all precipitating systems will have an optimally sharp appearance at the same moment. It is helpful therefore if the plate can be photographed several times during the precipitation stage. For this purpose the plate should be illuminated by several light sources, at an angle from below, using a dark background. All the immunoelectrophoretic figures in this chapter were obtained in this way. When the pattern has completely developed, the plate can be washed, dried and stained. The washing is done in several changes of physiological saline over a period of two days. This removes unprecipitated proteins from the gel. After a short final wash in distilled water the plate is covered with filter paper and dried down at 37°C. The paper is then removed, after wetting if necessary, and the arcs can be stained using the protein stains described earlier. A variety of stains and reagents demonstrating different antigenic species have been developed, (see J. Uriel, Chap. 2, Grabar and Burtin[1]) but are beyond the scope of this chapter.

High Voltage Agar Gel Electrophoresis

The method of electrophoresis in agar gel described above as part of the immuno-electrophoretic method, with circular starting wells, is not in itself capable of good resolution. The resolution can be improved by cutting thin starting slots, but it is then necessary to fill the slots with a mixture of the components to be examined and molten agar gel. If this is not done the starting slot may empty during the run and produce a non-linear field which distorts the electrophoretic pattern. However, the heating of sensitive components when mixing a sample with warm agar is unsatisfactory. A method for electrophoresis in agar gel was developed by Wieme,[7,8] which avoids these difficulties. It has also the advantages of requiring relatively little material and being fairly rapid since it employs higher voltage fields. This is possible because evaporation and undue temperature rise are prevented by immersing the plate in petroleum ether.

Apparatus. Fig. 6.8 shows the construction of the tank used, which is made easily from perspex. The electrode vessels A and D communicate with agar gel wicks B and C. During electrophoresis, the plate coated with agar gel, which carries the samples, lies in the centre section with its gel surface downwards, touching the wicks B and C. This central section is filled with petroleum ether, which, if not cooled, may rise several degrees during the run. If required, thermostatic cooling may be achieved by bubbling air through the ether as described by Wieme.[9] In our laboratory the equipment is placed in a room maintained at 4°C, and the petroleum ether, with some air, is circulated from the tank by a car petrol pump into a vented aspirator reservoir, from which it runs under gravity through a metal coil then back to the electrophoresis tank. Circulation rate is controlled by adjusting the reservoir height, and for lowest temperature runs the metal coil can be cooled in ice. However, considerable cooling takes place by evaporation in the reservoir. Precautions against the hazard of fire should, of course, be taken.

Method. Buffered agar gel is prepared as described above for immuno-electrophoresis. To pour the agar wicks the holes connecting the electrode compartments A and D to the agar compartments B and C are first stoppered. Molten agar is poured into sections B and C until the level rises just above the walls of the central compartment. When this agar has set the stoppers are removed and buffer of the same composition as used in the gel is added to the electrode vessels. Square plates or microscope slides are coated with the gel as for immuno-electrophoresis, on a levelled surface, and when set may be stored in a wet box. Slots are cut by pushing the end of a strip of 3 MM filter paper into the gel. The slots are filled from a Hamilton micro-syringe which can be rapidly washed out with buffer between each sample, about 2 μl.

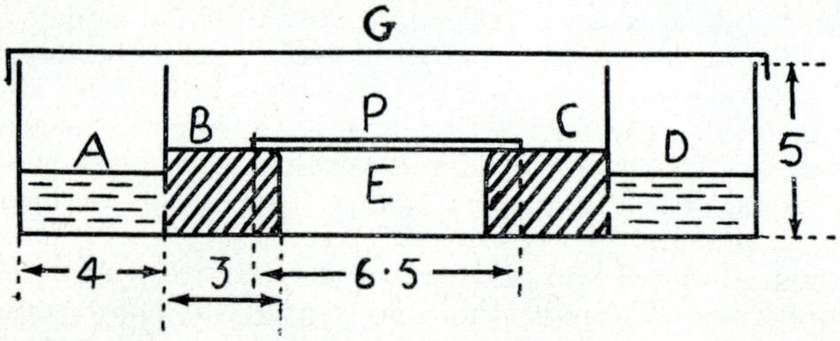

FIG. 6.8. Construction of high voltage agar gel electrophoresis tank. Compartments B and C filled with agar gel, A and D filled with buffer, E filled with petroleum ether to a level above that of the gel plate, P. The apparatus is enclosed by means of a cover, G.

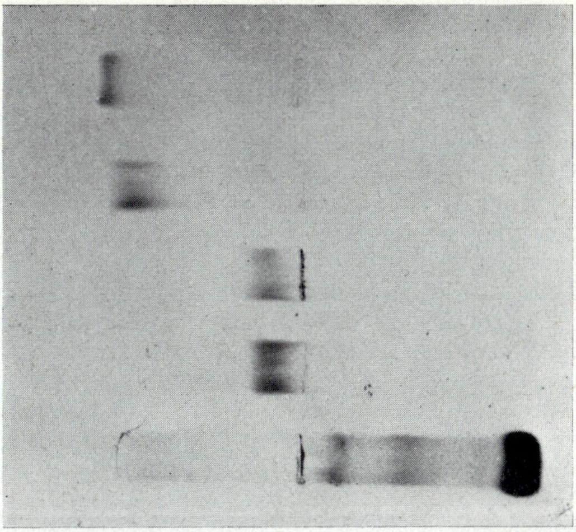

FIG. 6.9. Electrophoretic patterns of four myeloma proteins and normal human serum.

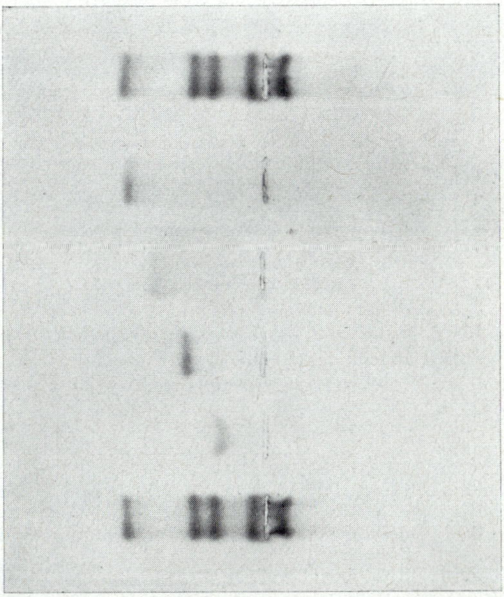

FIG. 6.10. Patterns of Bence-Jones proteins. Top and bottom patterns are of a standard B.J. protein mixture. Intermediate patterns show two B.J. preparations. Each pair shows the effect on electrophoretic migration of a single electronic charge increment, introduced by reduction and carboxymethylation.

of sample being required for a 1 cm. slot. The plates should be filled rapidly enough not to allow drying out of the slots first filled. The inverted plate is then placed across the agar wicks, under the petroleum ether which fills the centre section. After covering the tank a potential is applied across the electrodes, to give a field of 25 V/cm. across the plate. A satisfactory separation of serum components in this field is achieved in 30 to 45 minutes at 10°C. After the run the plate is removed, the petroleum ether is blotted off, and the plate is fixed, dried and stained, as described (for simple electrophoresis) in the section on immuno-electrophoresis.

Fig. 6.9 shows the stained patterns of human serum and four myeloma proteins of widely differing electrophoretic mobility. Fig. 6.10 shows the effect on the electrophoretic mobility of 2 urinary Bence-Jones proteins when a single negative charge increment was added to each by carboxymethylation. The patterns of the mixed Bence-Jones protein preparations show the reproducibility and resolution of which this method is capable.

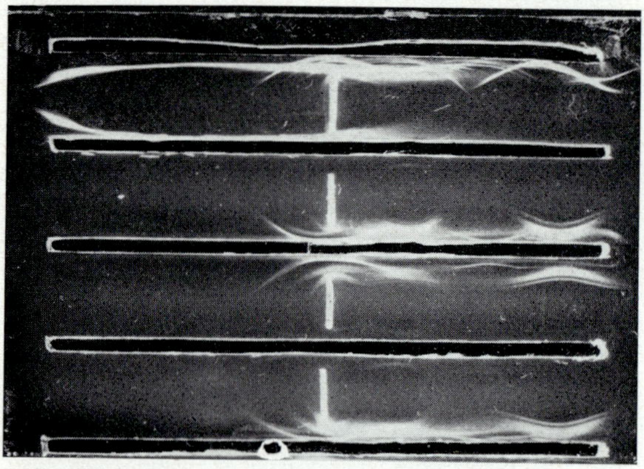

FIG. 6.11. High voltage immunoelectrophoretic patterns of an adult pig serum (top) and three fœtal pig sera.

The method may be used as the first step in immuno-electrophoresis, and is considerably faster than the usual method. Fig. 6.11 shows the immuno-electrophoretic patterns produced in this way of adult and fœtal pig sera.

REFERENCES

1. Immuno-electrophoretic Analysis (Eds. P. Grabar and P. Burtin), Elsevier, 1964.
2. Scheidegger, J. J. *Intern. Arch. Allergy Appl. Immunol.*, 1955, **7**, 103.
3. Osserman, E. F. *J. Immunol.*, 1960, **84**, 93.
4. Smith, H., Gallop, R. C., and Tozer, B. T. *Immunology*, 1964, **7**, 111.
5. Goudie, R. B., Horne, C. H. W., and Wilkinson, P. C. *The Lancet*, 1966, 1224.
6. Porter, R. R. *Biochem. J.*, 1959, **73**, 119.
7. Wieme, R. J. *Clin. Chim. Acta*, 1959, **4**, 317.
8. Wieme, R. J. "Agar gel electrophoresis," Elsevier, 1965.
9. Wieme, R. J. *Nature*, 1961, **190**, 806.
10. *Bull. World Health Org.*, 1964, **30**, 447.

CHAPTER 7

STARCH GEL ELECTROPHORESIS

SECTION 1—TECHNIQUES OF STARCH GEL ELECTROPHORESIS

Ivor Smith

ELECTROPHORETIC separations on paper, cellulose acetate and agar were based on a simple electrophoretic effect namely the migration of ions under the influence of an electric current. In general the medium acted as an inert support for buffer in which the separation occurred but if the medium did exert any effect at all then it was adverse and unfortunate, e.g. the slight denaturing effect of paper on proteins causing a small tailing. With the advent of starch gel electrophoresis the converse was the case. Not only did the gel affect the separation but it was so designed that this resulted in the separation of more proteins from any given mixture. Here the components of the mixture migrated through the pore structure of the gel which acted as a molecular sieve such that proteins of the same charge to mass ratio would separate if their molecular sizes were sufficiently different. The extreme example is of the single α_2 globulin band from paper electrophoresis which was resolved into about 10 sub-bands after gel electrophoresis, although all the globulin bands from paper yielded a number of bands after running in starch.

Previously a given population had showed one normal pattern, for humans this being one albumin and four globulins as found on paper. However due to the greater resolving power of the gel, this same population could be subdivided into a number of normal patterns and, for example, three genetically controlled normal haptoglobin patterns were found as well as a number of genetically variant transferrins.

"Starch-hydrolyzed" for gel electrophoresis is commercially available from the Conaught Laboratories and, more recently, frnm British Drug Houses Ltd; this starch is of a high quality. Nevertheless batches vary and it is usual to obtain large quantities from the same batch to avoid recalibration of small batches. Starch gel shows a slight electro-osmosis in the opposite direction to protein migration at alkaline *p*H's.

The original procedure of Smithies involved the gel lying horizontal as with horizontal paper, acetate and agar electrophoresis. This was basically an analytical procedure requiring no more than 10–15 μl. serum, and one in which the sample was applied on a strip of paper inserted into a cut in the gel. In an endeavour to use the procedure for larger quantities, a thicker sample was applied into a slot cut into the gel but due to electrodecantation this was not successful. The problem of using larger samples was overcome by standing the gel vertical. A fortunate corollary to this was due to the fact that the rate of protein migration is much greater in free solution than in the gel and consequently the protein band compacted greatly at the gel interface leading to much sharper bands. The albumen may be stained with bromophenol blue (BPB) as described elsewhere.

Acrylamide gel has largely displaced starch gel as a molecular-sieving medium for electrophoresis, for a number of reasons. It is an inert, synthetic chemical which is more easily prepared as a gel and leads to a much simpler and quicker technique. The amount of published work on starch gel decreases yearly although a number of major laboratories continue to use it for comparative purposes.

Buffers

Gels have been used successfully over a *p*H range of 2–11 but, in practice, most workers have used gels within the range of *p*H 8–10 for serum proteins together with an occasional run at *p*H 4–5. Buffer solutions can be prepared in bulk and stored in the cold.

For the buffers described below and elsewhere in this chapter, the quantities shown are weighed and dissolved as described; molarities shown in brackets following each substance refer to molarity of that substance in the final solution. However, it is not sufficient to take the weights shown and assume the *p*H of the final solution as substances both effloresce and deliquesce even though this may be too small to be visible. Final adjustment of the *p*H should be made using concentrated solutions of one or other of the components and titrating against an accurate *p*H meter reading in 0·01 units. Subsequently the final volume adjustment is completed.

Some useful molecular weights are:
Tris 121·1; boric acid 61·84; citric acid.1H_2O 210; EDTA.2H_2O 328·2; LiOH.H_2O 41·96; diethylbarbituric acid 184·2; NaAc.3H_2O 136·1; acetic acid 60·05; NaOH 40·01; urea 60·06.

The original buffer described by Smithies was a simple sodium borate buffer, *p*H 8·48, for the gel together with a more concentrated buffer at a somewhat lower *p*H for the reservoir; this resulted in a very wide albumen band although all other bands were narrow. Subsequently, Poulik described a discontinuous buffer system in which the gel buffer was a Tris-citrate system, approximate *p*H 8·7, coupled with the same reservoir buffer used by Smithies. The effect of the discontinuous buffer was that the proteins commenced migration in the gel buffer but completed migration in borate buffer due to the fact that borate ions have a higher migration than proteins in the gel and so over-run and pass them. The borate-Tris interface appears as a moving yellow line and serves as a natural front marker. The value of this procedure is that there is a voltage discontinuity at the buffer interface which results in a sharpening of the rear side of each band and, in particular, of the albumen band whose width may be half that found in the original procedure. Hence the sharpening is at an end and the experiment may be discontinued as soon as the yellow line has passed completely ahead of the albumen—BPB. Currently almost all work is done with discontinuous buffers, which are, more or less, modifications of the Poulik system. Another modification which has produced sharper bands is the use of lithium hydroxide instead of sodium hydroxide to prepare the reservoir buffer. A fair range of buffer concentration is possible. It should be noted that too dilute a reservoir buffer causes the anodic end of the gel to swell and too concentrated a buffer causes anodic shrinkage.

Buffer 1. Sodium borate buffer, *p*H 8·48 (original horizontal Smithies[1]); this refers to the *p*H of the gel mixture and is slightly lower than the *p*H of the original buffer as starch contains a few carboxyl groups in the molecule.

Gel:	Boric acid, (H_3BO_3, 0·03M), 1·855 g. plus	
*p*H 8·5–8·6	Sodium hydroxide, 12 ml. 1·0M (0·012M) in one litre.	
Reservoir:	Boric acid, 0·3M,	18·55 g.
	Sodium hydroxide, 0·06M,	60 ml. 1M solution
	Water to	1,000 ml.

Although this gel buffer is rarely used now, the reservoir or bridge buffer is in frequent use with discontinuous systems. Smithies noted that the ratio of boric acid to sodium hydroxide which resulted in best separations was that quoted above, namely 40 per cent neutralized acid. However as the starch varied slightly from batch to batch the exact concentration of gel buffer was determined by experiment. Gels were prepared using a series of buffers in which the original was diluted with water in a known manner and calibrated such that the front and back of the albumen band are equally sharp, and then the most suitable dilution of buffer was used for that batch. Frequently the dilution factor was such that two volumes of gel buffer above plus one volume of water was required, i.e. boric acid 0·02M and NaOH 0·008M. For vertical gels, the above ratios of boric acid and NaOH were maintained but a slightly higher concentration was required to produce a similar albumen band.

Using an 0·023M boric acid buffer, i.e. above diluted 77 ml. to 100 ml., and a constant current of 3 mA per 5 cm. wide horizontal gel, the voltage drop across a 15 cm. long gel is 34 V and across the power pack 80 V, and this remains unchanged overnight. Overday, 10 mA gives a comparable separation in about 5 hours. Vertical gels, 30 cm. long × 12·5 cm. wide, require a constant current of 15 mA overnight and the voltage figures are 110–150 also remaining relatively unchanged during the run. These figures are based on 6 or 8 wicks as described under "Apparatus and Method".

Buffer 2. Acetate buffer, *p*H 4·5.

Acetic acid, 1M	187·5 ml.
Sodium acetate, 1M (136 g. NaAc.$3H_2O$/l) .	150 ml.
Water to	3 litres

This solution is used for the reservoir but for the gel it is diluted with an equal volume of water.

Buffer 3. Barbitone buffer, *p*H 8·0.

Diethylbarbituric acid, 18·4 g. dissolved in hot water to about	600 ml.
NaOH, 1M	60 ml.
Water to	1,000 ml.

The acid is dissolved in hot water. After cooling the alkali is added and, again after cooling, the volume is made to one litre.

Buffer 4. Tris-citrate discontinuous buffer, horizontal or vertical, Poulik.[2] Stocks can be made 5 or 10 times concentrated and diluted as required. *p*H of concentrate is somewhat higher and must be allowed for, by checking a diluted sample; *p*H 8·65

Gel:	Tris, (0·076M)	9·2 g.
	Citric acid (0·005M)	1·05 g.
	Water to	1,000 ml.
Reservoir:	Boric acid (0·3M)	18·55 g.
	NaOH 1M (0·06M or 0·05M)	60 ml. or 50 ml.
	Water to	1,000 ml.

Horizontal gels can be run overnight at 3 mA constant current for a 5 cm. wide gel. The authors found this to correspond to a gel voltage of 20 V and a power pack voltage of 70 using a 15 cm. long gel at the commencement of the experiment and rising to 35–85 at the end. Comparable runs are obtained in 5–6 hours using 10 mA.

Vertical gels, 30 cm. long × 12·5 cm. wide, run overnight at 15 mA constant current. Starting voltages found are 65 V across gel and 100 V on the power pack rising to 160–230 V at the end of the run.

Buffer 5. Tris-citrate-Lithium, vertical.[3]

Gel:	A.	Tris (0·016M)	1·94 g.
*p*H 8·04		Citric acid. H_2O (0·003M)	0·63 g.
		Water to	1,000 ml.
	B.	LiOH. H_2O (0·02M)	0·84 g.
		Boric acid (0·0776M)	4·8 g.
		Water to	1,000 ml.

Mix 9 parts A plus 1 part B as required.

Reservoir:	LiOH. H_2O (0·1M)	4·2 g.
	Boric acid (0·38M)	23·5 g.
	Water to	1,000 ml.

12 mA for 16 hours were quoted with a voltage of 120. Alternatively 45 mA for 3–4 hours in the cold room at 550 V. This buffer has been used for serum proteins and 1 per cent Hb separations.

Buffer 6. Tris-citrate-Lithium, vertical.

Gel: A. Tris, 6·29 g., plus citric acid, 1·6 g. per litre
B. LiOH. H_2O, 1·2 g., plus boric acid, 11·89 g. per litre

For the gel, mix 9 volumes of A with one volume of B. For the reservoir use B alone.

Buffer 7. Tris-EDTA-boric acid, for *p*H's over range 9·5–8·0. For Hb's vertical.

Gel: Tris, 60·5 g., plus EDTA.$2H_2O$ 7·8 g. in approximately 2 litres. Titrate in concentrated boric acid solution to the required *p*H. Adjust final volume to 20 litres.

Reservoir: boric acid, 185·5 g., plus NaOH, 24 g., in 10 litres. *p*H 9·0

Run for 16–20 hours at about 175 V.

Tris-EDTA-Borate Buffer (for Hb's) For gels of *p*H 8·6 or 8·3, stock buffers are made as follows:

Tris	109 g. or 67·9 g.
disodium EDTA	5·84 g.
boric acid	30·9 g.
distilled water to 1 litre.	

The stock solution is diluted 1 to 20 (15 ml. to 300 ml.) with distilled water for making the gel and 1 to 7 (100 ml. to 700 ml.) for the buffer vessels. The samples are applied 5 cm. from one end of the gel. A second row of samples can be applied 7 cm. further along on the same gel. This Tris-EDTA-borate, *p*H 8·6, buffer gives the best overall resolution for the detection of abnormal hæmoglobins, and has several advantages over other buffers. The first is that hæmolysates can be applied without dilution, and a hæmoglobin concentration of *ca.* 10 g. per 100 ml. does not overload the gel. The system is therefore ideal for the detection of abnormal minor components as well as Hb-A variants. A second advantage is that these buffers consistently resolve Hb-F from Hb-A. Furthermore, it resolves methæmoglobin from oxyhæmoglobin, whilst the oxy, carbonmonoxy and cyanmet derivatives migrate together. Lastly, both "slow" and "fast" hæmoglobins are well resolved; "slow" and "fast" in this context refer to the rate of migration towards the anode compared with that of Hb-A.

The reason for using *p*H 8·3 as an alternative to *p*H 8·6 in some analyses is to alter the migration rate of slow minor components in relation to the origin. For example, Hb-α^4 (the α-chain hæmoglobin) found in β-thalassæmia and in association with some unstable hæmoglobins stays at the origin at *p*H 8·6 but migrates towards the cathode at *p*H 8·3. Both the *p*H 8·6 and 8·3 gels are run at 20 mA per gel for approximately six hours. Better resolution is obtained in the cold than at room temperature. In any case, it is useful to apply the sample to a gel which has been cooled in the refrigerator.

Sodium Phosphate Buffer Systems.[1,4] These have a *p*H range of 6·3–7·4 and are made up as follows. 40 ml. of 0·04M Na_2HPO_4 ($Na_2HPO_4.12H_2O$ 14·2 g. per litre) of a given *p*H (see Table 7.1) are diluted to 300 ml. with distilled water for making the gel. It will be seen from the table that for gels with a *p*H of 7·4 no adjustment of the *p*H of the sodium phosphate solution is necessary, and for gels of lower *p*H the *p*H of the sodium phosphate is adjusted according to Table 7.1. The electrode buffers for each run are made by adjusting a solution of

0·04M Na_2HPO_4 with phosphoric acid to the *p*H of the gel. As the actual *p*H obtained with different batches of starch may vary slightly, it is advisable to check this on an aliquot of gel poured into a beaker at the time of preparation. The gel is run at room temperature using a constant current of 50 mA for 2–5 hours according to the separation required.

TABLE 7.1

Composition of solutions for phosphate buffer starch gels (pH measured at 25°C)

*p*H to which 0·04M $Na_2HPO_4 . 12H_2O$ was adjusted with H_3PO_4*	*p*H of gel
8·9†	7·4
7·8	7·1
6·8	6·8
6·2	6·3

* 40 ml. of this solution diluted to 300 ml. with H_2O used for making gel.
† No H_3PO_4 added to this solution.

The phosphate buffer was originally used for the detection and resolution of Hb-β_4 (Hb-H) or Hb-γ_4 (Hb-Bart's). However, the Tris-EDTA-borate system resolves these almost as well. The phosphate buffer is, therefore, mainly used in the detection of the M-hæmoglobins as well as in screening for the unstable hæmoglobins (see later).

Apparatus for Starch Gel

Both horizontal and vertical apparatus will be described; in general these two types are based on the original designs of Smithies. The thickness of the gel is always 6 mm. as this is about the maximum that can be used without special cooling procedures; horizontal trays are made in 5, 10 and 15 cm. widths and lengths vary from 15–20 cm.

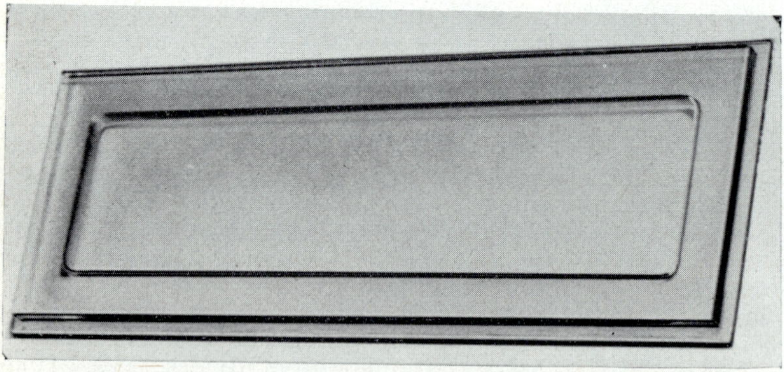

FIG. 7.1. Apparatus for horizontal gel preparation. The rectangular "sides" unit is silicone-greased and attached to the flat base plate. After pouring the gel, it is covered with a second flat plate which is lowered, starting at one short end, to cover the gel completely. Excess gel flows over the sides and the top plate is secured with some heavy weight.

whereas vertical gels are usually 12·5 cm. wide and 25–30 cm. long. The vertical method tends to give sharper, more discrete zones in general and is particularly useful for differentiation of closely moving bands such as the haptoglobins.

Horizontal Apparatus and Method. Subsequent to the electrophoresis, the gels must be sliced lengthwise for staining and the advantage of the apparatus to be described is that the sides can be removed without removing the gel which remains then on a plate such that it can be sliced directly.

The gel plate is laid on a horizontal surface, the "sides" unit is lightly covered with silicone grease on its bottom, top and inside surfaces and placed squarely on the base plate so that it is firmly attached and there

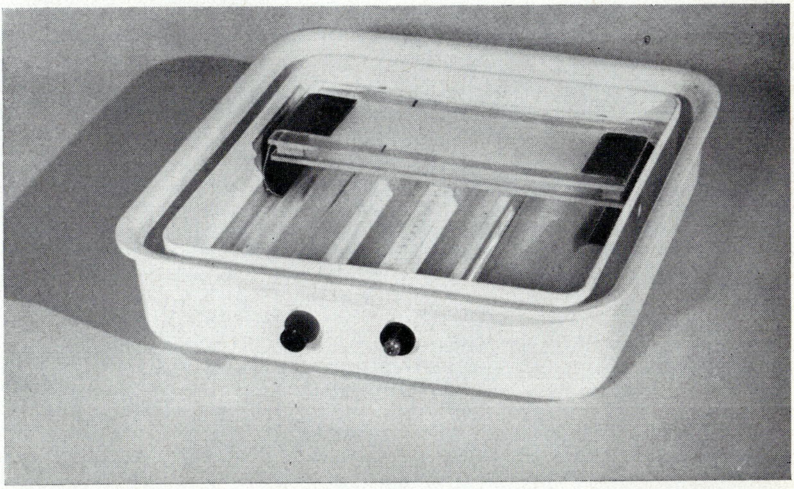

Fig. 7.2. The horizontal gel in position in the tank. Wicks lead from the edges of the gel (1 cm. overlap) down to the buffer and are held securely in place by means of the original top plate. The wicks are shown black to enhance the contrast.

is no chance of gel leaking out. Gel is prepared and poured into the unit so that the surface stands just proud of the top of the holder unit. Starting at one narrow end, place a second sheet of siliconed plastic onto the side unit and slowly flatten it down so that excess gel is squeezed out and the top-plate comes to rest freely without any entrapped air bubbles. Whilst holding it in position place a heavy weight such as brick onto the top plate so that it is held permanently in position with all edges held in contact with the sides. The gel is allowed to cool for a minimum of one hour but usually two hours at the temperature at which it is to be used.

When ready for use, trim excess gel from around the outer edges, insert a spatula edge between the top plate and the side and gently prise it off; it comes away cleanly without pulling or tearing the gel. The gel should be kept covered at all times to avoid evaporation and drying out and the simplest procedure is to use polyethylene film or bags. During the application of samples, two pieces of film are used leaving just the origin line free for working.

Whilst waiting for the gel to set, place 500 ml. buffer into the tank as described in Chapter 3, and level the buffer; reservoir buffer is used once only. Cut twelve wicks from thick paper such as 3 MM, each the width of the gel and about 6–8 cm. long so that they will reach easily from the gel surface to the buffer. Apply the sample, as described later, and cover the gel with film again. Place the gel horizontally in the tank and apply six wicks, as a single pad, to each short side of the gel. Wicks are wetted most of their length by immersion but the last one or two cms. by allowing the solvent to rise into the pad. Then raise 1 cm. of film and place the wicks squarely in position overlapping the gel by about 1 cm. at each end. The pads can be kept in position by covering with the original top plate which will rest on the slightly raised paper and not press onto the film below. Run the experiment for a suitable time as discussed under the various buffers and elsewhere.

Vertical Apparatus and Method. Assemble the gel tray by screwing the two end pieces to the main unit and grease lightly with silicone

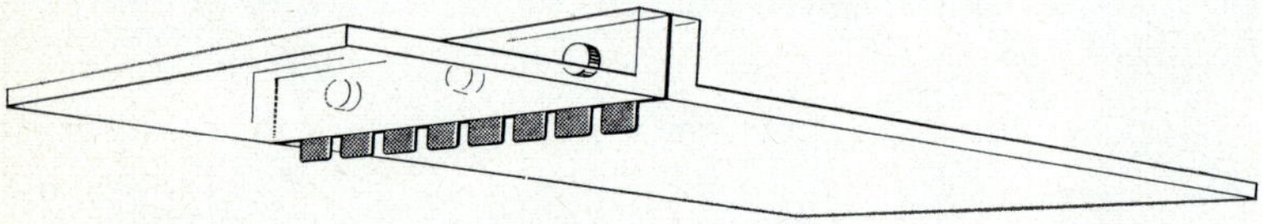

FIG. 7.3. Comb or slot-former for vertical gel. Note that the unit unscrews so that the two plastic plates can be removed separately from the slot-former to avoid damage to the gel. Combs may be used to provide eight individual slots for different samples or a single slot running the width of the gel for a preparative analysis.

grease or vaseline. Place on a flat sheet on a horizontal surface. Lightly grease the two parts of the top plate with the comb and screw together so that the teeth of the slot former project out from the plate; a small paint brush aids even application around the comb. Place this top unit in a warm oven, 50°C, until required.

Pour the gel into the tray taking care not to trap any bubbles and until the liquid surface is just proud of the tray sides. Take the top plate from the oven and align one long edge with a long edge of the gel tray so that the corners are square, ensuring that the slot-forming comb is near the top end of the tray as it will subsequently hang in the vertical position. Gently lower the top plate until it lies flat with no visible bubbles entrapped and immediately place heavy weights in position so that the whole plate is held down and ensuring that all corners remain closed. This operation requires care as it cannot be repeated with the same gel volume. Allow a minimum of 1·5 hours to elapse for cooling and setting at a temperature similar to that at which the experiment will take place.

Whilst the gel is setting, the electrophoresis apparatus can be set up. Place 400 ml. buffer into each trough, levelling each by tilting until the buffer liquid surface just crosses the mid-bridge. Connect the inner terminals to the electrodes, the upper being positive. Prepare two pads, each of six sheets 28 × 12 cm. Whatman No. 1 paper, place across the mid bridges of each trough and ensure that the pad in the lower reservoir squarely covers the whole floor as the gel will later rest on this. Cut 8 sheets of No. 1, each 12·5 cm. square, for each gel and place as a pad into the upper buffer. Fewer sheets but an equal thickness of thick paper such as 3 MM can be used or, for buffer wicks only,

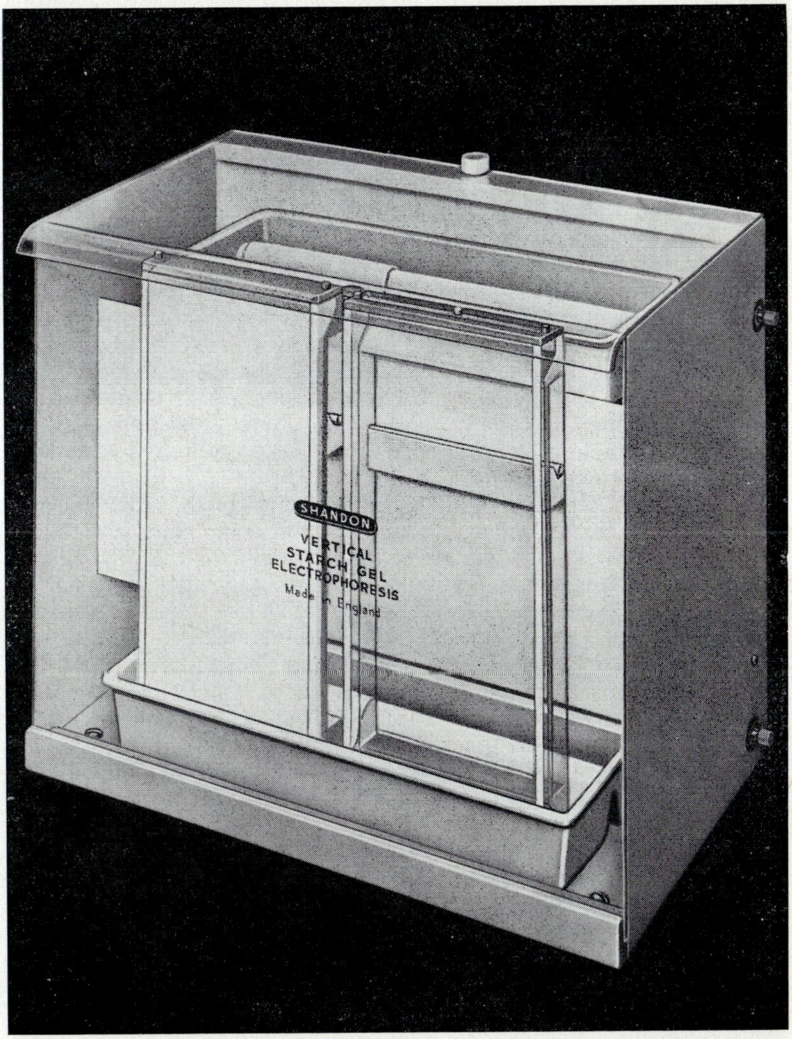

Fig. 7.4. Vertical gel set-up. The left hand gel unit contains gel whereas the right hand unit is empty to illustrate the method of suspension and the features of the gel tray. *Shandon Southern Instruments, Camberley, Surrey.*

some thicknesses of surgical lint or gauze. Replace the main cover on the apparatus until the tank is required.

After the sample has been applied and with the taped film in position (see sample application method C), unscrew the two small end plates, remove the lower one entirely but leave the upper one loosely in position. Place the gel tray vertically in position (check with a spirit level vial), resting evenly on the pad in the lower trough and hanging on the cross bar of the main apparatus. Take the gel pad, which is standing in the upper reservoir, and place it evenly so that it just covers the thick top surface of the gel and lightly screw down the Lucite plate so that the pad is held evenly in position but without any pressure being exerted on the gel.

Replace the outer cover, screwing in the contact breaker, and apply the current. Some workers prefer not to replace the outer cover, particularly if the experiment is to be performed in the cold room or if further cooling such as a fan blowing across the apparatus is to be used; if the cover is eliminated then remove the contact-breaking screw and insert separately into the apparatus. It is necessary to work without the outer cover when determining voltage drop across the gel by means of a pair of point-contact terminals connected to an Avometer (VOM). A current of 12–15 mA on a Poulik-buffered gel gives about 90 V across the gel but considerably more across the power pack.

Discontinuous systems show the yellow line which overtakes and passes the albumen and the run can be continued, if desired, until the yellow nearly reaches the lower edge of the gel. Alternatively the BPB method can be used to locate the albumen band.

Preparation of the Gel. The horizontal gel tray, 18 × 5 × 0·6 cm., requires about 100 ml. buffer plus starch and it is probably wise to gain facility by preparing a single gel at one time even if the vertical procedure is the main technique to be used. Subsequently sufficient for up to three horizontal trays can be prepared at one time. The vertical apparatus requires about 450 ml. buffer plus starch.

To prepare the gel use a long-necked round-bottom or conical flask with a wide standard joint; for one tray use a 250 ml. flask, for three trays a 750–1,000 ml. flask and for the vertical method a one litre flask. During the procedure the neck may get quite hot in which case it is useful to bind it with asbestos string so that it may be handled comfortably.

Attach a standard joint tube with a side arm to the water pump and set going before making the gel. The tube connection to the pump can have a three-way tap capable of letting air into the system and the tube should be stoppered at the upper end.

Weigh the prescribed amount of gel shown on the bottle, or other suitable amount, 11–15 g. usually, and pour it gently through a long-necked funnel into the flask already containing the buffer so that all the starch falls directly into the liquid and does not adhere to the sides of the flask. Swirl to suspend the solid evenly in the buffer and heat over a medium hot flame with a continuous rotary motion taking care that the starch does not char on the bottom of the flask. At the beginning, as heating progresses, the viscosity remains low and the solution swirls easily. Then there is a sudden large increase in viscosity and the suspension becomes almost opaque and semi-solid due to the bursting of the starch grains. Great care is necessary here in order to prevent charring as the flask appears to get much heavier due to the drop in swirling ability. Continue heating and swirling until the viscosity drops again and the liquid becomes semi-transparent and mobile as before. Remove the flask from the flame and rapidly connect to the pump. The solution swells up rapidly, the froth breaks and vigorous boiling is allowed to occur for a few seconds until de-aeration is complete. De-aeration may be continued up to 20 seconds but care should be taken not to boil off too much water to avoid volume changes. Disconnect the vacuum, remove the tube with side arm, pour the gel carefully but rapidly into the tray ensuring that no air-bubbles are trapped. Fill the tray until the liquid surface is just proud of the sides and cover with a top plate as previously described under "apparatus and method". After a setting period of 1–2 hours, the top plate is prised off by inserting a spatula at one corner and gently lifting; if necessary any excess gel adhering to the outer sides is first cut away. With the horizontal trays, the plate can be lifted slowly from the narrow end and slightly upwards along the length of gel when it should come away cleanly although a spatula can be used to assist separation if necessary. With the vertical apparatus, the screws are first removed so that the two portions of the cover and the slot maker are separate and the spatula inserted from the side to lift away the smaller part of the cover first. The comb is gently withdrawn in order to avoid damage to the slots and, finally, the longer cover is raised using a spatula to keep the gel down although this should not be necessary if the cover has been adequately greased and warmed in the oven as previously described. The gel is covered immediately with plastic film to prevent drying out and tearing away from the sides.

Gels may be used after the minimum cooling time or kept until the next morning or even up to two days. Although all such gels yield satisfactory separations, an "ageing" effect is observed with the result that an older gel shows separations similar to a younger one with less starch. Consequently, whatever procedure is used it should be adhered to fairly rigidly in order to get maximum comparability between results.

Application of the Samples to the Gel. Sample is applied to the horizontal trays either on a narrow strip of thick paper or as a slurried suspension in starch grains. With the vertical method, the sample is applied in free solution, e.g. whole or diluted serum is applied directly into the slot, and hence there is no effect due to possible adsorption on the paper or starch grains.

(*a*) Application on paper strips. Take a one-sided razor blade, prise open the holder, remove the blade and cut along the short dimension so that a piece 1 cm. wide is obtained, this requires a sharp instrument to avoid the blade splitting. Then insert the blade back into the holder and draw a scratch with a diamond exactly 5 mm. up from the sharp edge of the blade. Make two scratches 5 cm. along from the narrow edge, one on either side, of the gel tray. Place a ruler across

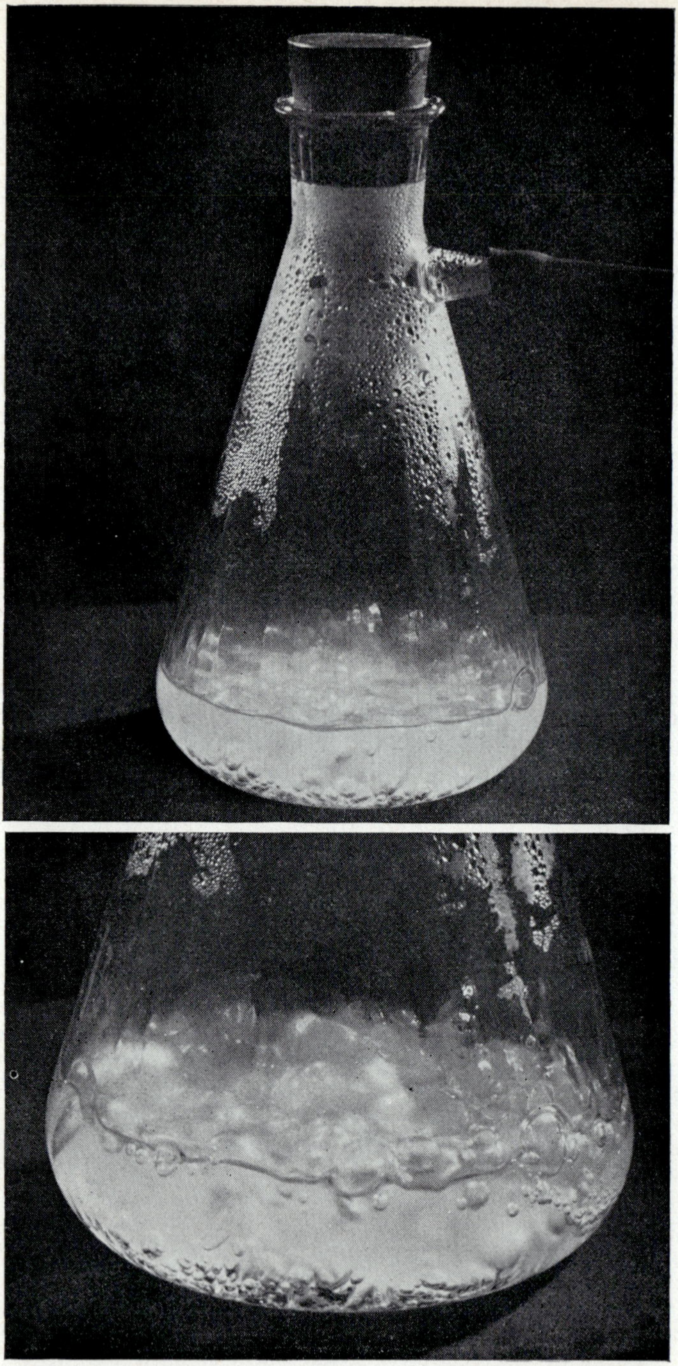

Fig. 7.5. During the process of making starch gel it is necessary, if the gels are to be reproducible in pore size—(and to get rid of air bubbles), to standardize absolutely the boiling effect under reduced pressure. This is fully described in the text, but these photographs clearly show exactly what should happen.

the scratches and use this to steady the blade whilst inserting it into the gel. Take a piece of precut paper, 10 × 5 mm. of 3 MM, apply 10–20 μl. sample coloured with bromophenol blue, BPB, if desired, and with a small pair of wide-end forceps insert it into the gel by very slightly pulling the blade backwards, placing the paper in front of it and edging it down into the gel. The blade must not cut down through the bottom of the gel and the top of the paper strip must be below the gel surface. Holding the paper in position by means of the forceps, carefully remove the blade and press the gel surfaces together again; no bubbles should be trapped. With practice three such inserts can be made, leaving a 5 mm. gap at each outer edge and about 3 mm. between strips on a 5 cm. wide gel. To compare minor differences, the strips can be placed even closer when similar bands will coalesce during the run and stain as uninterrupted lines. Even more samples can be compared by applying on 5 mm. widths of paper. Cover the gel with film.

(*b*) Application as a starch grain slurry. Two blades are prepared as above and returned to the holder with a 1 mm. rectangular separator sandwiched between them at the upper end. By inserting this into the gel, a slot can be cut and removed taking care that the two blades remain parallel during the operation and at right angles to the direction of migration. A bent spatula may be needed to clean out the slot. 20 g. ordinary potato starch is suspended in 100 ml. acetone and the supernatent is poured off after two minutes. The process is repeated three

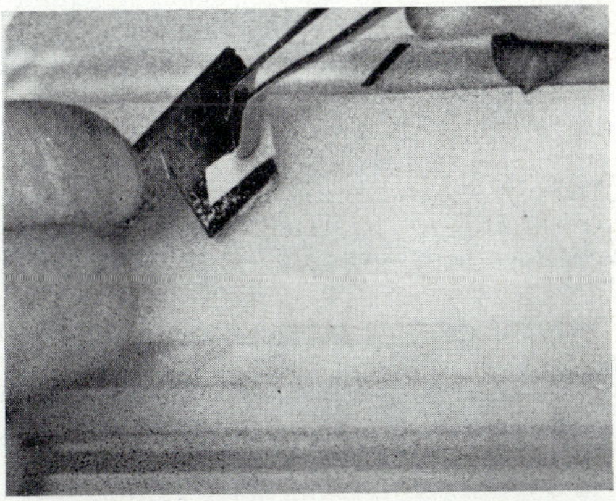

FIG. 7.6. Close up showing insertion of sample-on-paper strip. Note the permanent black lines on the apparatus indicating the origin, the cut-down razor, or blade in the razor holder, the 5 mm. indicator on the blade to ensure that the slit does not go through to the base plate.

times and the residual starch is dryed at 45°C. To 1 ml. of buffer add 1 g. starch grains, 0·2–0·5 ml. serum and mix into a slurry. Suck the slurry into a pipette, allow the grains to sink and separate from excess supernatent liquid and deposit the slurry carefully in to the slot without overflow. Cover the slot with just molten paraffin wax (petroleum jelly) using a wax which melts in the 30–50°C region so that the sample is not overheated. Cover the gel completely with plastic film and set up in the electrophoresis tank.

(*c*) Application of free solution sample for vertical electrophoresis. Determine the volume of the slots—usually about 50 μl. with a single comb. Mix 20 μl. serum with 30 μl. water containing a trace of BPB and apply to the slot without overflow or bubble entrapment. It is sometimes better to mix a larger volume so that the pipette can be removed with some residual solution to avoid bubble formation. The added water increases the resistance of the sample and helps to sharpen the protein "volume" before it enters the gel; the actual thickness of protein layer may be one fifth or even one tenth of the slot width and this can be observed if the BPB method is used. Pipette just molten wax, as described above, onto the slots so that they are completely enclosed. Cover with a sheet of film which slightly overlaps the width of the gel and fix this to the long sides with adhesive tape to hold the gel in position when vertical. Transfer to the tank and complete the experiment.

Power Supply and Current. The Vokam power pack described on p. 8 is satisfactory for starch gel techniques. Some figures have been recorded under specific buffers described previously but, in general 2 mA constant current per one cm. width of gel is suitable for 5–6 hour separation overday and about 0·7 mA/cm. for 18 hour overnight runs, the current referring to the power pack reading. Where buffers of quite different concentration to those previously mentioned are used, empirical observation is necessary to determine correct amperage which may be raised to a figure which just does not allow the gel to warm up above about 30–35°C. A draught of air can be blown over the gel to carry away excess heat although care should be taken so that separations remain straight within the gel.

Slicing the Gel. All starch gels, particularly when sliced to 3 or 1·5 mm. slabs or when prepared with urea, are fragile and easily broken. It is therefore always best to transport them on solid plastic sheets (not film) and, in transferring from reagent to reagent or wash solution we use a plastic (nylon or polyethylene) net such that the gel rests on the plastic sheet which is held in the net. The net should extend about 10 cm. out from each of the four sides of the plastic sheet so that the whole can be easily picked up and, if the net extends over the sides of the staining dishes, then the hands need not be wetted in the sequence of transfers. A variety of staining and washing dishes should be available such as pyrex rectangular pie dishes, photographic developing polyethylene trays, etc., slightly longer that the gel and at least twice as wide so that gel slices can be slid or floated sideways onto a new plastic sheet. A further aid to gel flotation and transfer is a series of wide palette or wall-paper scraping knives, 5, 10, and 15 cm. wide, which can be slipped between cut slices of gel to raise the upper slice slightly and just sufficiently for the solution to seep between the slices.

Due to the fact that top and bottom surface effects are often observed, it is usual to slice the gel through the 6 mm. thickness and to observe the stained bands from the inner surfaces of the two half-thickness gels. After practice it is possible to cut 3 × 2 mm. and even 4 × 1·5 mm. thicknesses thus giving three or four gels for use with different stains, enzyme location reagents, etc.

(*a*) *Slicing the Horizontal Gels.* The slicer board possesses an adjustable horizontal wire held at an angle to the two sides. The height adjuster allows for the thickness of the plastic sheet and for 3 and

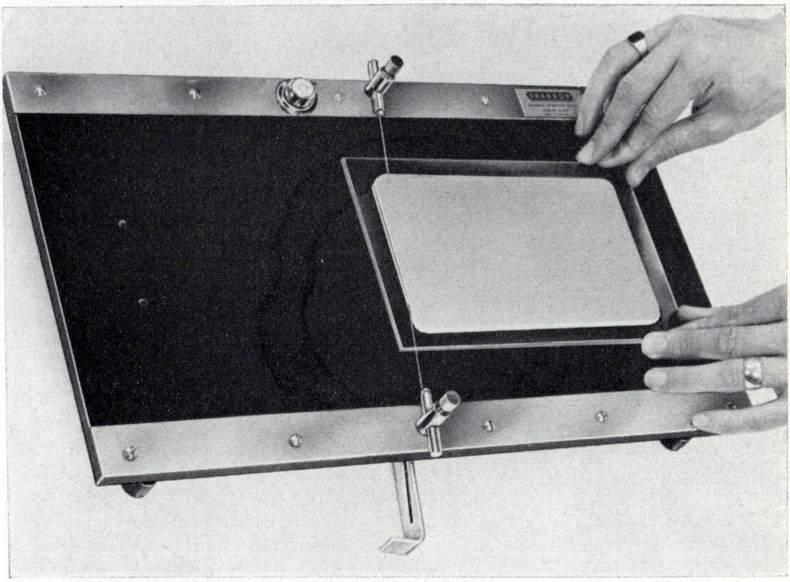

FIG. 7.7. Gel slicer. A wide gel still on its original base plate but with the "sides" unit removed is being cut. The adjustable wire allows slices of 1·5 and 3 mm. to be cut so that either two or four slices are obtained from the gel. Vertical gels are inverted onto a base plate before slicing. *Shandon Southern Instruments.*

1·5 mm. slices of gel to be cut. The gel unit is laid flat on the slicer board and the inserted sample strips of paper are carefully removed with forceps. The "sides" unit is gently prised up and away from the gel and should come away easily due to the previous greasing. If it does not, it can be loosened by running a spatula round between the unit and the gel to separate the two. The gel, still on its original base unit, is presented towards the wire, first making a cut at one corner of the gel and gently moved sideways so that it penetrates through the whole leading edge. Then, with a continuous, steady motion the plastic sheet is moved forward until the wire has sliced right through and comes out of the rear edge of the gel. If desired, the gel can be trimmed by applying one of the palette knives vertically down on the far side of the BPB and again about 2 cm. back from the slots towards the cathode direction or even at the slots themselves if interest is confined only to anodically migrating ions. The sliced gel still on the base is placed in a net and transferred to a dish containing solution, submerged and the top slice is floated off onto a second plastic sheet resting on a net, using a palette knife to aid flotation. The solution used in the dish depends on the subsequent location procedure. For chemical procedures involving the fixing of proteins, etc., 5 per cent

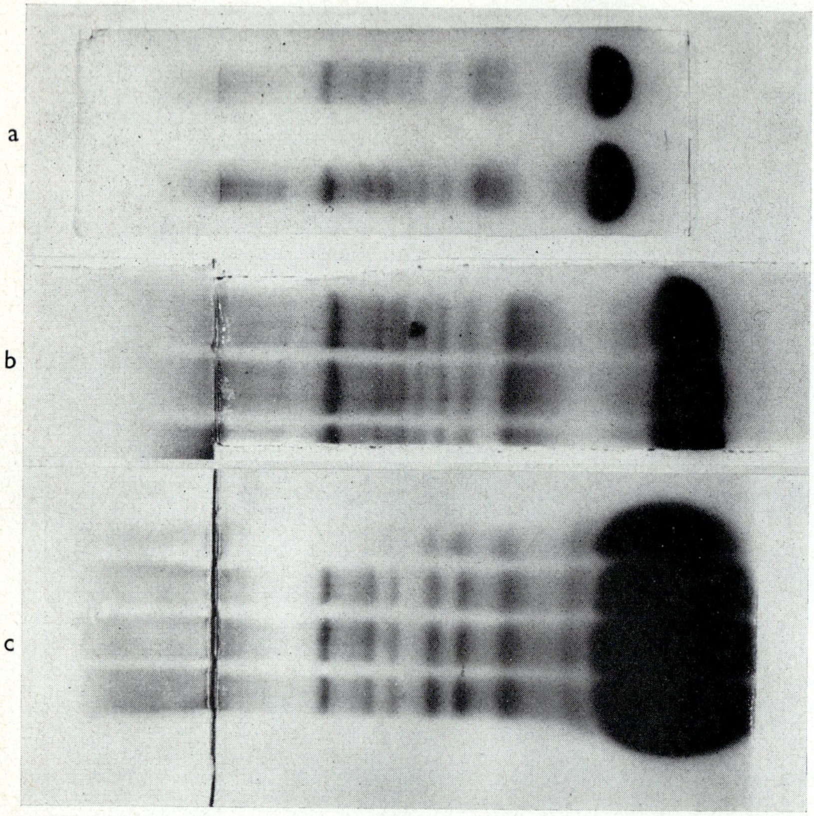

Fig. 7.8. Comparison of gels using normal serum and electrophoresis conditions as described in the text. (*a*) Horizontal Poulik buffer; (*b*) vertical Poulik buffer, (*c*) vertical Smithies buffer.

acetic acid may be used but for procedures depending on the retention of activity as with enzymes the solution can be the buffer in which the NAD, lactate, etc., is subsequently dissolved. An alternative procedure to this submersion is to raise the end of the upper slice with the knife and flood between the slices with the solution just mentioned. When completely wet, the upper slice can be gently slid off onto another plastic sheet. The slice is now transferred on its base sheet to the stain or locating solution; the main gel on the base plate is returned for further slicing, if desired, or stained.

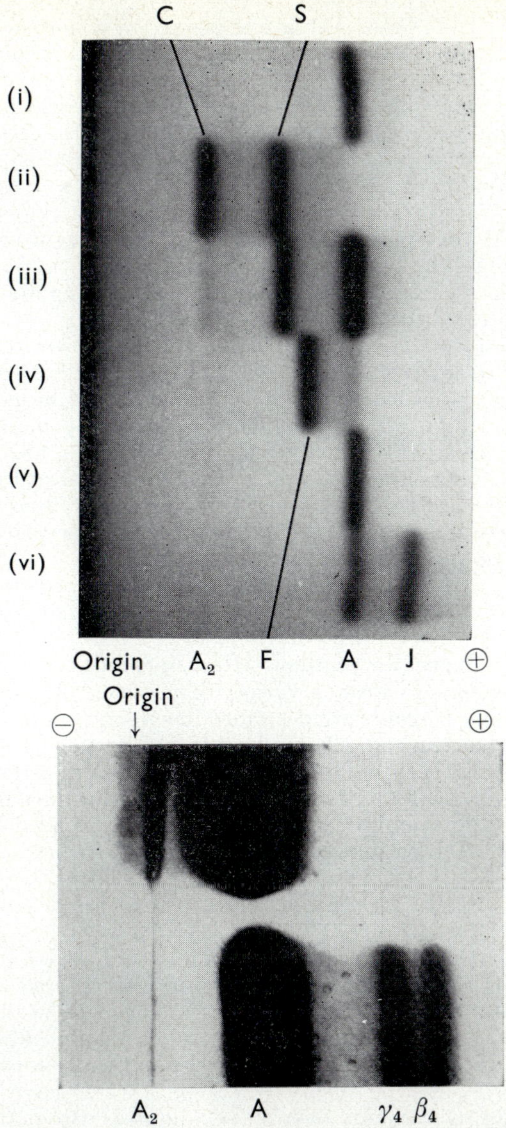

FIG. 7.9 (*a*) Starch gel electrophoresis in Tris-EDT-borate buffer, *p*H 8·6: (i) normal adult hæmolysate; (ii) Hb-S + Hb-C; (iii) Hb-A + S; (iv) normal cord blood; (v) normal adult; (vi) Hb-A + J. The slight difference in electrophoretic migration between Hb-S in samples (ii) and (iii) is due to the effect of the other hæmoglobins present in each hæmolysate, illustrating one of the difficulties encountered in interpreting slight differences in electrophoretic mobility. (*b*) Starch gel electrophoresis in phosphate buffer, *p*H 7·1: (i) normal adult; (ii) patient with Hb-β_4 (Hb-H) and Hb-γ_4. Note the absence of Hb-A_2.

(*b*) *Slicing Vertical Gels.* The gel is placed on a horizontal table, a spatula is run round the sides of the gel-unit interface and the gel is trimmed at the far side of the BPB or yellow line to remove the thick edge. Cathodic portions are also removed at a line which experience indicates is suitable; sometimes the gel is sliced at the slot position if only anodic materials are being investigated. A suitably sized sheet of plastic is placed over the remaining gel, the whole unit inverted and placed on the table. The gel container is slightly raised from one narrow end and the gel should come away and remain on the lower sheet; if it sticks, the wide knife is used to release the exposed edge and the unit slowly raised until free of the gel. Slicing and staining is carried out as described in (*a*) above.

Even when anode and cathode portions of the gel are to be stained it is often wiser to cut at the slot position and stain the halves separately but, in this case, the identity of the portions should be preserved by marking the gel. Another useful trick is to cut a small triangular section from one corner of the gel before it is sliced. This is particularly useful if three or four slices are to be cut and then realigned for comparison after different staining procedures, as it is sometimes difficult to be certain which way up the gel should be placed in order to have the same samples for comparison.

Stains

Gels may be stained for proteins, Hb's, lipids, etc., or they may be treated with reagents to locate a variety of enzymes. Apart from those listed below, many reagents are described in subsequent sections of this chapter. To avoid tearing, gels are best transferred on plastic or glass sheets. Stains are best used in large rectangular plastic, sandwich boxes with lids so that the whole can be kept clean during prolonged reactions and readily moved about the laboratory if need be. This is particularly satisfactory during destaining when the closed box can be placed on a rotating horizontal turntable to hasten the process.

Naphthalene Black. Naphthol blue black, amidoschwartz, for proteins.

Naphthalene black, 1 per cent, in methanol-acetic acid-water (5:2:5), or in 7 per cent acetic acid. The dye solution is filtered and the gel stained for 30–60 seconds. Excess dye is removed by soaking the gel in a number of changes of the solvent in which the dye was prepared. Destaining can be hastened by placing the gel in the destaining medium on a horizontal circulating turntable. Some workers have found that 0·1 per cent stain is adequate over 30–60 minutes. Bromophenol blue, light green and Ponceau S, as described in other chapters, have also been used.

Nigrosine. Used for weak protein bands or those which are not precipitated in aqueous acetic acid stains.

Nigrosine, 0·5 g., in 100 ml. of 50 per cent trichloracetic acid solution containing 5 per cent phosphotungstic acid. Dilute 1 volume with 4 volumes ethanol as required. Stain overnight and destain with 50 per cent ethanol containing 1 per cent PTA.

Lipid Crimson, for Lipoproteins. This stain can be used both as a prestain by adding it to the serum before the run or as a stain after electrophoresis.

Lipid crimson, saturated in ethanol, 60 ml., plus water, 40 ml., and filtered. The stain is best aged for a few weeks but this is not essential. For prestaining, add 10–20 μl. dye to 0·3 ml. serum before electrophoresis. For post-staining, transfer the gel to the stain overnight. Destain in 5 per cent acetic acid containing 5 per cent Teepol ($R.R'.CH.SO_3Na$; Shell Chemicals). Such gels can be counterstained for proteins using the above stains or the light green prepared as described for paper electrophoresis. Oil red O is used as an alternative to lipid crimson under the same conditions.

Hæmoglobin and Hæmoglobin-Haptoglobin Complexes. On paper electrophoresis at 8·6, haptoglobins migrate with the a_2 band which is faster than free hæmoglobin but slower than the Hb-Hp complex. On starch gel, the complexes move slower than either the corresponding free haptoglobins or hæmoglobin because of the filtration effect on the increased molecular weight. The basis of this determination is that Hb, either free or combined with Hp, catalyses the oxidation of leuco-dyes in the presence of hydrogen peroxide to form blue or brown dyes depending on the nature of the dye used. The free Hp's can, of course, be stained as for proteins.

The cells are washed four times with a 20-fold excess of isotonic saline. One volume of cells is then mixed with approximately 1·5 volumes of water and 1 volume of toluene and allowed to stand for half an hour in the cold. The solution is then centrifuged at 3,000 r.p.m. for 15 minutes. The clear intermediate layer is used for electrophoresis. Some of the unstable hæmoglobins, as well as Hb-β_4^A (Hb-H) and Hb-α^A, are denatured by toluene and when the presence of one of these is suspected the hæmolystae is cleared by ultracentrifugation. Chloroform was used at one time instead of toluene but, due to its adverse effect on a number of abnormal hæmoglobins, its use has now been discontinued. Prepare a 1·5 per cent Hb solution, and add 1 volume to 4 volumes serum to form a final concentration of 300 mg. per cent. Both the Hb and serum must be fairly fresh. Run, if desired, both original serum and serum-Hb on starch gel and stain one slice of each with naphthalene black. Stain one slice of the Hb gel with one of the stains shown.

Dianisidine Method.

Sodium acetate-acetic acid buffer, 1·5M, *p*H 4·6 .	35 ml.
Water	80 ml.
o-dianisidine, 1 mg./ml. in ethanol . . .	35 ml.
H_2O_2, 100 vols. per cent (added just before use) .	1 ml.

Pour the stain on the gel (or paper), close the tank and rock a few times. Leave until the brown bands are intense. Wash in 2 per cent acetic acid and store in 2·5 per cent Teepol for up to a few weeks. Ceruloplasmin does not stain except in very large amounts.

Benzidine Method.

Benzidine, 0·2 per cent, in 0·5 per cent acetic acid in water (warm to dissolve, store cold) . .	100 ml.
H_2O_2, 100 vols. per cent (30 per cent w/v) . .	0·2 ml.

Pour freshly mixed stain onto the gel and blue bands become visible within a few minutes. Wash free of peroxide with water and store. Excess peroxide must be avoided as it decolorizes the bands and also because the red cell catalase reacts with vigour and causes the gel to explode due to the rapid evolution of oxygen. The benzidine method is more sensitive than the dianisidine but a little less permanent.

P.A.S. This reagent cannot be used for glycoproteins as the starch also reacts positively.

Further Techniques

Preservation of Gels. Gel strips can be dehydrated and preserved almost indefinitely by this procedure which requires the preparation of a sandwich of the gel between two layers of dialysis membrane such as Viscose tubing.(5) Tubing commonly available in England seems to be thinner than that in the U.S.A. and the thinner membrane allows easier preparation with less trouble.

The gel is stained and destained in the usual manner. It is then immersed in a solution containing 15 per cent glycerol and 3 per cent acetic acid in water for 60 minutes. Two sheets of membrane, about 4 cm. wider than the gel are similarly immersed until soaked through although longer times do not harm. One sheet is laid flat on a piece of glass of similar dimensions and the bubbles smoothed out. The gel is laid flat on the membrane so that a 2 cm. width of membrane is left free on all four sides of the gel. The second membrane sheet is placed on top of the gel and rolled to remove air bubbles. The membrane sides are secured in position by means of bulldog or binder clips. Drying is allowed to proceed in the air although it can be hastened by blowing air over. During the course of about 24 hours drying is completed although the sandwich should be turned over occasionally to ensure uniformity of drying. When complete the sandwich is placed in a polyethylene bag or the membrane is simply peeled off the gel which is then transferred to a polyethylene bag, the latter being sealed with adhesive tape. Such gels can be placed in notebooks or stored elsewhere indefinitely (see Fig. 7.8).

Polypeptide Chains—Separations. A great deal of interest now centres around the chains of such proteins as the myeloma and macroglobulin multi-chain molecules. These chains are now split with mercaptoethanol, dissociated in the presence of urea and separated on starch gels. Although, in principle, the gel electrophoresis method is the same, the very useful review by Poulik(9) should be consulted first.

Scanning and Quantitation. Transparent gels are essential for good scanning. The gel should be washed completely free of background dye first and then rendered transparent by one of the following procedures. (1) The procedure described above for preservation using glycerol-acetic acid-water is used when a thin slice is obtained and, from which, the viscose sheet can be peeled off for this purpose. (2) The gel slice is placed in 90 per cent acetic acid which is brought to the boil and slowly allowed to cool when the strip obtained remains transparent for some hours but is very fragile. (3) The gel is heated in glycerol when a strong gel results. The first procedure is valuable in that a strong thin gel is obtained which can be preserved indefinitely.

The staining procedure has comparative value only but this is usually sufficient for determining normality. Biuret values of some of the bands so obtained from gels are shown below.

Two-Dimensional Electrophoresis. The first use of the two dimensional procedure was by Poulik and Smithies[6] in a successful attempt to correlate the zones found on paper electrophoresis with those found in the starch gel. Subsequently the method has been used a great deal to compare sera both within a given species, such as humans, and between various families such as those of the primate order. The method is a combination of the two procedures, namely paper electrophoresis first followed by insertion of the paper into the gel as already described. Occasionally the first run has been on acetate, in agar or even in starch at a different *p*H.

Paper electrophoresis is carried out as previously described in Chapter 2 except that approximately 15 μl. is applied per 5 mm. width of paper and BPB is used as a marker. A 4–5 cm. width of paper can be used and, after the run, cut with a sharp razor to provide a middle strip in the direction of migration of 5 mm. width, or the dry paper cut be cut first and then reassembled in the tank as if it were a single strip to avoid "side" effects often found on narrow strips. If only one strip is required for the gel, this should be flanked by spare strips to prevent undue evaporation or run in the Unikit apparatus. The length of separation is adjusted via voltage and time of run to provide a suitable length of paper for insertion into the gel. The position of the bands on the paper can be calculated or one of the outside portions can be rapidly stained to locate them. For the beginner, and often for better separation or if interest is centred in only one band, the paper can be cut into separate pieces each containing one protein band and these pieces inserted into individual but adjacent slots in the gel. With practice, the complete 8–12 cm. strip can be inserted into a single slot in the gel. The gel is then run and stained as previously described. To determine purity or traces of minor components, electrophoresis can be run first in starch at one *p*H, a slice or portion of a slice 5–10 mm. wide transferred to a second gel and run at a different *p*H. In this latter case, the buffer concentration should be lower in the first run to avoid double buffer effects.

Immuno-electrophoresis. The interpretation of precipitin lines becomes more certain as the number of these lines per unit length or area becomes smaller. Hence the greater the electrophoretic separation, the less ambiguous the interpretation becomes. Starch gel, particularly after two-way runs, provides maximum separation of proteins and readily lends itself to the immuno method. It must be remembered, however, that starch slowly adsorbs protein and, further, if diffusion is carried out in non-buffered agar there may be changes or even precipitation due to rapid removal of buffer ion from the protein areas.

One- or two-dimensional electrophoresis is carried out as previously described, the gel is sliced and rapidly stained to locate the bands. Depending on the area of interest, a small square or a long narrow non-stained slice about 2 mm. thick is cut from the unstained portion of the gel, placed on a sheet of glass, the surface just covered with a piece of polyethylene film, and surrounded with a 1 per cent, or similar, agar solution which is allowed to set.

The general procedure for preparing, pouring and immuno-diffusion in agar is described in Chap. 6.

Elution and Preparative Recovery. For this purpose, sample is applied along almost the complete width of the gel using a comb which makes a single slot in the original gel preparation. It is also probably better to use the vertical technique rather than the horizontal so that the sample can be applied without a support but, if the horizontal technique is used then the starch grain method is employed. The slot can also be made wider by using two of the comb-slot formers in the one holder to form a double width. Smithies has stated that the maximum volume of sample such as serum protein should not exceed 0·1 ml./cm.^{2} gel cross-section as otherwise definition falls considerably. To separate samples containing only a few types of protein molecule which run well separate it would appear that larger quantities could be used. The gel is prepared and run under the usual conditions. Subsequently a slice is cut from each long side, in the direction of the migration, stained and destained as rapidly as possible and realigned as guide strips but without touching the main section of gel in case protein denaturants are still present in the guides. Gels should be kept cold whilst awaiting the guides. If the staining procedure is an extended one then it can often be speeded up by reducing the times involved so that just weak bands are visible as this will be enough to use as the guide. It should be remembered that the guide strips often shrink a little during the staining process particularly if some of the solvents containing the stain are dehydrating reagents. At the present time it seems best to consider the use of acrylamide for full scale preparative isolation.

1. *Electrodialysis method of Koen and Shaw.*[(7)] This simple method appears to be the best so far described. It is equally applicable to starch, acrylamide discs and flat gels, agar, etc.

Electrophoresis is carried out as usual with slots up to 2·5 cm. long by 1·6 mm. wide, the latter being made with a number of slot-formers inserted between the two halves of the top-plate of the vertical gel cover. Similar gels up to 1 cm. deep can be used. A slot 2·5 cm. × 1·6 mm. × 5 mm. will hold about 0·2 ml. Subsequently the gel is notched at one corner for rematching, a 1·5 mm. slice is taken, stained for the appropriate substance as rapidly as possible and washed with water; fixing solutions must be avoided to prevent shrinkage. The untreated gel is stored covered in the cold room and the stained section is placed there also for about 20 minutes to come to equilibrium. The two portions are aligned and the desired bands are cut out by means of a scalpel cutting through both sections although only the lower, untreated section is used for recovery.

A piece of nylon or nylon acetate net, 1·5 mm. mesh, is cut into 4 × 4 cm. squares and two opposing sides are sewn together with nylon thread, leaving a long piece of thread trailing from one corner. The far, open end is sewn together leaving a second long thread trailing. The sections of starch are placed in the bag. A piece of dialysis tubing, slightly longer than the bag is wetted with buffer and the nylon bag, closed end first, is inserted into the tubing so that the thread protrudes a fair way through the far end. The open end of the bag is twisted with the end of the dialysis tubing as tightly as possible and secured with a rubber band, again with the trailing thread outside. A small volume of buffer, 0·5 ml. upwards, is added through the open end of the tubing and this is also twisted and closed with a rubber band.

The bag is then suspended by means of the lengths of nylon thread

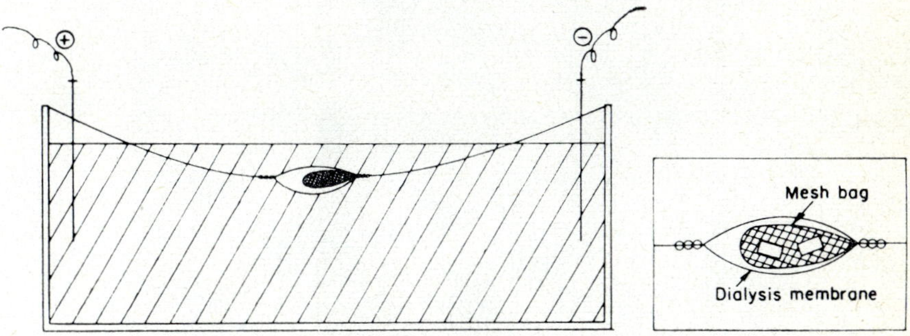

FIG. 7.10. Diagram of apparatus for recovery of protein from starch or acrylamide gel. Inset figure shows gel held in mesh bag within dialysis membrane bag as described in the text.

in a volume of the same buffer so that it lies longitudinally and is kept horizontal by means of small weights attached at each side. Electrodialysis is carried out and even the slowest ions are completely transferred within about 4 hours. Because of the low resistance, a high current flow at low amperage is possible, e.g. 100 mA at 80 V. At completion, the dialysis tubing is opened and the small amount of solution drained off. Gentle pressure on the gel bag will free a drop or two more and the gel can also be rinsed with 0·1 ml. buffer to obtain maximum yield. Up to eight samples can be processed simultaneously.

The method has a number of advantages. Because the buffer is the same inside and outside the membrane, no volume change occurs and a concentrated solution is obtained; this is particularly valuable for those enzymes which denature on dilution. The volume of buffer added to the bag is independent of the quantity of starch gel which also aids concentration and avoids any later concentration stage.

2. *Freezing and Thawing.* Freezing and thawing breaks down the gel structure leaving a sponge-like substance. This spongy material is supported on some sort of glass nail or golf tee fitting fairly tightly to

the centrifuge tube and spun at not more than 1,500 r.p.m. as otherwise the gel will spin down into the sample. Simple filterfuge tubes containing a filter medium can also be used at higher speeds. This must be done immediately after electrophoresis as otherwise the proteins will absorb irreversibly to the gel sponge. Recoveries are very variable depending on the type of protein, but poor for enzymes.

SECTION 2—MICRO STARCH GEL ELECTROPHORESIS

Connell L. Marsh and Carl R. Jolliff

The conventional starch gel technique of Smithies[1] has been modified[8] to make it simpler, more rapid and thus adaptable to routine clinical usage. The micromethod has been used for the electrophoretic analysis of serum from many species (Fig. 7.11): human, pig,

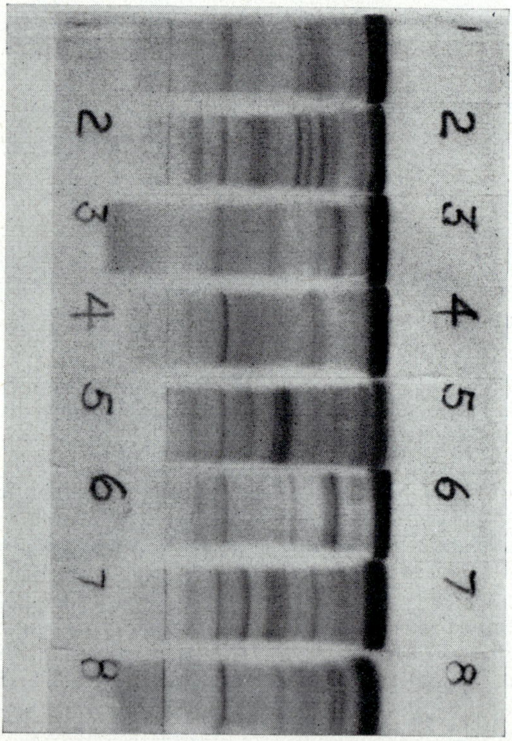

FIG. 7.11. Micro starch gel patterns of serum from different species: (1) human, (2) dog, (3) sheep, (4) cow, (5) chicken, (6) rabbit, (7) pig, (8) guinea pig.

rat, antelope, deer, chicken, turtle, and shark. Excellent patterns have been obtained with soluble protein extracts of brain, liver, heart, spleen, kidney, and gingival tissue. Other protein mixtures which have been analyzed are egg white, wheat proteins, saliva, perilymph, cerebrospinal fluid, urine and colostrum.

Reagents and Apparatus

Gel Buffer Stock, *p*H 8·7.

Tris (hydroxymethyl) aminomethane, (Tris) .	36·25 gm.
Ethylenediaminetetraacetic acid . . .	1·0 gm.
Hydrochloric acid 1·0N	48·0 ml.
Distilled water to a final volume of . . .	1 l.

Gel Buffer. One part of stock buffer is diluted with 4 parts distilled water to prepare the buffer for starch gels.

Electrode Chamber Buffer, *p*H 8·3.

Glycine	22·6 gm.
Tris (hydroxymethyl) aminomethane . . .	4·7 gm.
Distilled water to a final volume of . . .	1 l.

As a tracking dye, 1·0 ml. of 0·5 per cent Bromophenol Blue in 1 per cent acetic acid is added per litre of buffer.

Starch. Starch-hydrolyzed for gel electrophoresis may be obtained from Connaught Medical Research Laboratories, Toronto, Ontario, Canada.

Stains. Naphthalene black and benzidine as previously described.

Apparatus. The Shandon Universal electrophoresis apparatus is used as the basic electrophoretic unit, coupled with a suitable constant current—constant voltage power supply such as the Vokam* or the Spinco Duostat.†

Gel Trays. Gels are poured in perspex trays 20 cm. long and 10 cm. wide with 8 individual channels 2 cm. wide and 3 mm. deep. The trays are constructed with removable endgates and are supported at each end of the channels so that the tray will stand by itself without touching the centre partition of the electrode chamber (Fig. 8.11). For the Shandon Universal chamber the tray supports should be 5 cm. high.

Electrical connections to the gels are made with two thicknesses of filter paper cut to fit the gel tray. The wicks supplied with Durrum paper electrophoresis cells (Spinco-Beckman 319–329) are excellent for this purpose.

Gel Slicing Tray. The slicing tray for the micro gels is constructed of perspex, outside dimensions 4·5 cm. by 15 cm. The gel channel is 14 cm. long, 2·5 cm. wide and 3 mm. deep (Fig. 8.12). It is convenient to have several perspex shims just slightly narrower than the channel so that they slide easily, and about 2 cm. longer than the tray so they may be easily inserted and removed. The shims may be 1 mm. thick but it is convenient to have several which are 2 mm. thick. A 3 mm. thick piece the same length as the tray and 3 cm. wide is used to hold the gel in place while slicing.

The gel cutter is an ordinary cheese cutter with the original wire replaced by fine gauge piano wire, or, better still, recorder wire, which is very thin and quite strong.

* Shandon Southern Instruments, Camberley, Surrey; Sewickly, Pa. 15143, U.S.A.

† Spinco Division, Beckman Instruments, Inc., Belmont, California.

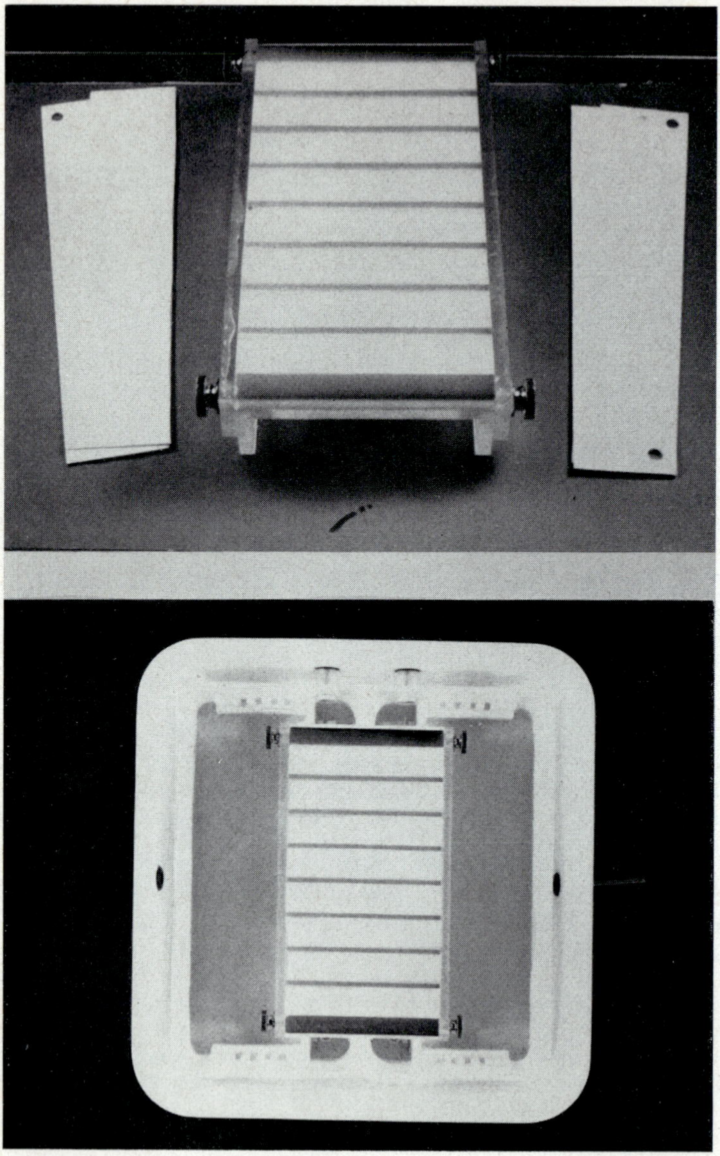

FIG. 7.12. *Top*: Gel tray and wicks. *Bottom:* Gel in migration chamber ready for electrophoresis.

Preparation of Gels. When a discontinuous buffer system such as Tris-Glycine is employed it is necessary to use slightly more starch than is recommended by Connaught for borate buffered gels. Usually 1·5 times the recommended amount is used, however, a good procedure is to do several trial runs to determine the optimum amount for each new batch of starch.

125 ml. of gel buffer and the proper amount of starch are thoroughly mixed in a 500 ml. Pyrex suction flask, cooked and degassed as described previously, p. 162. The hot, bubble-free starch is poured into the gel

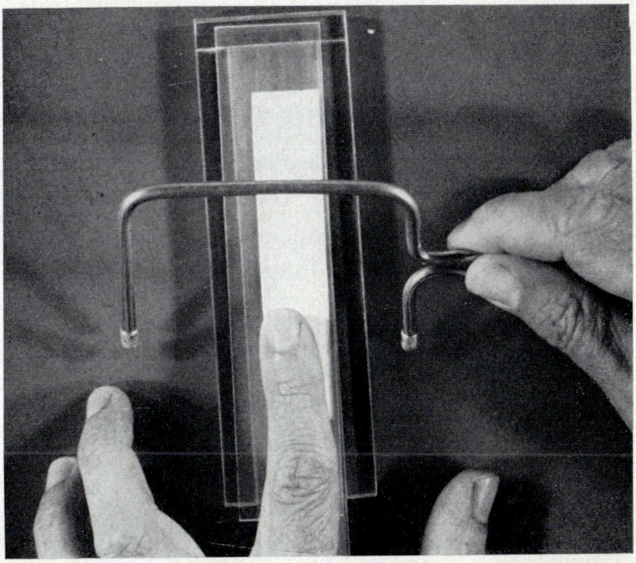

FIG. 7.13. Gel slicing tray, showing the procedure for holding the gel in place and slicing.

tray, making certain that all channels are filled to overflowing. The tray is then scraped with a straight-edged piece of plastic in the longer axis, i.e. at right angles to the gel channels. This removes excess starch and gives a smooth flat surface. When the gel has cooled to room temperature (about 30 minutes) it may be covered smoothly with a piece of Saran* Wrap a little larger than the tray. This is easily accomplished by starting the Saran at one edge of the tray and allowing it to settle slowly onto the starch. Gels may be used the day they are prepared but they are easier to handle and somewhat better resolution of patterns is obtained when they are poured, scraped, allowed to cool under a cover to exclude dust, and then placed in an airtight container and left overnight. The Saran is then applied just before the gel is to be used.

* The Dow Chemical Company, Midland, Michigan.

Procedures. Serum, hæmoglobin, or other protein extracts are introduced into the gels by wetting 2 × 15 mm. strips of Whatman 3 MM filter paper with the sample and inserting them in vertical cuts made with a razor blade broken to the proper width. The wet strip is laid on the blade and the cut is made about 3 cm. from the cathodic end of the gel. The starch is gently pressed back with the blade enough to allow the strip to be pushed down into the cut. The upper edge of the strip should be below the gel surface and care must be taken that no bubbles are left on the interfaces between strip and gel; otherwise streaked or distorted patterns may result. The endgates are now carefully removed so that the ends of the individual gels are not damaged. The filter paper wicks, cut to fit the tray are soaked in the migration chamber buffer and placed in position against the ends of the gels. The perspex end pieces are then replaced and the knurled nuts tightened sufficiently to hold the wicks snugly in place but not prevent buffer flow. The Saran, which has been laid back for sample insertion, is now replaced and the whole assembly is placed in the migration chamber. A constant current of 40 mA (about 200 V) is applied, and after 20 minutes the Saran is again laid back from the sample slots and, without turning off the current, the paper strips are carefully removed with fine tipped forceps. The cuts are closed with a glass rod, if necessary, making sure that all bubbles are removed before the Saran is replaced. The gel may be cooled for the remainder of the run by placing a clear plastic tray of ice water directly on the gels. Cooling is not absolutely necessary for serum or hæmoglobin separations but is desirable if enzymes or other labile proteins are to be subsequently examined.

As electrophoresis progresses a buffer front migrates from the cathode chamber through the gel and is clearly visible due to the added bromophenol blue. Albumin is stained by the dye, allowing one to observe the progress of the run. To ensure uniformity from run to run the stained albumin band is allowed to migrate a selected distance, usually 5·0 cm. in a gel of 10 cm. length. With the Tris-Glycine buffer this takes approximately an hour and a half.

At the conclusion of the run the gels are removed from their channels individually, placed in the slicing tray (Fig. 7.13) and cut with the wire cutter. Shims of the proper dimensions are placed under the gel in the slicing tray to give the desired thickness of slice. If several slices are to be cut from the same gel it is best to place several shims in the tray and remove them in succession after each cut, i.e. make cuts starting from the bottom of the gel. After cuts are all made the shims are replaced one at a time to raise each successive slice above the level of the tray for easy removal. The slices are transferred to a shallow tray or Petri dish of the proper size for staining. *Both* ends of each slice are identified *before* staining by marking it with a smooth-pointed applicator stick dipped in a little serum. The strips are stained for 5 minutes and are then destained with successive changes of solvent or by means of a continuous, circulating destainer. With the destainer patterns are clear enough for scanning in about 4 hours, otherwise destaining is not complete until the next day. A convenient destaining apparatus may be obtained from E.C. Apparatus Corporation.* Typical patterns obtained with the micromethod are seen in Fig. 7.14.

Hæmoglobin and Haptoglobin. Hæmoglobins may be run at the same time as serum samples and may be stained with either amido black or benzidine. The benzidine stain is more specific since it does not stain non-hæmoglobin proteins. Staining is complete with benzidine in about 5 minutes, (Fig. 7.15) after which the strips are washed in tap water to remove excess benzidine. If an opaque, white background is

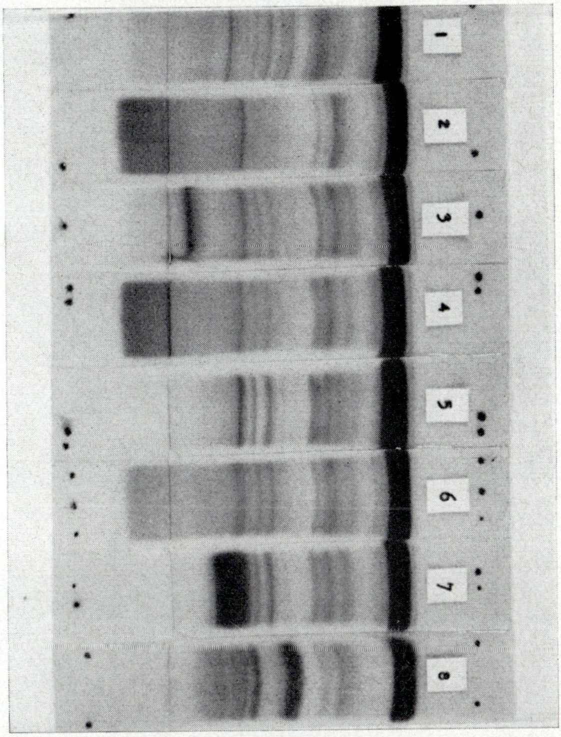

Fig. 7.14. Typical patterns obtained from human sera with the micro method: (1) normal human, (2) cirrhosis of liver, (3) Waldenstrom's macroglobulinæmia, (4) cryoglobulinæmia, (5) agammaglobulinæmia, (6) lupus erythematosus, (7) multiple myeloma (gamma G type), (8) multiple myeloma (gamma A type).

desired the stained gels may be immersed in the methanol-water-acetic acid destaining solvent for a few minutes.

Haptoglobins may be specifically visualized by adding a small amount of hæmoglobin to the serum before electrophoresis. 20 μl. of 5 per cent hæmoglobin solution are added to 1 ml. of serum, and a similar amount is added to water and used to determine the position of unbound-hæmoglobin in the patterns. After electrophoresis is complete the gels

* E.C. Apparatus Corporation, Swarthmore, Pennsylvania.

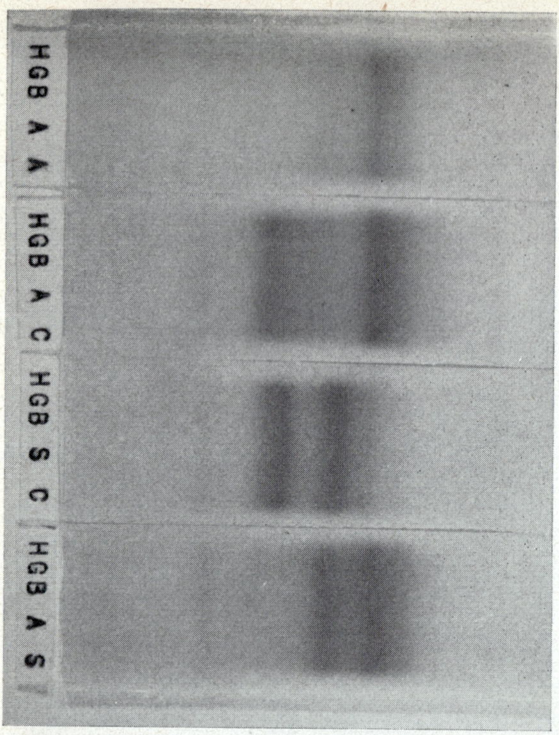

Fig. 7.15. Various hæmoglobin patterns, naphthalene black stain.

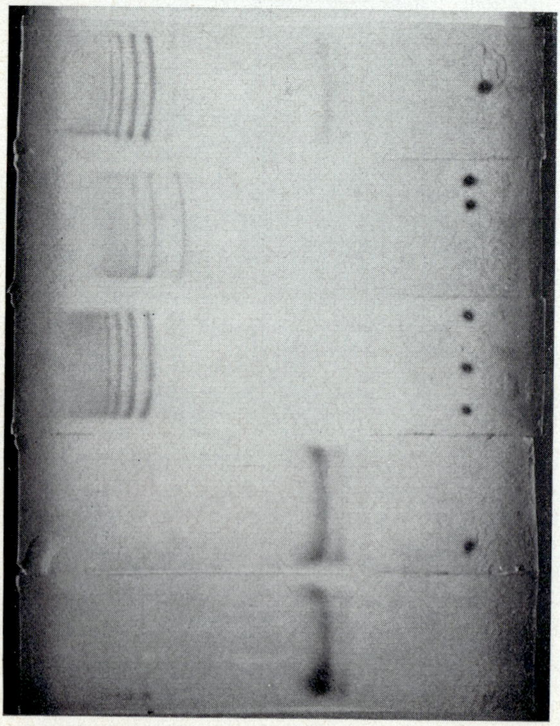

Fig. 7.16. Haptoglobin patterns, benzidine stain: (1) type 2–2, (2) type 2–1, (3) type 2–2, (4) type 2–2 (bands do not show in the photograph. This was a case of liver cirrhosis and the haptoglobin concentration was very low), (5) free hæmoglobin.

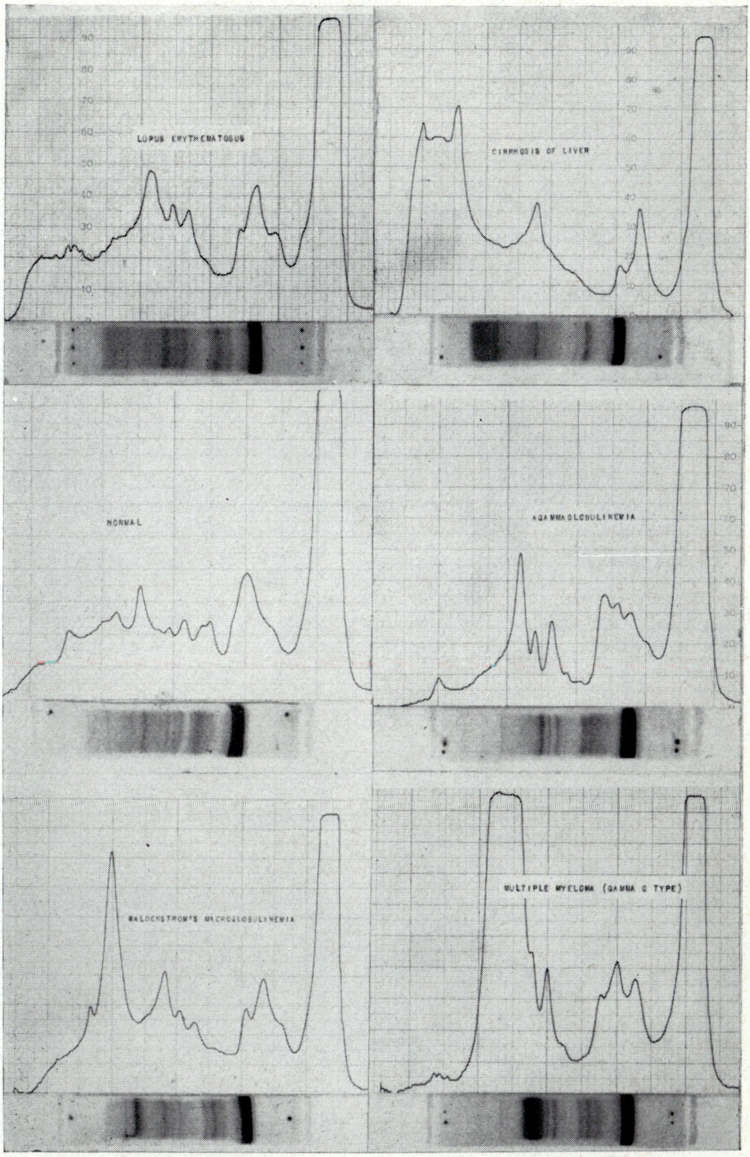

FIG. 7.17. Examples of densitometer patterns obtained with the Photovolt Densicord.

are cut as usual, the slices covered with the benzidine reagent, and allowed to stain until the different bands are well defined, but not smeared. The excess reagent is removed by washing with tap water. Immersion of the strips in destaining solvent improves the background for photography, (Fig. 7.16). When the benzidine stain is used gels are identified by marking with a little hæmoglobin just before staining.

Gel patterns may be scanned with a suitable densitometer such as the Densicord.* Scanning is performed on the destained gels without further clearing. Good resolution of the bands is obtained (Fig. 7.17), however exact quantitation is not possible since the gamma globulins extend from behind the point of sample application up to, or beyond the alpha 2 macroglobulins (slow alpha 2).

The starch gel patterns may be photographed for a permanent record if so desired. We have found the Polaroid MP-3 camera assembly to be most desirable for this purpose. Gel slices are placed on a white porcelain plate which has been previously moistened with several droplets of water to ensure easy positioning of the slices. Polaroid 3,000 speed/type 107 film is used with a 105 mm. (4″) f/4·5 lens. Exposures utilizing the 4 photoflood (75 w) bulbs have remained constant at 1/60 second at f22 for clear visibility of the starch gel bands. If additional copies of the patterns are desired at a later time copying of the photographs by the same unit is possible. Coloured photographs and transparencies for projection can also be made utilizing the same photographic unit except for the film type.

REFERENCES

1. Smithies, O. *Biochem. J.*, 1955, **61,** 629.
2. Poulik, M. D. *Nature*, 1957, **180,** 1477.
3. Ferguson, K. A., and Wallace, A. L. C. *Nature*, 1961, **190,** 629.
4. Smithies, O. *Arch. Biochem.*, 1962, Suppl. **1,** 125.
5*a* Dangerfield, W. G., and Faulkner, G. *Nature*, 1963, **200**, 388; 1964, **202,** 520.
5*b* Baur, E. W. *Nature*, 1964, **202,** 520.
6. Poulik, M. D., and Smithies, O. *Biochem. J.*, 1958, **68,** 636.
7. Koen, A. L., and Shaw, C. R. *Anal. Biochem.*, 1964, **9,** 495.
8. Marsh, C. L., Jolliff, C. R., and Payne, L. C. *Am. J. Clin. Path.*, 1964, **41,** 217.
9. Poulik, M. D. "Methods of Biochemical Analysis." 1966, **14,** 455.

CHAPTER 8

SEPARATION AND VISUALIZATION OF ENZYMES ON GELS*

M. J. Siciliano and C. R. Shaw

Early methods for the detection of enzymes after electrophoresis in semi-solid media such as starch gel and starch block involved cutting of the gel into small segments, eluting the enzyme from the segment, and assaying the eluate for enzyme activity by standard quantitative procedures. This was relatively laborious, depending upon the number of segments made, and resolution of the zones of activity was necessarily limited, so that slight differences in migration between two samples were probably often missed.

In 1957, Hunter and Markert first applied histochemical staining procedures to starch gels after electrophoresis for the *in-situ* demonstration of enzyme activities. The position of the enzyme is marked by a band or zone of stain directly in the gel. This visual display of enzymes in the gel has been termed a zymogram. The major advantages of the technique over earlier methods are speed of processing, and high resolution of the zones, given the appropriate gel condition. Thus, large numbers of enzyme extracts from many individual specimens can be processed in a single day (several hundred), and with the high resolution, slight differences in migration between two specimens can be easily detected by simple inspection of the gel. The high resolution has also facilitated the study of the multiple molecular forms of an enzyme, called isozymes if products of different genetic loci or allozymes if products of 2 different alleles at the same locus (Fig. 8.1). The zymogram technique has become a major tool in a number of the biological and biochemical fields, and has significantly advanced our knowledge in such areas as enzyme genetics, protein subunit structure, population genetics, molecular evolution, developmental biology, and clinical chemistry (see reviews by Markert and Whitt 1968, Shaw 1969, Siciliano *et al.*, 1974).

The two major limitations of the method are: (1) Only a relatively small number of enzymes can be studied, i.e. those for which histochemical staining methods are available; (2) results are not always accurately quantifiable. Of some one thousand enzymes which have been identified, fewer than fifty can now be demonstrated by the zymogram procedure. New staining methods are developed from time to time, but these have accumulated at a discouragingly slow rate. Techniques for quantitative evaluation of the gels have consisted

* Supported in part by Grants GM-19513 and GM-15597 from United States Public Health Service National Institutes of Health. Authors gratefully acknowledge the excellent technical services of Ms. Billie White and the helpful comments of David A. Wright, Ph.D., Biology Dept., M.D. Anderson Hospital and Tumor Institute, Houston, Texas.

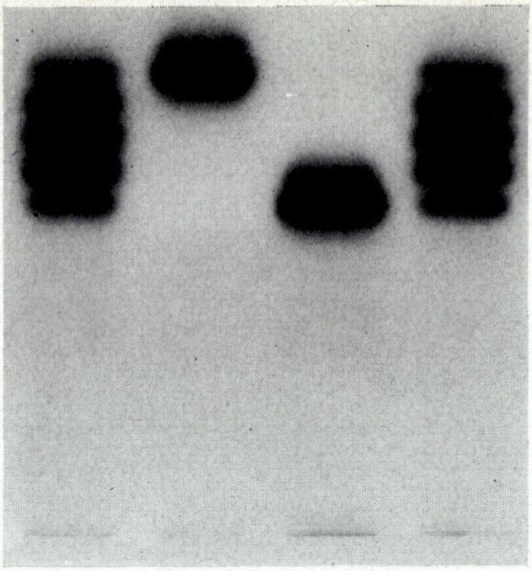

O

FIG. 8.1. Zymogram of glyceraldehyde-3-phosphate dehydrogenase (G_3PDH) following electrophoresis of brain homogenates from four platyfish originally placed in the 4 slots at the origin (O). Extracts in the middle 2 slots were taken from 2 fish homozygous for 2 different alleles—one specifying an anodally faster migrating G_3PDH than the other. The five banded patterns produced from homogenates in the end slots are from the offspring of the 2 fish in the middle slots and are typical of patterns from organisms heterozygous for electrophoretically different alleles which code for enzymes composed of 4 polypeptide subunits (explained in Wright, *et al.*, 1972).

mainly of some type of photoelectric scanning device, based on transmittance or reflectance. However, these have generally been little more accurate than simple visual inspection. The difficulty lies mainly in the lack of uniformity of the gel medium, and variable resolution of the bands. In this regard, polyacrylamide is usually superior to starch. Another approach to quantification is to excise the segment of the gel known to contain the enzyme, elute by a variety of methods, and carry out quantitative assay on the eluate. Again, there is difficulty in controlling the completeness of elution.

In this chapter we present the details of the methods of starch gel zymogram preparation currently in use in our laboratory. The histochemical staining procedures, though described particularly for application to starch gel slices could, upon modification, also be used to visualize enzymes run on other types of electrophoretic media, i.e. polyacrylamide or cellulose acetate.

Homogenate Preparation

Crude homogenates from any sort of biological material—body fluids, tissues, cell culture, or sub-cellular particles—may be suitable for electrophoretic analysis if properly prepared. In the preparation it must be remembered that, in most cases, visualization of the protein on the gel depends upon releasing it from its cellular component with sufficient enzyme activity and in sufficient concentration to give a histochemical reaction. Consequently all operations are carried out at not $>$ 4°C. For certain kinds of body fluids (e.g., urine), it is often beneficial to concentrate the sample. We have been successful using the A25 Minicon concentrator (Amicon Corp., Lexington, Mass., 02173, U.S.A.) for that purpose. Solid tissues as well as cells from culture after being washed in an isotonic solution, need to be mechanically disrupted. This is best accomplished by diluting the sample in a homogenizing medium (approximately 1:3, weight:volume) and rigorously grinding either manually in all glass or with a motorized teflon and glass homogenizer. Homogenizing medium contains 0·01 M tris-HCl, *p*H 7·5, 0·001 M β-mercaptoethanol and 0·001 M EDTA. Bacterial samples may be subjected to sonication after dilution. Homogenates are then cleared by centrifugation (30 minutes at 12,000 × g. is usually sufficient). Cleared supernatants may then be applied directly to gels.

A special problem is often presented by very small solid samples (<20 mg.) which are often difficult to recover after homogenization. Recovery can be facilitated by overdiluting for homogenization and after centrifugation, concentrating the supernatant as described above. Another problem is presented by extremely fatty material (e.g., fish livers, or yolky eggs). The lipids cause considerable "streaking" of the gels during electrophoresis. This can be avoided by taking care to decant only the aqueous phase of the sample after centrifugation.

Buffer Systems

The literature reveals an almost endless variety of both continuous and discontinuous buffer systems that can be used for electrophoresis. Since in our laboratory we are interested in examining many different proteins run on the same gel, it has been in our best interest not to have a different buffer system suited to each different protein studied. Instead, over the last several years we have worked to develop just a few buffer systems which would give clear resolution for many proteins. We have successfully reduced the number of buffer systems to the two described here. Every protein for which we list a histochemical staining reaction has been clearly resolved in one or the other of these systems. In each case the electrode buffer is used directly in the tanks at the anodal and cathodal ends of the gel. The gel buffer is mixed with the starch.

A. Tris–Citrate (TC)

1. *Electrode:* 0·13, M Tris–0·043 M Citrate, *p*H 7·0.

Tris	16·35 g.
Citric acid (granular monohydrate)	9·04 g.
Water	up to 1l.

Adjust *p*H upward with ION NaOH or downward with Conc. HCl for final *p*H of 7·0.

2. *Gel:* 0·009 M Tris–0·003 M Citrate, *p*H 7·0.

Dilute 40 ml of electrode buffer up to 600 ml with water.

B. Tris-Versene-Borate, (TVB)

1. *Electrode:* 0·5 M Tris–0·016 M Versene–0·65 M Borate, *p*H 8·0.

Tris	60·6 g.
Boric acid	40·0 g.
$Na_2EDTA–2H_2O$	6·0 g.
Water	up to 1 l.

Adjust pH upward with 10 N NaOH or downward with Conc. HCl for final pH of 8·0.

2. *Gel:* 0·05 M Tris–0·0016 M Versene–0·065 M Borate, pH 8·0. Dilute 60 ml. of electrode buffer up to 600 ml. with water.

Gel Preparation

Proper preparation of the gel is necessary for sharply resolved bands of enzyme activity after electrophoresis and for the production of a tough "rubbery" gel which allows itself to be sliced into as many as 10, 1 mm slices, each of which could be handled and stained for a different enzyme. In order to obtain these results we use 95 g. of Connaught starch and mix it with 600 ml. of gel buffer (one of two described above), in a 2000 ml. thick walled flask. This mixture is heated over a gas flame with constant shaking. First the mixture thickens and then begins to thin again as the mixture comes to a boil. At this point remove from the flame as overcooking produces fragile gels. If a cofactor (NAD, NADP, or ATP) is to supplement the gel, it is added here. The molten mixture is then degassed with a vacuum pump (to about 25 mm vacuum), 2 drops of β-mercaptoethanol are added and gently mixed in. The mixture is then slowly and carefully poured into the form avoiding the production of any bubbles (Fig. 8.2a). A variety of horizontal and verticle starch-gel forms are available (Buchler Instruments, Fort Lee, N. J., U.S.A.). The top of the form, containing the sample slot maker, is then pressed on, squeezing out the excess molten gel material (Fig. 8.2b.). After cooling, 30 minutes at room temperature followed by 30 minutes at 4°C, the ends and top (with the slot maker) of the form can be removed leaving the gel set in the remainder of the form.

Electrophoresis

Slots in the cooled gel may be better separated to prevent cross contamination of samples by laying a thin line of melted vasoline between them. Cleared supernatant samples are best filled into the slots with a Pasteur pipette that has been drawn out into a fine tip (Fig. 8.2c). After all the slots in the gel have been filled they can be sealed off by covering them with melted vasoline. The entire surface of the gel is then covered over with a clear plastic wrapping material (Saran Wrap, Dow Chemical Company, Midland, Michigan, 48640, U.S.A.). Gel is then placed in the verticle (or horizontal) apparatus (Buchler) located in a cold room or refrigerator (we use a sliding glass door, "beer cooler" type refrigerator). Filter paper wicks from the tanks containing the electrode buffer are applied to both ends (anodal and cathodal) of the gel (Fig. 8.3). In vertical systems, which we use primarily, the one end of the gel actually gets immersed in the anodal buffer tank. Electrodes are wired to a constant voltage power supply (e.g., Heathkit IPW-17, Heath Company, Benton Harbor, Michigan, 49022, U.S.A.) and a current of 400 V is applied. For most gels 1 cm thick and 30 cm long this will produce a current of approximately 35 milliamperes (ma.).

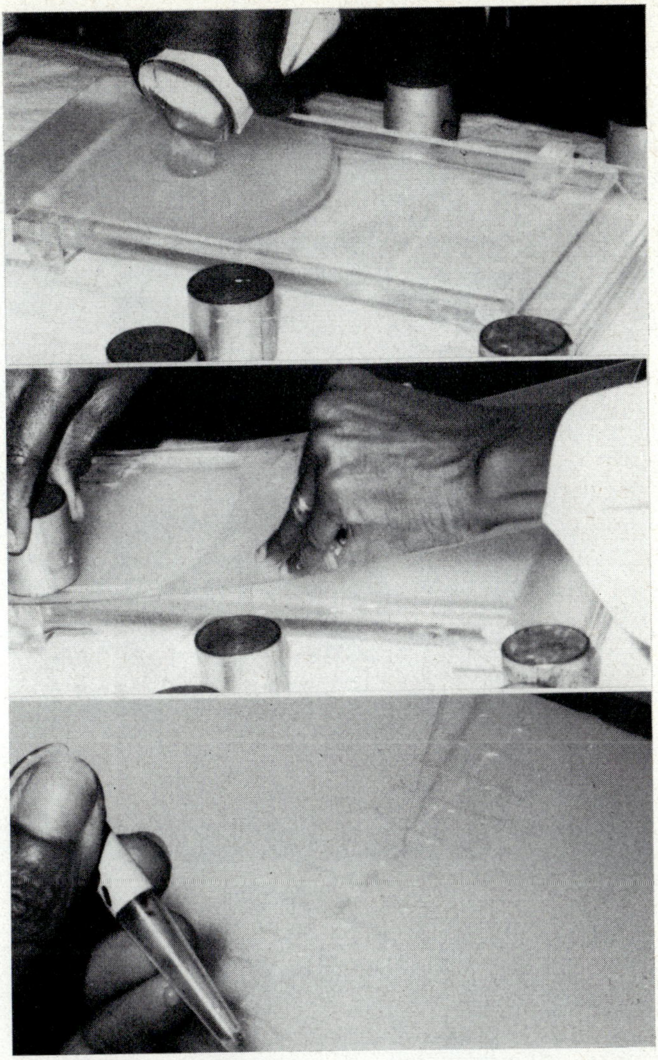

FIG. 8.2.a (*top*). Molten, degassed starch–buffer mixture is shown being poured into a gel form. Cylindrical objects around the edge are lead weights which will be placed on the form top after the gel is poured. b (*middle*). Gel form top containing the sample slot maker (held and hidden by the technicians left hand) is shown being pressed onto the freshly poured molten starch buffer mixture. Top is applied from one end and pressed on with the aid of lead weights toward the other end to squeeze out all excess molten gel material. c (*bottom*). After the top had been removed from the cooled gel, the sample slots, separated by vasoline strips, are shown being filled with a drawn out Pasteur pipette.

This current will be in proportion to the thickness of the gel and concentration of buffer each of which may vary in different cases. Therefore care should be taken not to exceed 35 ma. so as to avoid heating the gel. Where 35 ma. is exceeded the voltage should be reduced and the running time increased proportionately. We have gotten consistently good resolution of most separable proteins by running 5 hours at 400 V.

FIG. 8.3. Gel in its form (minus top) covered by clear plastic wrap has been placed vertically in the electrophoretic apparatus. Anodal buffer tanks, connected by filter paper wicks, are at the bottom of the apparatus and one end of the gel is resting in one of them. The cathodal buffer tanks, at the top rear, are connected to the top of the gel with filter paper wicks.

Slicing, Staining and Fixation

After electrophoresis the gel is removed from the form, trimmed and layed flat to a well washed, smooth surfaced table. Using piano wire strung tightly to a Buchler gel slicer, slices as thin as 1 mm. may be removed from the undersurface of the gel (Fig. 8.4a and b). As many as 10 slices may therefore be gotten from a gel 1 cm. thick. Each slice may then be put into a shallow staining dish and stained for a different protein (Fig. 8.4c).

The histochemical stains, recipes for which are listed below, are usually made up to 50 ml. in order to immerse the gel slice in the reaction mixture. However this is not always practical since for some stains, certain reagents, such as helper enzymes, are too expensive to be maintained at effective concentrations in large volumes. The classical way of approaching that problem has been to make up the reaction mixture

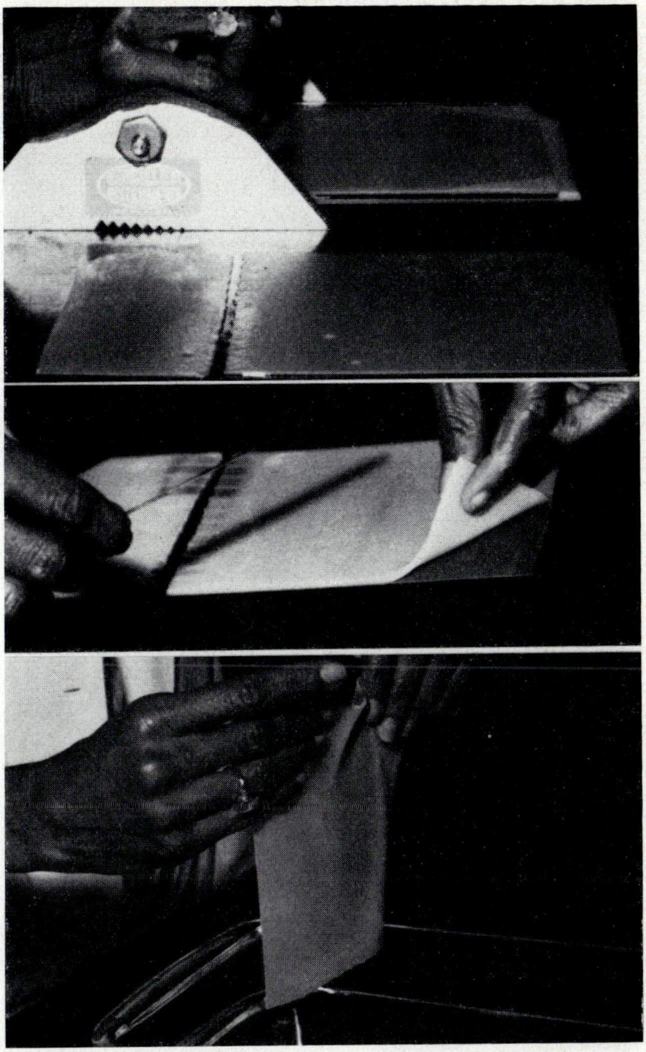

FIG. 8.4.a (*top*). Gel laid flat on a smooth, black formica, topped table and covered by a glass plate to add weight to help keep it pressed to the table top is shown being sliced by a gel slicer to cut slices of 1 mm thickness off the undersurface. In the foreground is such a slice prepared prior to the one shown being sliced. b (*middle*). After the slicer has been moved through the gel, the upper piece is shown being removed to be laid down on the table to be sliced again. c (*bottom*). A 1 mm. thick gel slice is shown being lifted into a pyrex dish for histochemical staining.

in 10 ml. of molten agar and pouring that over the cut surface of the gel. This brings the concentrated reagents into intimate contact with the proteins to be stained for and also prevents diffusion of the bands as the agar hardens on cooling. We have always had misgivings about dumping helper enzymes into molten agar and also in pouring hot material over the cut surfaces of our gels. Also the agar method is very messy. We have found that we can concentrate reagents in expensive stains and prevent band diffusion by making the reaction mixture up in a total of 6 ml. of fluid. This is just enough to flood the cut surface of the gel if applied dropwise with a Pasteur pipette. The mixture soaks directly into the gel where the reaction takes place and is easily visualized.

After staining, gels are fixed in one of the two fixatives listed below. Following fixation we have found it best to keep permanent records of the gels on 35 mm. film. Stains involving a fluorescent reaction cannot be fixed and must be photographed through a yellow filter (Wratten Gelatin Filter No. 2E, Eastman Kodak Company, Rochester, M. Y., 14650, U.S.A.). Our best all-round fixative is 50 ml. of the "alcohol gel wash" mixture shown below. It not only fixes the stain but toughens the gel and helps bleach out some of the background. For some reactions however it also dissolves the stained precipitate and is therefore unacceptable. In those cases the glycerin–water fix is used. It is good at preserving color but unfortunately does not toughen the gel as well as gel wash.

A. Alcohol Gel Wash

Ethanol	1000 ml.
Acetic acid	400 ml.
Glycerine	200 ml.
Water	800 ml.

B. Glycerine and Water

Glycerine	500 ml.
Water	500 ml.

We list here only those stain recipes which we use in our laboratory and are for proteins which are well resolved on the buffer systems indicated. If a reference for the method is listed in Shaw and Prasas (1970) it is not given here. If the protein requires a cofactor (NAD, NADP, ATP) in the gel, it is indicated. A protein not requiring a cofactor may be stained for on a slice off a gel containing a cofactor as long as the gel was made up in the appropriate buffer. Since cost is very often a factor in deciding which methods to include in a program, the approximate total price (in U. S. dollars) of the ingredients of each stain is included.

Most of the stain recipes call for a buffer to be used in the stain. We shall list the various stain buffers and their ingredients here. These buffers can be made up in amounts greater than needed in any one stain since they can be stored under refrigeration for long periods of time.

A. Stain buffer recipes

1. 0·2 M Tris–HCl, *p*H 8·0.

Tris	24·228 gms.
Water	up to 1 *l*.

Bring *p*H down to 8·0 with conc. HCl.

2. 0·02 M Tris–HCl, *p*H 7·0.

Same as above except keep adding HCl to *p*H 7·0

3. 0·05 M Citrate, *p*H 6·0.

Sodium citrate	1·47 g.
Water	100 ml.

Adjust *p*H downward with citric acid crystals or upward with 1N NaOH to *p*H 6.

4. 0·1 M Phosphate, *p*H 7.

$NaH_2PO_4.H_2O$	1·38 g.
Water	100 ml.

Adjust *p*H to 7·0 with 1 N NaOH or HCl.

5. 0·1 M Phosphate, *p*H 6·0.

$HaH_2PO_4.H_2O$ 27·8 g/l	460 ml.
$Na_2HPO_4.7\,H_2O$ 53·65 g/l	40 ml.
Water	500 ml.

Adjust *p*H to 6·0 with 1 N NaOH or HCl.

6. 0·5 M Phosphate, *p*H 7.

$NaH_2PO_4.H_2O$ (74·5 g./l.)	255 ml.
$Na_2HPO_4.7\,H_2O$ (131·7 g./l.)	245 ml.

Adjust *p*H to 7·0 with 1 N NaOH or HCl.

7. 0·05 M Citrate–0·05 M Phosphate, *p*H 4·0.

Na_2HPO_4 (anhydrous)	3·55 g.
Citric acid (anhydrous)	5·25 g.
Water	500 ml.

Adjust *p*H to 4·0 with 1 N NaOH or HCl.

With the exception of the soluble protein stain, all stains should be made up fresh—to be used within an hour. Dry ingredients of a stain can be put together first and kept in a freezer. They can then be combined with the wet ingredients just before use. Helper enzymes should not be added until the very last second (as the stain is being poured over the gel). Substrates solution (such as 1 M Na–DL–lactate, *p*H 7·0 for the LDH stain) as well as stain buffers may be made up as stock solutions and kept refrigerated for many months.

For some enzymes there are several staining methods which we have found effective—sometimes helping to clarify the differences between isozymes as well as underscoring the specificity of the bands observed.

B. Stain recipes

1. Acid phosphatase or AP–E.C. 3.1.3.2.

Buffer system: TC.

We use 2 methods for this enzyme:

a. (1) Combine in a 100 ml. flask:

Sodium α-napthyl acid phosphate	50 mg.
Black K salt	50 mg.
Water	50 ml.

(2) Pour above over cut surface of gel and incubate at 37°C for one hour. Red bands indicate zones of enzyme activity.

(3) Pour off stain, rinse once with water and fix in 50 ml. of glycerin–water (Fix "B").

(4) Cost: $0.25.

b. (1) Combine in 100 ml. flask:

Phenolpthalene mono- or di-phosphate	200 mg.
0·05 M Citrate, *p*H 6·0 (Stain buffer #3)	50 ml.

(2) Pour over cut surface of gel and incubate at 37°C for one hour.

(3) Pour liquid off gel and sprinkle with 7·4 N NH_4OH. Pink bands indicate zones of enzyme activity.

(4) Photograph immediately before bands diffuse away.

(5) Cost: $1.30.

2. Aconitase—E.C. 4.2.1.3.

Buffer system: TC.

a. Prepare substrate solution—0·1 M cis-aconitic acid, *p*H 8·0.

Cis-aconitic acid	1·74 g.
Water	100 ml.

Bring up to *p*H 8·0 with 1 N NaOH.

b. Combine in a 10 ml. tube:

Above substrate solution	2 ml.
NADP	15 mg.
NBT	15 mg.
PMS	1 mg.
$MgCl_2$	50 mg.
0·2 M Tris–HCl *p*H 8·0 (Stain buffer #1)	1 ml.
Water	2 ml.
Isocitrate dehydrogenase	72 units

c. Flood the above over the cut surface of the gel dropwise with a Pasteur pipette and let mixture soak into gel.

d. Incubate in the dark at 37°C and look for dark blue bands indicating zones of enzyme activity.

e. Rinse in water and fix in 50 ml. of alcohol-gel wash (Fix "A").

f. Cost: $3.00.

3. Adenosine deaminase or ADA—E.C. 3.5.4.4 (Spencer *et al.*, 1968)
Buffer system: TVB.
a. Combine in a 10 ml. tube:

Adenosine	40 mg.
NBT	15 mg.
PMS	5 mg.
Sodium arsenate	50 mg.
0·2 M Tris–HCl, *p*H 8·0 (Stain buffer #1)	1 ml.
Water	4 ml.
Xanthine oxidase	1·6 units
Nucleoside phosphorylase	5 units

b. Flood the above over the cut surface of the gel dropwise with a Pasteur pipette and let mixture soak into gel.
c. Incubate in the dark at 37°C and look for dark blue bands indicating zones of enzyme activity.
d. Rinse in water and fix in 50 ml. of alcohol gel wash (Fix "A").
e. Cost: $4.55.

4. Adenylate kinase or AK—E.C. 2.7.4.3.
Buffer system: TC.
a. Combine in a 50 or 100 ml. flask:

ADP	20 mg.
NADP	15 mg.
NBT	15 mg.
PMS	1 mg.
$MgCl_2$	20 mg.
Glucose	45 mg.
0·2 M Tris HCl, *p*H 8·0 (Stain buffer #1)	2 ml.
Water	18 ml.
Hexokinase	170 units
Glucose-6-phosphate dehydrogenase	80 units

b. Pour above over cut surface of the gel and incubate at 37°C in the dark. Look for dark blue bands indicating zones of enzyme activity.
c. Rinse in water and fix in 50 ml. of alcohol gel wash (Fix "A").
d. Cost: $5.70.

5. Alcohol dehydrogenase or ADH—E.C. 1.1.1.1.
Buffer system: TC or TVB.
a. Combine in a 100 ml. flask:

95% butanol, ethanol, octanol or any other alcohol	2 ml.
NAD	25 mg.
NBT	15 mg.
PMS	1 mg.
0·2 M Tris–HCl, *p*H 8·0 (Stain buffer #1)	7 ml.
Water	41 ml.

b. Pour above over cut surface of gel and incubate at 37°C in the dark. Look for dark blue bands indicating zones of enzyme activity.

c. Rinse in water and fix in 50 ml. of alcohol gel wash (Fix "A"). d. Cost: $0.70.

6. Aldolase—E.C. 4.1.2.7.

Buffer system: TVB.

a. Combine in a 100 ml. flask:

Fructose-1, 6-diphosphate (Tetrasodium salt)	275 mg.
NAD	25 mg.
NBT	15 mg.
PMS	1 mg.
Sodium arsenate	75 mg.
0·2 M Tris–HCl, *p*H 8·0 (Stain buffer #1)	10 ml.
Water	40 ml.
Glyceraldehyde-3-phosphate dehydrogenase	100 units

b. Pour above over cut surface of gel and incubate at 37°C in the dark. Look for dark blue bands indicating zones of enzyme activity.

c. Rinse in water and fix in 50 ml. of alcohol gel wash (Fix "A"). d. Cost: $2.35.

7. Catalase—E.C. 1.11.1.6.

Buffer system: TVB.

a. Make up solution "A"

3% H_2O_2	2·5 ml.
0·1 M Phosphate *p*H 7 (Stain buffer #4)	5 ml.
0·06 M $Na_2S_2O_3 . 5 H_2O$ (1·49 g./100 ml.)	3·5 ml.

b. Make up solution "B".

0·09 M KI (1·49 g./100 ml.)	25 ml.
Water	25 ml.

c. Soak gel in solution "A" at room temperature for 15 minutes.

d. Pour off "A" and rinse thoroughly in distilled water.

e. Add solution "B". Zones of enzyme activity are indicated by white bands on a blue background.

f. Rinse well in water and fix in 50 ml. of glycerine–Water (Fix "B"). g. Cost: pennies.

8. Enolase—E.C. 4.2.1.11 (Omenn and Cohen, 1971).

a. Combine in a 10 ml. tube:

2 Phospho-D-glycerate (Sodium salt)	25 mg.
NADH	15 mg.
ADP	2·5 mg.
$MgCl_2$	20 mg.
Fructose-1, 6 diphosphate (Tetrasodium salt)	1 mg.
0·2 M tris–HCl, *p*H 8·0 (Strain buffer #1)	1 ml.
Water	4 ml.
Pyruvate kinase	20 units
Lactate dehydrogenase	20 units

b. Flood the above over the cut surface of the gel dropwise with a Pasteur pipette and let mixture soak into the gel.

c. View under long wave UV (375 nm) and note even fluorescence over the entire surface of the gel. Zones of enzyme activity will appear as dark bands of defluorescence.

d. Photograph immediately through a yellow filter.

e. Cost: $4.00.

9. Esterase—E.C. 3.1.1.1.

Buffer system: TC or TVB.

a. Make up the following substrate solutions:

1. 1% α Napthyl butyrate

α Napthyl butyrate	1g.
Acetone	15 ml.
Water	85 ml.

2. 1% α Napthyl acetate

α Napthyl acetate	1 g.
Acetone	50 ml.
Water	50 ml.

3. 1% α Napthyl propionate

α Napthyl propionate	1 g.
Acetone	100 ml.

b. Combine 3 ml. of each of the first 2 substrates with 1·5 ml. of the third for final substrate solution.

c. Add to the final substrate solution the following:

Fast Blue RR	100 mg.
Water	38 ml.

Stain Buffer:

If TVB gel, 0·1 M phosphate, pH 6 (Stain buffer #5)	5 ml.
If TC gel, 0 5 M phosphate, pH 7 (Stain buffer #6)	5 ml.

d. Pour above over cut surface of the gel. Zones of esterase activity are indicated by dark brown bands.

e. Rinse in water and fix in 50 ml. of alcohol gel wash (Fix "A").

f. Cost: $0.50.

10. Fumarase—E.C. 4.2.1.2.

Buffer system: TC or TVB.

a. Combine the following into a 100 ml. flask:

Fumaric acid (Disodium salt)	385 mg.
NAD	40 mg.
NBT	15 mg.
PMS	1 mg.
0·2 M Tris–HCl, pH 8·0 (Stain buffer #1)	10 ml.
Water	40 ml.
Malate dehydrogenase	600 units

b. Pour above over cut surface of the gel and incubate in the dark at 37·C. Zones of enzyme activity are indicated by dark blue bands.
c. Rinse in water and fix in 50 ml. of alcohol gel wash (Fix "A").
d. Cost: $1.70.

11. α-Galactosidase—E.C. 3.2.1.22 (Beutler and Kuhl, 1972).
Buffer system: TC.
a. Combine:

4-Methylumbelliferyl-α-D-galactopyranoside .	10 mg.
0·05 M citrate–phosphate, *p*H 4·0 (Stain buffer #7)	2·5 ml.
Water	2·5 ml.

b. Pour above over cut surface of gel and incubate 45 minutes at 37°C.
c. Rinse gel with water and sprinkle 7·4 N NH_4OH over surface.
d. View under long wave UV (375 nm) light and note zones of enzyme activity as bands of fluorescence.
e. Photograph immediately through a yellow filter.
f. Cost: $1.90.

12. Glucose-6-phosphate dehydrogenase or G_6PDH – E.C. 1.1.1. 49
Buffer system: TVB + 20 mg. NADP added to gel mixture
a. Combine in a 100 ml. flask:

Glucose-6-phosphate (Disodium salt)	200 mg.
NADP	15 mg.
NBT	15 mg.
PMS	1 mg.
$MgCl_2$	50 mg.
0·2 M Tris–HCl, *p*H 8·0 (Stain buffer #1) . .	10 ml.
Water	40 ml.

b. Pour above over cut surface of gel and incubate in the dark at 37°C. Zones of enzyme activity appear as dark blue bands.
c. Rinse in water and fix in 50 ml. of alcohol gel wash (Fix "A").
d. Cost: $3.00.

13. Glucose phosphate isomerase or GPI (also called PGI)—E.C. 5.3.1.9.
Buffer system: TVB.
a. Combine in a 100 ml. flask:

Fructose-6-phosphate (sodium salt)	80 mg.
NADP	10 mg.
PMS	1 mg.
MTT	10 mg.
$MgCl_2$	40 mg.
0·2 M Tris–HCl, *p*H 8·0 (Stain buffer #1) . .	25 ml.
Water	25 ml.
Glucose-6-phosphate dehydrogenase	80 units

b. Pour above over cut surface of gel and incubate at 4°C. in the dark. Zones of enzyme activity appear as dark blue bands.
c. Rinse in water and fix in 50 ml. of glycerine–water (Fix "B").
d. Cost: $5.00.

14. β-Glucosidase—E.C. 3.2.1.21 (Beutler *et al.*, 1971).
Buffer system: TC or TVB.
a. Combine the following:

4-Methylumbelliferyl-β-D-glucopyranoside	10 mg.
0·05 M Citrate–phosphate, *p*H 4·0	2·5 ml.
Water	2·5 ml.

b. Pour above over cut surface of gel and incubate 45 minutes at 37°C.
c. Rinse gel with water and sprinkle 7·4 N NH_4OH over surface.
d. View under long wave UV (375 nm) light and note zones of enzyme activity as bands of fluorescence.
e. Photograph immediately through a yellow filter.
f. Cost: $0.20.

15. β-Glucuronidase—E.C. 3.2.1.31.
Buffer system: TVB.
a. Combine the following:

4-Methylumbelliferyl-β-D-glucuronide trihydrate	10 mg.
0·05 M Citrate–phosphate, *p*H 4·0 (Stain buffer #7)	2·5 ml.
Water	2·5 ml.

b. Pour above over cut surface of the gel and incubate 45 minutes at 37°C.
c. Rinse gel with water and sprinkle 7·4 N NH_4OH over surface.
d. View under long wave UV (375 nm) light and note zones of enzyme activity as bands of fluorescence.
e. Photograph immediately through a yellow filter.
f. Cost: $1.00.

16. Glutamate–oxaloacetate–transaminase or GOT—E.C. 2.6.1.1. (De Lorenzo and Ruddle, 1970).
Buffer system: TVB or TC.
a. Combine the following at room temperature, in the following sequence (each must be dissolved before adding the next) while stirring:

0·2 M Tris–HCl, *p*H 8·0 (Stain buffer #1)	50 ml.
Pyridoxal-5′-phosphate	1 mg.
L-Aspartic acid	200 mg.
α-Ketoglutaric acid	100 mg.
Fast blue BB	300 mg.

b. If necessary, adjust *p*H up to between 7·8 and 8·0 with 1 N NaOH

c. Pour above over cut surface of gel and incubate at 37°C until zones of enzyme activity appear as brown bands.
d. Rinse in water and fix in 50 ml. of glycerine–water (Fix "B").
e. Cost: $0.20.

17. Glutamate–pyruvate–transaminase or GPT—E.C. 2.6.1.2 (Chen *et al.*, 1972).

Buffer system: TVB.

a. Combine in a 10 ml. tube:

NADH	15 mg.
L-Alanine	20 mg.
α-Ketoglutaric acid	10 mg.
0·2 M Tris–HCl, *p*H 8·0 (Stain buffer #1)	1 ml.
Water	4 ml.
Lactate dehydrogenase	150 units

b. Flood the above over the cut surface of the gel dropwise with a Pasteur pipette and let mixture soak into the gel.
c. View under long wave UV (375 nm) and note even fluorescence over the entire surface of the gel. Zones of enzyme activity will appear as dark bands of defluorescence.
d. Photograph immediately through a yellow filter.
e. Cost: $0.60.

18. Glycerate-2-dehydrogenase or G_2DH—E.C. 1.1.1.29 (Siciliano, in prep.).

Buffer system: TC.

a. Combine in a 100 ml. flask:

DL-Glyceric acid (Hemi-calcium salt)	100 mg.
NAD	25 mg.
NBT	15 mg.
PMS	1 mg.
0·2 M Tris–HCl, *p*H 8·0 (Stain buffer #1)	10 ml.
Water	40 ml.

b. Pour above over cut surface of gel and incubate in the dark at 37°C until dark blue bands indicating zones of enzyme activity appear. LDH will also react with this substrate so the next slice off the gel should be stained for LDH as a control.
c. Rinse with water and fix in 50 ml. of alcohol gel wash (Fix "A").
d. Cost: $0.75.

19. α-Glycerophosphate dehydrogenase or α GPDH—E.C. 1.1.1.8.

Buffer system: TC or TVB.

a. Prepare substrate solution—1·0 M Na-α-glycerophosphate, *p*H 7·0 (21·6 g./100 ml. H_2O, adjust *p*H down to 7·0 with 1 N HCl).

b. Combine in a 100 ml. flask:

Above substrate solution	5 ml.
NAD	25 mg.
NBT	15 mg.
PMS	1 mg.
0·2 M Tris–HCl, *p*H 8·0 (Stain buffer #1)	10 ml.
Water	35 ml.

c. Pour above over cut surface of gel and incubate in the dark at 37°C. Zones of enzyme activity appear as dark blue bands.

d. Rinse with water and fix in 50 ml. of alcohol gel wash (Fix "A").

e. Cost: $1.05.

20. Glyceraldehyde-3-phosphate dehydrogenase or G_3PDH—E.C. 1.2.1.12 (Wright *et al.*, 1972).

Buffer system: TC or TVB with 40 mg. NAD added to gel mixture.

a. Prepare substrate solution fresh each day:

(1) Combine:

Fructose-1, 6-diphosphate (Tetrasodium salt)	270 mg.
0·2 M Tris–HCl, *p*H 8·0 (Stain buffer #1)	2 ml.
Water	3 ml.
Aldolase	100 units

(2) Incubate above at 37°C for 30–60 minutes.

b. Combine in a 100 ml. flask:

Above substrate solution	5 ml.
NAD	25 mg.
NBT	15 mg.
PMS	1 mg.
Sodium arsenate	75 mg.
0·2 M Tris–HCl, *p*H 8·0 (Stain buffer #1)	10 ml.
Water	35 ml.

c. Pour above over cut surface of gel and incubate in the dark at 37°C. Zones of enzyme activity appear as dark blue bands.

d. Rinse in water and fix in 50 ml. of alcohol gel wash (Fix "A").

e. Cost: $2.60.

21. Hexoseaminidase or HA—E.C. 3.2.1.30 (Okada *et al.*, 1971).

Buffer system: TVB.

a. Combine the following:

4-Methylumbelliferyl-2-acetamido-2-deoxy-β-D-glucopyranoside	10 mg.
0·05 M Citrate–phosphate, *p*H 4·0 (Stain buffer #7)	2·5 ml.
Water	2·5 ml.

b. Pour above over cut surface of gel and incubate 45 minutes at 37°C.

c. Rinse gel with water and sprinkle 7·4 N NH_4OH over surface.

d. View under long wave UV (375 nm) light and note zones of enzyme activity as bands of fluorescence.
e. Photograph immediately through a yellow filter.
f. Cost: $0.50.

22. Isocitrate dehydrogenase—E.C. 1.1.1.42.
Buffer system: TVB + 20 mg. NADP added to gel mixture.
a. Prepare 0·1 M Sodium isocitrate (trisodium salt), *p*H 7·0 (2·58 g./100 ml. water—bring *p*H down to 7·0 with 1 N HCl).
b. Combine in a 100 ml. flask:

Above substrate solution	8 ml.
NADP	15 mg.
NBT	15 mg.
PMS	1 mg.
$MgCl_2$	50 mg.
0·2 M Tris–HCl, *p*H 8·0 (Stain buffer #1)	10 ml.
Water	32 ml.

c. Pour above over cut surface of gel and incubate in the dark at 37°C. Zones of enzyme activity appear as dark blue bands.
d. Rinse in water and fix in 50 ml. of alcohol gel wash (Fix "A").
e. Cost: $3.40.

23. Lactate dehydrogenase or LDH—E.C. 1.1.1.27.
Buffer system: TC.
a. Prepare substrate solution 1 M Sodium-DL-lactate, *p*H 7·0.
(1) Dissolve 6·07 g $Na_2CO_3 . H_2O$ in 50 ml. of water.
(2) Set above in ice bath and while stirring slowly add (dropwise) 10·6 ml. of 85% DL-lactate.
(3) Add water to a total volume of 100 ml.
b. Combine in a 100 ml. flask:

Above substrate solution	5 ml.
NAD	25 mg.
NBT	15 mg.
PMS	1 mg.
0·2 M Tris–HCl, *p*H 8·0 (Stain buffer #1)	10 ml.
Water	35 ml.

c. Pour above over cut surface of gel and incubate in the dark at 37°C. Zones of enzyme activity appear as dark blue bands.
d. Rinse gel in water and fix in 50 ml. of alcohol gel wash (Fix "A") .
e. Cost: $0.70.

24. Malate dehydrogenase or MDH—E.C. 1.1.1.37.
Buffer system: TC.
a. Prepare substrate solution—1 M Sodium-L-malate, *p*H 7·0.

(1) Dissolve 12·15 g. $Na_2CO_3 . H_2O$ in 50 ml. H_2O.
(2) Set above in ice bath and while stirring slowly add 13·4 g. of L-malic acid.
(3) Add water to total volume of 100 ml.

b. Combine in a 100 ml. flask:

Above substrate solution	5 ml.
NAD	25 mg.
NBT	15 mg.
PMS	1 mg.
0·2 M Tris–HCl, *p*H 8·0 (Stain buffer #1)	10 ml.
Water	35 ml.

c. Pour above over cut surface of gel and incubate in the dark at 37°C. Zones of enzyme activity appear as dark blue bands.
d. Rinse in water and fix in 50 ml. of alcohol gel wash (Fix "A").
e. Cost: $0·75.

25. Malic enzyme or ME—E.C. 1.1.1.40.

Buffer system: TVB + 20 mg. NADP added to gel mixture.

a. Combine in a 100 ml. flask:

1 M Na-L-malate *p*H 7·0 as prepared for MDH (Stain 24)	5 ml.
NADP	15 mg.
NBT	15 mg.
PMS	1 mg.
$MgCl_2$	50 mg.
0·2 M Tris–HCl, *p*H 8·0 (Stain buffer #1)	10 ml.
Water	35 ml.

b. Pour above over cut surface of the gel and incubate in the dark at 37°C. Zones of enzyme activity appear as dark blue bands.
c. Rinse in water and fix in 50 ml. of alcohol gel wash (Fix "A").
d. Cost: $2.00.

26. Mannose-6-phosphate isomerase or MPI—E.C. 5.3.1.8 (Nichols *et al.*, 1973).

Buffer system: TVB.

a. Combine in a 100 ml. flask:

Mannose-6-phosphate (Disodium salt)	50 mg.
NADP	15 mg.
NBT	15 mg.
PMS	1 mg.
$MgCl_2$	50 mg.
02 M Tris–HCl, *p*H 8·0 (Stain buffer #1)	10 ml.
Water	40 ml.
Glucose (or hexose) phosphate isomerase	100 units
Glucose-6-phosphate dehydrogenase	80 units

b. Pour above over cut surface of the gel and incubate in the dark at 37°C. Zones of enzyme activity appear as dark blue bands. Prolonged staining will also bring up bands of 6 phosphogluconate dehydrogenase (6PGD) in material where that enzyme is strong. A second slice off the same gel should therefore be stained for 6PGD as a control for MPI.

c. Rinse with water and fix in 50 ml. of alcohol gel wash (Fix "A").

d. Cost: $9.30.

27. α-Mannosidase—E.C. 3.2.1.24 (Poenaru and Dreyfus, 1973).
Buffer system: TVB.

a. Combine the following:

4-Methylumbelliferyl-α-D-Mannopyranoside . .	10 mg.
0·05 M Citrate–phosphate, *p*H 4·0 (Stain buffer #7)	2·5 ml.
Water	2·5 ml.

b. Pour over cut surface of gel and incubate 45 minutes at 37°C.

c. Rinse gel with water and sprinkle 7·4 N NH_4OH over surface.

d. View under long wave UV (375 nm) light and note zones of enzyme activity as bands of fluorescence.

e. Photograph immediately through a yellow filter.

f. Cost: $2.00.

28. Nucleoside phosphorylase or NP—E.C. 2.4.2.1 (Spencer *et al.*, 1968).
Buffer system: TVB.

a. Combine in a 100 ml. flask:

Inosine	100 mg.
NBT	15 mg.
PMS	5 mg.
Sodium arsenate	100 mg.
0·2 M Tris–HCl, *p*H 8·0 (Stain buffer #1) . .	10 ml.
Water	40 ml.
Xanthine oxidase	1·6 units

b. Pour above over cut surface of gel and incubate in the dark at 37°C. Zones of enzyme activity appear as dark blue bands.

c. Rinse with water and fix in 50 ml. of alcohol gel wash (Fix "A").

d. Cost: $1.80.

29. Peptidase—E.C. 3.4.3.1. (Lewis and Harris, 1967).
Buffer system: TVB.

a. Combine the following in a 100 ml. flask and have all components dissolved before adding substrate:

0·2 M Tris–HCl, *p*H 8·0 (Stain buffer #1)	25 ml.
Water	25 ml.
$MgCl_2$	50 mg.
o-Dianisidine (dihydrochloride)	50 mg.
Peroxidase	1500 units
Snake venom (*Crotalus adamanteus*)	10 mg.

b. Add 80 mg. of a di- or tri-peptide as a substrate. We have been most successful in resolving the greatest number of peptidase loci products with the following 3 substrates on different gel slices:
   Glycyl-L-leucine
   L-Leucyl-L-alanine
   L-Leucyl-L-glycyl-glycine.

c. Pour the above over cut surface of gel and incubate overnight (or at least 6 hours) at room temperature. Zones of enzyme activity appear as brown bands.

d. Rinse with water and fix in 50 ml. of glycerine–water (Fix "B").

e. Cost: $3.00.

30. Peroxidase (E.C. 1.11.1.7) **and Catalase** (E.C. 1.11.1.6).

Buffer system: TVB.

a. Add 1 N HCl to 50 ml. of electrode buffer until *p*H reaches 6·0. Soak gel in this solution for 45 minutes at 4°C and pour off.

b. Pour on solution containing:

KI	1 g.
Water	50 ml.
Glacial acetic acid	1 ml.

c. Soak gel in above for 60 seconds and wash thoroughly (3 rinses) with distilled water and add the following solution:

Water	50 ml.
3% H_2O_2	1 ml.

d. Incubate at 37°C until peroxidase appears as dark blue bands and catalase as white bands on a light blue background.

e. Photograph immediately.

f. Cost: $0.20.

31. Phosphoglucomutase or PGM—E.C. 2.7.5.1.

Buffer system: TVB.

a. Combine in a 100 ml. flask:

Glucose-1-phosphate (Disodium salt)	300 mg.
NADP	15 mg.
MTT	20 mg.
PMS	1 mg.
$MgCl_2$	50 mg.
0·2 M Tris–HCl, *p*H 7·0 (Stain buffer #2)	10 ml.
Water	40 ml.
Glucose-6-phosphate dehydrogenase	80 units

b. Pour above over cut surface of gel and incubate in the dark at 37°C. Zones of enzyme activity appear as dark blue bands.
c. Rinse with water and fix in 50 ml. of glycerine–water (Fix "B").
d. Cost: $5.00.

32. 6-Phosphogluconate dehydrogenase or 6PGD—E.C. 1.1.1.44.

Buffer system: TVB or TC, 20 mg. NADP added to gel mixture.
a. Combine in a 100 ml. flask:

6-Phosphogluconic Acid (Trisodium salt)	100 mg.
NADP	15 mg.
NBT	15 mg.
PMS	1 mg.
$MgCl_2$	50 mg.
0·2 M Tris–HCl, *p*H 8·0 (Stain buffer #1)	10 ml.
Water	40 ml.

b. Pour above over cut surface of gel and incubate in the dark at 37°C. Zones of enzyme activity appear as dark blue bands.
c. Rinse with water and fix in 50 ml. of alcohol gel wash (Fix "A").
d. Cost: $7.00.

33. Phosphoglycerate kinase or PGK—2.7.2.3 (Omenn and Cohen, 1971).

Buffer system: TVB.
a. Combine in a 10 ml. tube:

3 Phospho-glyceric acid (Sodium salt)	10 mg.
NADH	15 mg.
ATP	10 mg.
$MgCl_2$	5 mg.
EDTA	1 mg.
0·2 M Tris–HCl, *p*H 8·0 (Stain buffer #1)	1 ml.
Water	4 ml.
Glyceraldehyde-3-phosphate dehydrogenase	100 units

b. Flood the above over the cut surface of the gel dropwise with a Pasteur pipette and let mixture soak into the gel.
c. View under long wave UV (375 nm) and note even fluorescence over the entire surface of the gel. Zones of enzyme activity will appear as bands of defluorescence.
d. Photograph immediately through a yellow filter.
e. Cost: $0.75.

34. 2,3-Phosphoglycerate mutase or PGAM—E.C. 5.4.2.1 (Omenn and Cohen, 1971).

Buffer system: TVB.

a. Combine in a 10 ml. tube:

2 Phospho-glyceric acid (Sodium salt)	12·5 mg.
NADH	15 mg.
ATP	10 mg.
$MgCl_2$	20 mg.
EDTA	1 mg.
0·2 M Tris–HCl, *p*H 8·0 (Stain buffer #1)	1 ml.
Water	4 ml.
Phosphoglycerate kinase	320 units
Glyceraldehyde-3-phosphate dehydrogenase	100 units

b. Flood the above over the cut surface of the gel dropwise with a Pasteur pipette and let mixture soak into the gel.
c. View under long wave UV (375 nm) and note even fluorescence over the entire surface of the gel. Zones of enzyme activity will appear as bands of defluorescence.
d. Photograph immediately through a yellow filter.
e. Cost: $2.70.

35. Soluble protein.

Buffer system: TVB.

a. Make up the following stain solution:

Ponceau S	2 g.
7·5% Acetic acid	150 ml.
Alcohol gel wash (Fix "A")	50 ml.

b. Pour 50 ml. of above over cut surface of the gel and incubate several hours (overnight is OK) at room temperature.
c. Pour off stain back into flask since it can be reused several times.
d. Cover gel with alcohol gel wash (Fix "A") and allow to soak several hours.
e. Pour off fluid and repeat "d" several times until background of gel is bleached white.
f. Zones of soluble protein can be seen as pink bands.
g. Cost: Pennies.

36. Pyruvate kinase or PK—E.C. 2.7.1.40 (Omenn and Cohen, 1971).

Buffer system: TVB + 66 mg. ATP added to gel mixture and 66 mg. ATP/500 ml. of buffer in the cathodal buffer box.

a. Combine in a 10 ml. tube:

Phospho(enol) pyruvate (Trisodium salt)	25 mg.
ADP	2·5 mg.
$MgCl_2$	5 mg.
NADH	15 mg.
Fructose-1,6-disphosphate	1 mg.
EDTA	1 mg.
KCl	25 mg.
0·2 M Tris–HCl, *p*H 8·0 (Stain buffer #1)	1 ml.
Water	4 ml.
Lactate dehydrogenase	150 units

b. Flood the above over the cut surface of the gel dropwise with a Pasteur pipette and let mixture soak into the gel.
c. View under long wave UV (375 nm) and note even fluorescence over the entire surface of the gel. Zones of enzyme activity will appear as bands of defluorescence.
d. Photograph immediately through a yellow filter.
e. Cost: $2.00.

37. Tetrazolium oxidase or TO.

Buffer system: TVB.

a. Combine in a 100 ml. flask:

NADP	15 mg.
NBT	15 mg.
PMS	5 mg.
$MgCl_2$	50 mg.
0·2 M Tris–HCl, *p*H 8·0 (Stain buffer #1)	10 ml.
Water	40 ml.

b. Pour over cut surface of the gel and incubate under a lamp at room temperature. Background of the gel will turn a light blue and zones of enzyme activity appear as white bands.
c. Rinse in water and fix in 50 ml. of alcohol gel wash (Fix "A").
d. Cost: $2.00.

38. Triosephosphate isomerase or TPI—E.C. 5.3.1.1.

Buffer system: TVB.

a. Prepare substrate (dehydroxyacetone phosphate) fresh:
(1) Combine:

NAD	25 mg.
1 M Sodium-α-glycerophosphate, *p*H 7·0 (as made up in α GPDH stain—#19)	5 ml.
1 M Sodium pyruvate (11 g./100 ml. water)	5 ml.
0·2 M Tris–HCl, *p*H 8·0 (Stain buffer #1)	2 ml.
Water	18 ml.
α Glycerophosphate dehydrogenase	20 units
Lactate dehydrogenase	72 units

(2) Incubate at 37°C for 2 hours.
(3) Adjust *p*H to 2·0 for 5 minutes with 1 N HCl to inactivate enzymes.
(4) Readjust *p*H to 7·0 with Tris crystals.

b. Combine the following in a 100 ml. flask:

Above substrate solution	30 ml.
NAD	25 mg.
NBT	15 mg.
PMS	1 mg.
Sodium arsenate	100 mg.
0·2 M Tris–HCl, *p*H 8·0 (Stain buffer #1)	10 ml.
Water	10 ml.
Glyceraldehyde-3-phosphate dehydrogenase	820 units

c. Pour above over cut surface of gel and incubate in the dark at 37°C. Zones of enzyme activity appear as dark blue bands.

d. Rinse in water and fix in 50 ml. of alcohol gel wash (Fix "A").

e. Cost: $4.00.

REFERENCES

Beutler, E., and Kuhl, W. "Biochemical and Electrophoretic Studies of α-galac. tosidase in Normal Man, in Patients with Fabry's Disease, and in Equidae. *Amer. J. Hum. Genet.*, 1972, **24**, 237–249.

Beutler, E., Kuhl, W., Trinidad, F., Teplitz, R., and Nadler, H. "β-Glucosidase Activity in Fibroblasts from Homozygotes and Heterozygotes for Gaucher's Disease," *Amer. J. Hum. Genet.*, 1971, **23**, 62–66.

Chen, S. H., Giblett, E. R., Anderson, J. E., and Fossum. B. L. G. "Genetics of Glutamic–pyruvic Transaminase: Its Inheritance, Common and Rare Variants, Population Distribution, and Differences in Catalytic Activity," *Ann. Hum. Genet., Lond.*, 1972, **35**, 401–409.

DeLorenzo, R. J., and Ruddle, F. H. "Glutamate Oxalate Transaminase (GOT) Genetics in *Mus Musculus*: Linkage, Polymerphism, and Phenotypes of the Got-2 and Got-1 Loci," *Biochem. Genet.*, 1970, **4**, 259–273.

Hunter, R. L., and Markert, C. L. "Histochemical Demonstration of Enzymes Separated by Zone Electrophoresis in Starch Gels," *Science*, 1957, **125**, 1294–1295.

Lewis, W. H. P., and Harris, H. "Human Red Cell Peptidases", *Nature*, 1967, **215**, 351–355.

Markert, C. L., and Whitt, G. S. "Molecular Varieties of Isozymes," *Experientia*, 1968, **24**, 977–1088.

Nichols, A., Chapman, M., and Ruddle, F. H. "Polymorphism and Linkage for Mannosephosphate Isomerase in *Mus Musculus*," *Biochem. Genet.*, 1973, **8**, 47–53.

Okada, S., Veath, M. L., Lerov, J., and O'Brien, J. S. "Ganglioside GM_2 Storage Diseases: Hexosaminidase Deficiencies in Cultured Fibroblasts," *Amer. J. Hum. Genet.*, 1971, **23**, 55–61.

Omenn, G. S., and Cohen, P. T. W. "Electrophoretic Methods for Differentiation Glycolytic Enzymes of Mouse and Human Origin, *In Vitro*," 1971, **7**, 132–139.

Poenaru, L., and Dreyfus, J. C. "Electrophoretic Heterogeneity of Human α-mannosidase," *Biochim. Biophys. Acta*, 1973, **303**, 171–174.

Shaw, C. R. "Isozymes: Classification, Frequency and Significance." International review of Cytology, 1060, **25**, 207 331.

Shaw, C. R., and Prasad, R. "Starch Gel Electrophoresis of Enzymes—A Compilation of Recipes," *Biochem. Genet.*, 1970, **4**, 297–320.

Siciliano, M. J., Wright, D. A., and Shaw, C. R. "Biochemical Polymorphisms in Animals as Models for Man," *Prog. in Med. Genet.*, 1974, **10**, In press.

Spencer, N., Hopkinson, D. A., and Harris, H. "Adenosine Deaminase Polymorphism in Man," *Ann. Hum. Genet., Lond.*, 1968, **32**, 9–14.

Wright, D. A., Siciliano, M. J., and Baptist, J. N. "Genetic Evidence for the Tetramer Structure of Glyceraldehyde-3-Phosphate Dehydrogenase," *Experientia*, 1972, **28**, 889.

CHAPTER 9

ACRYLAMIDE GEL DISC ELECTROPHORESIS

SECTION 1: TECHNIQUES OF DISC ELECTROPHORESIS

Ivor Smith

THE technique of disc electrophoresis is the most recent addition to the many variations of electrophoresis in current, routine use. It is so named because the bands or zones stack up as a series of concentrated discs at the beginning of the procedure (see later) and not because of any discontinuity of buffer systems which may be used in the technique or because the proteins are located as a series of discs at the end of the experiment.

Disc electrophoresis possesses a number of advantages over the starch gel technique which it has partially replaced. Amongst these advantages are: ease of preparation of the gel, relative inertness of gel components, greater simplicity of technique, much quicker running time and less man hours required per experiment. However, the greatest advantage of the technique is that the gel can be tailored so that the pore size of the gel can be varied in a known manner. Thus it is as simple to produce a gel of 3 per cent concentration as one of 5, 7·5, 10 or more per cent concentration. Such increases in gel concentration produce a decrease in pore size with a consequent exclusion from the gel of molecules of decreasing molecular size or dimensions. This facility enables the gel to be used not only as a routine analytical tool based on a fixed concentration but as a versatile molecular sieve in which larger but less interesting molecules can be screened out, molecular weights determined, etc. Until now, disc gels have been used mainly as routine, fixed concentration gels as, for example, in the analysis of sera from different diseases in order to establish disease patterns. As interest turns more to the examination of particular bands, i.e. abnormal protein or other components, then one can imagine that tailored gels with very fine differences in concentration will be used, particularly on the preparative scale to isolate such bands for further chemical study.

The technique of disc gel electrophoresis was originally described by Ornstein and Davis.[1,2] A brief discussion of the original technique and the underlying basis of its success will be given as this also permits the introduction of all the necessary terms.

The following mixture is prepared:

Monomer (acrylamide, $CH_2{=}CH.CONH_2$, 20–40 per cent in water) Co-Monomer (methylenebisacrylamide, BIS, about 3 per cent in the monomer solution)	2 vol.

Buffer (tris-hydroxymethylaminomethane-HCl, Tris-HCl, *p*H 8·9) Initiator (tetramethylethylenediamine, Temed, 0·23 per cent in buffer solution)	1 vol.
Water	1 vol.
Catalyst (ammonium persulphate, 0·2–0·4 g./100 ml.)	4 vol.

The mixture is placed in a vertical closed tube about 70 × 5 mm., and is covered with a layer of water. Polymerization is complete in 10–20 minutes and the gel so formed is known as the small pore or running gel in which the electrophoretic separation occurs. The overlying drop of water is removed and is replaced by a similar mixture made from a 3 per cent monomer solution with a Tris-glycine buffer of *p*H 6·7; this is the spacer or large pore gel and is usually about 3 mm. in depth. Finally, a few microlitres of serum or other sample is mixed with a further 0·2 ml. of large pore gel and again polymerized on top of the spacer gel to form the sample gel. In each case the function of the drop of water is to overlay the gel solution so that it polymerizes with a flat surface devoid of a meniscus distortion. Sample gel is often photopolymerized with riboflavin instead of persulphate.

The gel tube is now arranged to stand vertically in a Tris-glycine buffer of *p*H 8·3 and it is also overlaid with a further quantity of this buffer containing a trace of bromophenol blue which migrates more rapidly than any of the proteins and so serves as a "front" marker. Electrodes are connected to the two buffer solutions and current passed for about 25–30 minutes at an amperage of 5 mA per tube at which time the dye just reaches the far end of the tube. The gel is removed from the tube, stained, excess stain removed and the resultant gel shows a number of horizontal bands or discs of separated components.

It was calculated that the small pore gel would have a pore size of about 50 Angstroms which is within the range of dimensions of many plasma proteins and, hence, a sieving action based on molecular size would operate. With the 3 per cent gel, the pore size is so much larger that the proteins could be considered to be migrating with a velocity similar to that in free solution. Kohlrausch described a "regulating function" which occurs in the case of two solutions, one layered above the other. If the ions in the upper solution have a mobility less than the ions in the lower solution then, on passage of a current, a fixed and sharp boundary is maintained at the interface of these ions which both then move at the same rate down the column. As applied here, the glycinate ion is placed above the chloride ion. It can be shown that in the wide pore gel the proteins migrate with mobilities intermediate between those of the glycine and chloride ions. Therefore as they migrate through the sample gel and into and through the spacer gel, each protein forms a discrete disc and the separate proteins form a stack of discs at the interface of the small pore gel with the glycinate ion immediately behind the last protein disc. As the proteins migrate into the small pore gel their mobilities fall and the glycinate ion overtakes and passes them all to form a sharp boundary at the chloride ion interface. The proteins continue to migrate in a uniform linear field or

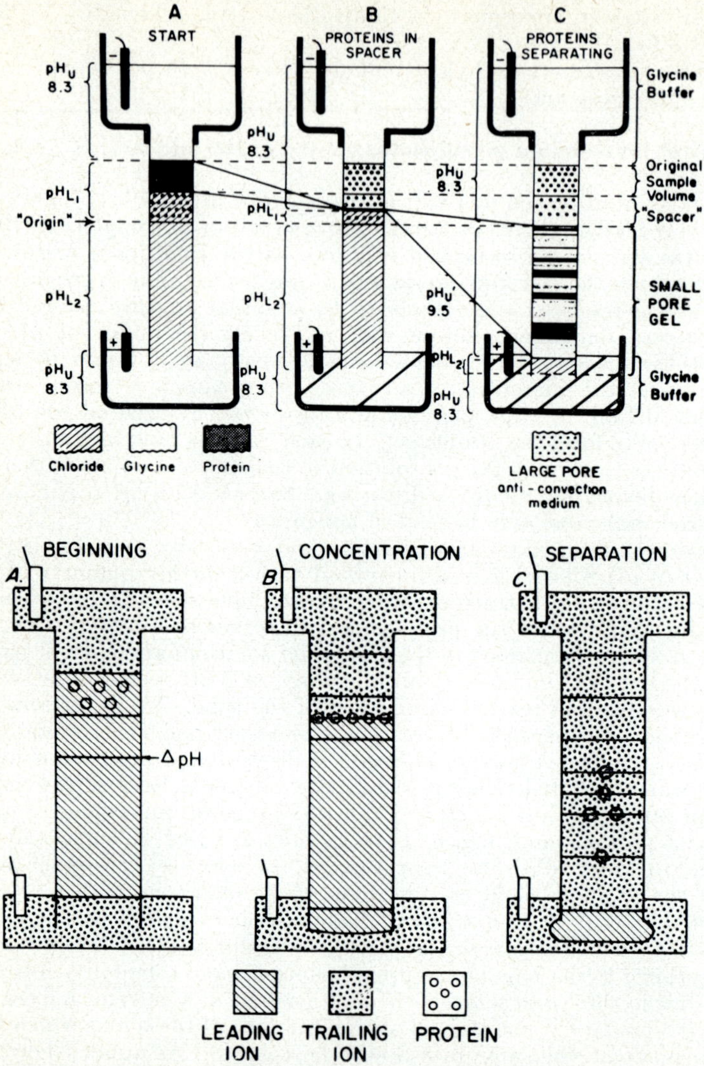

FIG. 9.1. This figure is included to illustrate the various terms and concepts used in disc electrophoresis as described in the text. Taken from Ornstein[(1)], and Williams and Reisfeld[(3)] with permission.

voltage gradient still as a series of thin discs. During this early part of the experiment the *p*H of the running gel changes. The original running gel buffer of Tris-HCl is transformed to a Tris-glycine buffer which has a *p*H of 9·5 and it is at this *p*H that the migration of proteins through the running gel occurs. Simultaneously, as the glycinate ion enters into the higher *p*H range its own mobility increases.

To summarize, the requirements for a satisfactory procedure, as originally described by Ornstein and Davis and further elaborated on by Williams and Reisfeld,[3] are:

1. The leading ion (i.e. the ion in running gel buffer and here the chloride ion) must have the highest mobility of all ions in the direction of migration.

2. The leading ion must be that of a strong acid and therefore unaffected by *p*H.

3. The trailing ion (i.e. the ion which eventually overruns the proteins and here glycinate) must be from a weak acid. This ion must show an

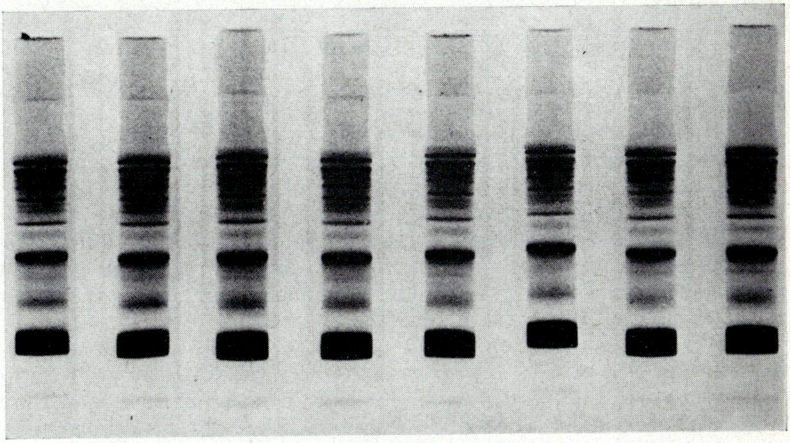

FIG. 9.2. A serum protein separation using the original three gel procedure. (Taken from Ornstein[1] with permission.)

increase in mobility as it passes from the spacer gel through to the running gel although this will follow from the fact that buffer within the running gel is changing *p*H as previously described.

4. The buffer counter-ions (i.e. the Tris migrating in the opposite or counter direction) must be capable of buffering both HCl and glycine and so must be a weak base capable of buffering within the *p*H of the running gel.

5. A sample gel is necessary to allow the layering of sample in a nondiffusing form.

6. A spacer gel is necessary to allow stacking of protein discs within the spacer and commencing at the spacer gel-running gel interface.

Theoretical arguments were used to show the correctness of all these six basic requirements. Without further elaboration, however, it can be said that empirical experimentation shows that many of these requirements are unnecessary. Clarke[4] showed that both spacer and sample gel could be dispensed with and many workers now use only the running gel. This is indeed often advantageous as some samples inhibit the gelling of sample gel and thereby create further problems. Hjerten, Jerstedt and Tiselius[5] argued that, as the analytical separation occurred at *p*H 9·5, it should be possible to use a continuous buffer system at this pH provided that the sample could be sufficiently sharpened at the gel interface. They then described previously known methods of obtaining narrow bands and showed that a single Tris-glycine buffer at pH 9·5 yielded an almost identical but equally satisfactory separation.

As with all techniques, and particularly this has been so in electrophoresis, continued use creates a further understanding of the actual process. Consequently both misunderstandings and misconceptions can be eliminated and correct theoretical requirements can be side-stepped provided that the final result is as satisfactory or better than the original procedure.

The Techniques and Apparatus of Disc Electrophoresis

Because of the number of different stages involved, it is unwise to attempt a discussion of apparatus and technique separately. Instead the individual stages of the procedure will be treated in sequence. For example, having discussed the tubes in which the electrophoresis occurs, the manner of producing the gel in the tube is next described. The order of the two subsequent stages, namely introduction of the tube into the electrophoresis apparatus and application of the sample, depends on the choice of sample application method. Thus, where only running gel is used, the sample may be applied before or after the tube is placed in the apparatus depending on the size of the sample and method of application but where spacer and sample gels are used the sample is placed in position and gelled before the tube is placed in the apparatus. The simplest procedure is that in which neither spacer nor sample gel is used and this will be described first. All other procedures, including the original Ornstein-Davis procedure will be described subsequently.

Gel Tubes. The tubes may be made of glass or rigid, cylindrical, transparent plastic tubing, length about 7·5 cm., internal diameter 0·5 cm., external diameter 0·7 cm. All tubes should be of the same length and scratches should be made, 1 cm. and 2 cm. down from one end. Tubes should be permanently numbered with a diamond scratch. Glass tubes should be permanently siliconized by dipping through 5 per cent solution of dimethyl-dichlor-silane in chloroform. Gels are most easily removed from permanently siliconized or plastic (Perspex, Lucite) tubes. The lower end of the tube is closed by fitting it into a tight plastic cap which surrounds it (see Fig. 10.3). Gels are very easily and cleanly removed from siliconized tubes as described later (see Figs. 9.6 and 9.7). Consequently only minimal washing of the tubes is required. Siliconising is not totally permanent and should be repeated occasionally, as this also aids even polymerisation.

Preparation of the Running Gel. As many tubes as required are stood vertically in a stand, the gel mixture is prepared and carefully filled into the tubes to just above one of the scratch marks. It should be run down the side of the tube, using a syringe or pipette, so that it completely fills the cap without trapping air which would otherwise slowly rise, form bubbles in the gel and inhibit polymerization. If desired, a length of narrow polyethylene tubing is affixed to the syringe so that the gel solution can be run right into the cap and allowed to displace the air in this way. Immediately 20 μl. of water to a depth of about 2–3 mm. is layered above the gel such that the interface, which is to become the running gel upper boundary, is seen to be horizontal. This can be seen easily as the gel is of different density and a schlieren line

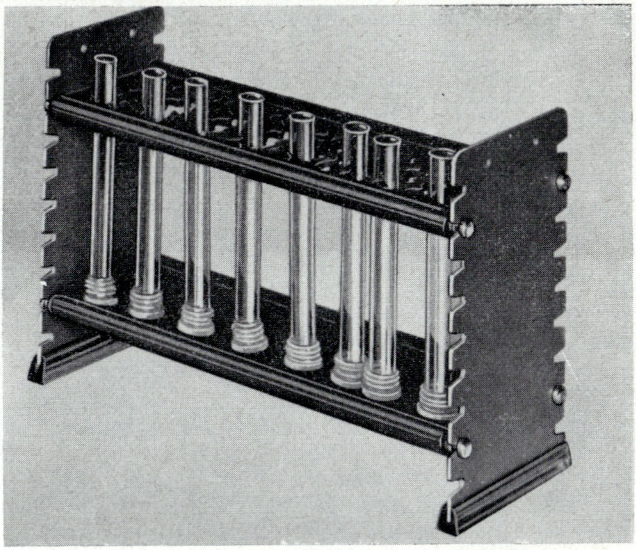

FIG. 9.3a. Gel tubes are inserted into plastic caps and stood vertically to ensure a horizontal meniscus.

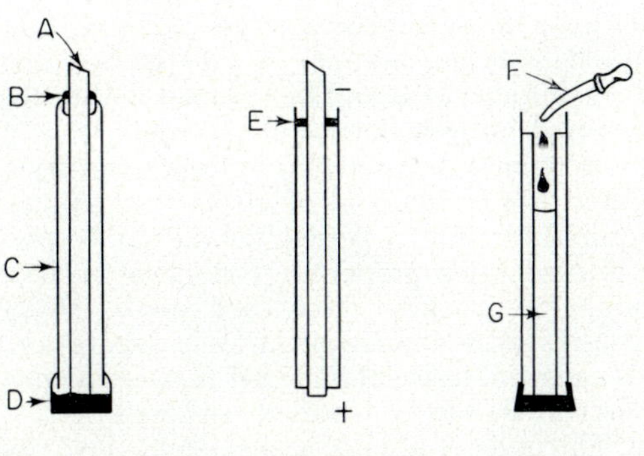

FIG. 9.3b. (*left*) A glass or plastic rod (A) is held in position in the usual gel tube C by plasticine bits B, and in the usual cap D. Liquid gel is inserted into the annular space. (*middle*) The annular gel ready for electrophoresis and showing flat meniscus E. (*right*) After electrophoresis, the central rod is removed, the gel tube is capped again and stain is added down the central well G using a pipette F.

can be observed at the interface. Great care must be taken to layer the water onto the surface and not to allow it to drop through and rise back as this may cause a density gradient to be set up in the first few millimeters or, worse, a gradient vortex in which case the discs migrate as cones rather than bands. Water can be applied via a 20 μl. microcap disposable micropipette with the end held to the side of the tube, and a few mm. above the gel solution, so that the water runs down the side and forms a layer; this method has the double advantage that the flow can be carefully controlled and a reproducable amount applied in one operation. Should drops fall deep into the gel solution the tube should be discarded, and so it might be worthwhile to prepare tubes in excess of requirements until satisfactory technique is attained. Alternatively, cotton or glass wool can be inserted into a syringe or pipette nozzle so that the water flows out slowly and smoothly onto the side of the tube. The gels set in 10–20 minutes but should be left at least 30 minutes although they can be left overnight in which case the water should be replaced by a quantity of buffer at the correct concentration added to ensure that the gel remains covered and does not dry out.

Zeineh *et al.*[43] have described an "immunocore electrophoresis" technique in which the solid gel is replaced by an annular gel prepared as shown in Fig. 9·3b. Although they used 12 cm long tubes, the technique is obviously equally applicable to the 6–7 cm tube. The gels can subsequently be stained *in situ* after removing the centering Perspex rod, or after removal in the usual way. They claim that increased sensitivity and resolution is observed Obviously gels can be removed, cut longwise and stained with a number of different reagent or the strips can be used for separation in a second direction as described in other chapters

Prior to electrophoresis the lower cap must be removed. This is best done by holding the tube at an angle and slowly turning the cap as it is pulled down and off. One edge of the cap should come free of the tube first in order to relieve any pressure which might detach the gel from the tube sides.

Stock Solutions

All stock solutions should be prepared from freshly distilled or boiled, cold water as it is essential to keep air out of the final gel mixture. The adverse effect of air is two-fold; oxygen inhibits polymerization and bubbles may form during the gelling process and be trapped within the gel. All solutions should be stored in dark bottles in the cold.

Stable solutions, such as buffers, may be prepared in bulk quantities. Monomer solutions have a shelf life in the cold of about two months and larger volumes than required for such a period should not be prepared; solutions should be filtered when prepared. Catalysts, such as ammonium persulphate, should be prepared fresh each week. Minor changes in these solutions can result in changes in protein separation-patterns, so great accuracy is essential in weighings, *p*H measurements, etc. Monomers are *neurotoxic* and should be handled with care.

Monomer Stock Solutions. The main monomer is acrylamide and this is used in conjunction with the co-monomer, methylenebisacrylamide. As previously explained, the concentration of the monomers forming the gel can be varied to suit experimental requirements but, in this case, as the monomer concentration is decreased the co-monomer concentration is increased to yield technically workable gels. Most common are the 7·5 per cent and 5 per cent monomer concentrations but up to 30 per cent has been used. The concentration referred to is that of final concentration in the monomer-buffer-catalyst mixture and the stock solutions are consequently of much higher concentrations. Unfortunately the definition of concentration is lax. Correctly, it would appear to be total monomers but some refer to acrylamide alone. Davis states that at monomer concentrations of less than 2 g. per cent plus BIS concentrations of less than 0·5 g. per cent gelation does not occur.

Addition of 0·5% agarose to the monomer solution results in the formation of good gels with even these low concentrations of acrylamide.

Although these gels are adequate for most work, they show a high ultraviolet absorption particularly in the 260–280 range. For work where this may be contra-indicated, the much cleaner, recrystallised monomers should be used.

The sample-spacer gel often detaches from the running gel during removal from the tube. If it is desired to maintain the connection then G replaces F below and concentrations are increased to monomer 15 g. and BIS 3·7 g./100 ml. This is valuable in determining whether proteins are retained in the sample-spacer gel due to non-penetration of running gel. Even at this *p*H some proteins may migrate cationically or retrograde and this can be investigated by using a somewhat shorter running gel and applying a further layer of running gel above the sample gel. Clarke has calculated that 10 per cent of serum protein does not enter the lower running gel.

Gels which can be solubilised. Gels crosslinked with BIS cannot be rendered soluble although this would be very valuable for some problems such as counting radioactivity in the gel. Two other crosslinking agents, which replace BIS, have been described. N,N′-diallyl tartardiamide (DATD)[6] can replace Bis in molar proportions and, subsequently, gel slices are dissolved by shaking for 20–30 mins in 2% periodic acid although highly alkaline gels should be neutralised first with acetic acid; DATD gels take about 1 hr to set. Ethylene diacrylate (EDA)[7] has been used at about double the BIS concentration and 1–2 mm slices are dissolved in 0·5 ml of solution containing 9 vol. 1 M pyridine and 1 vol. alcoholic Hyamine by shaking for 1–4 hr at 20–40°C.

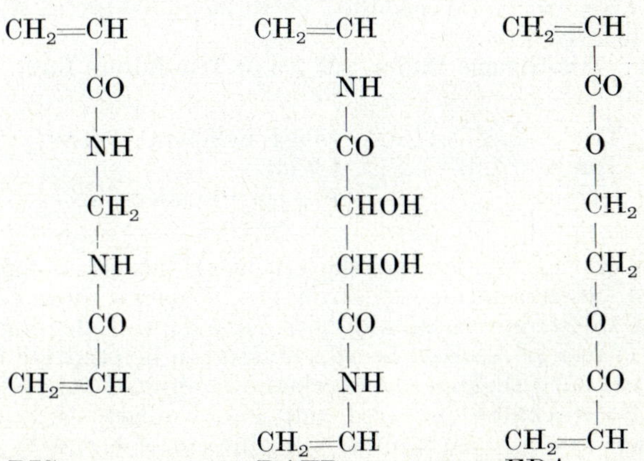

BIS DATD EDA

TABLE 9.1

Composition of monomer stock solutions

	Stock solution in water Acrylamide + BIS or DATD g./100 ml.		Final concentration in gel g./100 ml.
Running gel			
A	30·0	0·8–1·0	7·7
B	28·0	0·735	7·0
C	24·4	0·8	6·3
D	20·0	1·2	5·3 = 5·0
E	19·0	1·0	5·0
Sample and spacer gel			
F	10·0	2·5	3·0
G	15·0	3·75	4·7
Running gel			
H	30	2 EDA	8

Buffers. As previously mentioned, buffers are now used as continuous systems in which the same components (buffer salts, weak acids, etc.) are either in the same or different concentrations in the gel and the electrode vessel reservoirs, or as discontinuous systems where different buffers are used in the gel and reservoirs. Furthermore, where an intermediate spacer gel is used, a third buffer may be incorporated. Reservoir buffer should be used once only.

Ampholytes may run as anions or cations. In buffers 1–4 below, proteins run as anions in "anionic" gels. In buffers 5–6, they run as cations in "cationic" gels. For other buffers see subsequent sections of this chapter.

Urea can be incorporated into the buffer to depolymerize large units. A final concentration of 3–4 M in the gel is about as high as the gel can tolerate without becoming too fragile to be removed from the tube. If needs be, urea can be incorporated up to 36 g./100 ml. (6 M) in buffer and monomer solution.

Buffer 1: Tris-Glycine Buffer, *p*H 9·5 or Tris-Borate Buffer.

Tris . . .	45·5 g. in 900 ml. water }
Temed . . .	1·2 ml. }
Glycine . . .	20 g. in 100 ml. water

Titrate the glycine solution into the Tris to *p*H 9·5. Make final volume to one litre. Stock solution is 0·377 M Tris. This is a continuous buffer used at the same concentration in both gel and electrode reservoir. No sample or spacer gel is used. Stock solution can be prepared five times concentrated but *p*H should be checked on a diluted sample.

A Tris-Borate buffer prepared and used similarly is valuable for enzyme separations, particularly alkaline phosphatase[8] (see Fig. 9.8*b*).

Buffer 2[4]: Tris-Glycine, *p*H 8·3.

Tris	6·0 g.	per litre, for gel. Glycine is 0·387 M.
Temed	0·3 ml.	
Glycine	29 g.	

Tris, 6 g., + Glycine, 29 g., + N HCl, 5 ml., per litre. Diluted one in ten for reservoir, *p*H 8·1.

Buffer 3. This is the original buffer system described by Ornstein[1] and Davis[2] and requires three buffers, one each for small pore gel, large pore gels, and reservoir.

Small pore buffer *p*H 8·9	1 N HCl	48 ml.	
	Tris	36·6 g.	
	Temed	0·23 ml.	
	Water to	100 ml.	
Large pore buffer *p*H 6·7	1 N HCl	48 ml.	Accurate *p*H is ensured by titrating in the HCl to Tris and Temed dissolved in 50 ml. water.
	Tris	5·98 g.	
	Temed	0·46 ml.	
	Water to	100 ml.	
alternatively	Tris	5·7 g.	
	1 N H_3PO_4	25·6 ml.	
	Water to	100 ml.	
Reservoir buffer *p*H 8·3	Glycine	28·8 g.	Dilute ten times for use and use once only.
	Tris	6·0 g.	
	Water to	1,000 ml.	

Buffer 4.[3]

Small pore *p*H 7·5	1 N HCl	48 ml.
	Tris	6·85 g.
	Temed	0·46 ml.
	Water to	100 ml.
Large pore *p*H 5·5	1 M H_3PO_4	39 ml.
	Tris	4·95 g.
	Temed	0·46 ml.
	Water to	100 ml.
Reservoir *p*H 7·0	Diethylbarbituric acid	5·52 g.
	Tris	1·0 g.
	Water to	1,000 ml.

Buffer 5.

Small pore buffer	1 N KOH	48 ml.
	Acetic acid glacial	22·4 ml.
*p*H 4·2	Temed	4·6 ml.
	Water to	100 ml.
Large pore buffer	1 N KOH	48 ml.
	Acetic acid glacial	3·1 g.
*p*H 5·8	Temed	0·46 ml.
	Water to	100 ml.
Reservoir	Acetic acid glacial	2·4 g.
*p*H 4·0	Glycine	17·25 g.
	Water to	1,000 ml.

Buffer 6.[9]

Small pore buffer	1 N KOH	48 ml.
	Acetic acid glacial	17·2 ml.
	Temed	4·0 ml.
	Water to	100 ml.
Large pore buffer	1 N KOH	48 ml.
	Acetic acid	2·87 ml.
	Temed	0·46 ml.
	Water to	100 ml.
Reservoir buffer	β-Alanine	31·2 g.
*p*H 4·5	Acetic acid	8·0 ml.
	Water to	1,000 ml.

Buffer 7.

1 M propionic acid

Catalyst-Initiator Systems. Polymerization of the monomers is brought about by means of both a catalyst and an initiator. Catalysis may be both chemical via ammonium persulphate or photochemical via riboflavin; the running gel is invariably catalyzed with persulphate and the other gels, if used, with riboflavin or less frequently with persulphate. Temed is used most frequently as initiator although DMAPN, dimethylamino-propionitrile, has also been used. Where a running gel is used without sample and spacer gels and in a continuous buffer of constant composition, many prefer a preliminary $\frac{1}{4}$–1 hr passage of current to remove residual persulphate (or its decomposition product, sulphate) before applying the sample. The initiator is usually prepared as part of the buffer solution but the persulphate is kept separate until required because of the instability of the latter which therefore requires more frequent preparation of fresh solution. Rate of polymerization is controlled by the Temed concentration; increasing the Temed increases the rate, and the concentration is adjusted for polymerization to occur within the period of 10–30 minutes after mixing. The rate of polymerization is also increased in the cold and in the presence of oxygen as described below.

The riboflavin-photocatalyzed sample gel suffers from the disadvantage that the sample may often delay or completely inhibit polymerization. Hence some prefer to continue to use the spacer gel but to apply the sample above it in free solution, whereas others dispense with both sample and spacer gel, and apply the sample directly above the running gel in free solution as previously mentioned. Others polymerize the sample gel with persulphate providing that it does not affect the sample itself. Chemical catalysis with persulphate begins immediately the latter is added to the monomer. Photocatalysis with riboflavin requires bright light for the reaction to occur and is brought about as follows. The tubes containing the gel mixture, layered with water, are exposed to a daylight fluorescent tube placed as close to the tubes as possible. After about five minutes, schlieren effects indicate that polymerization has begun and this is complete in 30–45 minutes.

The procedure recommended below uses persulphate in preference to riboflavin. The persulphate procedure has been used successfully with many different types of protein and enzyme, including a large number known to contain labile groups such as disulphide bonds.

Persulphate Solution. Ammonium persulphate in water prepared fresh each week, kept cold and in a dark bottle.

0·14 g./100 ml.	solution L
0·20 g./100 ml.	,, M
0·40 g./100 ml.	,, N

Riboflavin Solution. 4 mg./100 ml.; refrigerated in a dark bottle.

Running Gel Mixtures and Reservoir Buffers—the Electrophoretic Set-up. When required, approximate volumes of each solution should be transferred into separate test tubes and warmed to room temperature. Exact volumes of monomer, buffer, catalyst, etc. are then pipetted into another tube or a small Buchner flask and mixed. If the stock solutions have absorbed oxygen, the fresh mixture should be degassed by stoppering the Buchner flask and attaching to a vacuum pump for a few seconds until all the gas has boiled out. The mixture is then rapidly transferred to the gel tubes by means of a plastics disposable syringe. If preferred, the mixture can be degassed in the syringe by closing off the exit, pulling the plunger further back and shaking to expel the oxygen from solution. The mixture begins to polymerize and must be transferred within about 5 minutes to the gel tube as, after this time, it becomes too difficult to transfer. It is best placed in the gel tube using either a pipette or a small syringe (2 ml.) and adjusted so that the lower part of the meniscus is level with one of the scratches on the tube wall. As soon as all the tubes have been filled with this gel solution, a volume of water is carefully layered onto the surface of the gel solution to a depth of about 2–3 mm. This is best done by running the layer down the side of the tube from a point only a few millimeters above the gel surface, using a microcap 20 μl. pipette. 20 ml. of mixture is adequate for 8 tubes.

Four examples will be given in detail, others can be arrived at by comparison.

A. Tris-glycine or Tris-borate buffer, *p*H 9·5, 5 or 7 per cent gel; running gel alone (no spacer or sample gel), continuous buffer.

Buffer 1	1 vol.
Monomer B or D	1 vol.
Persulphate M	2 vol.

The reservoir buffer is diluted with water, 1 volume plus 3 volumes, so that buffer concentration is the same in reservoir and gel. Both buffer and water are used cold. A 25 minute run is suitable.

B. Tris-HCl buffer, *p*H 8·9; 7 per cent running gel, 3 per cent spacer and sample gels, *p*H 6·7; reservoir Tris-glycine *p*H 8·3; discontinuous buffers; a similar procedure is used for buffers 4, 5 and 6.

The running gel is polymerized as previously described. The upper water layer is removed by flicking out or inverting and draining. The surface is then rinsed with spacer gel solution before applying a fresh quantity of about 0·15 ml. of spacer gel solution. Polymerization of this is brought about by placing a daylight fluorescent light about 0·5–1 inch above the gel surface or close to the gel tube for 0·5–1 hour; the beginning of polymerization can be seen after 5 or 10 minutes by a cloudiness and consequent light scattering in the gel. The process is repeated with sample gel.

Running gel:	Small pore buffer 3 *p*H 8·9	1 vol.
	Monomer B	2 vol.
	Water	1 vol.
	Persulphate L	4 vol.
Spacer gel:	Large pore buffer 3 *p*H 6·7	1 vol.
	Monomer F	2 vol.
	Riboflavin	1 vol.
	Sucrose, 40 per cent in water	4 vol.
Sample gel:	as for spacer gel. Add 2–5 μl. serum or other sample.	
Reservoir buffer:	Tris-glycine as under Buffer 3.	

C. Potassium acetate-Acetic acid buffer; 7 per cent running gel, 2·5 per cent spacer gel; reservoir β-alanine acetate *p*H 4·5. Cationic gel.

Running gel:	Small pore Buffer 6	1 vol.
	Monomer A (·4 g. BIS)	2 vol.
	Water	1 vol.
	Persulphate L	4 vol.
Spacer gel:	Large pore Buffer 6	1 vol.
	Monomer F	2 vol.
	Riboflavin	1 vol.
	Water	4 vol.
Sample gel:	as for spacer gel. Add sample before gelling.	
Reservoir buffer:	β-Alanine, acetic acid as under Buffer 6.	

D. Propionic acid buffer *p*H 2·1

Running gel:	4 M Propionic Acid	1 vol.
	Monomer H	1 vol.
	Persulphate 23 mg/100 ml	1 vol.
	Na hydrosulphite 35 mg/100 ml.	1 vol.
Reservoir buffer:	1 M propionic acid.	

This is a continuous system. Highly acid systems do not gel so well with persulphate, hence the use of hydrosulphite. Gel should be covered with hydrosulphite instead of water layer.

Sample Application. The methods of sample application depend to some extent, on the volume of the sample, and the use or not of a sample gel. In the case where sample gel is required, then the method described above is used. Where sample gel is not used then the other methods to be described are all satisfactory and, to some extent, alternative procedures.

Whatever media are used to contain the sample it should be remembered that they all possess certain common properties, namely:

(*a*) They are solutions of greater density than the buffer to prevent upward diffusion and dilution of sample.
(*b*) The conductivity is lower than that of the buffer as the resulting voltage discontinuity produces a sharpening effect on the zones before they enter the gel.
(*c*) The components of the medium do not absorb or otherwise affect the sample molecules so that they can be considered to be migrating as in free solution, i.e. with a migration rate considerably higher than that which they will possess after entering the gel.

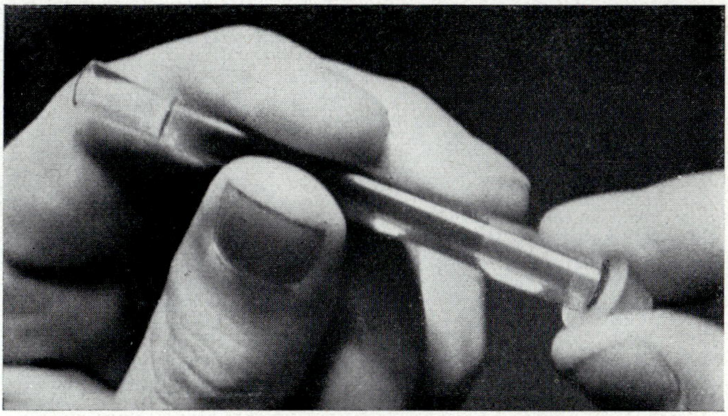

Fig. 9.4. Caps are removed from gel tubes by means of a turning motion so that the cap comes free from one side of the tube first. This ensures that the gel is not dislodged by a suction effect which might occur if the cap was simply pulled.

A valuable corollary, on which the method is based, follows from this. As the migration is so much faster in free solution than in the gel, the components of even a dilute solution will concentrate very many times just above the running gel surface. Further as each protein migrates at a characteristic rate, the components of a mixture stack up as a series of highly concentrated discs. Without an intermediate spacer gel there may be some slight overlap but this is eliminated in the early stage of passage through the running gel.

Although the leading albumen band can be seen as it migrates through the gel, due to the schlieren effect, it is much easier to see if a trace of bromophenol blue, BPB, is added as this stains the albumen blue. Any excess of BPB runs ahead of the albumen so that two blue bands can be seen. The BPB has a further advantage in that the stacking process can be seen occurring within the sample medium which is a check on correct method (for low *p*H buffers, methyl green has been used). Although no macromolecule should run faster than the BPB, such molecules may run between the BPB and the albumen band so the experiment should be stopped when the albumen-BPB complex is at least 10 mm. from the lower end of the gel. BPB does not affect any subsequent staining procedures and a reasonable excess is permissible.

Indeed there is an advantage in using sufficient to produce a leading band of free dye as the gel can be cut at this position prior to staining and this "front" is then a convenient point for the measurement of migration rates in a manner similar to Rf's in chromatography. The leading, free BPB has travelled about 1 cm. ahead of the blue human albumen-BPB band by the end of the experiment and in this lies its convenience as a marker; this is only true for buffers 1, 2 and 3 and running gel mixtures A and B above. Where the free BPB is not a useful marker, the albumen band is itself useful after checking the presence and position of pre-albumens. Of course, the relative migration of BPB and BPB-albumen complex will depend on the *p*H of the experiment. Before comparison of migration rates in gels prepared simultaneously but stained by different procedures or with chemical reagents, all gels should be given a final soak in the same inert medium, e.g. 5 per cent acetic acid in water, as gels often expand or shrink in different staining media especially if placed in a dehydrating solution such as concentrated alcohol.

In all these procedures it is usually best to remove the cap at the bottom of the gel tube and place the tube in position in the apparatus with the lower end of the gel dipping into reservoir buffer to avoid disturbing the applied sample layer subsequently.

If desired, a persulphate-catalyzed spacer gel can be used with these methods to act as a stacking medium but, as stacking seems to occur satisfactorily without the spacer, there may be no advantage in so doing. Solutions of value for sample application are as follows:

1. Sucrose, 40 per cent, in water[9] containing 0·5–1 per cent BPB; urea has been used instead of sucrose. Take up 2–5 μl. (or that volume containing a constant weight, e.g. 300 μg., of protein) in a microcap microlitre pipette, then twice as much sucrose solution and layer on the gel surface and under the buffer already in position in the complete apparatus. Alternatively mix one volume of sample with two volumes of sucrose and apply 6–15 μl.

2. Dilute sample with sucrose as above, pipette 6–15 μl. onto a precut circle of filter paper which will fit into the tube and place onto the gel surface, a drop or two of BPB-sucrose should be added followed by the buffer. This is a useful variant as circles can be punched from paper or acetate electrophoresis runs and the separated albumen and globulin fractions or bands can be examined individually by the gel procedure.[10]

3. Ficoll, 50 per cent, in buffer diluted 1 in 3 with water and containing BPB.[11] This is a thick, viscous solution like glycerol. The Ficoll method is most suitable for dilute solutions, such as parotid saliva, which may only contain about 1 per cent protein. Up to 200 μl. saliva can be mixed with an equal volume of Ficoll and the whole applied to the surface of the gel. Reservoir buffer is carefully layered onto the Ficoll. This volume of 400 μl. is similar to the total volume of sample plus spacer gel, acts similarly without the need to polymerize and yields identical results. It avoids the need to concentrate or liophyllize dilute solutions, such as saliva, as this often results in irreversible changes.

4. Sephadex G-200 or Dextran.[10] The material is expanded and washed, first in water then in buffer diluted 1 to 3 in water containing 20–40 per cent sucrose and BPB. Use as in methods 1 and 3 above. Hæmoglobin hæmolysates which inhibit gel polymerization, can be treated satisfactorily by methods 4 or 3.

5. Clarke[4] has described a number of other variations which have special value in a limited number of cases:

(*a*) Application of two samples to one gel. Gel is formed with the interface at the 1 cm. scratch. A rigid rectangular spacer, about 15 mm. long by a width equal to the diameter of the tube, is cut from heavy waxed paper (seam of a paper cup) and forced into the gel to about 1 mm. depth. Each side is filled separately with buffer and sample from a micro-cap. Electrophoresis is run as usual and it is recommended to remove the paper partition after the tracking dye has migrated 1–2 mm. into the gel although this may not be necessary.

(*b*) Ascending "retrograde" runs. Some proteins even at high *p*H's migrate cationically and are normally lost; this procedure allows a search for such proteins. Shorter gels are prepared as usual. Sample is applied under a small volume of buffer containing 20 per cent sucrose and a second gel is overlaid on the buffer, or sample is applied in a 3 per cent sample gel and overlaid with running gel. Short runs followed by staining indicate the possible presence of cationic proteins but an artefact band appears which is subsequently masked by the albumen.

(*c*) Where cationic bands are known to be present, clean separations can be obtained by means of the usual procedure except that the current flow is reversed by reversing the electrode connections.

6. Hjerten *et al.*[12] have used standard methods for electrophoretic "particle sieving" and have separated 30S from 50S ribosomes on analytical and preparative scales (see Chap. 10).

The Electrophoresis Assembly. The apparatus shown is made entirely of Lucite (Perspex) except for the platinum electrodes and glass gel tubes. The lower unit holds about 300 ml. buffer and has a ledge on which the electrode holder sits. The upper buffer reservoir is punched and silicone-rubber grommets are inserted to hold the gel tubes; unused holes are closed with rubber stoppers. The unit is gently lowered such that each gel tube slips through a corresponding hole of the electrode holder into the buffer solution and rests in position on three Perspex insets. About 250 ml. buffer is added and the upper electrode unit rests on the reservoir vessel by overlapping edges. Both electrodes are circular, and hence equi-distant from every gel tube, and sit submerged in buffer. A second reservoir with slightly larger holes is available for destaining gels held in slightly wider tubes as described under destaining.

Current Supply. At the beginning of the experiment a current of about 1–2 mA per tube is used in order to avoid overheating and

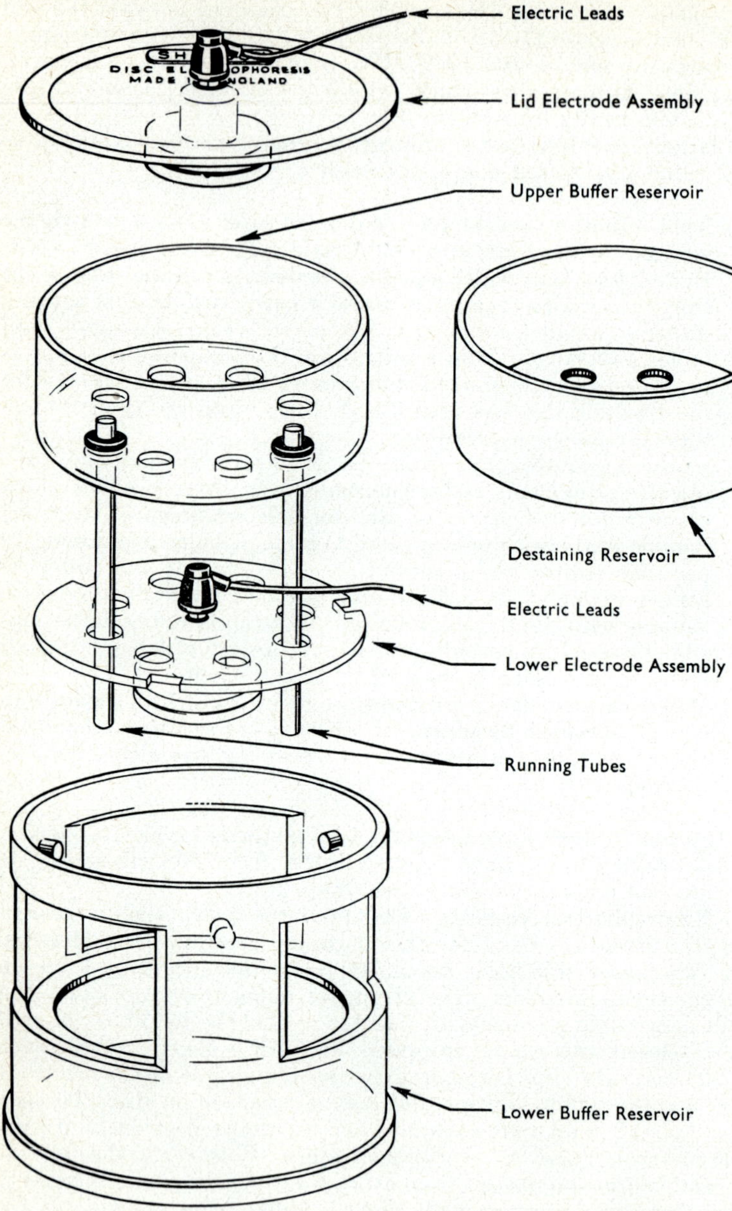

FIG. 9.5a. An exploded version of the gel electrophoresis set-up. Overall dimensions of the apparatus are 6 × 5 in. diameter.

diffusion. Within a short time the BPB enters the gel and current is increased to 5 mA per tube for the duration of the experiment. Depending on the conductivity of the gel, i.e. buffer concentration, etc., the run takes from 30–90 minutes. The tube should never feel more than just warm to the touch as this would indicate excessive heating within the gel and the need for lower amperage in repeat experiments. Constant current power packs should be used.

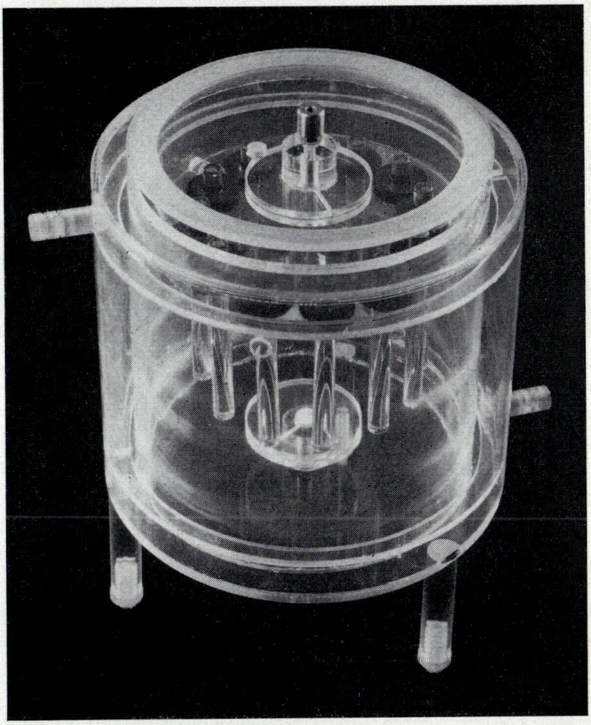

FIG. 9.5b. Photograph of a modified version of the apparatus in which the whole gel tube is emersed in cold buffer which can be circulated if desired.

The reservoir buffer together with water for the dilution of stock concentrates should be kept cold until required. The use of cold buffer in the apparatus helps to dissipate any heat evolved and this method is often adequate without further precaution in which case the tubes should not feel more than warm at the end of the experiment. Further cooling is sometimes effected by means of a fan and, for enzyme work, the experiment can be conducted completely in the cold. Alternatively a lowerage amperage, say 1–2 mA, is used for a longer time. However, excellent separations of enzymes including LDH, G6PD, alkaline phosphatases, etc., have been obtained working in the open laboratory, but starting with cold gels, buffer and water taken directly from the cold room.

Removing the Gel from the Tube. Before endeavouring to carry out a complete experiment, some practice in removing gels from the tubes is essential as a little skill is necessary in order not to ruin the gel at

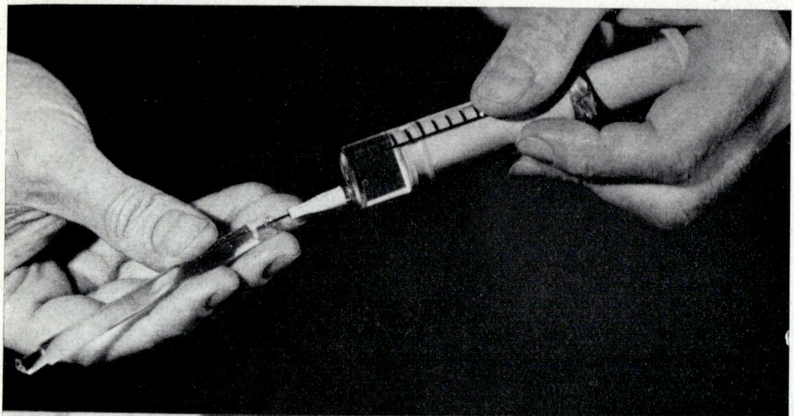

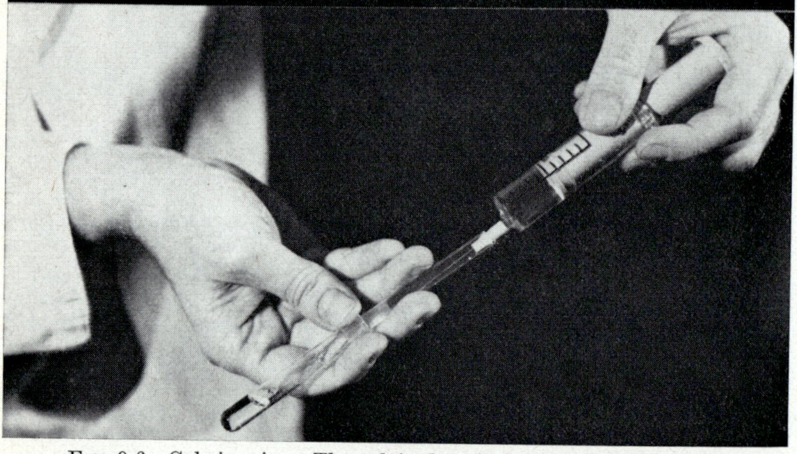

FIG. 9.6. Gel rimming. The gel is detached from the wall of the (siliconed) tube by gently rimming with a fine needle attached to a syringe. A small quantity of water, or ethylene glycol monomethyl ether, is slowly ejected to lubricate the gel-tube interface as it is freed by the needle. The needle is slowly advanced with a continuous motion while rotating the tube.

FIG. 9.7. The figure shows the rimmed gel sliding directly into stain or buffer as the tube is raised to the vertical position.

this advanced stage. With siliconized or plastic tubes, rimming presents no problem and the gels detach and slide out easily, using water as lubricant; this is the author's standard procedure for proteins and enzymes. The gel adheres strongly to the unsiliconized glass tube. Rimming of the gel is carried out (Figs. 9.6, 9.7) with a long thin needle to separate it from the walls. Each gel tube is removed from the upper reservoir and excess buffer shaken out. About 0·2 ml. of ethylene glycol monomethyl ether (may interfere with carbohydrate reagents) or liquid paraffin (a high boiling hydrocarbon fraction) is pipetted onto the gel surface and a fine wire or needle is inserted between the gel and the tube wall with the tube held at an angle. Rimming is carried out carefully in order not to scar or break the gel, turning the needle so that it passes right around the wall and simultaneously slowly down the gel allowing time for the lubricant to pass down with the needle. Should the needle not reach to the end of the tube, the same process is repeated from the lower end. The aim is simply to detach the gel from the walls and not to push it out of the tube. Aqueous solutions such as 7 per cent acetic acid or buffer can be used but require a greater skill and more patience.

A similarly numbered tube, of slightly larger dimensions and containing 3 ml. of the required reagent is taken and the end of the gel tube, held almost horizontal, is placed over its mouth. As the gel tube is now raised to the vertical position the gel often slides out directly into the lower tube. If it does not then a further loosening with the wire may start it off. Otherwise a small teat is applied to the upper end of the gel tube and a very slight pressure will send the gel slowly out. If it is desired to cut the gel at the BPB mark it should be allowed to slide into a suitable Petri dish containing the reagent and, after cutting, transferred to the tube. This step is essential as the BPB is lost by outward diffusion in the staining procedures. This method is suitable for staining with reagents prior to electrolytic destaining, or where stain is taken up only by the active band. However where a leaching step is required at any stage of the reaction, a much larger tube or Petri dish should be used and the solution agitated from time to time during the required period.

Stains and Staining Procedures

Staining is best carried out in small tubes, unless a leaching or washing stage is necessary. Such a procedure requires only small quantities of reagent which can be used once only and discarded. Also the gels are transferred from one tube to another similarly numbered through the sequence of reactions and cannot be mixed up. The time required for a particular staining step must be determined empirically as it is a function of stain concentration and rate of diffusion into the gel. The step can be judged complete when the band observed is continuous throughout the disc cross-section (as shown by cutting the disc if necessary) and without a clear zone of undyed middle portion. Longer staining times are required for high monomer concentrations.

Naphthalene Black (Naphthol Blue Black, Buffalo Black NBR, Amidoschwartz).

Naphthalene black, 0·1 per cent, in 7 per cent acetic acid in water (filter before use) or in methanol-water-glycerol-acetic acid-(50-50-20-1). Stain for 30–60 minutes, the acetic acid or methanol being used as protein precipitants. Occasionally 10 per cent TCA, 5 per cent sulphosalycylic acid and phosphotungstic acid have also been used. Destaining is carried out by soaking for prolonged periods in 7 per cent acetic acid or, better, by electrolytic destaining as described later. This is normally thought of as a protein stain but it has been shown by Hobbs to locate dextrans on cellulose acetate and by Weiss to locate neutral polysaccharides after paper electrophoresis and chromatography see p. 64.

Coomassie Brilliant Blue R 250. This stain is reported to be a much more sensitive and reproducible stain than naphthalene black and to obey Beers law over a range of 0·5 –20 μg. protein.

Coomassie blue, 0·25 g., in 100 ml. of methanol-water-acetic acid glacial (5:5;1 by vol.) or Coomassie blue, 0·25 per cent, in water containing 20 per cent sulphosalycilic acid or TCA. Stain for 20 minutes and destain electrolytically.

Not only is this stain more sensitive, thus allowing the use of smaller quantities of dilute protein solutions such as saliva fractions, but it yields different patterns to those obtained with naphthalene black procedures. The significance of this is unknown at present.

Ponceau S. Also used as a protein stain. Ponceau S, 0·6 per cent, in 3 per cent trichloracetic acid in water, stain for 1 hour. Wash in 4 changes of 7 per cent acetic acid over 48 hours. See section on human proteins. This stain is less sensitive than naphthalene black.

Benzidine for Haptoglobins.

Benzidine, 0·2 per cent, in 0·5 per cent acetic acid (warm to dissolve, store cold)	20 ml.
Ammonium chloride solid (dissolved just before use) .	30–50 mg.
H_2O_2, 100 vols. per cent (added just before use) . .	40 μl.

Immerse gels in fresh stain and blue bands become visible within 15 minutes. All bands appear within 50 minutes. The ammonium chloride preserves the blue bands for some hours as, without it, the bands rapidly fade to brown. Addition of ammonium chloride to the benzidine solution may cause precipitation hence the quantity of solid should be judged under local conditions and the solution filtered if necessary.

PAS (Periodic Acid-Schiff) for protein-bound carbohydrates. Clarke's method[(4)] seems by far the simplest of many variations in common use and will be described here—

(*a*) Periodic acid, 1 per cent, in 3 per cent acetic acid. Immerse for one hour.
(*b*) Leach in water for one hour.
(*c*) Schiff's reagent (commercial). Immerse for one hour.
(*d*) Sodium metabisulphite, 1 per cent in water. Store gel in this solution.

Brilliant red bands are seen. On keeping, there is a tendency for the storage solution to turn red although the bands do not seem to be diffusing out.

Note: Tris is oxidized by the periodic acid to an aldehyde which subsequently reacts with the Schiff reagent and the product is difficult to leach out. Hence gels containing large quantities of Tris should be given a preliminary leaching in a large volume of 7 per cent acetic acid for one hour prior to the oxidation in step (*a*).

Toluidine Blue for acid mucopolysaccharides and nucleoproteins. Toluidine blue, 0·1 per cent, in 1 per cent acetic acid.

Destaining. Many staining procedures involve chemical reactions in which colour is developed only at the site of the reactive substance. Thus, with the PAS location procedure for glyco-compounds, the coloured band is a direct result of interaction of stain with a reactive chemical group on the molecule being sought. Consequently, in those parts of the gel which are free of active compounds no colour is produced and no removal of excess dye (destaining) is necessary. However, with proteins, staining requires physical penetration of a dye into the total gel structure. Protein absorbs and fixes the dye but excess dye must be removed before the protein bands can be observed. Dye bound to the protein, which is precipitated during the staining procedure, is no longer free to diffuse but unbound dye can migrate out of the gel. Two procedures are available for removal of this excess dye. The gel may be transferred to a bath of dilute acetic acid when the dye passes out of the gel by simple diffusion but as diffusion is a slow process this procedure requires a number of changes of acid over 24 hours or more. The process may be hastened by applying a current across the gel thus causing electromigration of the dye out from the gel and this is known as electrolytic destaining.

Electrolytic Destaining. There are two variations of this procedure based on the same principle. The gel may be set up in somewhat wider tubes but in the same apparatus used for running the electrophoretic experiment and current passed down the length of the gel until it is seen to be free of excess stain, i.e. sharp protein-dye bands are seen with intervening water-clear areas free of dye. Alternatively, using a different apparatus, current is passed across the width of the gel until the same differentiation is observed. In the former case complete destaining may require 1–2 hours whereas in the latter case 15 minutes is adequate and the difference in time is in the ratio of length to width of the original gel for obvious reasons, providing comparable voltage gradients are used.

Destaining in the original apparatus is carried out as follows. A numbered tube of the same length but about 2 mm. greater in diameter than the original tube is taken. Two procedures are now available. First, a suitable size (00) "Viscap" (Visking Corporation) is taken and secured to the lower end of the tube with a rubber band. Alternatively the tube is capped as in the preparation of running gel, 1–2 ml. of gel mixture is added, allowed to polymerize until set and the cap is carefully removed. 3 ml. of water, 40 per cent sucrose or glycerol is added to the tube, the stained gel is transferred to the tube which is then placed in the apparatus with 7 per cent acetic acid in the lower reservoir. More water is pipetted into the gel tube just to the level of the surface of the gel. 7 per cent acetic acid is carefully layered onto the surface of the gel-aqueous sucrose to the top of the tube and the upper reservoir is filled with more acid solution. Current, at 10–15 mA per tube, is passed until the destaining is complete. Destained gels are then poured into small tubes containing 2–5 per cent acetic acid and can be examined or stored as such. With care, the destaining tube with its

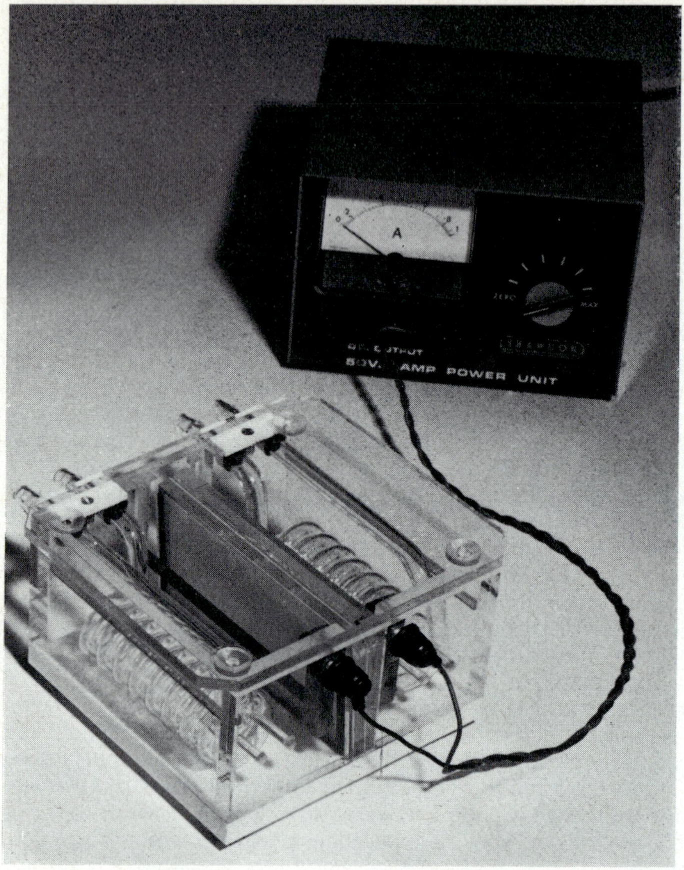

FIG. 9.8a. Destainer and power pack, showing cooling of both electrode chambers which enables high watt-heating to be dissipated.

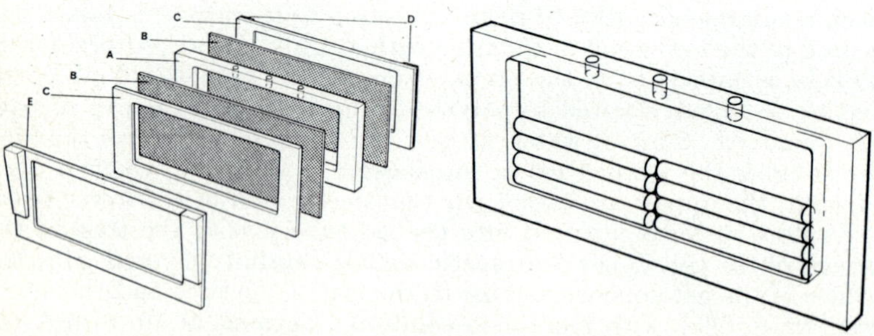

FIG. 9.8b. Expanded view of gel-holder. A = gel frame, B = porous plastics sheet, C = bezel, E = wedge-shaped bezel for tight fit, D = foam plastics sealing strip.

gel plug can be used many times before the plug becomes detached from the tube. The sucrose solution is a non-conducting support to hold the gels upright.

In the alternative procedure, gels are held horizontally between porous polyethylene sheets in a vertical, cooled Perspex unit.[44] Using 7% acetic acid, and a special power supply providing up to 1 A current. destaining is complete in not more than 30 minutes; see fig. 9.8a, 9.8b.

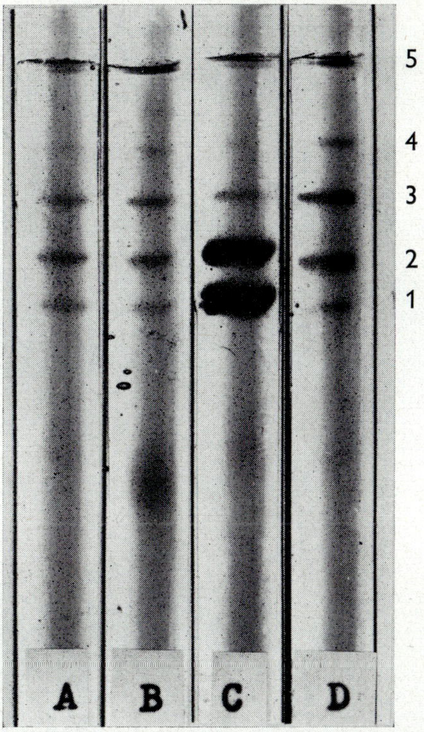

FIG. 9.9a. Enzyme separations. (*a*) LDH isoenzymes run in Tris-glycine continuous buffer *p*H 9·5. A, normal serum showing strong, fast bands 1 and 2 with weaker bands 4 and 5; B, normal plus faster moving pigment; C, myocardial infarction with greatly raised bands 1 and 2; D, unknown serum with raised bands 2 and 3 (this serum also showed raised alkaline phosphatase).

Enzyme Separations and Location. Methods used to locate enzymes depend on the enzyme activity and not on the particular buffer or separation technique. There seems little doubt that the location reagents described in Chapter 8 could be applied as described for starch gel. Procedures so described for LDH, G6PD, cæruloplasmin and alkaline phosphatase[(8)] have given good results after disc electrophoresis using the continuous Tris-glycine or Tris-borate *p*H 9·5 buffers.

It is generally assumed that the majority of assay procedures result in destruction of the enzyme or an entanglement with precipitate such that recovery is not possible. An excellent method for overcoming this is the methyl-umbelliferyl ester technique suitable for hydrolases. The location reagent, say methylumbelliferyl phosphate, is non-fluorescent whereas the product, methylumbelliferone, is fluorescent and so appears as a transient fluorescent band in the gel within 5–10 minutes. The fluorescent band is excised and the enzyme can be recovered from the small segment of gel using one of the methods

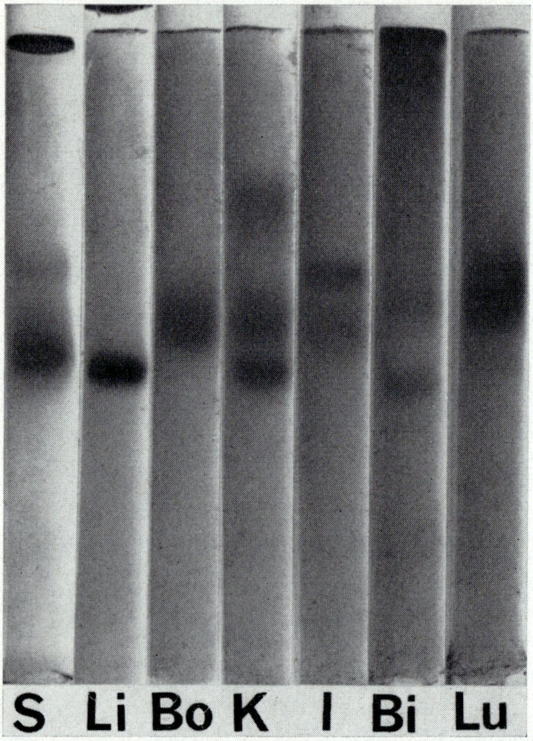

FIG. 9.9*b*

(*b*) Tissue alkaline phosphatase patterns[42] run in Tris-borate *p*H 9·5. S, serum; L, liver; Bo, bone; K, kidney; I, intestine; Bi, bile; L, lung.

described below, or the gel segment can be dropped on to an isoelectric focusing gel or implanted in a flat gel sheet for further separation using a different buffer or different electrophoretic principle. Other fluorescent agents, such as fluorescein esters, have been used. By means of coupled enzyme reactions, dehydrogenases, oxidases and other types of enzyme have been located. Fluorimetric methods have been well discussed by Leaback[13] and an application of the method to alkaline phosphatase gel electrophoresis has been made.[14]

Elution from Sample Gel. The sample gel is sliced at right angles to the direction of migration, sometimes after first freezing. The slices are homogenized with five volumes of buffer. Recovery can be well over 80 per cent. Alternatively the slices can be left for some hours or overnight in the refrigerated buffer and pass out by simple diffusion. The electrodialysis method described under "elution from starch gels" is probably the most valuable here (p. 236).

Immuno-diffusion after Gel Electrophoresis. Electrophoresis is carried out as usual and the gel released from the tube. Proceeding as rapidly as possible, a slice 1–2 mm. thick is cut along the length of the gel. The slice is cut again so that a section about 2 mm. square is obtained. This is embedded in agar as described in Chapter 6 and the normal agar technique is used for the diffusion study. This gives an overall gel-immuno pattern. Alternatively, individual discs can be sliced out and examined for particular antigens.

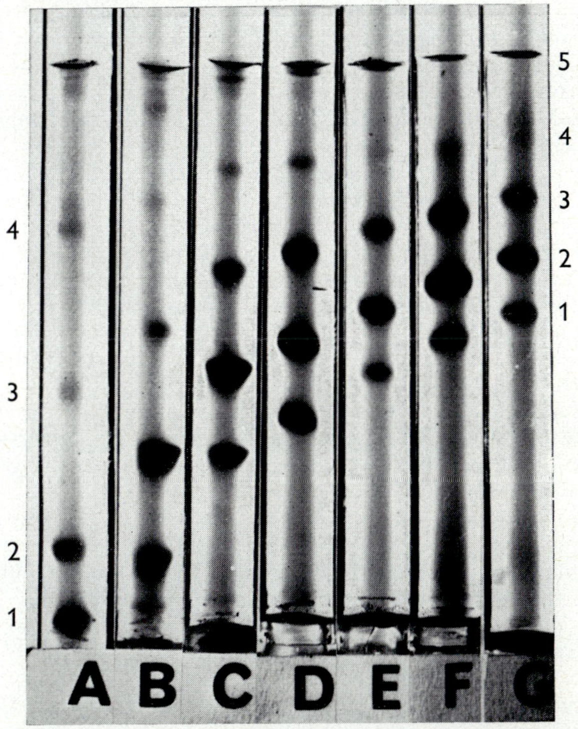

FIG. 9.10. The effect of increasing gel concentration on migration. A → G increasing from 3 → 9 per cent gel. In each case the albumin-BPB has run the full length of the gel which was then stained for LDH; all tubes were, of course, run simultaneously in the same apparatus. In tube A, the LDH's have run practically with the front but are held back increasingly with increase in gel concentration. Note that the hold back is of similar magnitude with all of the isoenzymes thus showing that the difference in running rates is not due to difference in molecular size.

The immunocore electrophoresis technique of Zaineh *et al.* (see p. 216) was specifically designed for immunoelectrophoresis. They claim that band sharpening during the immunodiffusion yields sharper resolution and, often, quicker results.

Setting Up the Experiment. As previously mentioned, the sequence of operations depends on the method of sample application to some extent. Hence the method outlined must be arranged according to the chosen method of the operator.

(*a*) Prepare the running gel in the tube with the lower end closed, using a stand in which the tubes are held vertically. If sample and spacer gels are used, prepare these with the tubes still in the same stand although tubes must be removed after each polymerization to flick out or drain the water layer.
(*b*) Place about 300 ml. reservoir buffer in the lower reservoir.
(*c*) Take each tube separately from the stand, flick out the water and remove the lower cap with the tube held at an angle and a circular turn of the cap so that it comes off one edge of the tube first. This is to negate any suction effect due to a straight pull of the cap which might damage the gel or pull it away from the side of the tube.
(*d*) Insert each tube into a grommet of the upper buffer reservoir, closing any unused places with rubber or silicone-rubber stoppers. Insert the reservoir into the apparatus.
(*e*) For sample application by method 1, place about 250 ml. buffer into the upper reservoir, ensuring that no air bubbles are trapped in the tubes. Bubbles may be dislodged with a syringe needle. Then apply the sample onto the surface of the gel and under the existing buffer when a thin, light blue layer should be observed at the interface.
(*f*) For sample application methods 2–4, apply the paper disc, Ficoll solution or Sephadex suspension as described. Cover with reservoir buffer gently to avoid mixing and eliminate any bubbles if necessary. Fill the reservoir.
(*g*) Apply the current and run until the free BPB band reaches to about 5 mm. from the lower end of the gel or the protein-BPB band is not less-than 10 mm. away from the end. For enzyme separations work in the cold or use a lower current.
(*h*) Disconnect the current and remove the gels from the apparatus by rimming with a wire or needle and allow the gel to slide into the staining solution.
(*i*) Stain or otherwise locate the bands for the requisite time.
(*j*) Destain, if necessary, and transfer the gels to storage tubes.

Notation. A number of methods for describing relative and absolute migration rates of individual protein bands are in use. Authors have described individual lines by relating their positions to identifiable proteins such as the haptoglobins, or by numbering the fractions of a pattern consecutively. This makes the discussion of findings equivocal or, at any rate, cumbersome. A standardized notation[15] overcomes this difficulty as far as serum proteins are concerned by placing a 0–100 scale along the pattern, from the top of the running gel to the advancing edge of albumin. The stained gels are photographed, and enlarged to the same final length of 100 mm. A millimetre scale is photographed next to the enlarged pattern. (See below for details of the photography, and Figs. 9.11 and 9.13–9.16 for the finished photographs). Bands are then identified by the relative distance they have travelled, in millimetres, i.e. by a direct reading taken from the millimetre scale. Thin lines are described by single numbers, wider bands by their back and front edges, and heavy bands are printed in bold type (e.g. five haptoglobin lines of a type 2–1 serum may be 8, 10, 12, 20, **46–50**).

The notation is reasonably reproducible and preferable to measuring from the centre of the albumin band where reproducibility is marginally better but errors in photography are greater. Taking the position of the free bromophenol blue marker gives more variable results but a more generally applicable procedure. The obvious limitation of the albumin notation is that the presence of albumin is required and its constant electrophoretic mobility assumed. It will work with serum except for rare cases of analbuminæmia and the homozygote state for the electrophoretically abnormal albumin variant in bisalbuminæmia, also very rare.

Photography.[16] The gels are photographed by the transmitted light of an X-ray viewing box. A Kodak X-Ray Illuminator is laid flat on the floor, the gel is placed on the opal screen of the box, and a Kodak half-plate camera is stood 18–24 inches (45–60 cm) above the gel. A strip of black paper is put next to the gel (for inclusion of a millimetre scale when printing). Up to four gels may be photographed on a sheet of quarter-plate film. The film used largely depends on the colour of the protein stain. Ilford N7E. 50 Line film is suitable for gels stained with Ponceau S; it is a high contrast film and gives good definition between the bands. Exposure will vary, but in the particular experimental conditions used the usual exposure is 18 seconds at f.22.

For printing, the negative is set up in the enlarger and the image is enlarged to a length of exactly 100 mm. A transparent millimetre rule is placed on the image over the projected clear film area next to the gel image. The size of this is adjusted so that the origin (the interface of the large and small pore gels) is at 0 mm. and the far end of the albumin band at 100 mm. Printing conditions vary with negative quality.

SECTION 2:

PROTEIN PATTERNS IN HUMAN SERUM: INTERPRETATION

A. L. Tárnoky

Disc electrophoresis is a versatile technique, many variations of method are possible, and the patterns obtained by different methods will also differ. This account of serum protein patterns assumes Ornstein and Davis[1,2] conditions in at least the composition of buffer solutions and preparation of the running and spacer gels.

The greater complexity of disc electrophoresis patterns makes it more informative in many ways than the cellulose acetate separations used in routine diagnostic work. It is, however, unlikely to replace these and its yield of information is not commensurate with the greater complexity, mainly because of the loss of high molecular-weight globulins which do not penetrate the 7 per cent running gel. The greater power of resolution brings its own problem, since it separates genetically determined protein variants: diagnosis by comparing the electrophoretic result with a normal pattern should now include at least mental selection of the appropriate genetic type; at least twelve are known. The problem is not peculiar to the disc technique and exists in starch gel electrophoresis and any other high-resolution methods; but this was the first time it arose in clinical biochemistry where other refined techniques had hitherto been too cumbersome for routine use.

In these conditions the method of assessing the results assumes added importance: (*a*) A computer-based comparison of very large numbers of normal and abnormal gel patterns or scans in the manner discussed by Ornstein(17) presupposes extremes in uniformity of technique and will be the longest in coming but should yield the most information utilizing even small details of pattern, without necessarily involving band identification. (*b*) Interpretation can depend on a comparison of stained gels; a general impression of the pattern is obtained rather in the way one looks at the general appearance of a chromatogram of aminoacids, observing the balance of the picture more than the individual spots; it is followed by comparing the size (width and depth of colour) of individual bands. Either the actual gels or their photographic enlargements can be used; photographs are preferable because they will show whether a doubtful band in the gel is really there or imaginary, and because enlargement of patterns to exactly 10.0 cm renders them uniform and allows direct comparison of sera whether by using the 0–100 mm notation (p. 236) or by putting the photographs alongside each other. Fig. 9.11 shows a normal adult pattern with the millimetre notation of all bands and names of those with an agreed identity. (*c*) Scanning of gels makes quantitative comparison possible. Densitometry of the stained pattern is affected by the differing dye uptake of different fractions(18). The method has been used by Mogi(19,20) whose scans, or their enlargements to 100 mm. can also be used for visual comparison without quantitation. Inspecting the scan will reveal changes in the size of fractions where inspection of the stained band would be misleading, and characteristic shapes of the scanned profile may themselves be diagnostic. Fig. 9.12, taken from Mogi, shows one type of normal and three disease patterns obtained in this way. Unstained patterns can be scanned in *UV* light.(21,22) Gels are held in quartz tubes during electrophoresis which is followed by direct scanning at 280 nm. This method involves fewer quantitative assumptions. (*d*) Gels may be cut into segments, the fractions eluted, read in a spectrophotometer, a colorimeter for stained bands or, with labelled proteins, a scintillation counter. This method is also the best for following changes in one particular protein fraction.

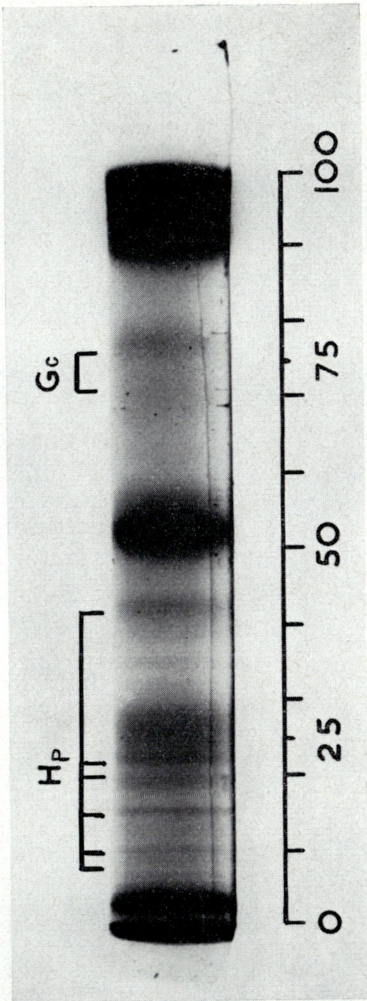

FIG. 9.11. Normal disc electrophoresis pattern using 300 μg. protein and Ponceau S. Male, 28 years, Hp type 2–1, Gc type 1–2. The protein fractions are at **1–4**, 5, 6, 8, 10, 15, 18, 20, **22–28**–31, 34, 37–**41–42**, 47–**48–54**–57, 62, 67–71, 75–77, 85–**87–100**. The main known fractions are 7S-γ-G 0–31; Hp 8, 10, 15, 20, **22**, **42**; main βG **48–54**; Tr **50**; Gc 71, **76**; A **87–100**.

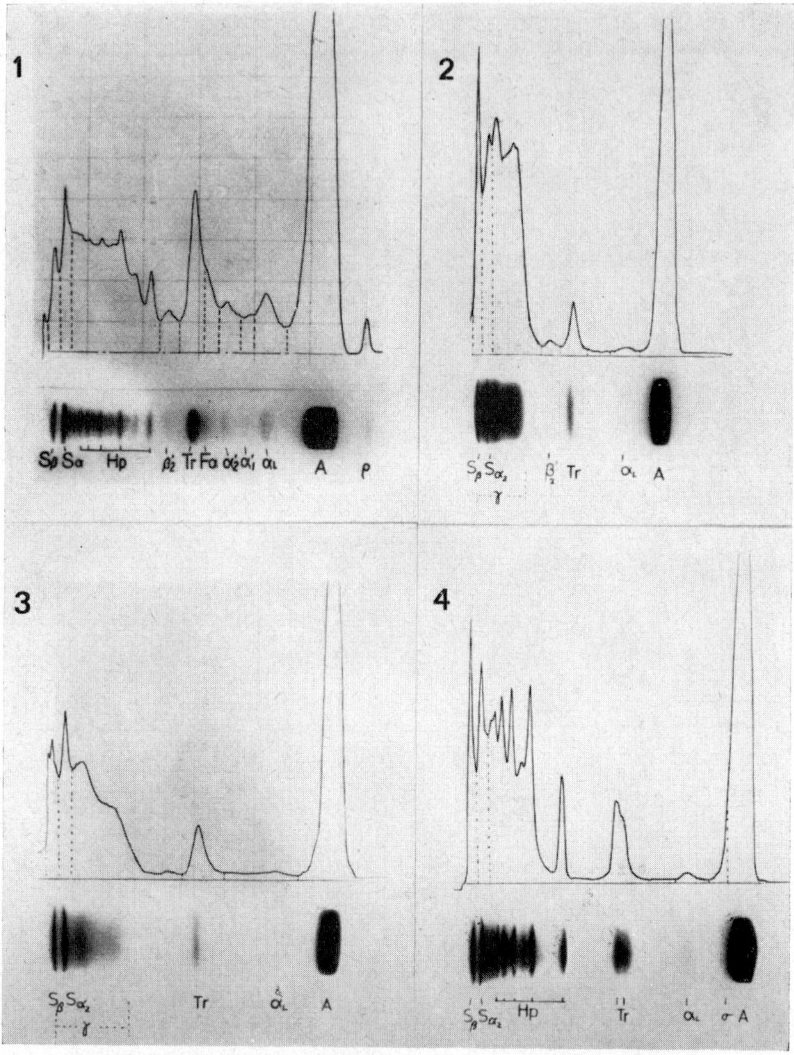

FIG. 9.12. Scanned disc electrophoresis patterns. (Mogi[19, 20]. Reproduced by permission from Seibutsu-Butsurikagaku.) (1) Normal, with fractions at 1, 4, **7·5–8**, 14, 16·5, 26, 31, 35, 41, **48–53**, 61, 73–75, **86–100**, 108; (2) cirrhosis of the liver: **1–6–10–18**–28, 36, 46–50, 72–75, **86–100**; (3) Banti's syndrome: **0–1·5–4·5–5·5–9–13**–23, 49–52, **92–100**; (4) ca. stomach: **1–5–10–11–14–23**–25, **32–34**, **50–53**, **54–57**, 75–76, **89–100**.

Any of these procedures may be supplemented (*e*) by special staining procedures for individual proteins or (*f*) by immunoelectrophoresis. The choice may eventually lie between detailed surveys of the (*a*) type and a *selective* use of the other procedures now available. Method (*b*) is the simplest for general purposes and will be followed here. Bands are referred to by their position on the 100 mm. pattern, though the name of identifiable fractions is sometimes also given.

Published work on general clinical interpretation is still scanty. This summary is based on the findings of Mogi,[19,20] Nishioka *et al.*,[23] Pastewka, Ness and Peacock,[24] Raunio, Ruoslahti and Krause,[25] Tárnoky *et al.*, [26,27] and of Zingale, Mattioli and their co-workers.[28,29,30] Some of these papers deal only with normal sera. Further clinical uses of the technique, not covered here, are indicated in the publications by Naito *et al.*[31] (lipoproteins), Smith *et al.*[14] and Goldberg *et al.*[32] (alkaline and acid phosphatase isoenzymes), Kleiner and Schram[33] (isoenzymes hydrolysing *l*-leucyl-β-naphthylamine) and Parish and Marchalonis[34] (molecular weight of proteins).

The Normal Pattern. This varies with age, and with genetically determined protein variants. Generally speaking there is a progressive change in pattern throughout life. The serum from normal cord blood shows bands at 0–2, **3–5, 51–54** (a thin main β-globulin band incorporating transferrin at 50), and albumin at **90–100,** understaining at the 0–30 end, and a few faint bands. Albumin decreases during the next months and then rises to the original level at 1 year. At this stage there are bands at **0–2, 3–5,** and a few thin bands in the 0–25 zone. The main β-globulin band around 50 tends to be wider than in later life. The number of lines increases with age, and the 0–30 zone fills in again during the second decade of life. Lines between 65 and the albumin develop somewhat later; haptoglobin lines darken and are pronounced bands in people in their sixties. The typical adult pattern contains 15–25 bands; their position is characteristic and varies only by a few millimetres on the 100 mm. scale. The monomeric albumin band is 12–20 mm. wide when enlarged to this scale, i.e. it is e.g. at **80–100** or **86–100.** Albumin oligomers are in the immediate post-albumin zone or travel with the main β-globulin band. The main β-globulin band, including transferrin, is around 50, e.g. at **48–54** or **50–57**. Transferrin appears at **50**, cæruloplasmin at 56. The majority of patterns contain lines at 35 and 42. The position of the haptoglobin bands does not seem to be agreed and varies with the technique (since adding hæmoglobin to the serum for identification affects the mobility of these bands); a simplified picture is that of a single Hp line near the main β-globulin band for the homozygous Hp 1–1 type several lines usually three to five in the main positions 8, 12, 14, 18, 20 (without the addition of hæmoglobin) in type Hp 2–2, and both types of line, with small changes in position, for the Hp 2–1 heterozygote. The lines of the Gc group-specific protein types are a line at **76** in homozygote Gc 1–1 (with faint lines around 68 and 70 which do not appear to be immunologically Gc lines), a strong **71** and weaker 76 line in Gc 2–2, and the reverse 71, **76** for Gc 1–2. Prealbumin bands have been reported at 122–125 and even at 133 but are often not found, and it is possible that in some techniques they run off the gels.

Hæmolytic Disease of the Newborn. A paucity of bands is typical, showing at 1–3 days only albumin, a thin main β-globulin band around 50–52 which seems fainter than in normal cord blood, and very faint haptoglobin bands. Sharp **0–1** and **3–5** bands may be seen.

Anæmia. Iron deficiency anæmias of some severity (Hb near 40 per cent, serum iron 10 and TIBC 500 μg/100 ml) show reduced haptoglobin bands but marked lines at **4–5, 5–6** or both positions. One case of severe megaloblastic anæmia (Hb 20 per cent) had a blurred pattern consisting of only:

0, 2, 2·5, 3, (faint 4–26 background), 50–53, trailing albumin.

In pernicious anæmia the haptoglobins tend to disappear, with a reduction of the main β-globulin band around 50 and of the 0–30 background in which, however, a number of sharp bands now appear.

Pneumonia. Though haptoglobin bands have been reported as increasing in infection, they are more often seen decreased in pneumonia; there are some not very marked decreases in the main β-globulin band around 50.

Prominent, unusually shaped, biconvex bands were seen in the pattern of a 4-year-old child with pneumonia; those at **0–2, 2–3, 3–4, 5, 6–7, 10–12, 16–18** were the most prominent, some clearly haptoglobins. The pattern contained a faint band at 108–110.

Septic Abortion. A case of a fatal *Cl. welchii* infection (methæmoglobinuria, heavy proteinuria) was remarkable in having the main β-globulin band, usually near 50, moved to 57–**60–66**–71. The haptoglobin lines were weak, and the 0–30 background was absent. All the changes added up to a foreshortened appearance of the pattern, with bands compressed towards the right (100).

Erythema Nodosum due to streptococcal infection shows a diminished main β-globulin band, fewer and blurred bands in the 25–80 zone but a number of clear bands at the 0–20 end.

Infectious Mononucleosis (glandular fever) may exhibit a normal pattern for the first fortnight. It changes to fewer bands than present normally, a smaller main β-globulin band, but often an increase in the **0–1** and **3–5** bands and a further stained band in the spacer gel immediately over the running gel boundary, a pattern which may persist for some months afterwards.

Mumps is accompanied by a rise in haptoglobin bands.

Duodenal ulcer patterns tend to be emptier than normal, with no very distinguishing characteristic.

Tuberculosis. A case of primary pulmonary tuberculosis shows a dark 0–28 background and a main βG band thinned down to 50–53.

Cholecystitis. This acute illness shows an essentially normal 0–100 pattern with heavy staining in the spacer gel.

Rheumatoid arthritis. The transferrin and haptoglobin bands show clearly over light, diminished slow β and main β areas.

Pulmonary Aspergillosis. The one case seen is of sufficient interest to be mentioned. It shows a rise in haptoglobin lines, a decrease in the

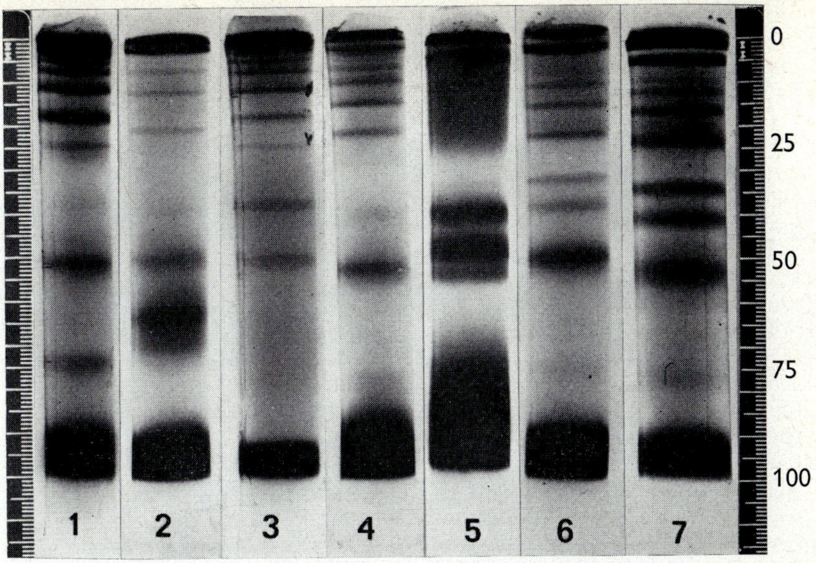

Fig. 9.13. Abnormal serum protein patterns obtained by disc electrophoresis: (1) pneumonia; (2) Cl. welchii infection; (3) erythema nodosum of streptococcal origin; (4) mumps; (5) sarcoidosis; (6) Crohn's disease; (7) adult nephrotic syndrome.

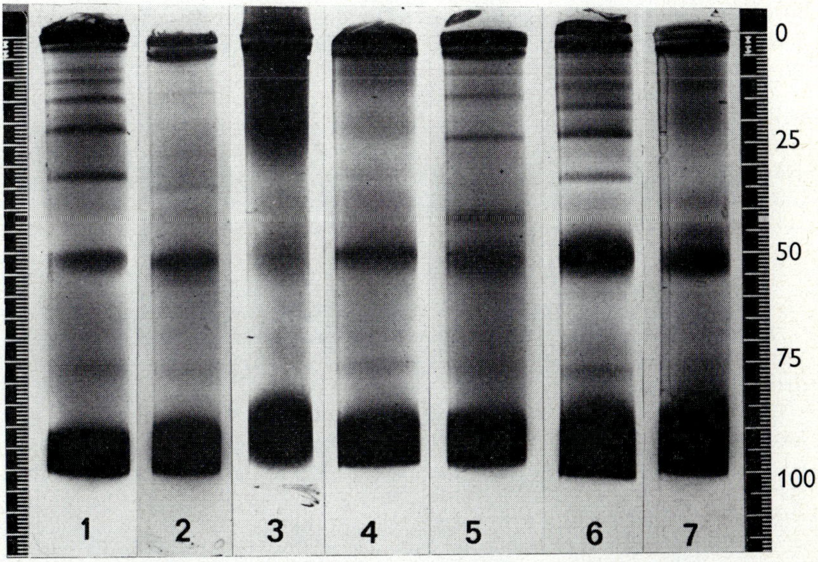

Fig. 9.14. Abnormal serum protein patterns obtained by disc electrophoresis: (1) early infective hepatitis; (2) recovering infective hepatitis; (3) hepatic coma; (4) lupoid hepatitis; (5) myxœdema; (6) essential hypercholesteræmia; (7) thyrotoxicosis.

main β-globulin band near 50, and a general lightness of background over most of the gel, unusual even for acrylamide.

Crohn's Disease. Though the **4–6** band is heavier in some cases and occasionally separates into **4–5–5–6,** the pattern is not consistent and largely mirrors the heterogeneous picture seen on cellulose acetate electrophoresis. It reverts to normal a few months after resection but may remain abnormal during remission in the unoperated patient.

Ulcerative Colitis. There is a general impression of sharpened bands, mainly of the main β-globulin, but the pattern is not uniform and seems unrelated to the changes in α_2- or γ-globulin on cellulose acetate.

Liver Disease. The early phase of infective hepatitis is marked by raised haptoglobin bands with or without a darkening of the 0–30 background. The transferrin and main β-globulin bands (both at 50) are likely to remain low throughout the illness. In the hepatocellular phase the serum of the recovering patient shows decreasing haptoglobin bands. If at this stage the 0–30 background darkens unusually, this suggests developing cirrhosis or liver failure. Chronic hepatitis, in which the IgM is high, develops a light 5–50 zone.

Active chronic hepatitis patterns give a general dark–light–dark–light–dark impression in which the first dark area is the increased background of IgG in the 0–30 zone, the centre is a normal main β-globulin band and the third, albumin.

Renal Disease. Children with acute glomerulonephritis have slightly diminished haptoglobins and an increase in the **4–6** band (α_2-macroglobulin). The number of bands is that of normal serum. The pattern in chronic nephritis is similar, with, in one case, a sharp prealbumin band at 121 and a band moving immediately ahead of the main β-globulin, making a composite band of 44–**45–50**–55.

Sera from nephrotic syndrome cases often (but not invariably) show raised transferrin bands (**50**). Neither child nor adult nephrotic syndrome gives a single pattern as typical as that on cellulose acetate. Bands in the 0–30 zone (which may understain) tend to appear stronger and those of the 30–50 zone weaker, and this shift of balance to the left seems to be the most consistent characteristic of an otherwise heterogeneous picture.

Uræmia shows the pattern of the disease that has caused it. A striking decrease in haptoglobins is seen in uræmia with hæmolysis.

Thyroid Disorders. Hyperthyroid serum shows a largely normal pattern, but the bands around 50 may appear sharper and more articulated than usual.

The tendency in hypothyroidism is towards emptier-looking gels. The band around 50 is simpler, besides often being weaker, and there appears to be a very approximate inverse relationship between its size and the serum cholesterol level. Thyroidectomy may give this picture.

Sarcoidosis. In this disease disc electrophoresis reflects the position of other electrophoretic methods: it may or may not give an abnormal pattern which may or may not appear typical. But where the pattern on acrylamide seems characteristic it gives several distinct types which may perhaps form the basis of further classification in terms of the mode of onset and stage of disease.

1. Heavy background staining of the 0–30 zone due to IgG up to about 18 and to IgA above it.

2. Heavy haptoglobin bands which may show through the background in 1;

3. Unusually heavy mid-region bands consisting of the usual, but more prominent, **46–52** with or without a strong **39–42** and weaker bands to give an articulated band, e.g. **39–43**–**46–51**–56;

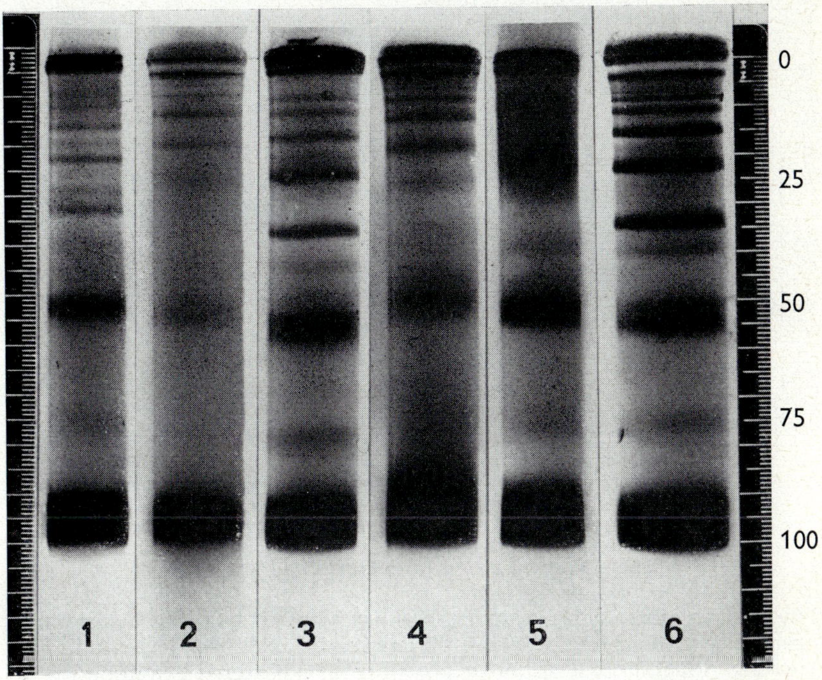

FIG. 9.15. Abnormal serum protein patterns obtained by disc electrophoresis: (1) ca. œsophagus; (2) ca. prostate, metastases in bone; (3) ca. liver; (4) ca. head of pancreas, liver metastases; (5) ca. ampulla of Vater; (6) hypernephroma.

4. Unusually elongated band or bands ending in albumin. Some or all of these characteristics may occur together.

Malignant Diseases. Carcinoma presents a varied picture, with a few fairly characteristic patterns. Disc electrophoresis may eventually provide the means of further differentiation of tumours and it may become possible with more experience to use the pattern to locate the site of the primary neoplasm where only the secondary sites are known and to relate patterns to histological appearances.

In general there is a reduction of bands between the main β-globulin and albumin, so that the 0–50 half of the patterns seems the more important. Splitting of the main β-globulin band has occurred in a few

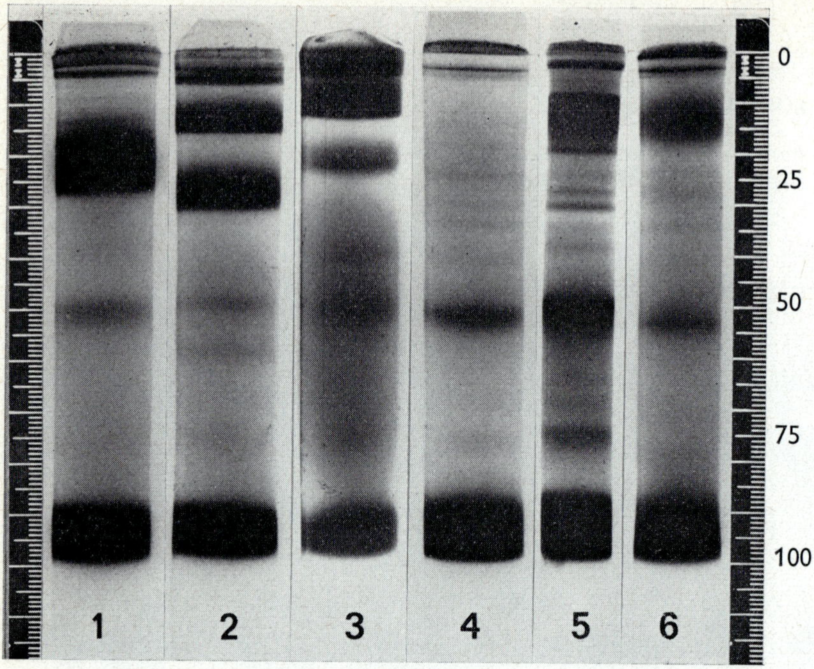

Fig. 9.16. Disc electrophoresis patterns of paraproteinaemia. (The letters in brackets give the position of the abnormal band on cellulose acetate electrophoresis.) (1) IgA (β_2G); (2) IgG and IgA (β_2G); (3) IgG ("M"-fraction); (4) IgM and immunosuppression (γG); (5) treated with Melphalan and (6) six months later, before the next course of treatment.

cases. Carcinoma of the œsophagus shows a decreased 0–30 background, generally thinner lines in the 0–50 half of the gel and a decrease in the main β-globulin band. There is probably no typical pattern in prostatic carcinoma, but the majority of cases with metastases in bone have a reduced 0–30 background and reduced haptoglobins while both these are raised in cases with liver secondaries. Bronchial carcinoma shows no obvious characteristics. Metastases in the liver are accompanied by a general darkening and/or increase of bands in the 0–35 zone. Primary carcinoma of the liver may show an α-fetoprotein band at 80–83. Carcinoma of the head of the pancreas or the ampulla of Vater both show, probably because of the biliary obstruction, a heavy left-hand side to the pattern, with raised haptoglobin lines, an increase in the main β-globulin band and cloud-like staining around some of the bands of the 0–50 area, not fully settled into discs.

One case of Cushing's syndrome (adrenocortical carcinoma) had strong **0–2** and **4–5** lines, with haptoglobin lines standing out well above the light 0–30 staining. One case of hypernephroma had definite bands, slightly biconvex in shape, in the 0–35 zone; the main β-globulin band had split into 50–53, **53–56.**

Myeloma patterns show the expected further differentiation of protein types as compared with paper or cellulose acetate separations. Depending on its concentration, IgG will be found at around 4–20, e.g. at 4–14 or 7–19, and IgA in the 17–28 zone. IgM does not enter the gel. An IgM myeloma or macroglobulinæmia pattern will appear empty. One case with IgM and κ chains shows four heavy adjoining bands, almost one articulated 24–36 area, with an almost completely unstained portion at 62–74. A heavy band up to 15 or 20 followed by a white zone is characteristic of an IgG myeloma with immunosuppression of IgA. Suppression of IgG in an IgA myeloma looks less striking. Cellulose acetate patterns with one abnormal band in the β_2-position have given one disc pattern with a heavy abnormal band at 16–28, a sharp, straight cut-off of the advancing edge at 28, and a smaller, slower band trailing into it from 14; the other gel contains bands at **5–7**, 10–**11–17** and 21–**24**–32, again with the unusually straight edge at 17. Sera with an "M"-band between the β_2- and γ-positions on cellulose acetate have given abnormal-type bands at **0–6**, **7–14** and 17–**19–25**, or only one band at **0–15**–16. The γ-myeloma pattern of cellulose acetate has given a single **10–20** myeloma band, or two at **0–5** and **7–11**, or none. Untreated myelomatosis generally gives an otherwise poorer pattern and —except for the abnormal bands—a blurred appearance.

Hypogammaglobulinæmia. The sera show almost no background at 0–30. There is no visible change in the (larger pore) spacer gel.

Hyperlipoproteinaemias. The electrophoretic lipoprotein patterns[31] of such cases will be be more characteristic, but some features of the protein pattern may be noted. A family with hypercholesterolaemia and raised serum total lipids (W.H.O. type II) had heavier 0–2 and 3–5 bands than usual, with a tendency to darker 0–30 staining for their age, and some blurring of lines up to 45. The raised background and blurring were seen also in the normocholesteraemic members of this family.

A patient with a high serum cholesterol and total lipid had a faint and somewhat blurred pattern in the 4–45 zone, and the spacer gel contained much protein of a lipoid appearance. When a fat-free diet brought the cholesterol into the normal range and reduced the lipid level (though not to normal) the appearance of the spacer gel persisted but the 4–45 zone became more normally differentiated. The patient's two children both had normal cholesterol but raised lipid levels; their patterns showed the same spacer gel protein, but the pattern in the running gel was normal, except for raised 20–22 bands.

Other cases of W.H.O. type IIa show blurring over 6–50, some with their main β-globulin band displaced towards 52–60. W.H.O. type IV sera show displacement but no blurring. A patient, W.H.O. type unknown, admitted with a myocardial infarct had heavy 0–2 and 3–5 bands in an otherwise normal pattern.

The above attempts a consensus of published opinion, some of it perhaps based on too few observations and some mutually contradictory. Electrophoresis on thin-layer acrylamide gel gives patterns broadly similar to the disc method; the present findings may be used for it in a general way, i.e. the descriptions of patterns apply but the

100-mm. positions do not. Thin-layer methods fairly close to Ornstein and Davis conditions (e.g. Nandi and Lewis[35]) are of course more comparable.

But the main application of disc electrophoresis still seems to be to individual problems rather than supplanting existing screening techniques in clinical chemistry. On present knowledge such problems are likely to be:

a more precise characterisation of myeloma types and a means of following the progress of treated cases;

classification of sarcoidosis;

improved diagnosis of carcinomas, aimed at locating primary sites;

following the effect of treatment in a number of conditions, e.g. when giving thyroxine in myxœdema;

classification into hereditary protein types (haptoglobin even in stored blood up to 2 years old,[42] group-specific protein, transferrin); and

the detection and measurement of selected protein bands by special methods (staining or immunochemical).

This limitation applies to the practice of fixed technique runs where only the subsequent handling of the gel is varied, and not to the great latitude of conditions in which the actual disc electrophoresis can be carried out.

Hæmoglobin Separations

Although acrylamide gel has been used for the electrophoresis of the hereditary hæmoglobin variants[36,37,38] disc electrophoresis has provided few data. Neerhout *et al.*[39] described a useful quantitative method for determining hæmoglobin A_2 with good differentiation between normals and patients with thalassæmia minor or heterozygotes for the sickle cell trait. They used Ornstein and Davis conditions with only minor changes. Good qualitative separation of at least some of the hæmoglobins is obtained under Ornstein and Davis conditions (A, A_2, C, F, J, S) but AF separations are not successful. A three *p*H system[40] in which the buffers are *p*H 9·1→8·9→8·6 (instead of Davis's 8·3→8·9 →8·3) gives somewhat sharper separations but, since it changes relative mobilities from $C<A_2<S<F<A$ under standard conditions to $A_2<C<S<F<A$, separation of SC is poor. Cellulose acetate, being simpler, is the better method for qualitative work, with the exception of A–A_2 separations for which disc electrophoresis is preferable. Smith and Evatt[41] achieved improved separations (though not of A and A_2) at *p*H 9·2, by a method further removed from Ornstein and Davis conditions.

REFERENCES

1. Ornstein, L. *Ann. N.Y. Acad. Sci.*, 1964, **121(2)**, 321.
2. Davis, B. J. *Ibid*, p. 404.
3. Williams, D. E., and Reisfield, R. A. *Ibid*, p. 373.

4. Clarke, J. T. *Ibid*, p. 428.
5. Hjerten, S., Jerstedt, S., and Tiselius, A. *Anal. Biochem.*, 1965, **11**, 219.
6. Anker, H. S. *Febs. Letters*, 1970, **7**, 293.
7. Choules, G. L., and Zimm, B. H. *Anal. Biochem.*, 1965, **13**, 336.
8. Smith, I., Perry, J. D., and Lightstone, P. J. *Clin. Chim. Acta*, 1968, **19**, 499.
9. Reisfield, R. A., Lewis, U. J., and Williams, D. E. *Nature*, 1963, **195**, 281.
10. Broome, J. *Ibid*, 1963, **199**, 179.
11. Colover, J. Personal communication.
12. Hjerten, S., Jerstedt, S., and Tiselius, A. *Anal. Biochem.*, 1965, **11**, 211.
13. Leaback, D. H. "An Introduction to the Fluorimetric Estimation of Enzyme Activities". publ. Koch-Light Laboratories, Colnbrook, Bucks., England.
14. Smith, I., Perry, D. J., and Lightstone, P. J. *Clin. Chim. Acta*, 1971, **35**, 59.
15. Dowding, B., and Tárnoky, A. L. *Proc. Assoc. Clin. Biochem.*, 1966, **4**, 18.
16. Williams, L. F. *Med. Biol. Ill.*, 1969, **19**, 105.
17. Ornstein, L. *J. Mount Sinai Hosp.*, 1965, **32**, 437.
18. Kruski, A. W., and Narayan, K. A., *Biochem. Biophys. Acta*, 1968, **168**, 570.
19. Mogi, G. *Seibutsu-Butsurikagaku*, 1965, **11**, 99.
20. Mogi, G. *Seibutsu-Butsurikagaku*, 1965, **11**, 171.
21. Dravid, A. R., Fredén, H., and Larsson, S. *J. Chromatogr.*, 1969, **41**, 53.
22. Watkin, J. E., and Miller, R. A. *Analyt. Biochem.*, 1970, **34**, 424.
23. Nishioka, M., Hironaga, K., and Fujita, T. *Clin. Chim. Acta*, 1971, **31**, 439.
24. Pastewka, J. V., Ness, A. T., and Peacock, A. C. *Clin. Chim. Acta*, 1966, **14**, 219.
25. Raunio, V., Ruoslahti, E., and Krause, U. *Acta path. et microbiol. Scandinav.*, 1966, **67**, 424.
26. Tárnoky, A. L., and Dowding, B. *Clin. Biochem.*, 1967, **1**, 48.
27. Tárnoky, A. L., and Vickers, M. F. (to be published).
28. Zingale, S. B., and Mattioli, C. A. *Medicina*, 1961, **21**, 121.
29. Zingale, S. B., Mattioli, C. A., Bohner, H. D., and Bueno, M. P. *Blood*, 1963, **22**, 152.
30. Zingale, S. B., Cedrato, A. E., Viscardi, E. B., and Mattioli, C. A. *Argentine Soc. Clin. Invest.*, Nov. 1964.
31. Naito, H. K., Wada, M., Ehrhart, L. A., and Lewis, L. A. *Clinical Chemistry*, **19**, 228.
32. Goldberg, A. F., Takakura, K., and Rosenthal, R. L. *Nature*, 1966, **211**, 41.
33. Kleiner, H., and Schram, E. *Clin. Chim. Acta*, 1966, **14**, 377.
34. Parish, C. R., and Marchalonis, J. J. *Analyt. Biochem.*, 1970, **34**, 436.
35. Nandi, M., and Lewis, G. P. *J. Clin. Path.*, 1970, **23**, 727.
36. Ferris, T. G., Easterling, R. E., and Budd, R. E. *Blood*, 1962, **19**, 479.
37. Curtain, C. C. *J. Clin. Path.*, 1962, **15**, 288.
38. Nakamichi, M., and Raymond, S. *Clinical Chemistry*, 1963, **9**, 135.
39. Neerhout, R. C., Kimmel, J. R., Wilson, J. F., and Lahey, M. E. *J. Lab. clin. Med.*, 1966, **67**, 314.
40. Dowding, B., and Tárnoky, A. L. *Proc. Assoc. Clin. Biochem.*, 1967, **4**, 167.
41. Smith, E. W., and Evatt, B. L. *J. Lab. Clin. Med.*, 1967, **69**, 1018.
42. Hilgermann, R. *Z. Rechtsmed.*, 1972, **71**, 222.
43. Zeineh, R. A., Mbawa, E. H., Pillay, K. G., Fiorella, B. J., and Dunea, G. *J. Lab. Clin. Med.*, 1973, **82**, 326, and *Biochem. Biophys. Acta*, 1973, **317**, 1.

CHAPTER 10

CONCENTRATION GRADIENT POLYACRYLAMIDE GEL ELECTROPHORESIS

D. H. Leaback

THE resolving power of methods for the electrophoretic separation of proteins was increased greatly by the introduction of agar, starch and polyacrylamide gels as media for the electrophoretic processes. These gels retard the migration of the proteins and there is little doubt that the increased resolving power is due to the associated molecular-sieving effects. However, there has been some confusion as to which model of the structure of the gels might best account for the observable facts. Thus, a "fixed-pore" model was originally suggested[(1–4)] for polyacrylamide and dextran gel matrices, and it was envisaged that such pores would form insuperable barriers to spherical macromolecules of appropriate sizes. The model of the gels which has emerged in recent years is essentially that of Ogston[5] in which the gels are best pictured as a random mesh of hydrophilic fibres with "pores" whose cross-sections are not typically circular; whose dimensions are scattered about mean values, and whose boundaries may even be deformable.[(6,7)] It has been known for some years that if a mixture of proteins migrating in a weak gel encounters a boundary of gel at higher concentration, then the proteins tend to concentrate or "stack" into sharper zones due to the retarding effect of the denser gel matrix. Some workers have employed gel cylinders with up to five zones of increasing polyacrylamide concentration for the electrophoretic separation of proteins and have claimed increased resolving power.[(8)]

During the period 1965–1969 several groups of workers[(9,15)] reported the preparation of polyacrylamide gels of continuously increasing concentration and also described their use in the electrophoretic separation of proteins. Such gel preparations are sometimes known as "gradient pore" gels and have been prepared in both gel cylinder and slab formats. Gels with linear[(10,11,13,15,16,17,18,76)] concave[(11,12,19)] and sigmoid[(20,21)] gradient contours have been reported with concentrations up to about 30% (w/v) of polyacrylamide. Slab gels with concave gradients of about 4–30 are available commercially[(22–25)] together with associated equipment for vertical slab gel electrophoresis and the subsequent processing of the products.

The present chapter will include a description of the preparation and use of slab gels with concentration gradients of 4–30% (w/v polyacrylamide), but the methods can also be used for the production and use of gels with lower gradients (including homogeneous concentration gels) or for the preparation of similar cylindrical gels.

Apart from the high resolving power, a principle attraction of the method is that proteins assume increasingly stable positions in the gels as they approach what is termed their "pore limits" and that their penetrations into the gels are relatable to the molecular weights of the individual proteins. Using linear gradient gels, empirical relationships between these quantities have been recorded[10,15] and these relationships have been rationalized on the basis of the theoretical treatment for gradient gel electrophoresis advanced by Rodbard *et al.*[6]

Theoretical Considerations

It is now well established[26,27] that the velocity dx/dt of migration of a globular protein in a gel of total concentration T is related to the velocity of migration $(dx/dt)_{T=0}$ in the absence of the gel (but otherwise similar conditions) by the Ferguson equation—

$$\frac{dx}{dt} = \left(\frac{dx}{dt}\right)_{T=0} \exp\,(-bT) \qquad (1)$$

where b is the "retardation" coefficient.

If the protein is migrating electrophoretically into a gel with a linear concentration gradient, the concentration of the gel at any point may be represented by

$$T = T_0 + a_2 . x \qquad (2)$$

By calling the total length of the gel unity, and by expressing the distance (x) penetrated into the gel by a protein as a fraction of the total length, then equations (1) and (2) reduce to—

$$\frac{dx}{dt} = \exp.\,[-b(a_1 + a_2 . x)] \qquad (3)$$

Using the boundary conditions $dx/dt = 1$ when $x = t = 0$, equation (3) gives—

$$\frac{dx}{dt} = \frac{1}{b . a_2 . t + 1} \qquad (4)$$

and solving equation (3) with the same boundary conditions gives[6]

$$x = \frac{1}{b . a_2} . \log_e(b . a_2 . t + 1) \qquad (5)$$

After very long time intervals $b . a_2 . t \gg 1$, and equations (4) and (5) become

$$\frac{dx}{dt} \doteqdot \frac{1}{b . a_2 . t} \qquad (6)$$

and

$$x \doteqdot \frac{1}{b . a_2} . \log_e(b . a_2 . t) \qquad (7)$$

It follows from equations (6) and (7) that the movement of a globular protein in the gel gradient is never actually zero for any finite values of time, t. This is in accord with the experimental findings of Slater[10] and others. Nevertheless high values of the retardation coefficient b (i.e. small pore gels) and for high values of a_2 (i.e. steep gel gradients)

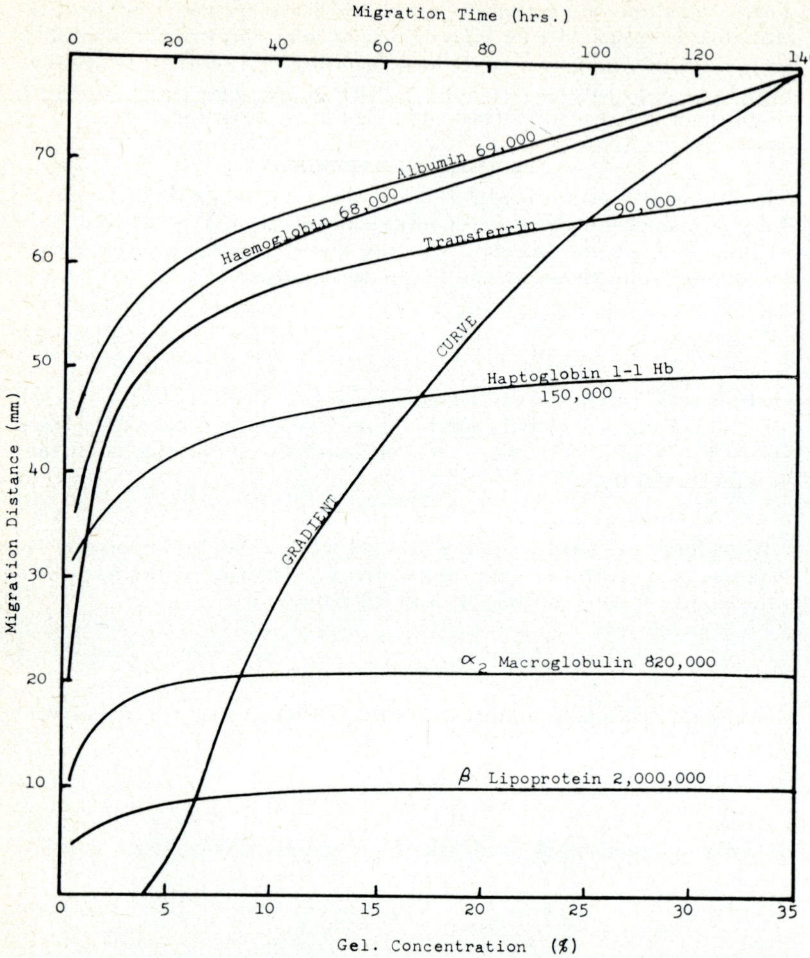

FIG. 10.1. Progress curves of Protein Migration into a concave 4–35% (w./v.) polyacrylamide gradient *p*H 8·3 and 23° using a potential difference of 100 V. (By courtesy of Dr. J. Margolis.)

would render many globular proteins of adequate size essentially stationary for all practical purposes (see Fig. 10.1).

It has been established[27] that the retardation coefficient b can be related to the radius of the protein molecule (R) and to that of the gel fibres (r) in the following way—

$$b \propto (R + r)^2 \qquad (8)$$

Moreover assuming similar partial specific volumes, the molecular weights (M) of globular proteins can be related to their radii by—

$$M \propto R^3 \qquad (9)$$

then if the radii of the gel fibres are small compared with that of the protein molecules i.e. $R \gg r$, then—

$$\log b = 2/3 \log . M + \text{const.} \tag{10}$$

Taking logarithms of equation (5) and combining with (10) we get for a homologous series of proteins (whose velocities of migration in the gel free system can be regarded as the same and equal to unit at $x = 0$) then—

$$\text{Log} . X = -2/3 \log . M. + \text{const.}$$

or

$$\log . M. = -3/2 \log . x + \text{const.} \tag{11}$$

Equation (11) is similar to the empirical relationships reported by Slater,[10] and by Kopperschlager *et al.*[15] (who obtained lines with slopes of $-1{\cdot}75$ and -2 respectively). It should be noted however, that from assumptions made above, the relationship (11) is unlikely to hold for non-globular proteins or for non-homologous series of proteins differing significantly in partial specific volumes or in net electric charge.

The above treatment can be extended to any gel gradient as defined by

$$T = f(x) \tag{12}$$

and in combination with equation (1) can, in principle, be solved to give relationships for the velocity of migration and the penetration distance into the gels. The signs are[6], that the solutions will be very complex for other than linear gel gradients.

The preparation of polyacrylamide concentration gradient gels

When it is necessary to prepare polyacrylamide concentration gradients in either cylindrical or slab gel forms, the most usual approach is to mix a gradient of two acrylamide polymerization solutions differing in concentration, and to introduce this polymerizing gradient into appropriate gel "moulds" by upward displacement of dilute (5% w/v) aqueous ethanol. The polymerizing solution is fed into the system such that the higher concentration (denser) acrylamide solution is at the lower ends of the gel "moulds". Some workers[11,40] have also incorporated gradients of sucrose and polymerization inhibitors respectively to provide extra density stabilization and to ensure the required polymerization sequence. Adequate stabilization is provided here during the polymerization without the sucrose gradient, and the required polymerization sequence is achieved by temperature differences in the component solutions and by suitable concentrations of polymerization initiator and accelerator. The required polymerization sequence is from top to bottom of the gel moulds and this is necessary in order to protect the potential polymer concentration gradient while large quantities of heat of reaction are generated at the bottom of the gel moulds.

Cylindrical concentration gradient gels can be formed in conventional "disc" electrophoresis tubes (say 70 mm. long, 5 mm. internal diameter) but the corresponding slab gels are usually formed in assemblies consisting of two glass plates (say 83 × 83 × 1 mm.^{3}) held together by waterproof adhesive tape but spaced 4 mm apart by two 83 × 4 × 2·7 mm.^{3} glass strips at the outer edges of the plates (see Fig. 10.3).

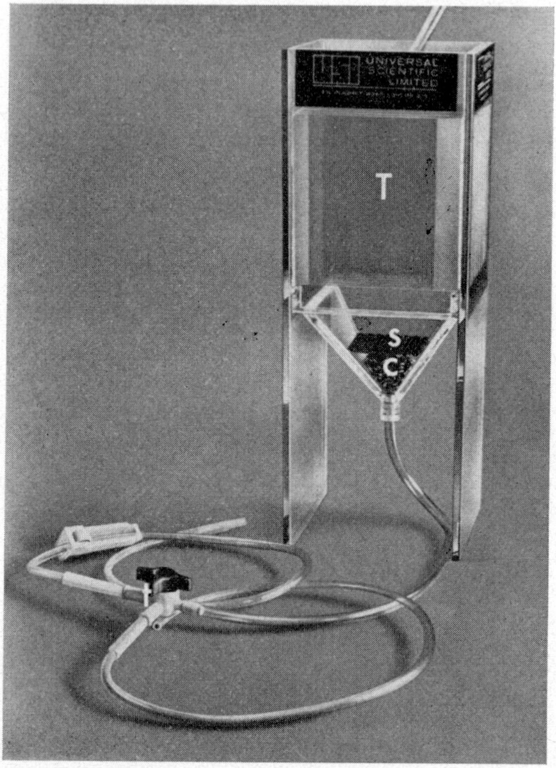

Fig. 10.2. Apparatus for Concentration Gradient Gel Formation. Transparant plastic tubing connects the gel forming tower (T) with the three-way tap (t). Within the tower is a coiled aluminium chain (c, to help stabilize the gradient as it passes up into the gel moulds) and a fine mesh metal screen (S, to trap any air bubbles that may get into the system). (By courtesy of Messrs. Universal Scientific.)

In the interests of reproducibility of results from gel to gel, there is a strong case for the simultaneous preparation of gel gradients in batches. Probably the best way to do this is to pack the gel tubes or slab assemblies compactly into a gel-forming "tower". Such a tower (Fig. 10.2) is available[22–24] commercially and is well suited for the

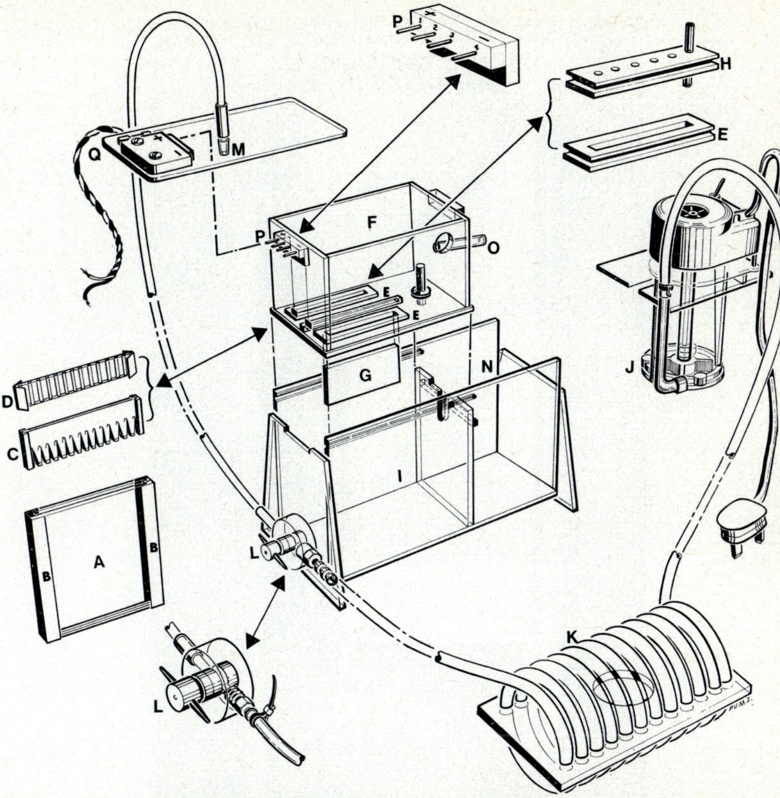

Fig. 10.3. Apparatus for cooled, vertical gel Electrophoresis. The slab gel mould assembly (A) consists of two 83 × 83 × 1 mm.[3] glass plates spaced 2·7 mm. apart by two glass strips and held together by waterproof adhesive tape. (B). Sample wells can be cast in the gels by means of the comb, (C) or samples can be added to up to 12 of the fourteen recessed compartments of the sample applicator (D) when the latter is in contact with the top surface of the gel. A slab gel assembly is inserted 1 cm. upwards into either of the two (silicone greased) neoprene gaskets (E) in position in the upper electrode chamber (F). If only one slab gel is to be run then the remaining gasket is plugged by the plastic plate (G). If cylindrical gels are to be used, gaskets of type (H) are substituted. The electrode chamber and the attached gel assemblies are lowered into position such that the latter are immersed in the buffer in chamber (I), and the buffer circulation proceeds via the pump (J) and the heat-exchanger (K; immersed in a cooled bath) through the flow regulator (L) into the upper and lower chambers at M. The buffer then returns to the pump containing chamber (N) through the outlet (O). A D.C. power supply suitable for the electrophoresis is applied to the plug (P) via the wired socket (Q). (By courtesy of Messrs. Universal Scientific.)

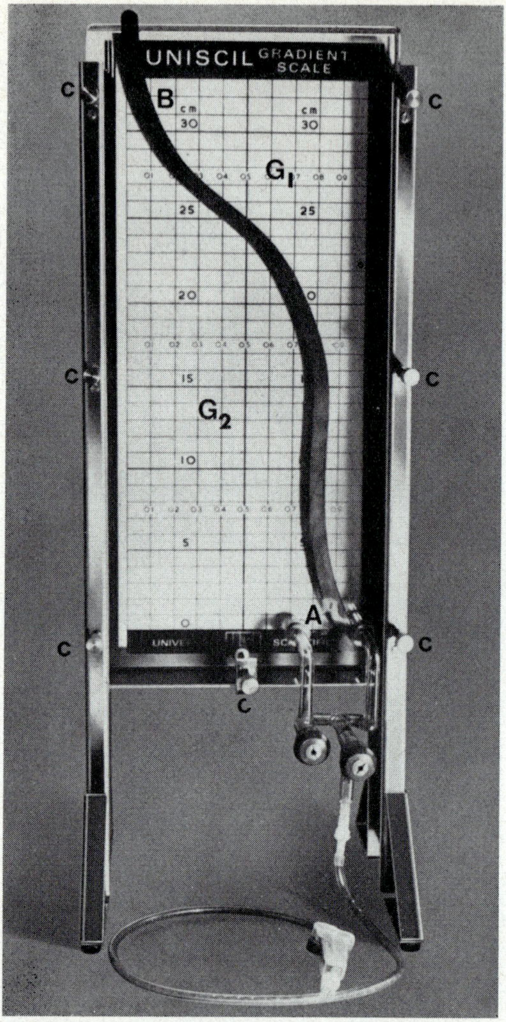

FIG. 10.4 A versatile Gradient Mixer.[22,31] The two component solutions (G_2 and G_2) of the proposed gradient are placed in the compartments indicated formed after adjusting the flexible partition AB to the required gradient contour and is secured by the clamps (C). The two solutions drain through tubes (H) and are mixed within the plastic tubing. As depicted in the figure, the first part of the gradient produced would contain more of the solution in the space G_1 than that in the space G_2. (By courtesy of Messrs. Universal Scientific.)

simultaneous preparation of up to about 15 slab gels with the above dimensions, but has also been used to form concentration gradient polyacrylamide gels in cylindrical form by packing the tower with appropriately sized tubes.[11]

Some types of concentration gradient can be formed using pumps and other means (see p. 276–7) but the versatile gradient forming device described by Margolis[31] (and available commercially)[22] is particularly convenient since it is simple, readily cleanable, and permits the production of gels with a wide variety of concentration gradient contours.

Requirements for Gradient Gel Formation

The procedure described here is designed for the preparation of batches of concentration gradient gel slabs, but it can be adapted to the preparation of fewer slab gels or to the production of cylindrical gels in batches, or even to the preparation of homogeneous concentration gels in either cylindrical or slab forms.

The experimental requirements are as follows:—

(1) **One gel forming tower** (Fig. 10.2). One of the kind described by Margolis and Kendrick[11a] is ideal for the formation of gel slab and cylinders, but procedures and apparatus devised specifically for gradient formation in gel cylinders in batches[12,28,29] and in individual gel slabs[20,30] have also been described.

(2) **An appropriate number of gel mould assemblies.** The procedure and tower are ideal for 14 gel mould assemblies of the kind described earlier, but if fewer are required, a block (say of weighted wood) of appropriate size and shape enclosed in a tight-fitting polythene bag, should be used to fill the volume not occupied by the gel mould assemblies (see A in Fig. 10.3 and Fig. 10.5; also Chap. 12).

(3) **One gradient mixer** (Fig. 10.4, 10.5 at G). This should be capable of producing a gradient of the required concentration contour and of up to about 500 ml. total volume. A variety of gradient forming devices have been used for this purpose,[12,13,43] but that described by Margolis[31] and illustrated in Fig. 10.4, will be employed here.

(4) **One three-way tap** (see "t" in Figs 10.2, 10.4 & 10.5; but note that a three-way tube connector together with appropriate tubing clamps can be used) is connected as shown in the diagram (Fig. 10.5a) using transparent plastic tubing (internal diam. say 3 mm.).

(5) **One glass funnel (F).**

The apparatus is assembled as in Fig. 10.5a.

The following solutions are required and would, for example, be suitable for the preparation of up to about 14 slab gels with concentration gradient 4–25% (w/v).

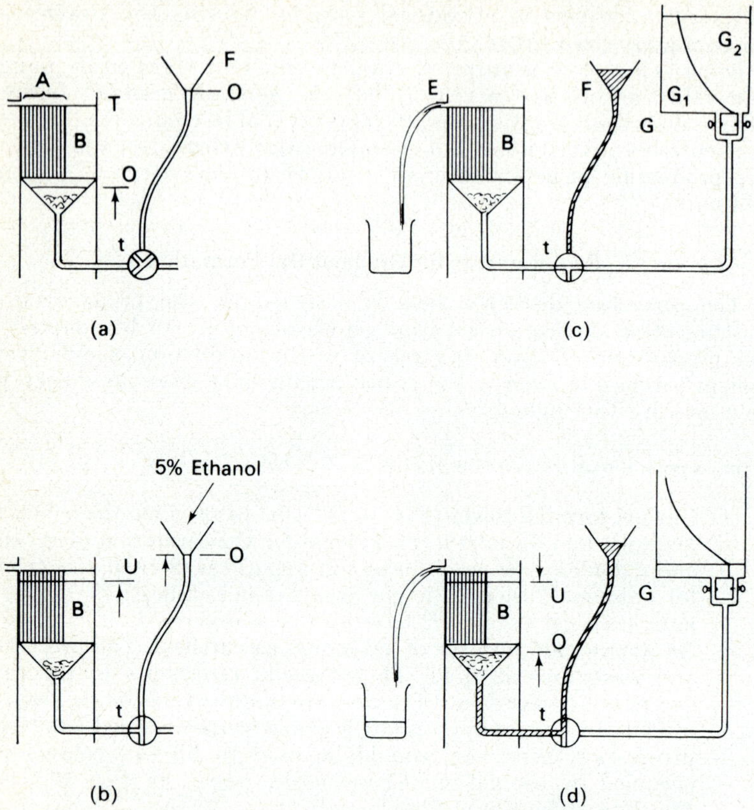

FIG. 10.5. Stages in the formation of the gradient Gels. (a) Shows position of the 5% ethanol solution at the zero (0) marks on the Funnel (F) and the Tower (T). That is, at about 1 cm. below the spacer block (B) and the gel mould assemblies (A). O is the first position of the surface of the 5% ethanol. The three-way tap (t) is turned to an intermediate position shown after adjustment of the solution level. (b) Shows position of the 5% ethanol solution required p in the gel mould assemblies for the upper limit of the gels. U is the second position of the ethanol surface and a measured volume of 5% ethanol is added to maintain level in funnel. (c) Shows tower fitted with overflow tubing (E) and the preparation of the gradient from solutions G_1/G_2 in the gradient former (G).The setting of the three-way tap (t) should be noted. (d) Shows final position of the gradient supported on the dense sucrose support solution (thus 1111) which has been added from the funnel (F). U is final position of gel gradient.

(A) Ethanol (25 ml.) made to 500 ml. with water. (note; the ethanol in this solution helps to free air bubbles from the gel slab assemblies; to form a stable interface with the underlying acrylamide solutions; and to give the gel assemblies a final de-grease). This solution is used at 25°.

(B) Sucrose (60 g.) is made to 200 ml. with water, coloured slightly by the addition of Methylene Blue, and cooled to 5° before use. (Note; this is the dense "chase" solution to support the gradient before polymerization: some workers do not use the dye)

(G_1) Acrylamide (60 g.); methylene bis acrylamide ("Bis"; 3 g.), Ammonium persulphate (0·4 g.) and Tetramethyl ethylene diamine (Temed; 0·2 ml.) made up to 200 ml. with buffer (see below). This solution is used at 15°.

(G_2) Ammonium persulphate (0·4 g.) and Temed (2·0 ml.) made to 200 ml. with the same buffer. This solution is used at 25°.

Procedure.

The above solution temperatures, catalyst and accelerator concentrations have been chosen to bring about the polymerization of the weak (4%) gel in 20–30 min., and the concentrated gel in 60–90 min. The reader should check in a preliminary experiment whether his solutions gel after similar intervals. It may prove necessary to de-gas some of the polymerization solutions by evacuation for 2 min. in a closed flask using a water pump.

It is then important to measure the volume of gel gradient necessary to fill the gel mould assemblies in the tower. This can be done by admitting the 5% ethanol solution to the tower such that the liquid surface is about 1 cm. below the bottom edges of the gel mould assemblies (see Fig. 10.5a). A "zero" mark (0) is then made at the surface of the liquid in the stem of the funnel (F) and a corresponding one on the side of the tower. A measured amount of 5% ethanol is then added to the funnel and the level of liquid in the tower is raised until it is about 1 cm. below the top of the gel moulds (see U in Fig. 10.5). The volume of 5% ethanol admitted to the gel moulds is then calculated after measurement of the volume residual in the funnel above the "Zero" mark (0).

An overflow tube (see E in Fig. 10.5c) is connected to the tower and further 5% ethanol solution is admitted to cover the gel moulds completely. Care should be taken that there are no air bubbles in the system. The funnel and tubing up to the tap (t) are filled with the dense solution (B). The solutions G_1 and G_2 are added to the gradient mixer after filling the tubing as far as the tap (t) with solution G_1. The required volume of freshly mixed gradient is then admitted to the tower and is adjusted to the correct position by means of the colour in the dense chase solution. (Fig. 10.5d. Note: the interface between the dense chase solution and the gradient can usually be seen without the need of the dye.) The mixing and a correct positioning of the polymerizing gradient should take about 10 min.

Gel polymerization is usually indicated by turbidity and heat generation, and should be complete in 90–120 min. After several hours the gel moulds can then be removed from the tower (gentle water pressure from the base is often helpful), trimmed free of excess gel and stored at 5°.

If homogeneous, single concentration gels are required the above gradient is simply replaced by a corresponding volume of appropriate polymerization mixture.

Measurement of the Concentration Gradient.

The most usual method for measuring the concentration gradient of a batch of gels is to remove the glass plates from one or two gels and to dry and weigh segments cut from the gels. Thus, 4 mm. discs are cut from the gel along the length of the gradient using a cork borer. The gel discs are placed in weighed glass tubes and heated to constant weight at 105°. The results from several of such determinations typically give averages within 5% of the theoretical values.[11]

Gel concentrations have also been determined by a refractometric method.[28] Commercial manufacturers of gradient concentration gels usually provide a concentration gradient curve with each batch of gels.

Buffers

Factors involved in the choice of buffers for the conduct of concentration gradient gel electrophoresis include, that—

(1) The buffering power should be adequate to maintain the *p*H of choice.

(2) The conductivity of the buffered gel should not be so great that large amounts of electrical power are dissipated in the gels in order to provide the necessary electric field strength.

(3) The *p*H of the gel should be such that the proteins of interest migrate rapidly into the gel gradient. This means that the *p*H should be several units away from the isoelectric points of the proteins concerned.[33] Since many proteins are isoelectric in the *p*H 4–6 region, this means that it is convenient to use buffers in the *p*H region 8·3–9·3, and for the dense end of the gel to be nearest the anode. On the other hand, basic proteins should be run at acid *p*H (say *p*H 4) and for the cathode to be near the dense end of the gel.[33]

Buffers that have been used with the gradient gels include:—

Tris–EDTA–Borate *p*H 8·3

Tris(hydroxymethyl)methylamine (Tris)	10·75 g.
Disodium ethylene diamine tetracetic acid (EDTA)	0·93 g.
Boric acid	5·04 g.

made to 1 litre with water.

Tris–EDTA–Borate *p*H 9·3

Tris	10·12 g.
EDTA	0·96 g.
Boric acid	0·37 g.

diluted to 1 litre with water

Sodium Acetate buffer *p*H 4[33]

Glacial acetic acid	25·95 ml.
Sodium acetate (trihydrate)	13·61 g.

made to 1 litre with water but the solution is diluted 1/10 before use.

In the electrophoresis apparatus commonly employed there is a relatively large amount of circulating buffer, so it is often convenient to polymerize the gel in the Tris–EDTA–Borate buffer *p*H 8·3 and to pre-run for a short period in the buffer of choice before adding the protein samples. (Note: some buffers inhibit the polymerization process.)

Apparatus

For conventional sizes of cylindrical concentration gradient gels typical apparatus for "disc" gel electrophoresis can be used. Special apparatus has been described for electrophoresis in gradient gels contained in capillary tubes.[34,35,79,80]

The use of horizontal concentration gradient gel slabs has been described[20] but most separations have been carried out with apparatus in which the gel slabs are used vertically. Such apparatus is commercially available[22,25] and is in principle very similar to that used for disc gel electrophoresis. Versions are available to take either two or four slab gel assemblies dipping into, and cooled by, circulated electrolyte buffer (usually maintained at about 5°; see Fig. 10.3).

Electrophoretic Procedure

One of the most widely used techniques for sample application employs a sample applicator (see "D" in Fig. 10.3) pushed firmly into the surface of the weak upper end of the gel. The samples (5–30 μl.—preferably containing 20% sucrose) are added to each of the wells in the applicator. The samples are then overlaid carefully with buffer and the gel assembly is inserted 1 cm. upwards into the (silicone greased) gasket (E) in the upper electrode chamber (F), and then lowered gently into the lower electrode chamber (I) containing cooled buffer. The circulation (approx. 200 ml./min.) of the buffer solution is brought about by pumping via tube M into the upper electrode chamber. Care must be taken that no air bubbles are trapped at either end of the gel slab.

Some workers prefer to form sample wells at the upper end of the gel slab by layering a fast-setting gel made from acrylamide (4 g.), methylene bisacrylamide (0·2 g.) in water (95 ml.) and ammonium persulphate (0·5 g.) in water (5 ml.), and inserting a sample "comb" of the kind shown at C in Fig. 10.3. Such wells are not already incorporated in commercial gels, partly because such gels do not transport satisfactorily but also because the gels may be required for two-dimensional work.

If a pre-run of the gel is desirable, the samples can be added later by layering under the buffer in the same way as is often used in disc gel electrophoresis. The D.C. supply unit is connected to the apparatus and the voltage applied. With the Tris–EDTA–Borate buffers 75 V at 30 mA. are typical starting conditions, but these often change to 75 V. at 10 mA. after 2–3 hr.

If a sample applicator has been employed, it is convenient to arrest the electrophoresis after about 30 min. and remove the applicator before recommencing the run. The total duration of the electrophoresis will vary greatly with the applied voltage, the size and electric charge of the proteins under examination, and the nature of the gel employed. With the above conditions serum proteins give good resolutions in 4–25% concave gradients after 8–32 hr. (see Fig. 10.6).

After the electrophoresis is complete, the adhesive tape is removed and the glass plates and spacer strips are removed carefully from the gel after passing a thin razor blade between the joins in glass surfaces. The gels can be fragile and should be kept moist and handled with a broad spatula.

Staining and Destaining

The most widely used stain to date for the detection of proteins in gradient gels is Amido Black (0·3%) in acetic acid–methanol–water (1;3;6). The gel is immersed in the staining solution for at least 6 hr. before removal of the staining solution and immersion in several changes of acetic acid (7% w./v.) in water. This takes typically 5 or more changes of washing fluid (with occasional agitation) over about 2 days for a satisfactory background colour. Alternatively, freshly stained gels may be rinsed with the washing solution and destained in purpose built, commercially available, electrolytic destainers.[22–25] In such apparatus the gel is placed in a plastic destaining cell,[22–25] the system filled with dilute acetic acid (5% w./v.) and the gel destained electrophoretically at about 12 V., 3 A. (a car battery charger is often convenient as a power pack) while the electrolyte is circulated by pumping through absorbent wool discs. Electrolytic destaining can take as little as 30 min. under these conditions, but much depends upon the degree of staining with the Amido Black solution.

Two protein staining procedures have been described which require little or no destaining. In the first of these,[36] the gradient gels are sealed (using adhesive tape) in small polythene bags containing 25 ml. of a filtered solution (W. No 1 then 0·45μ Millipore) of 0.04% Coumassie Brilliant Blue G250 in 3·5% (w./v.) perchloric acid. Dense protein bands appear within seconds and almost full intensity is reached after 45 min. at 37°. After several hours the staining solution is discarded and the gels may be resealed in the bags with a little fresh 3·5% perchloric acid solution. Destaining is not required and the background colour is not usually sufficient to interfere with the inspection of scanning of the gels.[30]

In an alternative procedure,[37] proteins are stained for 30 min. in 0·25% (w./v.) aqueous solution of Coumassie Brilliant Blue G250 diluted (1/20) in aqueous 10% (w./v.) trichloroacetic acid. Stained gels are rinsed in 7% acetic acid.

Lipoproteins[13,38] have been stained with Sudan Black B whereas nucleic acids have been located in gradient gels using Methylene

Blue[29,39] Acridine Orange[39] and Pyronine Y Stains[40] (see also Chapter 12). Tritium labelled glycoproteins have been detected in concentration gradient gels by scintillation counting in extracts of 0·6 mm slices[41] cut from the frozen gels after electrophoresis. It is worth noting that proteins have also been detected,[42] without staining, by U.V. excited fluorescence after cooling the gels to −70°.

Sometimes it is desirable to use specific methods for protein location (e.g. for enzyme activities) as well as staining for protein. This may be done by cutting the gel longitudinally into two thin gel sheets using a gel slicer. The latter is available commercially[22] and consists of a plastic tray to contain the gradient gel; some plastic shims of appropriate thickness and a saw fitted with a fine, stainless steel wire. To cut the gel it is placed in the tray on a shim of the chosen thickness, and the taut wire of the saw is drawn with a slight sawing action through the gel (from the dense to the less dense end) while keeping the wire against the guide-rim of the gel tray. The two slices of gel can then be used for different zone localization techniques (see Chapter 8).

Preservation of the gels.

Commercially produced gels have 50 p.p.m. sodium azide present,[22–24] and can be preserved without deterioration at 5° for at least a year in a sealed container. Unlike thin homogeneous polyacrylamide gels, those with concentration gradients cannot be dried after use due to differential shrinkage. Gels can be preserved in sealed plastic bags in the presence of a little of the acid medium (see ref. 36) but probably the best plan is to photograph the gels for permanent records.

Some common Practical Difficulties.

One of the most common difficulties in gradient gel electrophoresis is that there is a certain amount of leakage and intermixing of samples due to an inadequate seal between the sample applicator and the top of the gel. The applicator should be pushed firmly into the top of the gel slab. This problem can often be solved by the addition of a very thin (say 1 mm.) layer of a very soft (2·5% w./v.) polyacrylamide gel on the surface of the gradient gel itself. In any case, care should be taken not to disturb the sample applicator during the addition of the samples.

On occasions the gel slabs come away from the glass plates while in the electrophoresis apparatus. This can be disastrous if it happens before the samples are fully within the gel slab. The most frequent causes of gel detachment are inadequate cleanliness of the glass plates prior to gel formation (see recommended clearning procedure on p. 300) or due to excessive heat dissipation in the gel.

The gels also show differential volume changes during processing and frequently assume a trapezoid shape. Some solvent mixtures are particularly bad in this respect. The new gels with cross-linkage gradient (see p. 268) are said to be less prone to this effect (private communication —Messrs. Universal Scientific Ltd.).

Factors involved in the choice of buffers and operating pH are discussed on pp. 260, 266.

Some applications and Conclusions.

Four relatively short accounts of concentration gradient gel electrophoresis have appeared recently,[43–46] and these contain some references on the subject. Only particularly novel applications of the approach or those illustrating interesting principles will be mentioned here.

High resolving power is probably the most recurrent theme in the literature on concentration gradient gel electrophoresis. At least part of this resolving power must be attributable to the zone sharpening which occurs as migrating proteins meet increasing resistance to movement through the gels.

The resolution possible by the gradient gel method for serum and plasma proteins can be seen in Fig. 10.6, and there are numerous reports in the literature that particular serum proteins, iso-enzymes, glycoproteins and membrane proteins can be separated more readily by means of gradient gel electrophoresis than by other procedures.[41,47–53] The serum haptoglobins are especially well resolved for instance, so that the gradient method is the one of choice for phenotype determina-determinations.[96,53] Another example is the resolution of the plasma lipoproteins, for following the demonstration by Pratt and Dangerfield[13] that the high molecular weight species can be readily separated using the gradient gels; the method has been extended to a variety of plasma lipoproteins.[16,35–38]

The high resolving power of the method is attractive in other clinical applications in the search for significant differences in various normal and pathological states. Examples include the detection of a "pregnancy" protein characteristic of pregnant women or those taking contraceptive pills[11d]; differences in haptoglobins in Alzheimer's disease[51]; differences in the patterns of proteins present in normal and pathological urines[59]; and possible differences in the proteins present in the sputum of normal and asthmatic individuals.[58]

Procedures have been described for the combination of the gradient concentration gel method in two dimensional techniques—either to increase yet further the resolving power, or in order to produce a separate display of the molecular size and electric charge characteristics of the individual proteins in a mixture.[60–64] This is usually carried out by electrophoresis or isoelectric focusing in a gel cylinder of uniform concentration, followed by laying the gel cylinder across the upper (low density) edge of the concentration gradient gel, and then electrophoresis of the partly separated proteins into the concentration gradient gel. Examples of such separations are shown in Fig. 11.12, p. 296.

That the gradient gel method might give an indication as to the relative molecular sizes of the separated proteins is a great attraction. It is now well established that indications of this kind can already be obtained from results in homogeneous concentration gels provided that the relative velocities of the proteins of interest can be measured in at least two gels of known concentration.[26,27] Thorun and Mehl[65] have

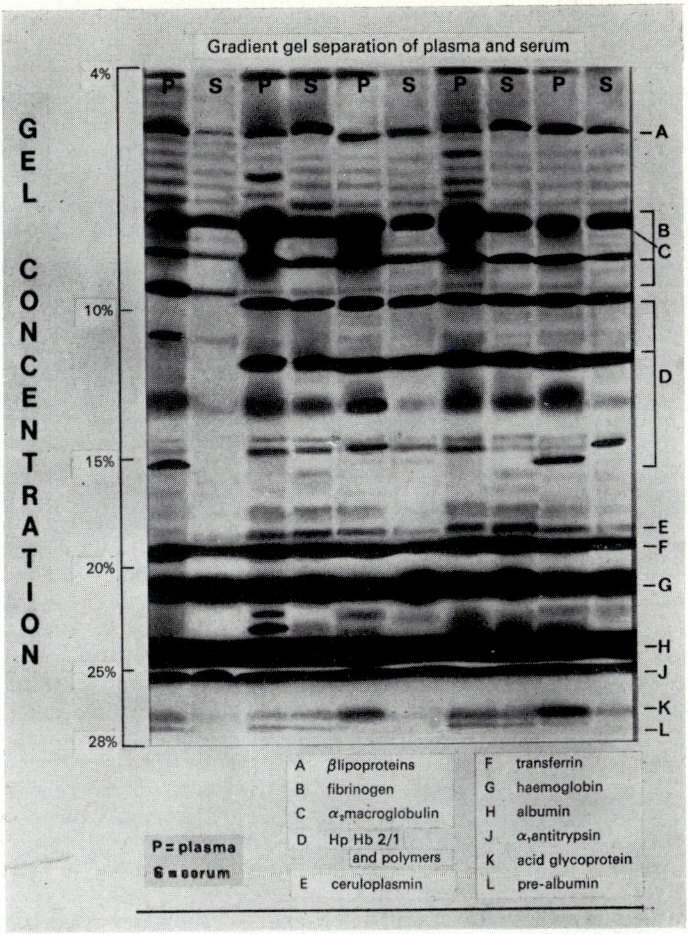

Fig. 10.6. Results[45] of the electrophoresis of corresponding plasma (P) and serum (S) samples into a 4–28% (w./v.) concentration gradient gel and stained with Amido Black. The protein types indicated at the foot of the figure have been allocated to the zones A–L. (By courtesy of Dr. J. Margolis.)

shown that this can be done in an ingenious way by measurements on the electrophoretic migration of a mixture of proteins into a slab of polyacrylamide gel at right angles to the direction of the concentration gradient (i.e. at right angles to the usual direction of migration into such a gel). Most globular proteins yielded satisfactory estimates of molecular size with this approach.

Two groups of workers[10,15] have now shown that the penetration of globular proteins during electrophoresis into a linear concentration gradient polyacrylamide can be related to their relative molecular sizes by a double logarithmic relationship of the kind which has been rationalized[6] theoretically as outlined earlier (see p. 253). What has not yet been rationalized is the report that the penetration of globular proteins into gels with concave (but mathematically otherwise unspecified) concentration gradients, is relatable to the molecular sizes of the proteins involved.[66] Of the proteins employed in these studies, only fibrinogen and the γ-globulins behaved really anomalously. The anomalous behaviour of fibrinogen has been attributed to the asymetric nature of the molecule.[10] Leaback[33] has shown that human IgG monoclonal γ-globulins differ considerably in isoelectric point, and that this can affect profoundly the migration of these proteins in concentration-gradient gel electrophoresis. Thus, it is hardly surprising that such of these proteins that have isoelectric points within about one *p*H unit of the operating *p*H do not migrate rapidly into the gels (see samples J & K, Fig. 10.7). What was surprising, however, was that the globulins with differing isoelectric point, but 4–5 *p*H units away from the *p*H of the electrophoresis, penetrated the gels to different extents (see samples A & B, Fig. 10.7). Experiments with other homologous proteins (bacterial L-asparaginases)[67] suggest that this phenomenon is not confined to the γ-globulins. It seems likely therefore that the anomalous streak reported by Slater[10] for γ-globulin is due to the presence of species with a range of isoelectric points and/or with isoelectric points too close to the *p*H of the electrophoresis. Leaback[33] has suggested that the success of the quantitative results of Kopperschlager *et al.*[15] might be explained by the fact that the calibrating proteins employed probably had typical, and closely similar, isoelectric points in the *p*H region 4–6. There would appear to be a good case for monitoring the isoelectric points of proteins before drawing conclusions on relative molecular sizes of proteins from results of this kind in gradient gels. It should be pointed out however, that the quantitative approach of Thorun and Mehl[65] using concentration gradient gels is independent of the net electric charge of the proteins provided that they give discrete zones and show measureable velocities at the gel concentrations concerned.

Nucleic acids (both DNA and RNA species) show very sharp bands on electrophoresis[29,39,40] in linear concentration gradient gel electrophoresis, and moderately satisfactory quantitative results have also been obtained[40] analogous to those obtained with proteins. Parish and Marchalonis[54] have expressed the belief that random coil molecules would electrophorese as approximately spherical particles such that their mobilities too would be related to their molecular weights. The electrophoresis of proteins with extended conformations has been carried out into concentration gradient gels using 8 M. urea,[18] sodium dodecyl sulphate[41,68,69] or phenol–acetic acid–water[65] in the electrophoretic medium. Satisfactory quantitative results have been claimed for estimates of the size of proteins down to about 20,000 daltons.[18] Other workers[15] put the lower limits of the gradient gel method at about 50,000 daltons and it is clear from the results in Fig. 10.1, that proteins of about this size are anything but stationary in a gradient gel of up to 35% (w./v.) polyacrylamide and with 5% cross-linkage.

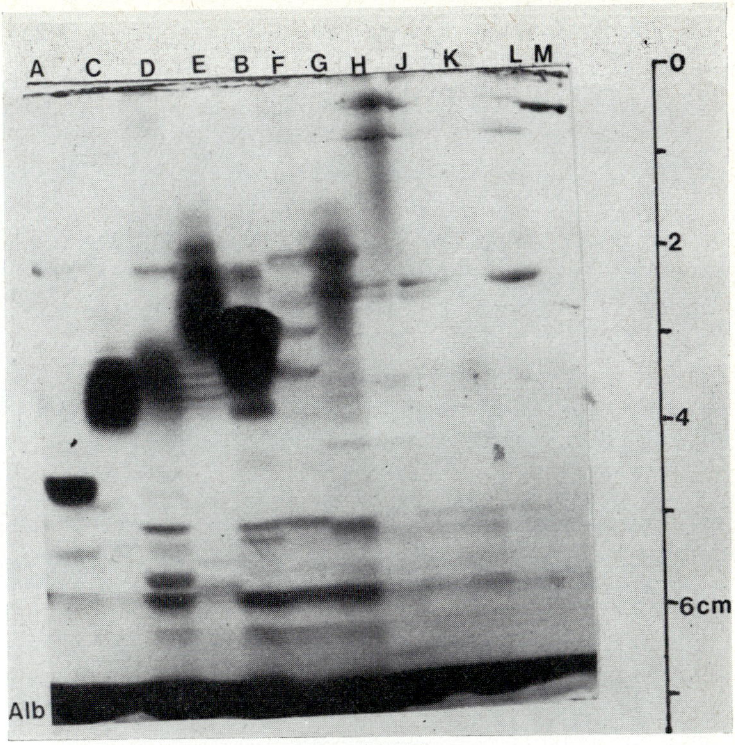

FIG. 10.7(a)

FIG. 10.7. Results of the electrophoresis of the same ten human IgG (A–K) myeloma sera as seen in Fig. 11.16, p. 305 together with a corresponding human IgA myeloma serum (L) and an IgM serum (M) into commercial 2–28% "Gradipore" gels after visualization with tho Amido Black staining/destaining procedures. In both cases the gel concentration increased from top to bottom of the illustrations and the penetrations of the proteins into the gels are indicated approximately by the centimeter scales. (a) Results of electrophoresis for 24 hr. at 5° towards anode at 70 volts, 20 mA in the Tris–EDTA–borate buffer *p*H 8·3. Note the series of IgG myeloma immunoglobulin zones from samples A–H which show variable penetrations of 0–4·5 cm. whereas immunoglobulins from samples J and K do not appear to have entered the gel. (b) As above but electrophoresis carried out in *p*H 9·3 buffer. Note that samples J and K have now entered the gel but the degree of penetration of even samples A and B is different. The polymeric series of IgA and IgM proteins in samples L and M is noteworthy. Such differences were not eliminated even after 92 hr. electrophoresis but could be on electrophoresis towards cathode in similar gels at *p*H 4 in the acetate buffer.[33] The position of the erum albumin is indicated as "Alb" in both cases. (By courtesy of Messrs. Butterworths, London.)

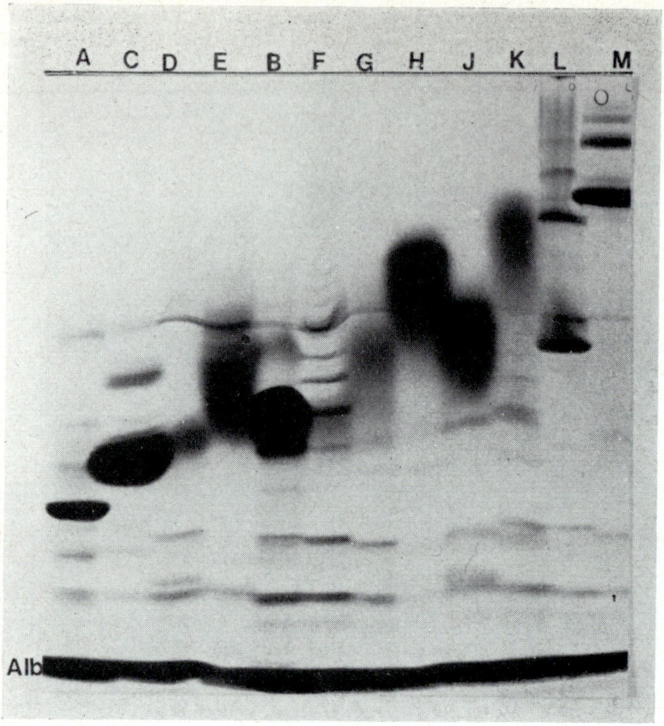

FIG. 107(b)

Maurer[69] has discussed the potential practical advantages in the use of different concentration gradient contours, and in the use of varying acrylamide and/or Bis concentration gradients. Indeed, there are indications[70] that the incorporation of a 3–6% concentration gradient of Bis into standard 4–26% concave gradient gels produces superior handling properties and greater retention for the smaller globular proteins.

Another significant technical innovation is the recent development of "direct" staining techniques[36,37] for the detection of proteins in the gels without extensive destaining procedures. These improved techniques should not only eliminate the need for lengthy and tedious destaining procedures but also avoid the need of expensive, purpose-built electrolytic destaining apparatus.

As with other techniques for the electrophoretic separation of proteins (e.g. gel isoelectric focusing, see p. 297) it is usually important to be able to make close comparisons of multiple samples after separation, so that there are considerable advantages in the use of gradient gels in the slab format. The fact that the cylindrical gradient gels can often be used in pre-existing (disc–gel) electrophoresis apparatus, has been advantageous, but is decreasingly so as equipment for the electrophoresis of slab gels becomes more common. The cylindrical gel format lends itself most readily to gradient gel electrophoresis of exceedingly small samples of proteins in capillary tubes,[79,80] but a procedure has been described for the preparation and use of corresponding micro-scale slab gels.[74]

Thin-layer gel filtration[72] is the only other simple technique available at present in which multiple samples of proteins can be compared after separation according to molecular size. The resolving power of the gel filtration technique is not very high and is particularly poor above about 150,000 daltons. These disadvantages in the gel filtration technique are precisely the strong features of the gradient gel method, and examples of highly-resolved, polymeric series of proteins with molecular weights greater than 150,000 daltons can be seen from samples L & M in Fig. 10.7. Similar results for polymeric series of serum albumin,[18,79] immunoglobulins[33,73] and haptoglobulins[74] have been reported. Homologous series of proteins of this kind are most likely (but not absolutely necessarily)[75] to have components similar in partial specific volume, shape and net electric charge, and should therefore be satisfactory for molecular weight calibration purposes. Provided these factors are taken into account, the prospects are that the relative molecular sizes of other, non-homologous, proteins should be deducible from the results of the gradient technique. Thus, Leaback[33] has presented evidence that the behaviour on cellulose acetate electrophoretograms of certain immunoglobulins can be rationalized in a qualitative way from electric charge and molecular size data obtained from the results by gel isoelectric focusing and by concentration gradient gel electrophoresis respectively.

The future of concentration gradient gel electrophoresis seems assured—partly because of its proven ability to resolve complex mixtures of proteins or nucleic acids, but also because of its potential for the determination of the relative sizes of the biological macromolecules. In this respect the method is in accord with the current trend towards simple, high-resolution procedures which can give insight into important molecular properties of the separated substances.[46]

REFERENCES

1. Ornstein, L. *Ann. N.Y. Acad. Sci.*, 1964, **121,** 321.
2. Porath, J. *Pure Appl. Biochem.*, 1963, **6,** 233.
3. Squire, P. G. *Arch. Biochem. Biophys.*, 1964, **107,** 471.
4. Ackers, G. K. *Biochemistry*, 1964, **3,** 723.
5. Ogston, A. G. *Trans. Faraday Soc.*, 1968, **54,** 1754.
6. *See* Rodbard, D., Kapadia, G., and Crambach, A. *Anal. Biochem.*, 1971, **40,** 135.
7. Ogston, A. G. *Brit. Med. Bull.*, 1966, **22,** 105.
8. Wright, G. L., Farrell, K. B., and Roberts, D. B. *Clin. Chem. Acta*, 1971, **32,** 285.
9*a*. Slater, G. G. *Fed. Proc.*, 1965, **24,** 225.
9*b*. Slater, G. G. *Anal. Biochem.*, 1968, **24,** 215.
10. Slater, G. G. *Anal. Chem.*, 1969, **41,** 1039.
11*a*. Margolis, J., and Kendrick, K. G. *Anal. Biochem.*, 1968, **25,** 347.

11*b*. Margolis, J., and Kendrick, K. G. *Nature*, 1967, **214,** 1334.
11*c*. Margolis, J., and Kendrick, K. G. *Biochem. Biophys. Res. Commun.*, 1967, **27,** 68.
11*d*. Margolis, J., and Kendrick, K. G. *Austral. J. Exp. Biol. Med. Sci.*, 1969, **47,** 637.
12. Vidal, J. C. *Anales Asoc. Quim Argentina*, 1971, **59,** 353.
13. Pratt, J. J., and Dangerfield, W. G. *Clin. Chim. Acta*, 1969, **23,** 189.
14. Epstein, E., Houvras, Y., and Eak, B. *Clin. Chim. Acta*, 1968, **20,** 335.
15. Kopperschlager, G., Dierzer, W., Bierwagen, B., and Hofmann, E. *FEBS Lett*, 1969, **5,** 221.
16. Brumby, E., and Welch, V. A. *J. Dairy Res.*, 1970, **37,** 121.
17. Kidby, D. K. *Anal. Biochem.*, 1970, **34,** 478.
17. Kidby, D. K. *Anal. Biochem.*, 1970, **34,** 478.
18. Anderson, L. O., Borg, H., and Mikaelson, *FEBS Lett.*, 1972, **20,** 199.
19. Sheridan, J. W., Kendrick, K. G., and Margolis, J. *Biochem. Biophys. Res. Commun.*, 1969, **35,** 474.
20. Arcus, A. C. *Anal. Biochem.*, 1970, **37,** 53.
21. Wein, J. *Anal. Biochem.*, 1969, **31,** 405.
22. Available from Messrs. Universal Scientific Ltd., London, E13 0QU, U.K.
23. Available from Messrs. Isolab Inc., Akron, Ohio, U.S.A.
24. Available from Messrs. Gradient Pty. Ltd., Lane Grove, Australia.
25. Available from Messrs. Pharmacia Fine Chemicals A.B., Uppsala, 1, Sweden.
26. Ferguson, K. A. *Metabolism*, 1964, **13,** 985.
27. Rodbard, D., and Crambach, A. *Anal. Biochem.*, 1971, **40,** 95.
28. Komm, L., and Mees, J. *J. Chromatog.*, 1971, **62,** 383.
29. Caton, J. E. ,and Goldstein, G. *Anal. Biochem.*, 1971, **42,** 14.
30. Laulhere, J. P., Lambert, J., and Berducou, J. *Analysis*, 1972, **1,** 234.
31. Margolis, J. *Anal. Biochem.*, 1969, **27,** 1969.
32. Kitchin, F. D. *Proc. Soc. Exp. Biol. Med.*, 1965, **119,** 1153.
33. Leaback, D. H. In "Isoelectric Focusing" by Arbuthnot, J. P., and Beeley, J. A. (eds). Butterworth, London, 1974, 201.
34. Smeds, S., and Bjorkman, U. *J. Chromatog.*, 1972, **71,** 499.
35. Ruchel, R., Mesecke, S., Wolfrum, D. I., and Neuhoff, V. *Z. Physiol. Chem.*, 1973, **354,** 1351.
36. Reisner, A., Nemes, P., and Bucholtz, C. *Anal. Biochem.* 1975, **64,** 509.
37. Diezel, W., Kopperschlager, G., and Hofmann, E. *Anal. Biochem.*, 1972, **48,** 617.
38. Green, J. *Scand. J. Clin. Lab. Invest.*, 1972, Suppl. **126,** 29.
39. Schaup, H. W., Best, J. B., and Goodman, A. B. *Nature* (*Lond.*), 1969, **229,** 864.
40. Jeppesen, P. G. N. *Anal. Biochem.*, 1974, **58,** 195.
41. Voyles, B. A., and Moskowitz, M. *Biochem. Biophys. Acta*, 1974, **351,** 17.
42. Patterson, B. D., Atkins, C. A., Graham, D., and Wills, R. B. H. *Anal. Biochem.*, 1971, **44,** 388.
43. Lamy, J., Lamy, J., Comping, S., and Weill, J. *Ann. Biol. Clin.*, 1971, **29,** 125.
44. Crambach, A., and Rodbard, D. *Science*, 1971, **172,** 440.
45. Margolis, J. *Lab. Practice*, 1973, p. 107.
46. Leaback, D. H. *Chemistry in Britain*, 1974, **10,** 376.
47. Hochberg, M. L., and Sargent, M. L. *Anal. Biochem.*, 1973, **51,** 516.
48. Cornell, N. *Proc. Austral. Assoc. Clin. Biochem.*, 1973, **5,** 176.
49. McIntosh, C. H. S., and Plummer, D. T. *Biochem. J.*, 1973, **133,** 655.
50. Blake, N. M., Kirk, R. L., McDermid, E. M., Case, J., and Burshir, H. *Austral. J. Exp. Biol.*, 1973, **51,** 209.
51. Velde, W. O. D., and Stam, T. C. *Brit. J. Psychiat.*, 1973, **122,** 331.
52. Velde, W. O. D., and Stam, T. C. *Humangenetic*, 1973, **20,** 25.
53. Clifford, B. J. In "The Examination and Typing of Blood Stains in the Chemistry Laboratory". U.S. Department of Justice publication, Washington, 1973.
54. Parish, C. R., and Marchalonis, J. J. *Anal. Biochem.*, 1970, **34,** 436.
55. Widman, M., Nilsson, I. M., Nilsson, J. L. G., Agurell, S., Borg, H., and Granstan, B. *J. Pharm. Pharmacol.*, 1973, **25,** 453.

56. Melish, J. S., and Waterhouse, C. *J. Lipid Res.*, 1972, **13,** 193.
57. Bautovich, J., Dash, M. J., Hensley, W. J., and Turtle, J. R. *Clin. Chem.*, 1973, **19,** 415.
58. Hellpern, S., and Rebuck, A. S. *Thorax*, 1972, **27,** 26.
59. Vincent, D., and Magron, J. *Clin. Chim. Acta*, **50,** 1974, 181.
60. Kendrick, K. G., and Margolis, J. *Anal. Biochem.*, 1973, **33,** 204.
61. Wright, G. L. *Clin. Chem.*, 1971, **17,** 430.
62. Wright, G. L. *Amer. J. Clin. Path.*, 1972, **57,** 173.
63. Wright, G. L., Affronti, L. F., and Reich, M. *Infection & Immunity*, 1972, **5,** 482.
64. Wright, G. L., Farrell, K. B., and Roberts, D. B. *Biochim. Biophys. Acta*, 1973, **295,** 396.
65. Thorun, W., and Mehl, G. *Biochim. Biophys. Acta*, 1968, **160,** 132.
66. Robinson, H. K., and Leaback, D. H. Unpublished results.
67. Moon, K. E., and Thompson, O. P. *Austral. J. Biol. Sci.*, 1971, **24,** 755.
68. Anderson, M., Cawston, T., and Cheezeman, *Biochem. J.*, 1974, **139,** 653.
69. Maurer, H. R. *Ann. Biol. Clin.*, 1971, **29,** 1971, 205.
70. Margolis, J. and Wrigley, C. W. *J. Chromatog.*, 1975, **106,** 204.
71. Maurer, H. R., and Dati, F. *Anal. Biochem.*, 1972, **46,** 000.
72. Johansson, B. G., and Rymo, L. *Acta Chem. Scand.*, 1962, **16,** 2067; Morris, C. J. O. R. *J. Chromatog.*, 1964, **16,** 167; Roberts, G. P. *J. Chromatog.*, 1966, **22,** 90.
73. Fine, J. M., Lambin, P., and Frommel, D. *Biomedical*, 1973, **18,** 145.
74. Jarrett, I. G. *Austral. J. Biol. Sci.*, 1972, **25,** 941.
75. Blatter, D. P., and Reithel, F. J. *J. Chromatog.*, 1970, **46,** 286.
76. Foebsy, H., *J. Chromatog.*, 1975, **106,** 51.
77. Sprinzl, M., Wolfrum., and Neuhoff, V., FEBS Lett., 1975, **50,** 54.
78. "Electrophoresis and Isoelectric Focusing in Polyacrylamide Gel." by R. C. Allen and H. R. Maurer (eds.), de Gruyter, Berlin, 1974.
79. Ruchel, R., see ref. 78. p. 215.
80. Dames, W., and Maurer, H. R., see ref., 78, p. 221.
81. O'Farrell, P. H., *J. Biol. Chem.*, 1975, **250,** 4007.
82. Chua, N. H., and Bennoun, P., *Proc. Nat. Acad. Sci. U.S.A.*, 1975, **72,** 2175.

Notes added in proof.

Work has continued on the application of concentration gradient electrophoresis to exceedingly small samples(79,80) and to the separation of neucleic acids.(77) The use of the gradient gels in the presence of sodium dodecyl sulphate (SDS) is a notable recent development in the separation of transfer RNA species,(77) in the separation of cell membrane polypeptides,(82) and in the spectacular separations of E. coli protein chains described by O'Farrell.(81)

CHAPTER 11

ISOELECTRIC FOCUSING OF PROTEINS

D. H. Leaback and C. W. Wrigley

Proteins, being ampholytes, can behave either as acids or bases. Thus at high *p*H, they are negatively charged and migrate towards the anode in an electric field, and at acidic *p*H they are positively charged and migrate towards the negative electrode (cathode). Between these extremes, there is a characteristic *p*H for each protein, its isoelectric point (*p*I), where it does not migrate since oppositely charged sites are balanced.

Principles of isoelectric focusing

The technique of isoelectric focusing takes advantage of these ampholytic properties. The method involves applying an electric field to proteins contained in a *p*H gradient (*p*H increasing from anode to cathode). Protein molecules in a basic region migrate anodically until they lose their net electric charge, cease to migrate and are said to be isoelectric. Similarly, molecules of the same protein in an area of low *p*H move in the opposite direction until they too are focused to the region of their isoelectric point. The derivation of the name isoelectric focusing is thus evident. The contraction "electrofocusing" is a common synonym.

Isoelectric focusing is distinct in principle from electrophoresis. Thus, electrophoresis involves a dynamic separation according to electric charge at particular *p*H (and according to size if the medium has a sieving effect), whereas isoelectric focusing is an equilibrium method which fractionates on the basis of isoelectric point—giving a measure of this characteristic for each of the proteins separated. Secondly, electrophoretic resolution depends on the sharpness of starting zones since protein boundaries continue to spread, due to diffusion, as separation proceeds. Isoelectric focusing, however, offers the advantage that the applied electric field also acts to sharpen the zones as equilibrium is approached and thus minimises diffusional spreading. As a result, sharply focused zones can be obtained even from a very dilute sample, since the whole *p*H range can, in principle, be used for loading the sample.

Forming the pH gradient

The general principle of isoelectric focusing has long been recognised, but its useful application to protein fractionation has been made possible only relatively recently by producing means for establishing a satisfactory *p*H gradient. In early applications of the principle, "artificial" *p*H gradients were prepared by layering different buffer solutions one upon another. However, such a gradient was disrupted when an electric field was applied due to migration of the buffer ions, and so fractionation could be pursued for only a brief time.

This difficulty was overcome by introducing the concept of "natural" *p*H gradients. This type of *p*H gradient is actually produced and maintained by the electric field (in contrast to "artificial" *p*H gradients).

In exploratory studies, application of an electric field to mixtures of small ampholytic molecules or protein hydrolysates caused the ampholytes to arrange themselves within the field in order of their isoelectric points, thereby creating a *p*H gradient which remained essentially stable while the voltage was maintained. However, these "carrier ampholytes" were poorly suited to practical fractionation of proteins. Some of the properties desirable in useful carrier ampholytes, as envisaged by Svensson(1,2) can be summarized as follows: i) Good buffering capacity and conductivity at their isoelectric points (to create good *p*H gradients and uniform electric fields). ii) Low molecular weight (so that they can subsequently be separated from proteins). iii) Good solubility in water at their isoelectric points (so they do not separate from solution at isoelectric focusing equilibrium). iv) Low light absorption above 260 nm. (to permit the detection of proteins at 280 nm.). v) Chemical properties different from proteins (to enable proteins to be detected in their presence by chemical means).

Synthetic carrier ampholytes

So far the best "carrier ampholytes" for production of natural *p*H gradients have been those synthesized especially for the purpose.(3,4) Heterogeneous mixtures of these carrier ampholytes are commercially available under the name "Ampholines" (LKB—Produkter A.B., S-161 25 Bromma 1, Sweden), and have been reported(5) to have the general structure (see also footnote Table 11.3, p. 308)

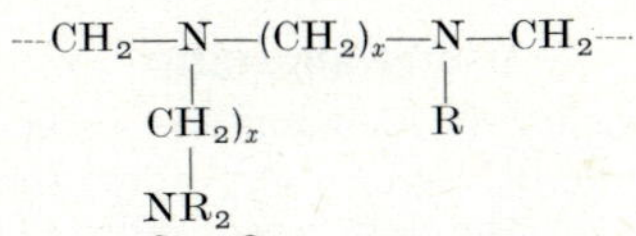

$x = 2$ or 3

$R = H$ or $-(CH_2)_x-COOH$

These "Ampholine" carrier ampholytes i) Typically can produce *p*H gradients of variable linearity and electrical conductivity(5) for two *p*H units or more over approximately the *p*H range 3·5–9·5. ii) Have good solubilities in water. iii) Have molecular weights mostly in the range 300–1,000 daltons but 1% may be as large as 1,000–5,000 daltons.(6) iv) Have some species which absorb light fairly strongly at 254 nm. but less so at 280 nm. (see ref. 5 and p. 287, 310). v) Having structural similarities to proteins often undergo similar reactions (e.g. dye binding). The chief disadvantages of the Ampholines is that they are relatively expensive. Cheaper alternatives have been described.(6,7)

In practical use the carrier ampholytes are usually protected from possible modification at the electrodes by using a solution of a strong base (or acid) at the cathode (or anode), respectively.

Stabilizing media

For isoelectric focusing separations to be useful, they must be performed in a medium stabilized against remixing due, for example, to electric heating, and in a medium suited for examining the final results of fractionation. In the main, two types of stabilizing media have been used for isoelectric focusing: a density gradient of sucrose or a gel of polyacrylamide. The former tends to be better suited to preparative fractionation, and the latter to analysis, but this is not exclusively so. The chapter deals mainly with methods based on these two media, but others have been used successfully (notably granulated gels.[107,108])

Further background information, details of early experiments, the use of other stabilizing media, and the mathematical treatment of isoelectric focusing are available in recent reviews[5,6,9,10,11] and in the proceedings of conferences on the subject. [12, 14, 109] A full list of literature references is available from LKB-Produkter, Sweden.

ISOELECTRIC FOCUSING IN DENSITY GRADIENT COLUMNS

The use of a density gradient as a stabilizing medium for isoelectric focusing offers the main advantages that it is uncharged (the *p*H gradient will thus be stable), and it is fluid (the medium can thus be drained and fractions separated following isoelectric focusing). The apparatus must be designed so that there is adequate temperature regulation as well as, provision for the escape of gaseous products of electrolysis and so that the solution can be drained with a minimum disturbance of separated zones. Three types of density gradient apparatus are described below: the U-tube, the Vesterberg and Svensson column and the Isco upward displacement columns.

U-tube apparatus

Density gradient isoelectric focusing can most simply be performed in a U-shaped glass tube of about 1 cm. diameter[15] (fig. 11.1). Fractionation takes place in the left arm only. The balancing arm is necessary so that electrolysis gases escaping from the "lower" electrode do not disturb the density gradient. The U-tube apparatus is useful in exploratory or analytical studies, in gaining familiarity with the methods in general, or for demonstrating the principle to laboratory classes.[16] In the latter applications, use wide range carrier ampholytes initially (*p*H 3–10), with 1 mg. coloured protein (native haemoglobin, myoglobin, cytochrome c, 0·2 mg. dye (Haematein, Fast Green, Orcein[10]) or even black tea.

Procedure (for 15 ml. side arm). Fill the U-tube half-way up the straight length of the side arms with about 15 ml. 1% ethanolamine solution, containing 40% sucrose. Form the density gradient in the left arm by carefully adding to it eleven 1 ml. solutions as follows. First add 1 ml. "dense" solution (1·5% carrier ampholytes in 30% sucrose). (Ampholine carrier ampholytes are generally supplied as 40% solution w./v.). The last addition is 1 ml. light solution (0·5% carrier ampholytes in 5% sucrose). The nine intermediate 1 ml. portions contain decreasing volumes of dense solution (0·9, 0·8 . . . 0·1 ml.) and increasing volumes of light solution (0·1, 0·2 . . . 0·9 ml., respectively). Finally, one or two millilitres of 2% sulphuric acid are layered on top of the density gradient.

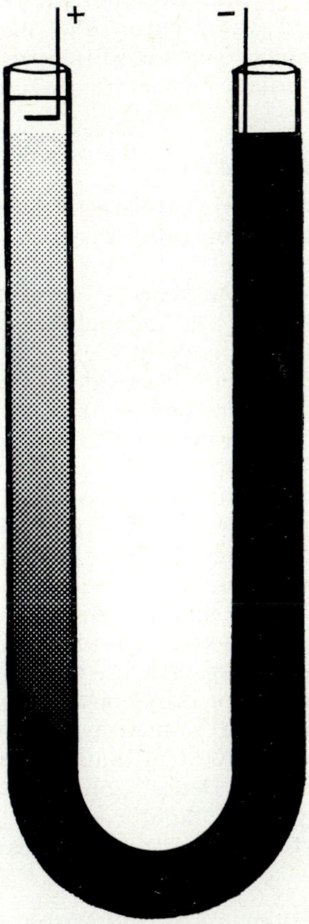

FIG. 11.1.

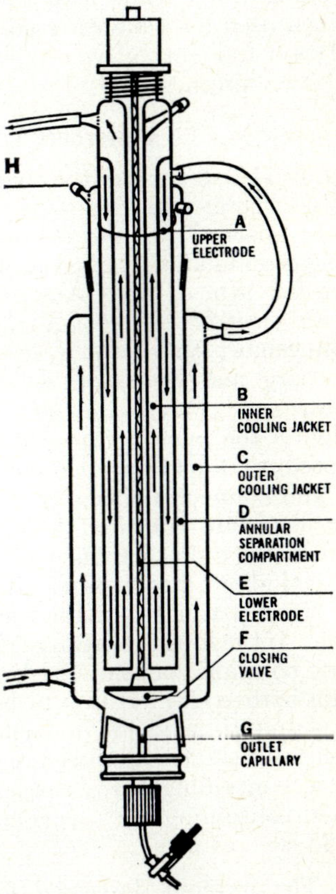

FIG. 11.2.

FIG. 11.1. U-tube apparatus for density gradient isoelectric focusing. Dense electrolyte solution in the right arm balance as density gradient in the left arm where fractionation takes place.

FIG. 11.2. Vesterberg and Svensson column for isoelectric focusing in a density gradient (*by courtesy of LKB Produkter A.B.*).

The protein sample (1 to 3 mg.) can be dissolved in one or several of the intermediate solutions of the density gradient.

Insert the anode into the sulphuric acid solution, and the cathode well into the ethanolamine solution. Increase the voltage gradually to a final maximum of 500 volts. Focusing should be complete after 4 to 10 hours.

The results of fractionation are evaluated after the run by pumping 40% sucrose into the right arm to force the density gradient up the left arm, out through a fine tube and into a fraction collector or U.V. absorbance monitor.

The temperature during the isoelectric focusing can be regulated by immersing the U-tube in a thermostatted water bath. The method has been used for a 33-ml. gradient,[(15)] but problems with temperature regulation and convective mixing of the gradient restrict its use to these smaller volumes.

Vesterberg and Svensson Column

Figure 11.2 illustrates the commercially available apparatus designed by Vesterberg and Svensson.[(3)] Fractionation takes place in the annular space, D, of the column, between concentric glass cylinders, which accommodate cooling jackets, B and C. The space D corresponds to the left arm of the U-tube in Figure 11.1. The balancing column of dense electrolyte solution (right arm) corresponds in Figure 11.2 to the innermost tube which contains the "lower" electrode, E. The Vesterberg and Svensson column is filled similarly to the U-tube apparatus, but it is drained from the bottom (through capillary G) after sealing off the inner compartment by raising valve F.

Procedure (for 110 ml. LKB column)* 1. Set up the apparatus vertically and connect a thermostatted water supply (say at 4°). 2. Pour lower electrode solution down the central tube to fill the bottom chamber, the outlet capillary G, and the inner and outer compartments to a level about 2 cm. above the bottom of the inner jacket. Operate valve F a few times to make sure no air bubbles are trapped and leave it open. If the lower (central) electrode is to be negative, use 40% W./V. sucrose solution containing 1% ethanolamine (or ethylene diamine or sodium hydroxide). If it is to be positive, use 40% sucrose containing 0·5% sulphuric acid (or phosphoric acid). 3. Next, a linear gradient of sucrose must be produced, extending from "Dense" solution (35% sucrose, containing 1·5% carrier ampholytes to "light" solution (5% sucrose, containing 0·5% carrier ampholytes) at the top. This can be done in several ways.

(*a*) The method described for the U-tube apparatus requires no specialized equipment. Measure 0, 0·1, 0·2 . . . 3·2 ml. of light solution into a series of 33 tubes. Add in the same order, 3·2, 3·1, 3·0 . . . 0 ml. of dense solution, so that the 33 tubes each contains 3·2 ml. Pour the mixed contents of the tubes carefully into the outer separation chamber D in the same order (densest first). As chamber D fills, the balancing column of lower electrode solution rises in the inner tube.

(*b*) Alternatively, if a pump with three channels is available, pump the light solution (at about 2 ml./min.) into a container in which 55 ml. dense solution is being stirred thoroughly. At the same

* This is a commercial version of Vesterberg and Svensson's column.

time, pump out of the container at twice the input rate and into the separation chamber (D) of the isoelectric focusing column. A schematic diagram of this type of gradient mixer[17] is shown in Figure 11.3.

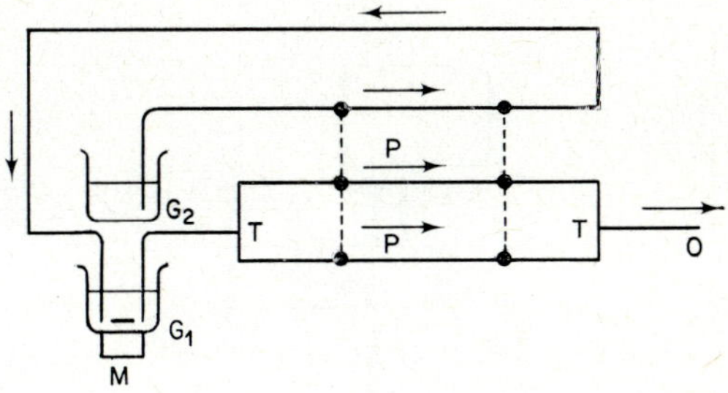

FIG. 11.3. Experimental arrangement[17] for the production of a linear density gradient. G_2, Beaker containing high-density solution. G, Beaker containing low density solution. M), Magnetic Stirrer. P), Three-channel peristaltic pump. T), T-junctions. O), Outflow of gradient. For a truly linear gradient it is important that each of the three channels should be of similar length and flow-rate. The vertical dotted lines indicate the coupling of the channels in the peristaltic pump.

(*c*) A further alternative is offered by the series—coupled mixing device (a commercial version is available through Messrs. L.K.B.-Produkter and is illustrated in figure 11.4 (cf. Svensson and Pettersson).[18] For this a pump is not necessary.

4. The Protein sample can be incorporated in a portion of the gradient solution (preferably close to the region where it will focus) or in either or both of the light and dense solutions. In the latter case it is advisable to put a layer of dense or light solution between the sample and the respective electrode solutions.

5. Finally, add a layer of upper electrode solution (0·5% sulphuric acid for anode, or 1% ethanolamine for cathode at top) so that it rises about 1 cm. above the upper electrode.

6. Turn on the current, and gradually increase the voltage, keeping below a maximum power (volts × amps) of 1 to 2 watts. Initially the *p*H thoughout the density gradient is approximately the average of the nominal *p*H range chosen. As the carrier ampholytes are focused by the electric field to their isoelectric positions, the resistance rises and the applied voltage can be increased without producing excessive heating. Apply a final voltage of between 300 and 900 volts depending on whether a wide or narrow *p*H gradient was used, respectively. Carrier ampholytes should be focused in about 8 hours at 500 volts.

Focusing of the protein zones takes longer, perhaps 30 to 60 hours. It is faster for wider *p*H ranges, higher field strengths and higher temperatures, and also depends on the mobilities of the particular proteins.

Completion of focusing is partly indicated by no further drop in curent (measured with a sensitive instrument). Within reason, resolution is not adversely affected by maintaining the electric field longer than the minimum time.

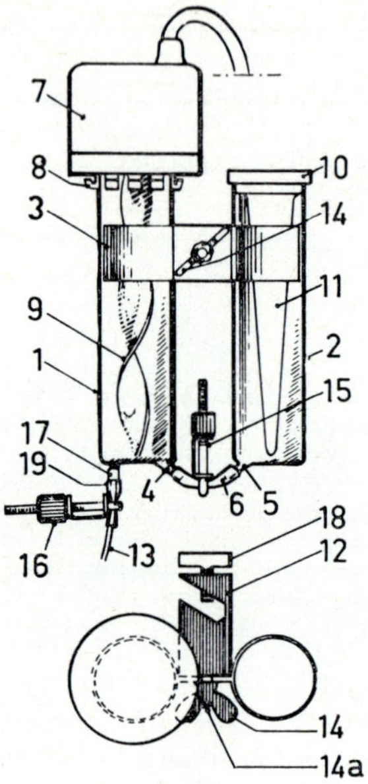

FIG. 11.4. Side and plan views of an apparatus for a linear gradient mixer. To use, the two equi-diameter cylindrical vessels (1, 2) are supported above the density gradient column using clamping facilities (3, 12, 14, 14a and 18). Connecting tubes (6 and 19) are sealed by clamps (15, 16) and the more dense and less dense solutions added to vessels (1 and 2) respectively. The electric motor/paddle assembly (7, 9) and plastic insert (11), designed[18] to compensate for the differing densities of the two solutions) are located accurately (8, 10) in the two vessels. Stirring is then commenced and clamps 5 and 16 are opened carefully and the density gradient column is filled slowly through the outflow tube 13. (*Diagram by courtesy of LKB Produketer A.B.*)

7. At the end of the run, turn off the power, close valve F and suck out the inner electrolyte solution. Remove the upper electrolyte solution, to prevent its diffusion into the upper layers of the gradient Open the capillary outlet tube and drain the column* into a fraction collector, perhaps through a U.V. absorbance monitor or even a continuous *p*H detector.[18a] The outflow should be adjusted to 1–2 ml./min. In an alternative procedure, the column contents can be collected at the outlet G (Fig. 11.2) after fitting tubing to the upper inlet H and pumping slightly tinted water on top of the upper column contents

using a non-pulsatile pump working at 1–2 ml./min. It is important to keep the direction of flow downwards to take advantage of the stabilizing effect of the density gradient. The possible trapping of bubbles in an absorbance cuvette can often be avoided by degassing solutions under vacuum before filling the column.

Isco density gradient electrophoresis Columns

In 1970, Grant and Leaback[19] showed that Isco density gradient electrophoresis columns Models 210 and 630 (which had been designed for electrophoretic mobility determinations[20]) could be used with advantage for the isoelectric focusing of proteins.

The result of this experience led to the development of the Model 212 apparatus. These three models (Fig. 11.5) are constructed almost completely of synthetic plastic materials and, in each case, comprise a central column surrounded by a jacket for circulating coolant water. The column contents are in contact with relatively large electrode chambers through semi-permeable cellulose membranes. Density gradients are formed in the central columns on top of a dense sucrose solution. By means of a peristaltic pump or a motor-driven syringe the latter dense "chase" solution can be pumped into or out of the central column and this allows the gradient either to be moved steadily past an optical unit (if fitted) or to be removed from the column by upward displacement.

Some potential advantages of these columns are as follows:—

1. If an optical flow cell is to be used during the removal of the column contents, the flow through this must also be upward to avoid inverting the density gradient. An upward flow minimizes the possibility of any gas bubbles in the system accumulating in the flow cell.
2. The fact that the sample is isolated from the electrodes by membranes and by relatively large volumes of electrode solutions should minimize the possibility of chemical changes in the central column due to electrode effects.
3. The contours of the central columns are particularly smooth and simple and this should minimize the risk of disturbing focused zones during their removal from the columns.
4. The model 210 has a built-in optical unit which operates at 254 nm. A similar facility can be offered with the Model 212 column by the fitting of an extension tube and an optical unit operatable at 254 or 280 nm. (see Fig. 11.5c). These optical units permit the progress of isoelectric focusing runs to be monitored without necessitating the removal of the gradients from the columns.

Some salient features of the three types of column are listed in Table 11.1 and are illustrated in Fig. 11.5. It can be seen that the model 210 and 212 columns are of relatively small capacity and should be regarded as primarily analytical (before and/or after removal of column contents), whereas the model 630 is primarily preparative.

TABLE 11.1

Some salient features of the Isco Models 210, 212 and 630 Density Gradient Electrophoresis Apparatus.

	MODEL 210	MODEL 212	MODEL 630
Ultra-violet Light Analyser	Built into centre of electrophoresis column	Added to top of column with or without extension tube (see Fig. 11.5)	Added to top of column
Principal material of central tube	Teflon	Cellulose acetate	Cellulose acetate/butyrate
Internal diameter of central tube	1 cm.	1 cm.	2·5 cm.
Convenient length and volume of density gradient	10·5 cm.* 8·5 ml.	10·5 cm.* 8·5 ml.	20 cm. 100 ml.
Volume of upper electrode solution layered on top of density gradient†	16 ml.	16 ml.	30 ml.
Volume of solution required for each electrode compartment	about 200 ml.	about 200 ml.	about 500 ml.

* It has been found possible to extend the length of the gradient in the Isco column to 15 cm. and a corresponding gradient volume of 13 ml. Volume of upper electrode solution is then reduced to 12 ml.

† Whereas the contents of the Model 210 must be displaced upwards after the completion of the electrophoresis, it has been found that the central column and end fittings of the Model 630 can be reversed, so that the contents of this column can, if required, be displaced downwards provided that it is assembled appropriately beforehand.

Equipment Required

In addition to either the Model 210 or 212 columns the following equipment is required:—

1. One Isco Model 182 Motor driven syringe or/A reversible, non-pulsatile peristaltic pump operatable at 0–5 ml./min.
2. One stabilized D.C. power supply operatable at say 0–1,000 V., 0–30 mA.
3. One Isco Model UA-2 Ultra-violet Analyser.
4. One circulating water system operating at 0–5°C.
5. Apparatus to make a density gradient.
6. One Isco A cm. flow cell and 254–280 nm. optical monitor unit.
7. One 0–10 mV., 1 second response chart recorder with a chart speed of 1 in./min.

Note: Only items 1–4 are necessary if analysis within the columns is not to be carried out, whereas items 1, 2, 4, 5, and 7 are required for internal analysis using Model 210 and items 1–7 required for Model 212.

Solutions for Isco Columns

Models 210 and 212

(*a*) Dense sucrose solution; 500 g. sucrose + 750 ml. distilled water (this makes about 1,100 ml. of stock solution of which about 350 ml. would be required for one run with the Isco Model 210 or 212 columns).

(*b*) Lower electrode solution; 300 ml. dense sucrose solution + acid or base. The acid and base are usually 0·5% phosphoric and 1% ethylene diamine respectively and should be added freshly before use use since such solutions develop colour on prolonged storage.

(*c*) Upper electrode solution; 270 ml. distilled water + acid or base.

(*d*) Dense sucrose solution for density gradient; 5 ml. dense sucrose solution + 200 μl. 40% W./V. carrier ampholyte solution.

(*e*) Less dense solution for density gradient; 5 ml. distilled water + 50 μl. 40% W./V. carrier ampholyte solution.

Note: that the final concentration of carrier ampholytes must be lowered to 0·4% (W./V.) if scanning at 254 nm. is contemplated.

Model 630

(*a*) Dense sucrose solution; 500 g. sucrose + 750 ml. distilled water.

(*b*) Lower electrode solution; 600 ml. dense sucrose solution + acid or base.

(*c*) Upper electrode solution; 600 ml. distilled water + acid or base.

(*d*) Dense solution for gradient; 50 ml. solution a + 2·0 ml. carrier ampholyte solution (40%, W./V.).

(*e*) Less dense solution for gradient; 50 ml. distilled water + 0·5 ml. carrier ampholyte solution (40%, W./V.).

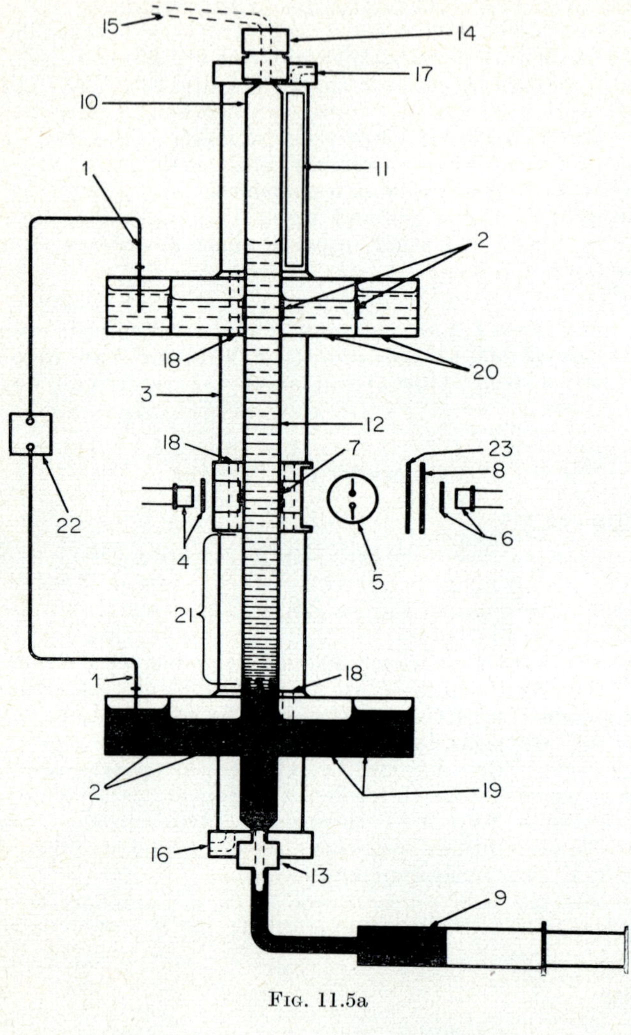

FIG. 11.5a

FIG. 11.5. Schematic diagrams of (a) The Isco Model 210 density gradient Electrophoresis column. (b) The Isco Models 212 and 630 density gradient columns set up for isoelectric focusing. (c) Section of the flow cell (H)/U.V. monitor (M) attachable to the tops (C) of the Models, 210, 212 or 630 and the extension tube (E) suitable for certain application of models 210 or 212 (see text). Note—that the extension tube E is replaceable by a simple screw-in cap of the kind shown in Fig. 11.5a. L = light source; D = light detector (*by courtesy of Shandon Southern Ltd*).

Key to Fig. 11.5a.

1. Platinum wire electrode
2. Semi-permeable membranes
3. Coolant jacket
4. Filter and measuring photocell
5. Low-pressure mercury vapour lamp
6. Filter and reference electrode
7. Quartz windows
8. Adjustable shutter
9. Motor-driven syringe
10. Transparent central tube
11. Centimetre scale
12. Teflon central tube (1 cm. I.D.)
13. Lower column support
14. Removeable cap for insertion of gradient and sample
15. To sample collector
16. Coolant inlet
17. Coolant outlet
18. Coolant passage
19. Lower electrode chamber
20. Upper electrode chambers
21. Gradient column
22. Constant current power supply
23. UV absorbance standards

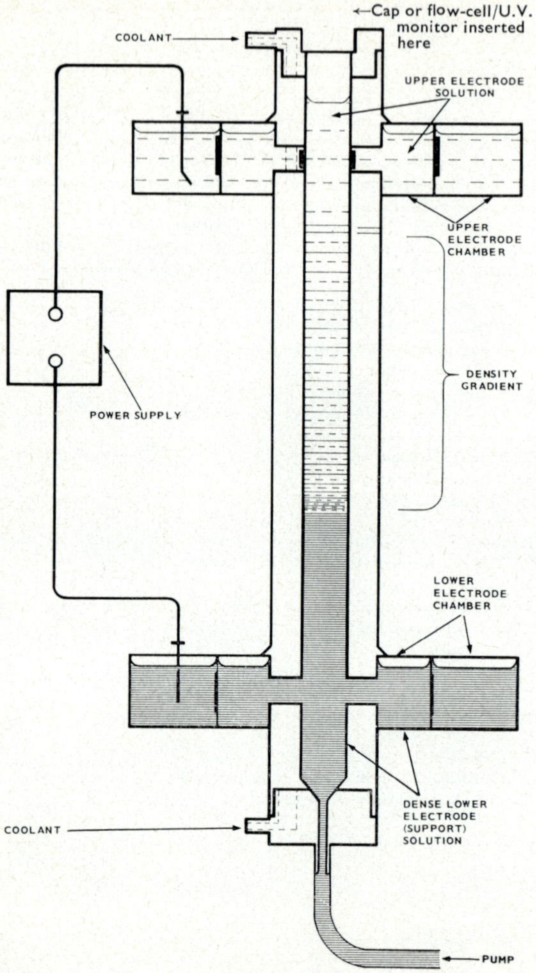

←Cap or flow-cell/U.V. monitor inserted here
COOLANT
UPPER ELECTRODE SOLUTION
UPPER ELECTRODE CHAMBER
DENSITY GRADIENT
POWER SUPPLY
LOWER ELECTRODE CHAMBER
DENSE LOWER ELECTRODE (SUPPORT) SOLUTION
COOLANT
PUMP

Fig. 11.5b

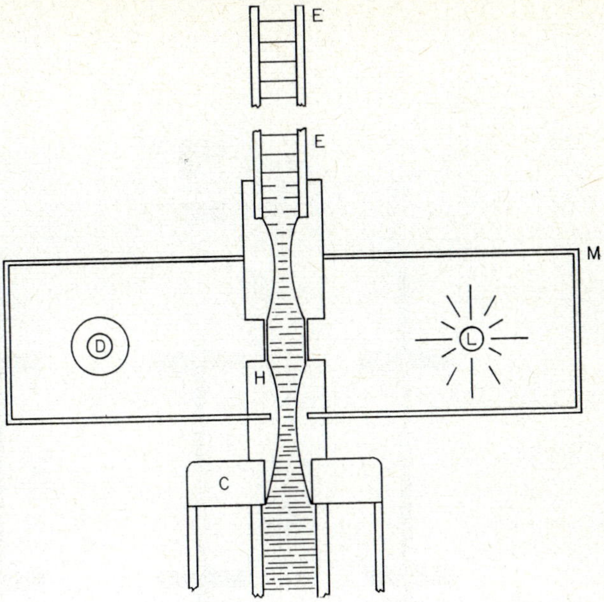

FIG. 11.5c

Operation of Models 210, 212 and 630 without "internal" analysis

(i) After connecting coolant jacket to supply of circulating water at 0–5°C., dense lower electrode solution is pumped upward until the column is almost full (see Fig. 11.6a).

(ii) A density gradient is then established on top of the dense support solution using appropriate volumes of mixtures (d) and (e) (see p. 281) and mixing methods described earlier (pp. 276–7).

(iii) When about half the sucrose gradient has been added the sample is added and the gradient is completed while the dense support solution is slowly lowered.

Note: Protein samples of up to 25 μl. can usually be added to the gradient without special procedures, but larger volumes call for the addition of sucrose (20% W/V) to the samples.

iv) After the sucrose gradient is complete (see Fig. 11.6b) the upper electrode solution is layered carefully on the top of the density gradient and then lowered to a position similar to that in Fig. 11.6c.

(v) Appropriate solutions are added to the electrode chambers and the isoelectric focusing commenced. Up to about 5 W (say 1000 V, at 5 mA with a typical focusing time of 2–4 hr.) can be dissipated in models 210 and 212 with efficient cooling whereas the requirements for the model 630 are much the same as for the Vesterberg and Svensson column described earlier (p. 277).

vi) After completion of the focusing process, the column contents are removed by upward displacement from the columns (often through

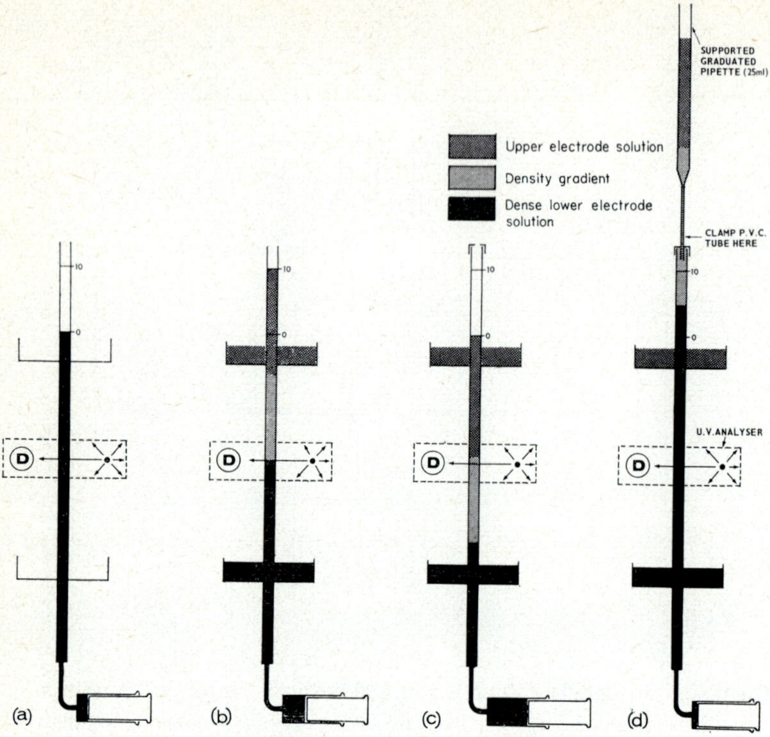

FIG. 11.6. Schematic diagram showing various positions of electrode solutions and density gradient in Isco model 210 column when 'internal' analysis is required. (a) Prior to addition of density gradient. (b) After addition of density gradient and electrode solution. (c) Density gradient is at its lower limit of travel. (d) Positions of solutions for addition of sample to preformed density/pH gradient (*by courtesy of Shandon Southern Ltd*).

a UV monitor) and fractions are collected and analyzed. Typical elution patterns from mixtures of human carbonmonoxy haemoglobins[22] A and F, after isoelectric focusing in Isco Models 210 and 630, are shown in Figs. 11.7 and 11.8.

Operation of Models 210 and 212 with "internal" analysis

Since the built-in UV monitor of the Isco model 210 only operates at 254 nm and since the carrier ampholytes absorb significantly at this wavelength, it proved important with this model to determine first the absorption characteristics at 254 nm. of the focused pH gradient itself. Half the pH gradient is then pumped at 2·5 ml./min. into a graduated pipette (as shown in Fig. 11.6d) and the connecting tubing clamped off. The sample is added to the top of the lower part of the pH gradient and the whole gradient carefully returned to the column

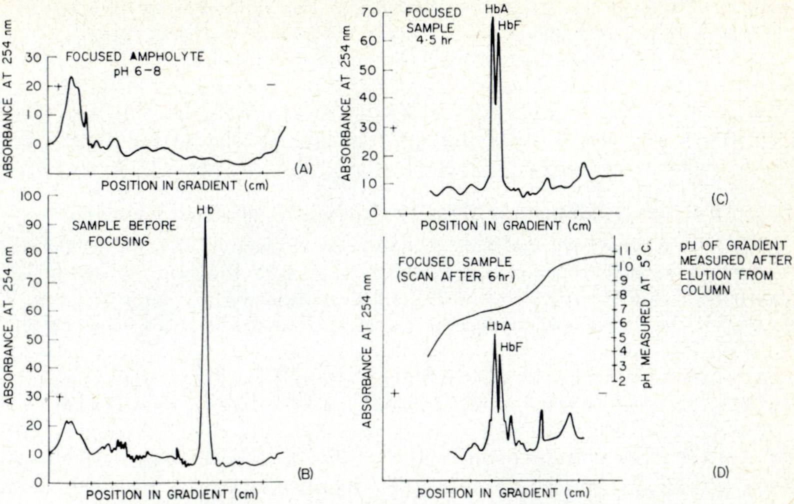

FIG. 11.7. Absorption characteristics at 254 m. of (a) pH 6–8 carrier ampholyte solution after isoelectric focusing. (b) As (a) but immediately after the addition of mixed carbonmonoxy-haemoglobin A and F samples. (c), As (b) but after 4·5 hr. isoelectric focusing. (d) As (c) but after 6 hours isoelectric focusing. In model 210 column (*by courtesy of Shandon Southern Ltd*).

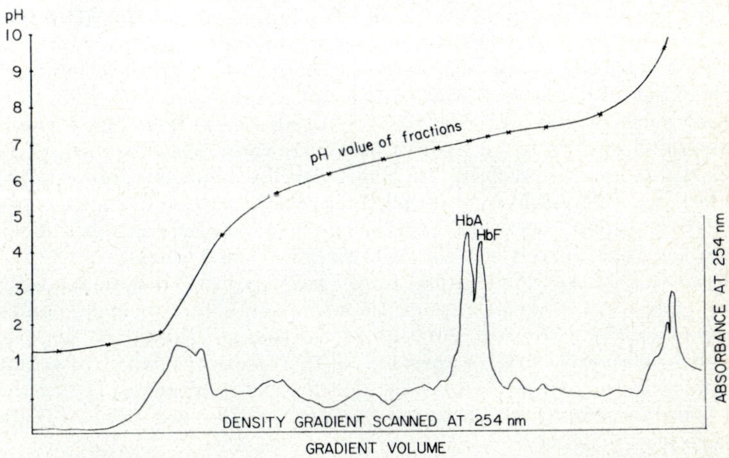

FIG. 11.8. Ultra-violet absorption (254 nm.) and pH (—x—x—) characteristics of fractions from the isoelectric focusing of carbonmonoxy-haemoglobin A and F using the Isco Model 630 column. Isoelectric points of haemoglobins A and F were found experimentally to be pH 7·1 and 7·2 respectively. These compare favourably with those of pH 7·0 and pH 7·07 for haemoglobins A and F, in the literature ref. 45 (*by courtesy of Shandon Southern Ltd*).

for re-focusing and the determination of the new absorption pattern. The results of one such separation are represented in Fig. 11.8.

A UV monitor operating at 280 nm. and an extension tube (Fig. 11.5c) can be fitted to the top cap of Model 212 column and since the absorption of the carrier ampholyte is low at this wavelength. In this case the above procedure involving pre-focusing of the carrier ampholyte is not usually necessary.

Further Guidance for Density Gradient Isoelectric Focusing

Choice of carrier ampholytes. A common approach with an uncharacterized protein mixture is to start with isoelectric focusing in a wide *p*H gradient (usually *p*H 3–10) and to follow this with a separation in a narrower *p*H gradient designed to give maximum resolution of the protein of interest.

Conditions for isoelectric focusing should be arranged carefully so that there is a reasonably uniform field strength throughout the column. If a zone of very low conductivity develops within a *p*H gradient a large voltage drop occurs in this zone with consequent local heating and convective disturbance: moreover in such cases, there is often little electric field strength remaining to produce focusing in the rest of the column.

The low conductivity of strong sucrose solution poses such a problem. For this reason, more of the carrier ampholyte solution is included often in the denser rather than in the less dense solution in order to raise initial conductivity at the bottom of the column. Carrier ampholytes with isoelectric points of about 6 also have relatively low conductivity, so that the electrode polarity should be chosen so that the *p*H 6 region will lie near the top of the column where sucrose concentration is low. For example, the anode should be located at the top of a column for a *p*H range of 6–9 and at the bottom for a *p*H range of 3–6.

Low conductivity near neutrality can also arise from the formation of a zone of pure water in this region, since water is itself an ampholyte. This difficulty can usually be overcome by routinely incorporating *p*H 6 to 8 carrier ampholytes at a level of about 10% of the total content of carrier ampholytes. This precaution is particularly important when operating with narrow *p*H ranges away from neutrality.

Protein Loading. As charged buffer ions introduced with the sample must be removed by electrophoretic migration before focusing can take place effectively, the salt content of the sample should be kept low, say, below about 5 mM. (expressed as the concentration if distributed throughout the whole volume of the density gradient). If necessary the sample should be dialysed against a 1% solution of carrier ampholytes of glycine.

For a gradient volume of about 100 ml., a loading of 5 to 25 mg. protein per zone is often possible. Much less sample would usually be applied for an analytical application requiring high resolution than for crude preparative fractionation. Decisions on loading, *p*H range, final voltage and time of focusing are related. For example, choice of a narrow *p*H range permits higher loading per zone, but probably warrants a higher field strength for longer time. If loading and zone sharpening are extreme, the density of the zone of protein might exceed the carrying capacity of the density gradient. In this case droplets of protein solution sometimes fall from the zone to remix with lower layers.

Detection of focused zones. Focused protein zones are usually detected by their absorbance at 280 mn. At this wavelength, absorbance due to carrier ampholytes is low but not uniform. Procedures and apparatus have been described above in which the absorbance of focused carrier ampholytes can be determined prior to the addition of the protein sample; but in other apparatus a control experiment (no protein) should be performed to established the baseline of absorbance. As we have seen, such procedures are particularly important if the absorbance is measured at a lower wavelength e.g. 254 nm.

Carrier ampholytes interfere strongly with the Lowry colorimetric assay for protein, they do so only slightly with the Biuret method (see ref. 10).

While some carrier ampholyte species are known to fluoresce in the ultra violet,[6] Penner and Cohen[23] reported ultra violet fluorescence to be a sensitive assay of protein involving little interference from carrier ampholytes.

Enzyme assays on isoelectric focusing fractions are particularly convenient if they permit adequate dilution of the fractions in order to minimize interference by the carrier ampholyte due to varying *p*H, buffering power, etc. Fluorimetric enzyme assays are therefore particularly attractive.[24] Addition of a metal co-factor might be necessary in view of the ability of carrier ampholytes to complex metals.[6]

Removal of carrier ampholytes. Contamination of fractionated proteins with carrier ampholytes and sucrose is a sometimes serious disadvantage of isoelectric focusing. Dialysis is satisfactory for removal of sucrose but relatively slow for removing carrier ampholytes, requiring over 20 hours.[9] Harsh methods, such as protein precipitation with trichloroacetic acid are often successful but usually inappropriate. Milder conditions are offered by precipitation with ammonium sulphate.[25]

Satisfactory separation of most proteins from carrier ampholytes is obtained with gel filtration, e.g., using Sephadex G-50.[26] Separation on the basis of size has also been obtained using ultra-filtration membranes.[27] Specific absorption of carrier ampholytes on an ion exchange column is a promising method,[29] but the suitability of this or other procedures above depends largely on the properties of the particular proteins concerned.

Applications of density gradient isoelectric focusing. The example in Fig. 11.9 illustrates the resolution obtainable by performing two consecutive runs to obtain a very shallow *p*H gradient (p. 489 in reference 12). The plot of radioactivity indicates microheterogeneity in the serum β-globulins with a high affinity for steroids. In a separate experiment, isoelectric focusing was carried out on non-labelled serum and individual fractions were checked for binding activity, with similar results. The heterogeneity did not seem to be an artifact of the procedure (e.g. possible interaction with carrier ampholytes), since a

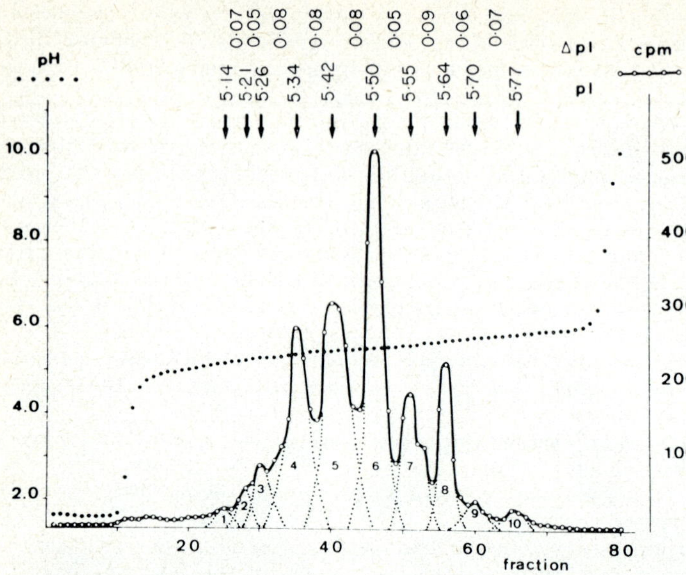

FIG. 11.9. Density gradient isoelectric focusing of a steroid-binding β-globulin of human pregnancy serum (p. 489 in ref. 12). Steroid depleted total serum was first focused in the range pH 4 to 6 (final voltage of 600; time 45 hours; anode on top; temperature of 4°). The region containing radioactively labelled steroids was submitted to isoelectric focusing a second time, with a shallower pH gradient (final voltage of 1000; time: 65 hours; cathode on top; temperature of 4°). The solid line indicates radioactivity (c.p.m.) of fractions drained from the column. The pH gradient is shown as closed circles.

refocusing of an individual peak again produced a single zone with the same isoelectric point. Fractionation after neuraminidase treatment suggested that the heterogeneity was connected with variation in the sialic acid content of the proteins.

The use of carrier ampholytes which have become available recently for isoelectric focusing in the pH range 2·4 to 4 has been demonstrated by the fractionation of pepsin isoenzymes from gastric juice.[29] Alternatively, mixtures of weak acids can be used[30] but in this case, the problem of low conductivity in the neutral pH region must be overcome (see p. 48 of reference 5). Proteins with high isoelectric points (including cytochrome c and lysozyme) have been focused using carrier ampholytes synthesized to produce the pH range 9 to 11.[31] For lysozyme, two components (isoelectric points 10·0 and 11·0) were separated from the urine of leukemic patients.[32] Mixed amines have also been used for isoelectric focusing up to pH 11.[33,34] For these high pH ranges, Vesterberg[31] has recommended replacing sucrose, as solute in the density gradient, with 0 to 50% sorbitol (or glycerol), since sucrose is partially dissociated at high pH.

A change of solute might also be justified to overcome protein precipitation during fractionation. Ethylene glycol (up to 33%) has proved a suitable replacement to prevent precipitation of serum lipoproteins.[34] In other studies on these proteins (and the corresponding apoproteins) solubility has been maintained by increasing the concentration of carrier ampholytes (to 2%, or even to 8%) and by adding urea (to 6 M or 8 M; see p. 311 in reference 13). Precipitation of proteins as they focus can be avoided simply in some cases by measures that decrease the protein concentration. This can be effected by decreasing the amount applied; by decreasing the final applied voltage; by selecting a narrower *p*H range, or perhaps by stopping the experiment when zone sharpening is sufficient but not quite complete.

However, for many proteins, solubilizing agents must be added to prevent precipitation. The most commonly used, urea, has been reported to give an overestimate of isoelectric point[55,36] and to affect assessments of microheterogeneity.[36] Urea has also been used effectively for studying subunits.[37] Non-ionic detergents can also be used for maintaining solubility. Triton X-100 has been used up to 1% with haemolysins (p. 465 in reference 12), and 0·5% Brij 35 has also proved effective.[38]

Addition of mercaptoethanol (with urea) has been used for the dissociation of proteins prior to the isoelectric focusing of the protein subunits.[39] On the other hand, thiol compounds at a very low concentration (·01 mM dithioerythritol) have been used to improve recovery of labile enzymes.[40] An alternative approach to the isoelectric focusing of sensitive enzymes is to prefocus carrier ampholytes before applying the sample directly to the region of the column where the enzyme is expected to focus.[23] In this way exposure to extremes of *p*H is avoided and focusing of the protein is quicker.

Density gradient isoelectric focusing can even be applied to particulate material. For example, it has been used with membranes of human red blood cells, platelets and lymphocytes.[41]

ISOELECTRIC FOCUSING IN GEL MEDIA

In 1968 several groups of workers[42–46] independently described procedures for carrying out isoelectric focusing in either rods or thin layers of polyacrylamide gel. The subsequent rapid development of the technique is reflected in the number of publications on the subject since 1968 (see references 5, 6, 9–11). This in turn reflects the importance of the experimental approach and the need that there was for an analytical technique by which multiple samples of proteins could be compared after separation according to the isoelectric focusing principle.

For most applications the principal requirements of the gel medium are that it should provide an effective protection against convective mixing and have satisfactory chemical and mechanical properties for the subsequent detection and analysis of the separated proteins. Because of the low endosmotic flow, the adequate mechanical and chemical properties, the high optical clarity and the excellent anti-convection properties of the polyacrylamide gel matrix, this has been the medium of choice for the vast majority of the applications of gel isoelectric focusing so far. There is little doubt that the potential of isoelectric focusing for very high resolving power is realized more nearly in the gel techniques since some resolution is usually lost in density gradient experiments while removing the column contents, whereas polyacrylamide is extremely effective in preserving resolved protein zones both before and after the termination of the focusing process.

The chief disadvantage of polyacrylamide as a medium for isoelectric focusing is that gels with reasonably good optical and mechanical properties have porosities which limit the migration of the larger proteins (greater than about 700,000 daltons). Satisfactory optical and mechanical properties are obtainable in gels of high porosity composed of agarose or of agarose/polyacrylamide mixtures, but these gels typically have the disadvantage of a significant endosmotic flow due to a net electric charge on the agarose. There are indications however that this difficulty may be overcome.(47,48)

Thin layers of granulated gels like Sephadex have poor mechanical properties, but Radola(8,49) has shown that they can be used to stabilize isoelectric focusing in flat beds and that they permit the free migration and isolation of even very large molecular weight proteins.(107,108)

Many experimental variations of gel isoelectric focusing procedure have been reported,(5,6,11,50,51) and although only established techniques using polyacrylamide in the form of gel cylinders or thin layers will be described in detail here, reference will be made to common difficulties and variations of the techniques.

Isoelectric focusing in Gel cylinders

Apparatus

The use of gel cylinders for isoelectric focusing became popular quickly because of the ready availability of disc electrophoresis apparatus, which can be used without modification. This section is complementary to Chapter 9 on disc electrophoresis, where a full description of apparatus and handling techniques is given.

On the other hand, because of differences in principle between isoelectric focusing and electrophoresis, more versatility in sample application is possible with gel isoelectric focusing. In further contrast to gel electrophoresis, zones become sharper as isoelectric focusing proceeds so that provided the same voltage gradient is applied, equally sharp zones can be obtained in longer gels than is normal for electrophoresis.

Figure 11.10 shows schematically a disc electrophoresis apparatus, set up for gel isoelectric focusing. A dilute acid is used for the anode vessel and a dilute base serves as catholyte. The gel tubes typically (5 mm. internal diameter by 60–120 mm. long) contain polyacrylamide and carrier ampholytes. The protein sample can be spread throughout the gel polymerization mixture (permitting the use of very dilute samples) or else layered on top of the polymerized gel under a layer of carrier ampholyte solution (which protects the sample from the extreme *p*H of the electrode solution).

Gel preparation. The following gel formula is sufficient for about twelve normal length gels (60 mm.), or eight double-length gels, containing 2% carrier ampholytes and 6% acrylamide monomer. Note that the acrylamide monomers are potentially toxic and that contact with skin, eyes, etc. should be avoided by careful handling and the use of plastic gloves.

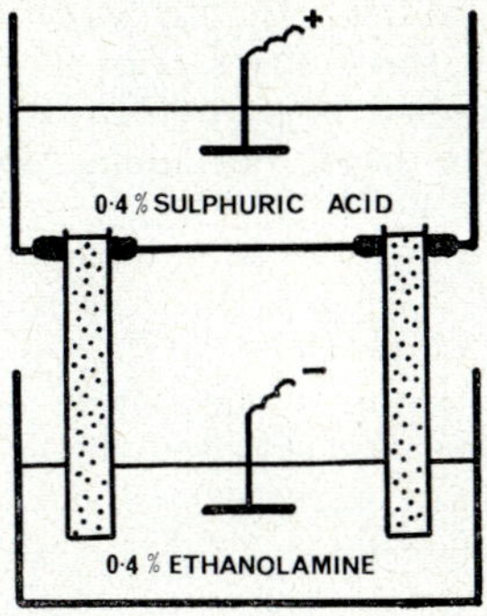

FIG. 11.10. The use of disc electrophoresis apparatus for isoelectric focusing in gel columns.

Acrylamide stock solution (see below) . .	4·0 ml.
Carrier Ampholytes (Ampholine, 40%) . .	0·5 ml.
Riboflavin (10 mg./100 ml.)	1·0 ml.
Water, to make a final volume of . . .	20·0 ml.

Acrylamide and riboflavin stock solutions should be stored at 5° in a brown bottle and renewed at least monthly. The acrylamide stock solution consists of:

Acrylamide	30 g.
N,N′-Methylene bisacrylamide . . .	1 g.
Water, to a final volume of	100 ml.

Dissolved air is removed from the gel mixture by applying a partial vacuum for a few minutes. The running tubes are filled with gel mixture to within about 5 mm. of the top and overlaid with water following the general instructions for disc electrophoresis (p. 214). The gel is photopolymerized by exposing the tubes to a strong light for 20 to 30 minutes at approximately 20° (i.e. sunlight, or at about 10 cm. from a fluorescent tube). Photopolymerization does not occur under normal levels of room lighting, so that shielding of the solution during tube filling is usually unnecessary. Photopolymerization is more difficult at higher *p*H values. In such cases, the riboflavin should be of particularly good quality.

If photopolymerization is unsatisfactory (e.g. because of inhibiting substances in a protein sample to be set in the gel, there is the alternative of chemical polymerization. In this case the riboflavin solution is replaced by a fresh solution of potassium persulphate (10 mg./ml.).

Polymerization occurs after about 30 minutes at 20°. The resulting gels are usually stronger than photopolymerized gels. but residual persulphate has been reported to modify sample proteins.

Sample Applications

(*a*) If the protein sample is to be incorporated into the gel, use some of the water of the gel mixture for dissolving the sample, and the balance is incorporated into the water in a concentrated gel mixture which is added to each sample. In this way, very dilute samples can be examined since nearly three quarters of the gel volume is available for each sample. The amount of protein to be applied depends on the means of detection to be used, but about 30 μg per component is normal.

(*b*) As an alternative to incorporation in the original gel mixture, the sample (in 10% sucrose) may be layered on top of the gel, and under a protecting layer (2 to 5 mm. thick) of 2% carrier ampholytes in 5% sucrose, either before switching on the current or after running for, say thirty minutes to establish the *p*H gradient. The following solutions are needed:

Ampholyte layer:

Carrier ampholytes (40%)	0·1 ml.
Sucrose	100 mg.
Water, to	2 ml.

Sample layer (**apply 0·05 ml.**)

Protein in sample	1 to 5 mg.
Sucrose	100 mg.
Water, to	2 ml.

Isoelectric focusing procedure

Gel tubes are assembled in the apparatus, as recommended for disc electrophoresis, and the electrode solutions are poured in. Sulphuric acid (0·4% v/v) is used for the anode compartment and ethanolamine (0·4% v/v) for the cathode. It is important to make sure that no air bubbles are trapped above or below the gel tubes. If the second sample-loading method is adopted, the upper electrolyte is introduced carefully through a funnel—to avoid mixing of electrolyte and the ampholyte layer.

Various other acids and bases have been chosen as electrode solutions, but the effect seems to be to alter slightly the *p*H limits at the end of the gels.[10]

The power supply is switched on and the voltage across the electrode is raised gradually. Initially, a maximum current of 2 mA per tube is maintained (1 mA for each double-length tube), but as the carrier ampholytes focus, the conductivity drops and a higher voltage can be applied whilst the same power is maintained. To avoid excess heating, the power gel is kept throughout the run below 0·3 watt/gel. A final voltage of about 400 volts for 60 mm. gels (700 volts for double-length gels) is usual. Normally, thirty minutes is ample time to reach maximum voltage and to establish the *p*H gradient. However, the time required to focus proteins depends upon the voltage gradient, the *p*H range, the tube length, the method of application, the salt content of the sample, the size of the protein molecules and their electric charge near the isoelectric points. For example, ovalbumin is focused (*p*H 3–10 gradient) in about 20 minutes if applied to the top of the gel, compared with about 40 minutes if set in the whole gel. While other proteins can require several hours, and in exceptional cases recommendations have been made to continue focusing overnight, many proteins are brought close to equilibrium after 90 min. If in doubt, samples should be run for various times and the results compared. When focusing is complete, the gels are removed from the tubes as described for disc electrophoresis (page 228). Methods for the determination of the *p*H gradient and for the detections of the focused zones etc. will be described in detail below (p. 306–7).

Special techniques using the results of isoelectric focusing in gel cylinders

Close comparison of two samples in the same gel tube can be made using a duplex tube. As illustrated in Fig. 11.11 (centre), such a tube

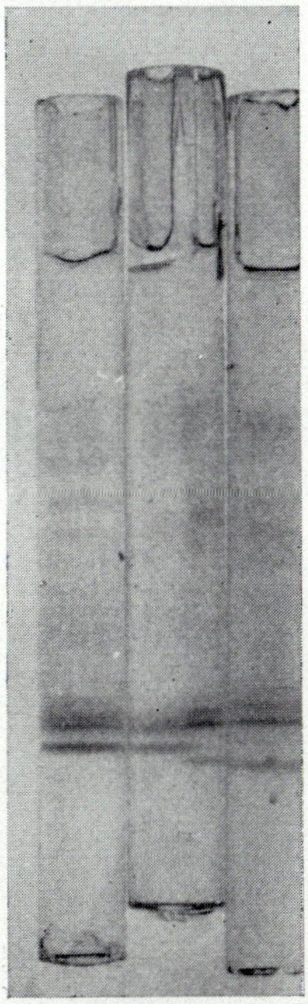

FIG. 11.11. Comparison of haemoglobin preparations from two different sources in separate tubes, and (centre) in the two halves of a duplex tube. The tubes were photographed immediately after isoelectric focusing with the anode on top for 40 minutes in the range *p*H 3 to 10 (*from Wrigley*, *p*. 417 *in ref*. 12).

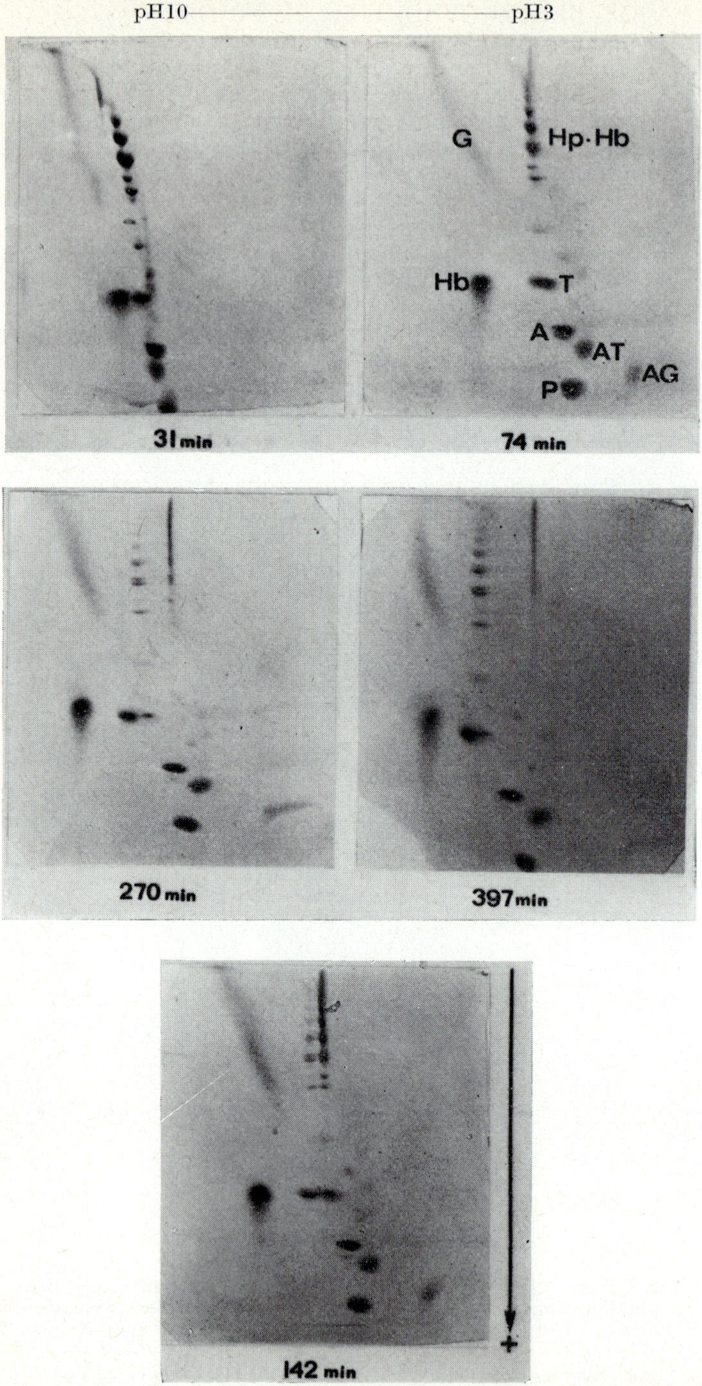

pH10
pH3
G
Hp·Hb
Hb
T
A
AT
P
AG
31 min
74 min
270 min
397 min
142 min
+

has the top 5 to 10 mm. divided into two compartments by a transverse wall of glass, fused in position, or stiff plastic sheet, wedged across the tube. Samples for comparison are applied to the top of the gel on either side of the partition. This method is particularly suitable for comparing a purified fraction with the original crude extract.

Resolution of various protein mixtures has been further increased by combining the resolving capacities of isoelectric focusing and electrophoresis. Smith *et al.*,[52] have done this quite simply by cutting a gel after isoelectric focusing into sections and applying each of the resultant discs to the top of another gel of the same diameter for electrophoresis.

Alternatively, a two-dimensional map can be obtained by placing the focused gel in the starting slot of a slab gel for electrophoretic separation (see pages 154 and 281 in reference 13). Crossed immuno-isoelectric focusing is an extension of this approach.[53] The combination of the two methods often has the advantage that carrier ampholytes are removed in the electrophoretic step, thus avoiding difficulties with staining.

Two-dimensional separation[54] comprising gel isoelectric focusing and gradient gel electrophoresis of a mixture of serum proteins is shown in Fig. 11.12. In particular, combination of the two methods has permitted separation of the immunoglobulins from the haptoglobin–haemoglobin series, and of free haemoglobin from transferrin. The time sequence of Fig. 11.12 illustrates the gradual focusing of the components (applied to the basic end of the gel) to give optimal focusing at between one and two hours but followed by a slight drift of the *p*H gradient towards the cathode. Combinations of gel isoelectric focusing with sodium dodecyl sulphate (SDS) electrophoresis have also been reported (see later; Table 11.4).

ISOELECTRIC FOCUSING IN THIN LAYERS OF POLYACRYLAMIDE GELS

Since the first reports of the use of thin layers of polyacrylamide gels for the isoelectric focusing of proteins by Leaback and Rutter[45] and by Awdeh, Williamson and Askonas,[42] this experimental approach has assumed great importance and many variations in apparatus and in experimental procedure have been reported.

The chief advantages of this form of gel isoelectric focusing are that multiple samples of proteins can be compared after separation in the same gel and that this gel format offers ready access to all the gel surface before, during and after the focusing process.

FIG. 11.12. Combined gel isoelectric focusing (*p*H 10–3 left to right and electrophoresis (second dimension, vertical, in 4·5 to 26% acrylamide gradient gel, *p*H 9·3) of serum proteins: Hp–Hb, Haptoglobin–haemoglobin complex; Hb, free haemoglobin; T, transferrin; A, albumin, AT, α_1—antitrypsin; P, prealbumin; G, Immunoglobulin (IgG); AG, α_1—acid glycoprotein. Isoelectric focusing was performed for 31, 74, 142, 170 and 397 minutes at 120 volts (*from Kenrick and Margolis*[46]).

Apparatus

In Fig. 11.13 are schematic diagrams of the principal experimental arrangements that have been employed for thin-layer polyacrylamide isoelectric focusing. In the original arrangement fig. 11.13b by Leaback and Rutter[45] the gel was cast in a shallow "home-made" glass tray and the focusing was carried out by dipping platinum electrodes into electrode troughs containing acid and base electrode solutions. The latter were subsequently found to be unnecessary[56] and the electrodes were replaced by cheaper flat graphite electrodes placed in direct contact with the gel (Fig. 11.13c). Awdeh *et al.*,[42] cast a thin layer of polyacrylamide gel onto a clean glass plate, placed the gel in a plastic box and carried out the isoelectric focusing with the plate inverted, and with the gel in direct contact with cheap, cylindrical, carbon electrodes (Fig. 11.13a).

The experimental arrangements in Fig. 11.13a–c have been shown to give satisfactory results and have the virtue that they can be conducted in apparatus which can be put together with a minimum of expertise and expense. However, if there is a strong case for obtaining quick results there are indications that it is advantageous to carry out the isoelectric focusing in a gel on a thin glass plate in contact with a cooled plate (see pp. 97 and 114 in ref. 14).

Commercially available thin-layer electrophoresis apparatus fitted with cooling plates can be used for this purpose. For example, the thin-layer electrophoresis apparatus supplied by Shandon-Southern (of Camberley, Surrey, U.K.), and which is fitted with a 150 × 230 mm. earthed metal cooling plate, has been used very successfully[57] for isoelectric focusing (see Fig. 11.13d) after simple modification—that is, by the use of two 5 mm. diameter, 250 mm. long carbon electrodes (Type WLO: Morgan Crucible Co., Battersea, London) gripped by metal "crocodile" clips and insulated wires to the terminals within the apparatus (see Chapter 4). The experimental arrangement shown in Fig. 11.13e incorporates many of the elements involved in Figs. 11.13a–d and illustrates the essentials of the LKB "Multiphor" apparatus set up for isoelectric focusing (Fig. 11.14). The apparatus utilizes a cooled 1 mm. × 260 mm. × 125 mm. glass plate on which is cast a 1–2 mm. gel layer. Alternative lids bearing platinum wire electrodes enable the isoelectric focusing to be carried out along the short or long dimensionsof the gel plate.

Other equipment required for gel isoelectric focusing include means for casting the gel layer (see below); a constant voltage/constant current power supply for up to 50 mA at 500 or perhaps 1000 volts. A supplyof circulating coolant water at 5° is also important for certain applications.

Gel Preparation*

The most common form of this polyacrylamide layer currently in use for isoelectric focusing is that cast between a glass "running" plate (to which the gel must adhere firmly) and another "gel former" plate (to which the gel must not adhere) spaced apart, and sealed by a

* LKB have recently introduced Ampholyne PAG plates, 245 × 110 × 1 mm layers of polyaerylamide gel on plastic sheets ready for the separation of proteins in a wide pH range.

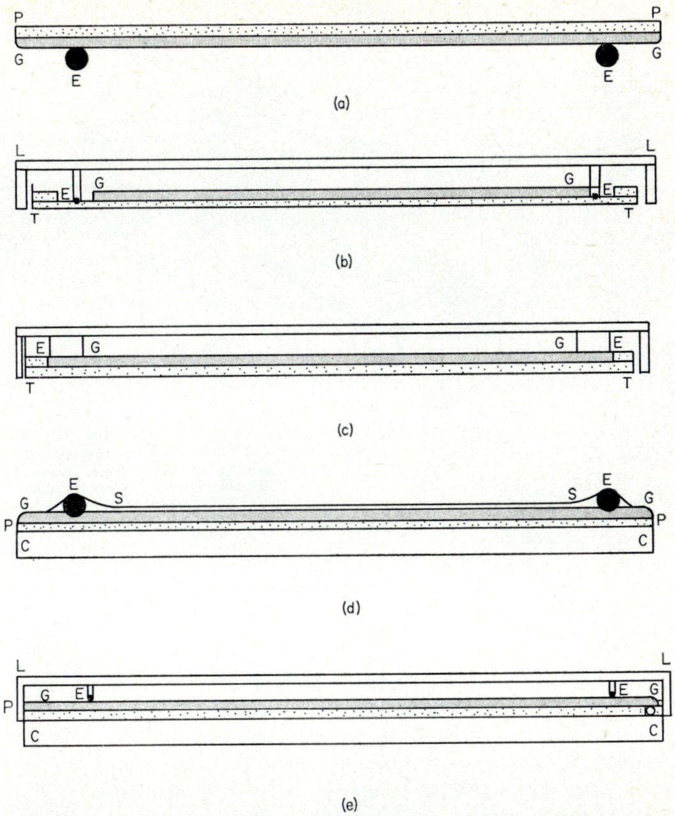

FIG. 11.13. Schematic sections of various Experimental Arrangements for Thin-Layer Gel Isoelectric Focusing. (*a*) Glass Plate (PP) with adhering gel layer (G) inverted on cylindrical carbon electrodes (E).[42] (*b*) Glass tray (T) containing gel layer (G) and a fabricated acrylic lid (L) bearing supported platinum electrodes (E) dipping into electrode wells containing acid or base.[45] (*c*) Later development of (*b*) in which the electrode wells have been eliminated and flat carbon block electrodes (E) substituted.[56] (*d*) Glass plate (PP) bedded onto underlying cooling platten (C). Cylindrical carbon electrodes (E) are laid on the thin-layer gel (G) and the assembly is overlaid by a thin sheet (S) of polythene.[57] (*e*) GlassP late (PP) bearing a thin gel layer (G). A fabricated acrylic lid (L) bearing supported platinum electrodes (E) placed on the gel (G). 1 cm. strips of filter paper wetted with either acid or base are often placed between electrode and gel. The glass plate (PP) is bedded on the cooling platten (C). The latter are the essentials of the LKB "Multiphor" Apparatus (see Fig. 11.14).

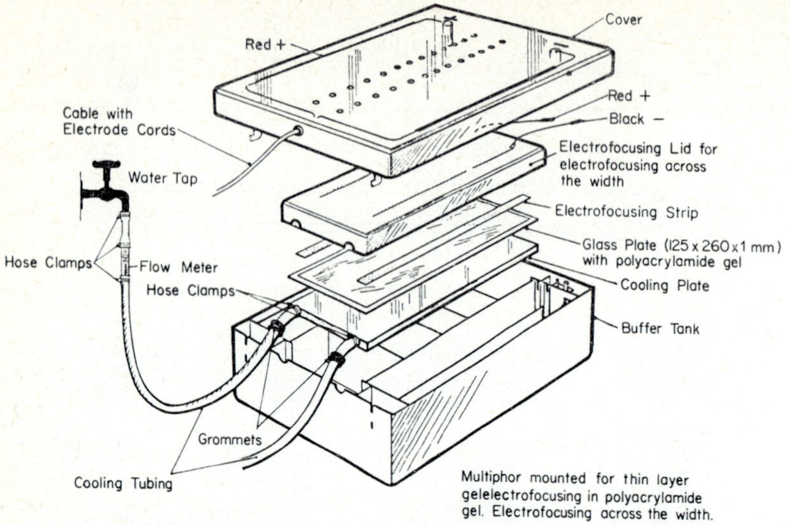

FIG. 11.14. Exploded diagram of LKB Multiphor Apparatus for thin-layer isoelectric focusing in polyacrylamide gels. The arrangement shown is for isoelectric focusing. across the width (100 mm.) of the gel, whereas another lid bearing electrodes is also available for isoelectric focusing along the long dimension (250 mm.) of the gels (*by courtesy of LKB Produkter A.B.*).

silicone rubber or polyvinylchloride gasket to form a thin, watertight mould to take the gel polymerization mixture. The whole assembly is held together firmly by 'Bulldog' clips for casting the gel (Fig. 11.15).

The most important requirements in the *running plates* are that they should be thin and made of glass (to help dissipate heat); that they should be of appropriate size for the number of samples contemplated, and for the length of the *p*H gradient required. Some workers believe that glass plates coated ("subbed") with a layer of chrome-hardened gelatine are advantageous in providing a key for strongly adhering gels. Clean glass plates of about 1·4 mm. thickness and subbed in this way on one side and in various sizes (approximately 250 × 210 mm.; 215 × 165 mm., 200 × 100 mm.) are available from Messrs. Ilford Ltd., Ilford, Essex, U.K. Other workers prefer clean, uncoated glass plates. Vesterberg[50] recommends the following procedure for plates of uncertain cleanliness: First clean with a detergent, then immerse in alcoholic KOH for 20 min. followed by rinsing in 0·1 N HCl and then distilled water.

The essentials of the gel former plate are that it should not be readily deformable and that it should not adhere to freshly polymerized polyacrylamide gels. Three types of casting plate are in use: (i) A thicker glass plate siliconed on the face to be used nearest the gel. A 260 × 125 × 3 mm.^{3} plate of this kind is supplied with the Multiphor apparatus for use with a 2 mm. thick moulded polyvinylchloride gasket during gel formation.

(ii) A sheet of acrylic polymer (perspex or Plexiglass) 6 mm. thick.

(iii) A sheet of acrylic polymer (perspex or Plexiglass) 6 mm. thick and the same size as the running plates with a 1·5 cm. ridge on three sides and channelled out to take a 2 mm.diam. 1 mm. bore silicone rubber tube (Esco Rubber Co., London)[(57)].

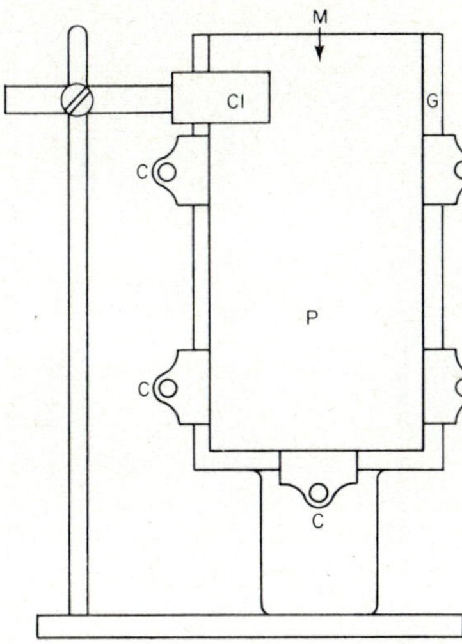

FIG. 11.15. Schematic diagram of arrangement for casting thin-layer polyacrylamide gels. The glass "running" and "casting" plates (P) and a silicone rubber or polyvinyl chloride gasket (G) are held firmly together by strong Bulldog clips (C) and the assembly is supported on a beaker by a clamp (Cl) before filling the mould with polymerization mixture at M and exposure to a suitable source of light. Note that it is usually necessary to employ more clips than is shown in the diagram in order to get a leak proof seal.

The following solutions would be required for a 5% photopolymerized gel with 3% cross-linkage (that is T = 5, C = 3 according to the nomenclature of Hjerten[(58)]).

Solution A. Acrylamide (29·1 g.) is made to 100 ml. with distilled water (measuring cylinder).

Solution B. N,N′-methylene-bisacrylamide (0·9 g.: also known as "Bis") is made to 100 ml. with distilled water (measuring cylinder).

Solution C. Riboflavin (4 mg.) is made to 100 ml. with distilled water.

Note: It is not normally necessary to recrystallize the above reagents from B.D.H., Poole, Dorset, U.K., but it is sometimes better to filter the solutions to remove dust particles before storage at 5° in dark bottles. Precuations (see p. 293) concerning the handling of the monomer solutions should be observed.

The *gelling procedure* described here will be for a photopolymerized golyacrylamide gel layer (T = 5, C = 3) incorporating a *p*H 3·5–9·5 radient suitable for either the Multiphor (260 × 125 × 1 mm) plates or

Table 11.2

Electrode and Carrier ampholyte solutions recommended(59) for thin-layer isoelectric focusing gels.*

Required pH range in gel	Volume of 40% carrier ampholyte (Ampholine) of nominal pH range (ml)							Anode Solution	Cathode Solution
	3·5–9·5	2·5–4	4–6	5–7	7–8	7–9	9–11		
3·5–9·5	2·8	—	0·2	0·2	—	—	0·4	$M \cdot H_3PO_4$	M·NaOH
2·5–6	—	1·2	1·2	0·6	—	—	—	$M \cdot H_3PO_4$	0·5% pH 5–7 Ampholine
5–8·5	—	—	—	1·5	1·5	—	—	$M \cdot H_3PO_4$	M·NaOH
7·5–10·5	—	—	—	—	—	0·2	3·6	0·1% pH 7–9 Ampholine	1·0% pH 9–11 Ampholine

* These volumes of Ampholines are those recommended for the 125 × 260 mm.[2] "Multiphor" plates but can be reduced by 33% for use with the 215 × 165 mm.[2] glass plates.

the 215 × 165 × 1·4 mm.^{3} plates described above. For other sizes the total volumes should be adjusted accordingly and for other *p*H ranges guidance will be given below.

The clean glass "running" plate is assembled with one of the gel former plates mentioned above and spaced with either a 2 mm. silicone tube (Esco Rubber Co., London) or a p.v.c. gasket (L.K.B. Instruments) to form a seal round the edges of the two plates. The assembly is secured by spring 'Bulldog' clips and when the worker is satisfied that the seal is watertight, the whole is mounted vertically and clamped in a stand as illustrated in Fig. 11.15. Using the Multiphor 260 × 125 mm. plates, 10 ml. of solution A; 10 ml. of solution B; 36·6 ml. of water and 3·0 ml. of 40% (w/v) carrier ampholytes are added with mixing to a 100 ml. evacuatable flask (these volumes can be reduced to 6·6; 6·6; 24·4 and 2 ml. respectively using the 215 × 165 mm. plates). The flask and contents are "evacuated" for 2 min. and, after carefully releasing the vacuum and adding 0·4 ml. solution C, the mixture is transferred (by means of a 25 ml. pipette or syringe but taking care not to aerate) to the gel moulding assembly illustrated in Fig. 11.15.

The polymerization is then carried out in front of an appropriate fluorescent lamp (or in sunlight) and should be complete in one hour (use fresh riboflavin solution if illumination is poor).

The 'Bulldog' clips and the rubber tubing or gasket are removed and the gel former plate is removed by easing with the edge of a spatula. Sometimes the glass gel former is difficult to remove but this difficulty can often be overcome by the storage of the gel assembly without the clips for 15 min. at 5°. If the gels are not required for immediate use they can be stored horizontally at 5° in a closed box for several weeks without significant deterioration.

In Table 11.2 is a list of carrier ampholyte combinations recommended by the manufacturer for different *p*H ranges (these volumes should be reduced by 33% for the 215 × 165 mm.^{2} plates). Acrylamide does not polymerize so readily at basic *p*H (e.g. in the presence of *p*H 7·5–11 gradients). In such cases the 0·4 ml. riboflavan solution is replaced by 0·8 ml. of a 1% (w/v) ammonium persulphate solution. The latter reagent however, is known to cause chemical changes in proteins,[61] but two approaches have been devised to reduce this hazard. Firstly, the persulphate polymerized gel can be pre-run for a period (thereby removing charged residues) before the addition of the protein sample. A second approach is to polymerize the gel with persulphate (replacing the carrier ampholytes by 0·1 ml. of Temed, i.e., N,N,N′,N′, tetramethylethylenediamine) and wash out soluble materials from the gel overnight in a large volume of distilled water. Leaback and Rutter[45] used this procedure and introduced the carrier ampholytes into the washed gels by diffusion, but it proved difficult to obtain economical and uniform entry of the carrier ampholytes in this way. An improved alternative procedure[60] is to dry a washed gel overnight at 37° (to about 10% of the original water content) and to introduce onto the dried gel any chosen carrier ampholyte mixture by means of the Perspex "gel former" described earlier (iii, see p. 301).[57]

Sample application

Sample application in isoelectric focusing is in principle non-critical, but in practice this is not always so, since proteins often deteriorate at high dilution or at certain *p*H. The most common method of application is to add the protein sample to a strip of 1 cm. wide filter paper (e.g. Whatman 3 MM or No. 542). The strips of appropriate size are held in forceps, wetted evenly with the sample (e.g., 1 cm. of Whatman No. 542/10 μl. sample) and placed lengthwise along the gel 5 mm. from adjacent samples. The best position for the sample application will vary with the *p*H stability, with the molecular size, and with the isoelectric points of the proteins of interest. It is generally good advice to add a sample at diffierent points along the length of the gel but preferably at sites away from the expected focusing region of special interest (see for example Fig. 11.16), since some stainable material usually stays at the point of application wherever this might be.

Staining the paper strips used for sample application almost always shows that some protein also remains attached to the paper. Other methods of sample application have been used including spreading the mixtures on the surface; adding samples to troughs in the gel, and adding samples to the surface in pieces of polyacrylamide gel.[45,51] A piece of centimetre squared graph paper placed temporarily beneath the gel plate during sample application can help greatly with the correct positioning of the samples and can be used later to record the *p*H gradient of the gel.

Isoelectric focusing procedure

Experience has shown that it is important to avoid as much as possible uneven conductivity along the length of the gels during isoelectric focusing—both because the resultant uneven electric field makes it difficult to bring proteins to a focus in areas of high conductivity and because local heating tends to cause trouble in regions of particularly low conductivity. The recommendations embodied in Table 11.2 are designed to overcome these difficulties, and it should be noted that areas of low conductivity are equally to be avoided in the region of the electrodes. In consequence when carbon electrodes are to be used these are steeped in the recommended electrode solutions for 10 min., wiped dry, then laid parallel to each other at the ends of the gel. When platinum electrodes are to be used, these are placed on 1 cm. wide strips of Whatman 3MM paper which have been wetted in the respective electrode solutions, blotted lightly, and then placed parallel to each other at the ends of the gel.

A polyacrylamide gel matrix can be so effective in preventing the convective transfer of heat generated within the gel during isoelectric focusing that localized heating and evaporation by water etc. can take place. Such difficulties have been overcome by a combination of limiting the electrical power dissipated in the gels; by keeping the layer of gel thin; and by ensuring good thermal contact of the gels through the glass "running" plate to an underlying cooled plate. It has been found that the gels described above can be focused in 18–20 hrs. using no cooling but by limiting the power dissipated to 1 watt: that

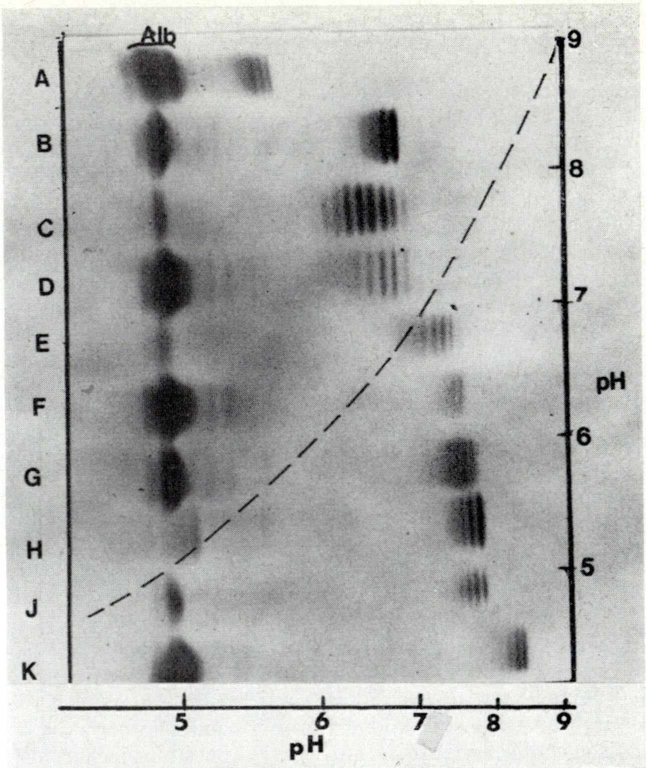

FIG. 11.16. Results of the Fast Green staining (see text) of samples (A–J) of human sera containing monoclonal IgG immunoglobulins after isoelectric focusing for 18 hr. at 320 V. on a 215 × 165 mm.[2] thin-layer gel, and employing a pH 3–10 gradient. The samples (10 μl) were added to pieces of Whatman 3MM paper (5 × 10 mm.,[2] and seen on left of figure). The resultant pH gradient was measured with a combined glass/reference electrode. Two methods of representing the measured pH gradient are shown in in the figure. The first method involves the non-linear pH scale shown horizontally, and the second uses the linear scale shown on the right which is not an ordinate but should be employed in conjunction with the broken curve. There are arguments for and against either method of representation. The characteristic banding pattern of monoclonal IgG immunoglobulins is seen on the right of each track (*by courtesy of Messrs. Butterworths*).

is, by starting the run at say 125 V and increasing the voltage from time to time (to about 350 V) as the current drops—but keeping the wattage at or about unity. The use of a layer of aqueous non-ionic detergent between the gel running plate and the underlying cooling plate has ensured good thermal contact and permitted the dissipation of up to about 40 watts in cooled isoelectric focusing gel sof the kinds described above. Such high wattages are not recommended for beginners.

FURTHER PRACTICAL ASPECTS OF GEL ISOELECTRIC FOCUSING

After the focusing process is complete, the usual requirements are that the gels should be prepared for the determination of the *p*H gradient and for the detection of the separated proteins. This usually involves the removal of gel cylinders from the "running" tubes, and the removal of any paper strips from the surface of the thin layer gels. By reflected light the surface of freshly focused thin-layer gels invariably show lateral contours which follow lines of equal *p*H and low electrical conductivity. The depth of the surface contours and their straightness can give a useful early indication of how ideal the run has been.

The Determination of the *p*H gradient

For gel cylinders, the *p*H gradient can be determined by cutting the focused gels laterally into about 20 segments of equal length using a set of razor blades held on threaded rods.[(62)] The resultant gel discs are steeped in 1 ml. aliquots of water for 2 hr. and the *p*H of the extract measured. The equivalent procedure for the thin-layer gels entails cutting 4 mm. diameter discs at 5 mm. intervals along the length of the *p*H gradient using a cork borer, and extracting the discs separately in water (0·75 ml.) for 2 hr. prior to *p*H measurement.

The possibility of using "marker" proteins of established isoelectric point to indicate certain *p*H values in the gradient and interpolating for other values was recognized in early work.[(45)] After painstaking work, Bours[(63)] has recommended the following commercially available proteins (isoelectric points in parentheses) as possible marker materials—bovine serum albumin (4·93); β-lactoglobulins A and B (5·35, 5·45); bovine carbonic anhydrase (6·38); and the two components of equine myoglobin (6·71, 7·56).

While such markers might be of use in the thin-layer gel format, they seem of very limited utility in gel cylinders since the proteins remain in the gels during the fixing and staining of the sample proteins. In this context the low molecular weight dyes can be attractive, because they focus rapidly, they are visible to the eye, and are washed out during gel processing. Their positions can be marked in the gels with a needle dipped in Indian ink. Routine inclusion of Bromophenol Blue has been recommended by Smith *et al.*[(52)] as a check that focusing is progressing normally. About 5 μg. of this dye or Fast Green FCF provide a useful indication of the *p*H 3 region in the gels.

The most commonly used method for the determination of *p*H gradients in focused gels involves direct measurement on the gels using a combined glass and reference microelectrode system. Due to the low but variable conductivity of the different regions of the focused gels, it is important that the measuring electrodes are small and are held at a fixed (and small) distance apart. The Pye Ingold combined electrode with a working diameter of 6 mm. has been used by a number of workers. Complications in the measurement of the *p*H gradient can arise due to the uptake of atmospheric carbon dioxide, or to the variation of *p*H with temperature. The latter effect can amount to a difference of 0·1–0·3 of a *p*H unit in the isoelectric point of a protein estimated at 4° and 25° (with the higher value at the lower temperature).[(67)] The *p*H gradient is therefore best measured on thin-layer gels soon after the focusing process is terminated and, with appropriate temperature compensation, while the gel layer is still in contact with the cooling plate.

The Detection of Proteins in Focused Gels—Staining, etc.

Since the carrier ampholytes currently in use for isoelectric focusing have some properties in common with proteins, they sometimes interfere with conventional methods for the detection of proteins in gels. The dye Amido Black is very commonly used in conventional gel electrophoresis for the staining of separated proteins, but this dye interacts so strongly with the carrier ampholytes that special procedures are necessary if it is to be used for isoelectric focusing gels.

The procedure which has been adopted is to fix (precipitate) the proteins into the gel using a denaturation reagent, then wash out the carrier ampholytes using conditions that do not remove the denatured proteins, and then stain with Amido Black or other dyes in the conventional way.

A typical procedure is as follows:

The focused gel is immersed in 5% (w/v) trichloroacetic acid solution. At this point many proteins appear as white bands in the gels and can, if required, be examined or photographed against a dark background using side illumination. The carrier ampholytes can then be removed from the gels by occasional agitation in at least five changes of the trichloroacetic acid solution.

When the gels are substantially free of carrier ampholytes, normal staining can proceed by immersion in Amido Black (Naphthalene Black: 10B 1% w/v in 7% acetic acid) for one hour followed by washing with several changes of 7% w/v acetic acid to remove excess stain.

With gel cylinders electrolytic removal of the carrier ampholytes is an attractive alternative. This may be carried out after the initial fixing of the proteins by reducing the concentration of the trichloroacetic acid to 0·5% and carrying out (preferably transverse) electrophoresis as described in the chapter on disc electrophoresis (page 231). Note that dilute acetic acid should not be used in such a process since it does not fix proteins adequately.

Riley and Coleman[(64)] were the first to introduce a "direct" staining technique for isoelectric focusing gels—that is, a stain mixture that could be applied directly to the gels without prior removal of the carrier ampholytes. Since then a number of direct staining procedures have been used including those involving the Bromophenol Blue, Fast Green FCF and Coumassie Blue recipes listed in Table 11.3. It can be seen that the direct staining mixtures listed all include fairly large proportions of an alcohol. It seems that under these conditions the carrier ampholyte/dye complexes are weaker or are more readily dissociated. The Bromophenol Blue and Fast Green procedures entail the

TABLE 11.3

*A Summary of Recipes and Procedures for the Direct Staining of Proteins in Isoelectric Focusing Gels**
The quantities indicated below are approximately those required for one typical thin-layer gel.

Stain (concentration)	Convenient Quantity of Stain	Stain solution made up in	Staining Conditions	Destaining Solution	Reference Number
Bromophenol Blue (0·2% w/v)	0·5 g.	ethanol, 125 ml. water, 112 ml. acetic acid, 12 ml.	1 hr. at 20°	ethanol, 300 ml. water, 650 ml. acetic acid, 50 ml.	65
Fast Green FCF (0·2% w/v)	0·5 g.	ethanol, 100 ml. water, 100 ml. Glycerol, 40 ml. acetic acid, 2 ml.	3 hr. at 20°	ethanol, 250 ml. water, 700 ml. acetic acid, 50 ml.	66
Coumassie Brilliant Blue R250 (0·1% w/v)	0·3 g.	Methanol, 75 ml. water, 180 ml. containing sulphosalicylic acid (9 g.) and trichloroacetic acid (30 g.)	15 min. at 60°	ethanol, 250 ml. water, 650 ml. acetic acid, 80 ml.	67
Coumassie Brilliant Blue R250 (i) 0·05% w/v (ii) 0·01% w/v	(i) 10 ml. of aqueous 1% w/v stain (ii) 2 ml. of aqueous 1% w/v stain	Methanol, 65 ml. water, 160 ml. acetic acid, 25 ml. containing $CuSO_4$ (0·25 g.)	4 hr. at 20°	ethanol, 100 ml. water, 800 ml. acetic acid, 100 ml.	6

* It appears (personal communication Messrs Bio-Rad) that at least some of these direct staining techniques are applicable to gels containing the new polyaminopolysulphonic carrier-ampholytes[106] of Bio-Rad; these are not yet available in England (August 1975).

immersion of the gels in the staining mixture for the time indicated, followed by immersion with occasional agitation in several changes of the appropriate "wash" solutions listed. These procedures have the advantage of simplicity, but show different affinities for some proteins and are generally not as sensitive as those techniques utilizing Coumassie Brilliant Blue.

In 1971, Spencer and King[(68)] reported a direct staining technique based on the Coumassie Brilliant Blue procedure of Crambach *et al.*;[(67)] (which needed no destaining) and which incorporated trichloroacetic and sulphosalicylic acids as protein denaturants. At about the same time Vesterberg[(67)] described a similar direct stain but one operating at 60° and containing ten times more stain. The staining solution is made up by dissolving the Coumassie Brilliant Blue R250 in the methanol and water mixture indicated in Table 11.3 and then adding the stated amounts of sulphosalicylic and trichloroacetic acids. This amount of stain solution is prewarmed to 60° in a metal, glass or plastic tray supported in a water bath at this temperature and the gel plate immersed in the staining solution for 15 min. After removal of the staining solution, it is recommended that the destaining mixture (see Table 11.3) should be agitated gently and changed at regular intervals.

The chief disadvantages of the latter method are that a staining procedure at 60° (in a fume cupboard) is not convenient in every laboratory, and that the procedure leads to high background colour which only clears slowly. Righetti and Drysdale[(6)] have adapted further the approach of Spencer and King to a procedure which operates throughout at room temperature and which, they claim, combines high sensitivity with low background staining.

Other methods for the detection (and quantitation) of the Separated Proteins

Separated proteins can be detected and perhaps estimated in gel cylinders by densitometric scanning at 280 nm. provided that quartz "running" tubes are employed.[(70,71)] Since individual carrier ampholytes show significant ultraviolet absorption (see Fig. 11.17) it is advisable in these applications to scan a control gel which has been focused simultaneously but without the addition of the protein sample. The prohibitive cost of even moderate sized quartz plates makes the scanning of thin-layer gels impractical at 280 nm., but densitometric scanning at higher wavelengths using coloured or stained proteins has been carried out on both cylindrical[(72)] and thin-layer gels[(68)] and can give good quantitative results. Thus, Spencer and King[(68)] excised pieces of gel (approximately 3×10 mm.^{2}) containing focused protein bands stained with Coumassie Blue, and extracted each with dimethylsulphoxide (1 ml.) for 24 hr. at 25° before estimating the extracted dye. They obtained good agreement in this way with integrated densitometric scans of the original stained gels. Robinson[(74)] has eluted excised portions of a focused gel with water (1 ml.) for 48 hrs. at room temperature and found enzyme (L-asparaginase) activities which agreed with the intensity of bands observable in the gel after staining.

Enzymes can sometimes be detected in focused gels without prior adjustment of the pH gradient provided that sufficiently sensitive enzyme detection procedures are employed (such as those based upon the fluorogenic methylumbelliferyl substrates)(75), but it is preferable to adjust the pH by short immersion (5 min.) of the gel in a fairly strong (say 0·5–1·0 M) buffer of appropriate pH containing the enzyme detection system. Alternatively the gel can be overlaid with filter paper or agar gel containing the buffer and the elements of the enzyme detection system. Zymogram methods along these lines have been listed by Wadstrom and Smyth(76) for amylase, deoxyribonuclease, Elastase, Hemolysin, Penicillinase, phospholipase and protease activities.

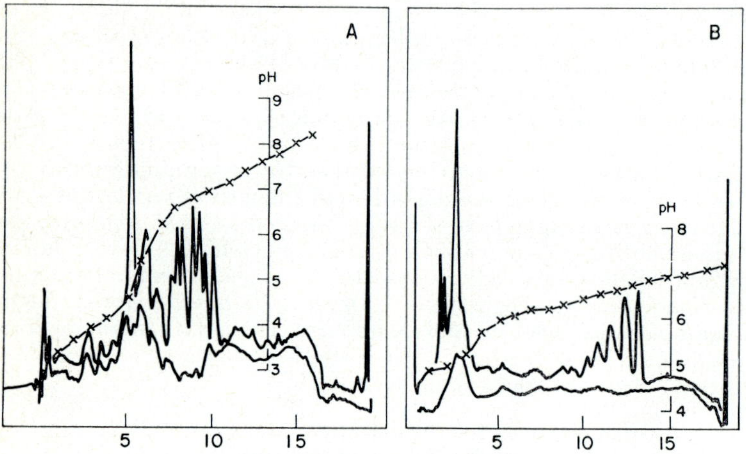

FIG. 11.17. Densitometric scans (280 nm.) of gels after isoelectric focusing of myeloma serum (50 μg. protein) in the pH gradient ranges (shown x – – – – – x) 3 to 10 (A) and 5 to 7 (B). The lower trace in each case shows the absorbance due to carrier ampholytes alone. (From Righetti and Drysdale, page 161 in reference 13.)

The characterization of focused zones by immunological means can most readily be performed by placing gel sections in the wells of buffered immunodiffusion (Ochterlony) plates. The slow diffusion of immunoglobulins in polyacrylamide gels(56) has led to the use of agar gel for isoelectric focusing in order to facilitate immunological assay.(77) In addition, many varied forms of "immuno-isoelectric-focusing" have been described and summarized by Catsimpoolas (page 144, in reference 13). "Crossed" electrophoresis after isoelectric focusing has proved a particularly powerful method for the identification and quantitation of the separated proteins.(78)

The estimation of radioactively-labelled proteins after isoelectric focusing has been carried out by digestion of excised portions of the gels for 5 hr. at 50° with 30% H_2O_2 followed by emission measurements with scintillation fluids.(68) The autoradiographic detection of proteins separated on thin-layer gel isoelectric focusing has been reported, but special precautions should be taken to ensure good contact between the dried gel and the sensitive film.[51]

Preservation of the Gels

After staining, isoelectric focusing gels are sometimes photographed for record purposes, but it is usually possible to store the gels themselves almost indefinitely after suitable processing.

Gel cylinders are usually stored in a stoppered tube containing 7% w/v acetic acid, and while the corresponding thin-layer gels can be stored moist in airtight containers, it is usually more convenient to dry such gels for storage. Thin-layer gels crack and peel from their glass plates if dried completely, but after drying at 20–37° to about 10% of their original water content (or drying in the presence of 5 w/v glycerol; see p. 97 of ref. 14) they can be transferred to polythene bags for storage.

Some common difficulties in Isoelectric Focusing

The difficulties of focusing proteins in gradients with non-uniform conductivities have been mentioned earlier, but can be illustrated from the results of Robinson.[74] Thus, Fig. 11.18 shows that samples (F, G, H and I) of a protein mixture appear to have focused when applied to a pre-run gel at *p*H values less than about 7, whereas similar samples applied at more basic *p*H clearly have not. Closer examination however, reveals that even samples F, G, H and I have not reached their respective isoelectric points (about *p*H 8·4),[74] and it can be deduced that these samples must have migrated rapidly at acid *p*H but then slowed greatly on reaching the basic region. Moreover, samples G–I must have undergone zone sharpening during electrophoresis and not as the result of isoelectric focusing. These results emphasize the importance of establishing as uniform as possible electrical conductivity throughout the *p*H gradient; of applying samples at various positions along the length of the gels, and of interpreting even the results of such experiments with caution.

Bringing basic proteins to a true focus can sometimes be difficult due to the contraction of the basic region of the *p*H gradient during prolonged isoelectric focusing. This drift of the *p*H gradient towards the cathode (see Fig. 11.18) is also known as the "plateau" effect,[6] In spite of intense investigation, this effect is still poorly understood.[79] There are some indications that running gels in an inert atmosphere of argon (and thus minimizing the chance of carbon dioxide absorption) can improve the basic region of the *p*H gradient a little,[80] but at present the best advice is to apply the sample as close as possible to the expected focusing point of the protein of interest.

The attainment of isoelectric focusing equilibrium in gels by very large proteins can be difficult in almost any region of the *p*H gradient. For example, Leaback *et al.*[56] showed, by the application of samples of an IgM immunoglobulin at more than one position along the length of a gel, that samples were far from focused after 16 hrs. isoelectric focusing in a 3% polyacrylamide gel.

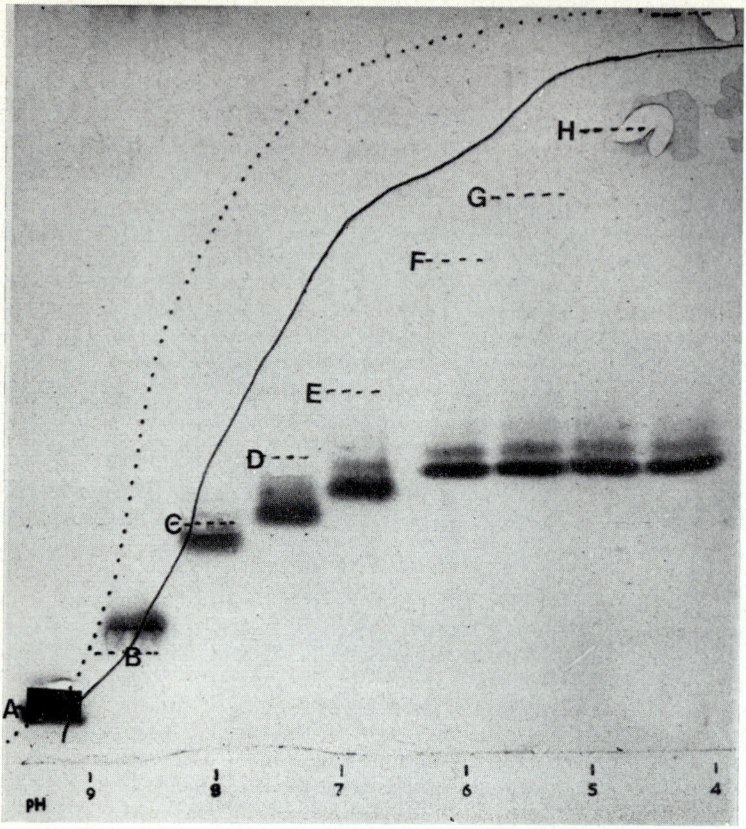

FIG. 11.18. A thin-layer gel pre-run for 6 hr. (pH gradient .....) and identical samples (A–I) of Erwinia, caratovora L-asparaginase applied at the positions indicated (– – – – –) and focused for a further 24 hr. before determination of the pH gradient (———); and staining with Bromophenol Blue procedure. The contraction of the basic part of the pH gradient between 6 and 24 hr. is particularly noticeable (*by courtesy of H. K. Robinson*).[57,74]

Some proteins undergo reversible or irreversible changes at certain pH so it is particularly fortunate that simple experimental procedures are available to check whether banding patterns obtained after isoelectric focusing may or may not be the consequence of the focusing procedure itself. Thus, the characteristic banding patterns of human monoclonal and polyclonal IgG immunoglobulins (see the right-hand sides of Fig. 11.19; samples a and i respectively)[45,46] are clearly not an artefact of the isoelectric focusing process since on excising the pattern from one such gel (e.g. Fig. 11.19b positions A, B, and C) and re-running the parts of the excised pattern separately (Fig. 11.19, c, d, e, and Fig. f, g, and h) the bands re-run true. In contrast, when

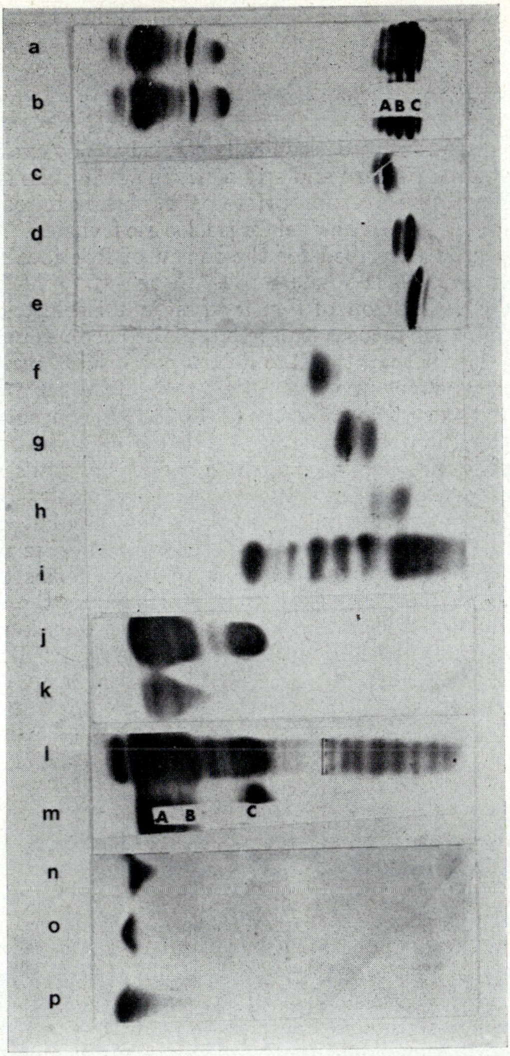

FIG. 11.19. The Detection of Isoelectric Focusing Procedural Artefacts.(56) The portions of the gels shown correspond approximately to pH 3–10 (*left to right*) with (a) 10 μl of a human IgG myeloma serum. (b) as (a) but with the monoclonal IgG proteins excised and sections A, B and C re-run on tracks (c), (d) and (e); (f), (g) and (h) are re-run portions of the polyclonal human IgG immunoglobulin pattern shown in (i); (j) and (k) are 10 μl and 2 μl respectively of a 3·5% human serum albumin solution; (l) and (m) are 10 μl and 2 μl respectively of a normal human serum showing similar focused albumin patterns similar to those in (j) and (k): Excised portions A, B and C from track (m) were then run on tracks (n), (o) and (p).

the same procedure was carried out on the banding pattern shown by human serum albumin (See Fig. 11.19, j and m) the excised bands did not run true (Fig. 11.19, n, m, o, and p) and this suggests that the characteristic pattern obtained for albumin is a reversible procedural artefact.[56]

This re-running procedure is typically carried out by excising portions of a protein banding pattern and setting them aside while the rest of the gel is stained for protein. Appropriate parts of the excised gel are then laid on separate tracks of a new gel for isoelectric focusing. An analogous procedure has been described for the re-run of fractions from density gradient columns.[81]

A frequent complication of density gradient isoelectric focusing is that as a protein zone focuses it precipitates or becomes unstable as the zone becomes denser than the stabilizing zone below. Such events can have a disastrous effect upon the other focusing zones and can sometimes be averted by certain additions to the column contents (see p. 291), or by increasing the density of the stabilizing medium in the vicinity of the offending zone. The precipitation of focusing proteins in the gels can also disrupt the *p*H gradient and this is usually countered by decreasing the protein loading. On the other hand it is sometimes possible to work satisfactorily with precipitating proteins. Thus successful gel isoelectric focusing of up to seven forms of insulin, as precipitated zones, has been reported.[82] However, the sample must be soluble at other *p*H values to permit ready focusing in the Gel. Urea, at high concentrations, has been commonly used for this purpose.[83,84,85] Other common causes of irregular protein zones in the gels include the addition of high levels of salt in with the protein sample, or local heating due to non-uniform electrical conductivities.

Some Applications and Conclusions

In this final section consideration will be given only to particularly novel applications, or to those which highlight some interesting aspect of isoelectric focusing techniques. For other applications the reader is referred to previous sections and to the numerous reviews on the subject (see p. 274).

All methods based upon the isoelectric focusing principle have two attractive features in common; that a wide variety of techniques for the application of the sample should yield the same final result, and that, for a given *p*H gradient and electric field, the result should be an equilibrium state dependent only upon the diffusional and charge properties of the ampholytes in question.

We have already seen how these non-critical requirements for sample application can be exploited in facile techniques for the detection of procedural artefacts, but they also make possible the addition of very dilute samples of proteins such as those from urine[86] or sweat.[87] Of course, isoelectric focusing has its limitations. For instance the method is not really applicable to proteins which undergo irreversible denaturation at their isoelectric points. It follows that the *p*H stability of a protein of interest should be investigated before attempting isoelectric focusing. We have also seen that there are

TABLE 11.4

Gel Isoelectric focusing as a component in some two-dimensional or quasi-two-dimensional techniques for the separation of proteins.

First Dimension	Second Dimension	Reference Number
Gel Isoelectric focusing	Crossed Immuno-electrophoresis	78
Gel Isoelectric focusing	Gel electrophoresis	52, 88
Gel Isoelectric focusing	Gradient gel electrophoresis	54 (see also Fig. 11.12)
Gel Isoelectric focusing	SDS-gel-electrophoresis	89
Gel filtration	Gel isoelectric focusing	24, 90
Gel Isoelectric focusing	SDS-gradient gel electrophoresis	110

still difficulties in bringing some (particularly basic) proteins to a focus, but isoelectric focusing procedures have been described here which can produce near-ideal results for a great number of proteins. Moreover the procedures are generally of very high resolving power and can separate very large homologous proteins with differences of but one or two units of electric charge.[56] When particularly complex mixtures of proteins are involved, it may be necessary to combine some other separation technique with isoelectric focusing. In Table 11.4 are listed some two-dimensional or quasi-two-dimensional techniques which employed gel isoelectric focusing and which were developed either to increase revolving power or to provide information on both the molecular size and the electric charge characteristics of proteins in mixtures.

The fact that isoelectric focusing permits the estimation of a very fundamental characteristic of proteins (the isoelectric point) has led to many applications including—helping to understand the behaviour of serum proteins in cellulose acetate and gradient pore electrophoresis;[92] in the detection and identification of haemoglobin variants;[45,93–95] in the demonstration of enzymatic and chemical modifications of proteins;[45,96,97] as an aid to guiding the choice of conditions for ion-exchange chromatography;[24] in the study of enzyme substrate and enzyme co-factor complexes (see Fig. 11.20) and of other proteins with ligands.[98]

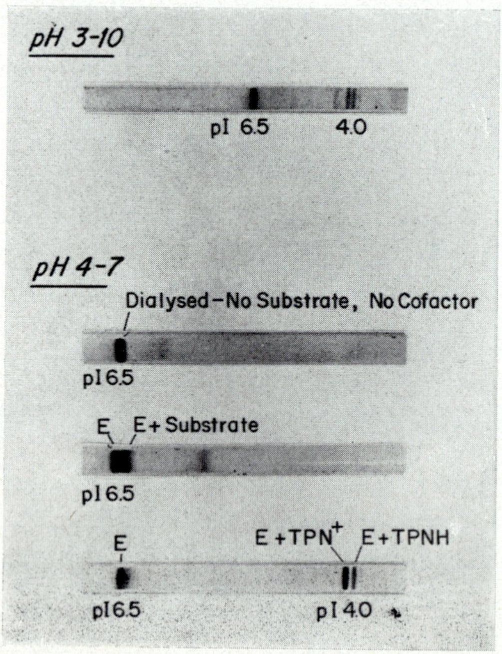

FIG. 11.20. Gel isoelectric focusing of enzyme, enzyme substrate, and enzyme coenzyme complexes (from page 163 of reference 13).

Density gradient stabilized isoelectric focusing is most usually carried out in columns of about 100 ml. capacity and is suited to relatively small-scale preparative purposes. Although columns of about four times this capacity are also commercially available, and continuous flow methods for the focusing of proteins have been developed,[91] the cost of the carrier ampholytes involved in the larger scale preparations of proteins by isoelectric focusing is impractical for all but very special applications. The column techniques are also well suited to analytical problems—especially in the determination of the isoelectric point distribution (the "isoelectric spectrum") of proteins with a particular assayable (e.g. enzymic) activity; for this kind of application small analytical columns that can give a result quickly are preferable. In contrast, isoelectric focusing in gels offers a technique for the determination of the "isoelectric spectrum" of all the proteins present in a mixture and this is not practicable by the density gradient procedures. Considered together, the results of the two kinds of isoelectric focusing techniques on a crude mixture of proteins can give a guide to the purification of a given component, and can be useful in monitoring the progress of the purification.[24,111]

The isoelectric spectra of the immunoglobulins can give great insight into the heterogeneity of these proteins. Williamson and his colleagues[81,99] have used the gel technique to investigate the origin of the heterogeneity and have ascribed differences to structural genes and to post-synthetic modifications. Most of the heterogeneity characteristic of IgG monoclonal immunoglobulins (see patterns in Fig. 11.16) is probably due to post-synthetic loss of amide groups from the proteins. Similar patterns are not shown by the corresponding IgA monoclonal proteins and this may be due to microheterogeneity in the attached carbohydrate moieties of the immunoglobulins or to the propensity of these proteins for polymerization.[56,91,100] Microheterogeneous carbohydrate moieties may also account for the striking gel isoelectric focusing banding patterns obtained for snake venom L-aminoacid oxidase isoenzymes.[101]

In spite of considerable interest and obvious potential, isoelectric focusing techniques have, rather surprisingly, not yet found applications in routine clinical biochemistry. Thus, Latner and his colleagues[43,88,102] have developed some interesting isoelectric focusing procedures and have employed them in the search for useful correlations between pathological states and the patterns of focused serum or tissue proteins. Monoclonal IgG immunoglobulins almost invariably show banding patterns of the kind illustrated in Fig. 11.15 and these can sometimes be used in the early detection of such pattern proteins in human sera;[103] unfortunately there is as yet no certain corrielation between the monoclonal protein patterns of this kind and the malignancy of the clone of plasma cells involved. Perhaps the most likely clinical applications of isoelectric focusing are in the detection of genetic variants of proteins. Thus the technique has been used to detect aberrant enzymes in storage diseases,[104] or in the investigation of pathological haemoglobin variants,[93–95] but it is probably true to say that the isoelectric focusing procedures have not as yet offered adequate advantages over existing methods to gain widespread acceptance for these purposes.

Part of the popularity of isoelectric focusing must be due to the fact that it is now possible, using established procedures, to get excellent results without much expertise or without recourse to elaborate apparatus. But the experimental approach is also popular for the information it can give on the electric charge properties of the separated proteins, and in this respect, is in accord with the present trend towards relatively simple, high-resolution electrophoretic techniques which offer insight into important molecular characteristics of proteins.[105]

O'Farrell[110] has resolved 1,100 different proteins from extracts of E.coli using an isoelectric focusing technique, and Miller and Elgin[112] have indications that focusing in 8 M urea may help in the difficult problem of removing sodium dodecyl sulphate from proteins. A soft lazer densitometer has been described for scanning of isoelectric focusing gels, and apparatus has been marketed by Desaga for the preparative focusing of proteins in granulated gels. Constant wattage D.C. power supplies are now available and are to be recommended for maximum efficiency in isoelectric focusing.

REFERENCES

1. Svensson, H. *Acta Chem. Scand.*, 1961, **15,** 325.
2. Svensson, H. *Acta Chem. Scand.*, 1962, **16,** 456.
3. Vesterberg, O., and Svensson, H. *Acta Chem. Scand.*, 1966, **20,** 820.
4. Vesterberg, O. *Acta Chem. Scand.*, 1969, **23,** 2653.
5. Haglund, H. In "Methods of Biochemical Analysis" (D. Glick, ed.) Vol. 19, p1. Interscience Publishers, New York, 1971.
6. *See* Righetti, G., and Drysdale, J. W. *J. Chromatog.*, 1974, **98,** 271.
7. Blanicky, P., and Pihar, O. *Coll. Czech. Chem. Commun.*, 1972, **37,** 319.
8. Radola, B. J. *Biochim. Biophys. Acta*, 1969, **194,** 335.
9. Vesterberg, O. In "Methods in Enzymology" (W. B. Jakoby, ed.) Vol. 22, p. 389. Academic Press, Inc., New York, 1971. Also Wrigley, C. W., p. 559.
10. Wrigley, C. W. In "Techniques in Amino Acid, Peptide and Protein Analysis" (A. Niederwieser and G. Pataki, eds.) p. 291. Ann Arbor Science Publishers Inc., Ann Arbor, Michigan, 1971.
11. Catsimpoulas, N. *Separation. Sci.*, 1970, **5,** 523.
12. Peeters, H. (ed.) "Protides of the Biological Fluids." Seventeenth Colloquium. Pergamon Press, Oxford, 1970.
13. Catsimpoulas, N. (ed.) "Isoelectric Focusing and Isotachophoresis." *Ann. N.Y.Acad. Sci.*, 1973, **209,** 1.
14. "Isoelectric Focusing" (Arbuthnot, J. P., and Beeley, J. A., eds.), Butterworth, London, 1975.
15. Weller, D. L., Heaney, A., and Sjogren, R. E. *Biochim. Biophys. Acta*, 1968, **168**, 576.
16. Heaney, A., and Weller, D. L. *J. Chem. Educ.*, 1970, **47,** 724.
17. Ayad, S. R., Bonsall, R. W., and Hunt, S. *Sci. Tools*, 1967, **14,** 40.
18. Svensson, H., and Pettersson, S. *Separation Sci.*, 1968, **3,** 209.
18a. Strongin, A. J., Baldvev, A. P., and Levin, E. D. *Science Tools*, 1973, **20,** 34.
19. Grant, G. M., and Leaback, D. H. Instrument Applications No. 31, 1970, Shandon Southern Instrument Co., Camberley, U.K., and Isco Applications Research Bulletin No. 3, 1970.
20. Brake, M. K., Allington, R. W., and Langille, F. A. *Anal. Biochem.*, 1968, **25,** 30.
21. Allington, W. B., and Aon, C. G. Isco Applications Research Bulletin No. 4, 1971.
22. Samples prepared according to Weiss, J. B., in Chromatographic and Electrophoretic Techniques" by J. Smith (ed.) Vol. II, p. 43, Heinemann Medical Books, London, 1968.

23. Penner, P. E., and Cohen, L. H. *J. Biol. Chem.*, 1971, **246,** 4261.
24. Robinson, H. R., and Leaback, D. H. *Biochem. J.*, 1974, **143,** 143.
25. Nilsson, P., Wadstrom, I., and Vesterberg, O. *Biochim. Biophys. Acta*, 1970, **22,** 146.
26. Vesterberg, O. *Sci. Tools*, 1969, **16,** 24.
27. Chang, P. C., and Dickie, N. *Biochim. Biophys. Acta*, 1971, **236,** 367.
28. Brown, W. D., and Green, S. *Anal. Biochem.*, 1970, **34,** 593.
29. Vesterberg, O. *Acta Chem. Scand.*, 1973, **27,** 2415.
30. Pettersson, E. *Acta Chem. Scand.*, 1969, **23,** 2631.
31. Vesterberg, O. Symposium on Chromatography and Electrophoresis. Publ. Press. Acad. Europeennes, Brussels, 1973, p. 81.
32. Lundblad, G., Vesterberg, O., Zimmerman, R., and Lind, J. *Acta Chem. Scand.*, 1972, **26,** 1711.
33. Flatmark, T., and Vesterberg, O. *Acta Chem. Scand.*, 1966, **20,** 1497.
34. Kostner, G., Albert, W., and Holasek, A. *Hoppe-Seyler's Z. Physiol. Chem.*, 1969, **350,** 1347.
35. Josephson, R. V., Maheswaran, S. K., Morr, C. V., Jenness, R., and Lindorfer, R. K. *Anal. Biochem.*, 1971, **40,** 476.
36. Ui, N. *Biochim. Biophys. Acta*, 1971, **229,** 567.
37. Bjork, I. *Acta Chem. Scand.*, 1968, **22,** 1355.
38. Friesen, A. D., Jamieson, J. C., and Ashton, F. E. *Anal. Biochem.*, 1971, **41,** 149.
39. Catsimpoolas, N. *Fed. Eur. Biol. Soc. Lett.*, 1969, **4,** 259.
40. Wadstrom, F., and Hisatsune, K. *Biochem. J.*, 1970, **120,** 725.
41. Jamieson, G. A., and Groh, N. *Anal. Biochem.*, 1971, **43,** 259.
42. Awdeh, Z. L., Williamson, A. R., and Askonas, B. A. *Nature* (*Lond.*), 1968, **219,** 66.
43. Dale, G., and Lather, A. L. *Lancet*, 1968, **1,** 847.
44. Fawcett, J. S. *FEBS. Lett.*, 1968, **1,** 81.
45. Leaback, D. H., and Rutter, A. C. *Biochem. Biophys. Res. Commun.*, 1968, **32,** 447.
46. Wrigley, C. *Sci. Tools*, 1968, **15,** 17.
47. Quast, F. *J. Chromatog.*, 1971, **54,** 405.
48. In "Isoelectric Focusing and Isotachophoresis" by Righetti, P. G. (ed.) Elsevier (Amsterdam) 1975, in press.
49. *See also* p. 184, ref. 14.
50. Vesterberg, O. *Biochim. Biophys. Acta*, 1972, **257,** 11.
51. Vesterberg, O. *Sci. Tools*, 1973, **20,** 22.
52. Smith, I., Lightstone, P. J., and Perry, J. D. *Clin. Chem. Acta*, 1971, **35,** 59.
53. Jirka, M., and Blanicky, P. *Biochim. Biophys. Acta*, 1973, **295,** 1.
54. Kenrick, K. G., and Margolis, J. *Anal. Biochem.*, 1970, **33,** 204.
55. Barrett, T., and Gould, H. J. *Biochem. Biophys. Acta*, 1973, **295,** 165.
56. Leaback, D. H., Rutter, A. C., and Walker, P. G. *Protides of Biol. Fluids*, 1970, **17,** 423.
57. Robinson, H. K. *Analyt. Biochem.*, 1972, **49,** 353.
58. Hjerten, S. *Arch. Biochem. Biophys.*, 1962, Suppl. **1,** 147.
59. Karlsson, C., Davies, H., Chman, J., and Andersson, U. B. LKB Application Note (March 1973).
60. Robinson, H. K., and Leaback, D. H.—unpublished results.
61. Mitchell, W. M. *Biochem. Biophys. Acta*, 1967, **147,** 171.
62. Matsumura, T., and Noda, H. *Anal. Biochem.*, 1973, **56,** 571.
63. Bours, J. *Science Tools*, 1973, **20,** 29.
64. Riley, R. F., and Coleman, M. K. *J. Lab. Clin. Med.*, 1968, **72,** 714.
65. Awdeh, Z. *Sci. Tools*, 1969, **16,** 42.
66. Leaback, D. H., and Grant, G. M.—unpublished results.
67. Vesterberg, O. *Biochim. Biophys. Acta*, 1971, **243,** 345, also p. 145 in ref. 109.
68. Spencer, E. M., and King, T. T. *J. Biol. Chem.*, 1971, **246,** 201.
69. Crambach, A., Reisfeld, R. A., Wykoff, M., and Zachari, J. *Anal. Biochem.*, 1967, **20,** 150.
70. Fawcett, J. S. P. 23 in ref. 14.
71. Catsimpoulas, N., and Wang, J. *Anal. Biochem.*, 1971, **39,** 141.
72. Fawcett, J. S. In ref. 12, p. 409.

73. Righetti and Drysdale. In ref. 6, p. 161.
74. Robinson, H. K. Symposium on Chromatography, Electrophoresis. Publ. Press. Acad. Europeennes, Brussels, 1973, p. 221.
75. *See* Leaback, D. H. "An Introduction to the Fluorimetric Estimation of Enzyme Activities," 2nd Edition, 1975—Available on request from Koch-Lights Ltd., Colnbrook, Bucks., U.K.
76. Wadstrom, T., and Smyth, C. J. *Sci. Tools*, 1973, **20,** 17.
77. Catsimpoulas, N. *Clin. Chim. Acta*, 1969, **23,** 237.
78. Jirka, M., and Blanicky, P. *Biochim. Biophys. Acta*, 1973, **295,** 1.
79. *See* contributions of A. Crambach and J. S. Fawcett in "Isoelectric Focusing and Isotachophoresis" by P. G. Righetti (ed.). Elsevier, Amsterdam, 1975, in press.
80. Radola, B. J. *Personal Communication*, 1974.
81. Williamson, A. R., Salaman, M. R., and Kreth, H. W. *Ann. N.Y. Acad. Sci.*, 1973, **209,** 210.
82. Kohnert, K. D., Schmid, E., Zuhlke, H., and Fiedler, H. *J. Chromatog.*, 1973, **76,** 263.
83. Flavini, F. R., Brivio, R. P., and Batlelle, B. A. M. *Anal. Biochem.*, 1971, **40,** 345.
84. Suginaka, H., Blumberg, P. M., and Strominger, J. L. *J. Biol. Chem.*, 1972, **247,** 5279.
85. Miner, G. D., and Heston, L. L. *Anal. Biochem.*, 1972, **50,** 313.
86. Rotbol, L. *Clin. Chim. Acta*, 1970, **29,** 101.
87. Kirka, M., and Blanicky, P. *Clin. Chim. Acta*, 1971, **31,** 329.
88. Dale, G., and Latner, A. L. *Clin. Chim. Acta*, 1969, **24,** 61.
89. Barrett, T., and Gould, H. J., *Biochim. Biophys. Acta*, 1973, **294,** 165.
90. Leaback, D. H., and Robinson, H. K. *FEBS Lett.*, 1974, **40,** 109.
91. Fawcett, J. S. *Ann. N.Y. Acad. Sci.*, 1973, **209,** 112.
92. Leaback, D. H. In ref. 14, p. 201.
93. Sonnet, J., and Deroyette, J. P. *Sci. Tools*, 1971, **18,** 12.
94. Drysdale, J. W., Righetti, P., and Bunn, H. F. *Biochim. Biophys. Acta*, 1971, **229,** 42.
95. Bunn, H. F. *Biochemistry*, 1974, **13,** 988.
96. Salaman, M. R., and Williamson, A. R. *Biochem. J.*, 1971, **122,** 93.
97. Bobb, D., and Hofstee, B. H. J. *Anal. Biochem.*, 1971, **40,** 209.
98. Akedo, H., Mori, Y., Kobayashi, M., and Okada, M. *Biochem. Biophys. Res. Commun.*, 1972, **49,** 107.
99. Awden, Z. L., Williamson, A. R., and Askonas, B. A. *Biochem. J.*, 1970, **116,** 241.
100. Leaback, D. H., and Walker, P. G. *Proc. Roy. Soc. Med.*, 1971, **64,** 15.
101. Wellner, D., and Hayes, M. B. *Ann. N.Y. Acad. Sci.*, 1973, **209,** 34.
102. Latner, A. L. *Ann. N.Y. Acad. Sci.* 1973, **209,** 281.
103. Leaback, D. H.—Unpublished results.
104. Sandhoff, K. *Z. Physiol. Chem.*, 1968, **349,** 1095.
105. Leaback, D. H. *Chemistry in Britain*, 1974, **10,** 376.
106. Cf. Pogacar, P., and Jarecki, R., p. 153 in ref. 109.
107. Radola, B. J., *Biochim. Biophys. Acta*, 1975, **386,** 181.
108. Delincee, H. and Radola, B. J., Eur. *J. Biochem.*, 1975, **52,** 321.
109. "Electrophoresis and Isoelectric Focusing in Polyacrylamide Gel" by R. C. Allen and H. R. Maurer (eds.), de Gruyter, Berlin, 1974.
110. O'Farrell, P. H., *J. Biol. Chem.*, 1975, **250,** 4007.
111. Cf. ref. 108.
112. Miller, D. W. and Elgin, S. C. R., *Anal. Biochem.*, 1974, **60,** 142.
113. Zeineh, R. A., Nijm, W. P., and Al-Azzawi, F. H., International Laboratory, March 1975, p. 25.

CHAPTER 12

ELECTROPHORESIS OF PROTEINS ON SODIUM DODECYL SULPHATE POLYACRYLAMIDE GELS

J. W. Payne

POLYACRYLAMIDE gel electrophoresis in the presence of sodium dodecy sulphate (SDS*) is now the most widely used method for determining the molecular weights of proteins. The negatively charged SDS molecules bind extensively to the polypeptide chains of proteins causing them to unfold and dissociate. The molecular weights of the constituent polypeptide chains can be determined by comparing their electrophoretic mobilities on SDS gels to the mobilities of well-characterized marker proteins of known molecular weights. The essential procedure is simply and rapidly carried out and requires only microgram amounts of proteins. While the principles remain the same, there do exist certain differences in methodology that provide advantages in particular casess i.e. use of gel rods or slabs, and use of single (phosphate) or discontinuous buffers, and it is useful to consider these aspects separately. In particular, the combination of discontinuous buffers and gel slabs is finding increasing application in the resolution of membrane proteins.

Experimental Procedures

Chemicals

Acrylamide, N,N′-methylenebisacrylamide (bis), and N,N,N′,N′-tetramethylethylenediamine (TEMED) are available from a number of commercial sources and may generally be used without further purification. If required, bis and acrylamide can be recrystallized from acetone[60]. Sodium dodecyl sulphate, in a specially purified form, can now be obtained from several suppliers (e.g. BDH).

Preparation of Reagents

1. Acrylamide—bis acrylamide (30:0·8): This stock solution is used for preparing gels in the range 5–20% acrylamide. It is prepared from 30 g. of acrylamide and 0·8 g. of N,N′-methylenebisacrylamide dissolved in 100 ml. water. Insoluble material should be removed by filtration through Whatman No. 1 filter paper. The solution is stable indefinitely when kept at 4° in a dark bottle. Great care should be taken in handling unpolymerized acrylamide as it is a poison and skin irritant (use automatic pipettes and gloves).

2. Sodium dodecyl sulphate (10%): stock solution prepared by dissolving 10 g. SDS in 100 ml. distilled water. The clear solution is stable at room temperature.

3. Ammonium persulphate (1·5%): 150 mg. of ammonium persulphate are dissolved in 10 ml. water. The solution is unstable and is preferably made up just before use, although it has been reported that a 10% solution is usuable for up to 2 weeks if stored under refrigeration[1].

4. TEMED: used directly. It is stable when kept at 4° in a dark bottle.

5. Continuous gel buffer: Sodium phosphate (1 M): 81·0 g. of NH_2HPO_4 and 59·3 g. of $NaH_2PO_4 \cdot H_2O$ dissolved in water and made to 1 litre, pH 7·2. Stable at room temperature.

6. Stacking gel buffer: Tris-HCl: Dissolve 6 g. of tris in water, adjust to pH 6·8 with 1N HCl and dilute to 100 ml. with water. Stable when kept under refrigeration.

7. Discontinuous gel buffer: Tris-HCl: Dissolve 36·3 g. of tris and 48 ml. 1N HCl and dilute to 100 ml. with water, pH 8·8. Stable when kept under refrigeration.

8. Electrode buffer: Tris-glycine: Dissolve 6 g. of tris and 28·8 g. of glycine and dilute to 1 litre with water, pH 8·3. Stable when kept under refrigeration.

Apparatus

Cylindrical Gels. A great many designs based on that described by Davis[(52)] have proved satisfactory. A commonly used one is supplied by Shandon, and a detailed description of its construction and use is given early in Chapter 9.

Flat-Slab Gels. Use of flat-slabs is comparately more recent but a number of different designs are available. A simple design was described by Adams *et al.*[(55)]. Studier[(53)] has given detailed information about an improved version of a design of Reid and Bieleski[(56)] that he and others[(2,48)] have used with great success. A number of commercial

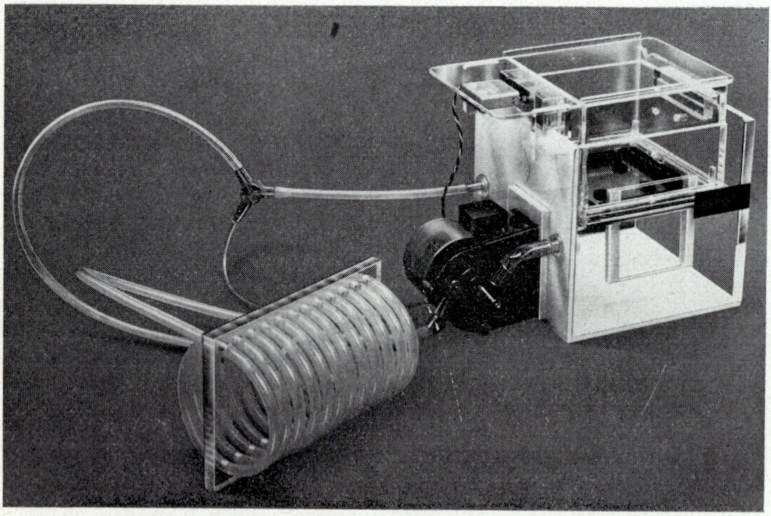

FIG. 12.1. An apparatus (Uniscil/Gradipore) for flat-slab electrophoresis. A gel sandwich is shown held vertically in one of the four available positions. This design is equipped with a buffer circulating pump and a cooling coil.

designs are available (Pharmacia, Uniscil/Gradipore) including a flat-bed system (LKB); see also Chapter 5.

In the vertical systems the principle is similar to that for rod gels. An example (Uniscil) of a system for holding four slab gels is shown in Fig. 12.1. The system has a built in buffer circulation and cooling system (±0·2°C), that is useful for maintaining buffer ionic strength when continuous buffer systems are used. The geometry of the electrodes is symmetrical to ensure uniform current density across the gels. The unit will accept either slabs or rods or a combination of both simply by changing the type of gasket used in the four slab positions. In Fig. 12.2 is shown a gel-slab used in this equipment. When preparing the gels (see below) the edges are sealed with an adhesive tape (Sellotape No. 1607, 25 mm.) that can withstand immersion in most buffers containing SDS for about 3 months. Also illustrated in Fig. 12.2 are a plastic "comb" (top) used for forming sample wells (see below) and a sample

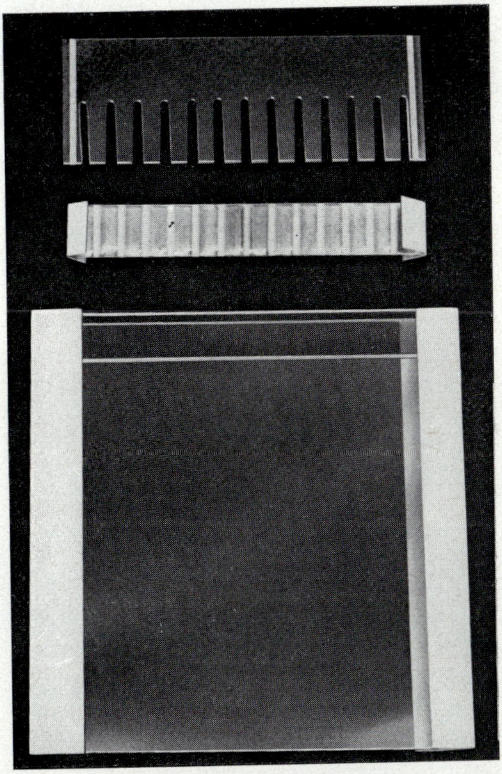

Fig. 12.2. A flat-slab gel sandwich for use in the apparatus shown in Fig. 12.1. Also illustrated is (top) a plastic "comb" for making sample wells in the top of the gel slab, and (centre) a sample applicator for adding spaced samples to the flat-slab gel when sample wells are not preformed on top of the gel.

applicator (centre) which offers an alternative mode of adding sample to the use of sample wells (see below).

Power supplies. Simple and inexpensive equipment can be built quite easily that can be entirely satisfactory[1,52]. It is useful to have separate power supplies for electrophoresis and for destaining (see below). An adequate source for electrophoresis supplies about 300 V. and 100 ma. Many are available commercially equipped with constant voltage and constant current regulation but this facility is not very important for SDS gels[1]. However, a common way to describe an electrophoretic run is by quoting a constant current per gel for a given time. For destaining electrophoretically any power supply (e.g. a car battery charger) with a high output current (0·25–1 ma.) is suitable.

Preparative Apparatus. The technique has been used almost exclusively at an analytical level, but in principle it should be relatively easy to scale up the gel sizes to accommodate larger protein loads (see below)[1]. In this connection, techniques have been described for the removal of bound SDS from proteins and enzymes, and for their subsequent renaturation.[56,57]

Gel Preparation

After the gel type has been selected, (see Section "Criteria for Selecting Experimental Procedure") it is important that all the ingredients listed in Tables 12.1 and 12.2 for stacking and separating gels should be mixed in the order given. Reaction of persulphate with TEMED causes initiation of polymerization, which in turn depends on their concentrations, and it is therefore important that persulphate is added last with

TABLE 12.1

Reagents for preparation of reservoir buffer and separating gels of various porosities for use in continuous SDS-phosphate system

	Final acrylamide gel concentration (%)[a]						Reservoir Buffer[b]
Reagent[c]	20	15	12·5	10	7·5	5·0	
Acrylamide (1)	20	15	12·5	10	7·5	5·0	—
Sodium phosphate (5)	3·0	3·0	3·0	3·0	3·0	3·0	100
SDS (2)	0·3	0·3	0·3	0·3	0·3	0·3	10
TEMED (4)	0·03	0·03	0·03	0·03	0·03	0·03	—
Water	5·2	10·2	12·7	15·2	17·7	20·2	890
Persulphate (3)	1·5	1·5	1·5	1·5	1·5	1·5	—

(a) Amounts of reagents (ml.) are those needed to make 30 ml. of gel mixture, sufficient to prepare 12–15 gel rods.

(b) Amounts of reagents (ml.) are those needed to make 1 litre of reservoir buffer.

(c) Numbers in parentheses are reagent numbers described in section "preparation of reagents".

Table 12.2

Reagents for preparation of reservoir buffer, stacking gel and separating gels of various porosities for use in SDS-discontinuous buffer system

Reagent[c]	Stacking Gel (5%)[d]	Final acrylamide concentration in separating gel[a]						Reservoir buffer[b]
		20	15	12·5	10	7·5	5·0	
Acrylamide (1)	1·67	20	15	12·5	10	7·5	5·0	—
Stacking gel buffer (6)	2·50[e]	—	—	—	—	—	—	—
Discontinuous gel buffer (7)	—	3·75[f]	3·75	3·75	3·75	3·75	3·75	—
Electrode buffer(8)	—	—	—	—	—	—	—	500[g]
SDS (2)	0·10	0·30	0·30	0·30	0·30	0·30	0·30	10
TEMED (4)	0·01	0·03	0·03	0·03	0·03	0·03	0·03	—
Water	5·22	4·42	9·42	11·92	14·42	16·92	19·42	490
Persulphate (3)	0·50	1·50	1·50	1·50	1·50	1·50	1·50	—

(a) Amounts of reagents (ml.) are those needed to make 30 ml. of gel mixture, sufficient to prepare 12–15 gel rods.
(b) Amounts of reagents (ml.) are those needed to prepare 1 litre of reservoir buffer.
(c) Numbers in parentheses are reagent numbers described in section "preparation of reagents".
(d) Amounts of reagents (ml.) are those needed to make 10 ml. of gel mixture, ample for the above gel rods.
(e) Final concentration 0·125 M Tris-HCl, pH 6·8.
(f) Final concentration 0·375 M Tris-HCl, pH 8·8.
(g) Final concentration 0·025 M Tris and 0·192 M glycine, pH 8·3.

rapid mixing so that a homogeneous solution is added to the tubes or slabs in which polymerization takes place. The gels should form within 10–30 min. depending on particular reagents and ambient conditions. If gelling occurs prematurely, (before all gels are cast) the amounts of initiators may be decreased.

Cylindrical Rods: Separating Gel Only. These are used with the continuous buffer system, and are prepared as itemized below.

1. Convenient glass tubes are 10 cm. long 0·6 cm. inside diameter, (but longer tubes (gels) are often used to provide improved resolution). Ensure tubes are first soaked in cleaning solution, rinsed and oven-dried. Mark each tube to ensure constant gel length (7 cm. for standard 10 cm. tube).

2. Seal one end of each tube with parafilm (twisted back on itself twice) or with a rubber stopper (e.g. a serum cap) and place the tubes in a vertical position. An appropriate rack may be used, or a thick block of perspex (plexiglass) or wood may be drilled with rows of holes into which the tubes fit snugly.

3. At room temperature, mix the required acrylamide solution, sodium phosphate, SDS, TEMED and water in the amounts indicated in Table 1, in a round bottomed flask and deaerate for 1 min. using a water pump. Add the ammonium persulphate solution as indicated in Table 12.1 and mix thoroughly.

4. Fill each tube up to the mark (approximately 2 ml. of mixture for standard gel tube) using a 10 ml. pipette (with manual attachment, e.g. pro-pipette).

5. As soon as the gel tubes are filled, overlay the gel solution with a small volume of water, or buffer of the same composition as that in the gel mixture, to a height of about 0·5–1 cm. This procedure prevents an air-gel meniscus and ensures a flat gel surface, but it must be done carefully or the gel surface will be disturbed. The water may be added from a syringe fitted with a needle bent at 90°, or using a peristaltic pump to deliver the water through a fine plastic tube at very slow flow rate and allowing the water to flow down the side of the tube, or by spraying a fine mist from a hand operated sprayer[(1)]. In the last case it is important to use a sprayer that atomizes not by a flow of air but by direct pressure on the liquid, and to spray directly on top of the tubes from several feet away.

6. Solidification of the gel leads to a visible interface. The gels may be used immediately, or may be kept for considerable periods if dehydration is prevented by storing in a plastic bag or some other sealed container.

Cylindrical Rods: Separating and Stacking Gels. These are used with the discontinuous buffer systems. (Table 12.2).

The separating gels are prepared in an identical manner to that described in steps 1–6 above, except that in step 3 the required reagents are those given in Table 12.2. After completion of steps 1–6, the overlayer of water is removed completely and the following steps are carried out to prepare the stacking gel.

7. At room temperature, mix the acrylamide, stacking gel buffer, SDS, TEMED and water as indicated in Table 12.2 deaerate as in step 3, add the ammonium persulphate as indicated in Table 12.2 and mix thoroughly.

8. Add the stacking gel mixture to give a height of about 0·5 cm. (about 0·15 ml. of mixture).

9. Overlay the gel solution as in step 5.

10. Solidification of the gel gives a visible interface. These gels are generally used as soon as possible after preparation, but storage for 24 hours seems not to affect results.

Cylindrical Rod Gels General Comments. Plastic tubes (plexiglass) may be used in place of glass, the gels adhere less strongly to plastic, and this may be advantageous in certain cases (see below). An alternative, rapid procedure for preparing a number of cylindrical gels is to pack the tubes upright in a vertical cylinder, and pour gel into the cylinder until it rises to the required height in the tubes. After overlaying with water and polymerization, the tubes are removed from the cylinder, separated and excess interstitial gel is discarded. The cylinder may be made from a section of plastic measuring cylinder fitted with a rubber bung, or a small evaporating dish or beaker may be suitable.

Sometimes the gel may crack, contain bubbles or pull away from the tube surface, particularly when high acrylamide concentrations are used. The cause may be dirty tubes, or too much heat produced during rapid polymerization; the latter case may be alleviated by ensuring that reagents are at room temperature, by decreasing the amounts of initiators, or siliconizing the surface of glass tubes or using plastic ones.

It is frequently recommended[(1)] that the height and volume of stacking gel should be approximately 1·2 times that of the sample, but this is not experimentally convenient under the usual conditions when sample volumes of about 20 μl are employed. A stacking gel of about 0·5 cm. is a convenient compromise.

Flat Slab Gels: General Comment. It was indicated in the section on "apparatus", that a number of flat-slab electrophoretic units are available commercially or can be made relatively easily, and consequently flat slabs may be made in many different sizes and thicknesses depending on the dimensions of the particular glass plates and plastic spacers respectively. Irrespective of this variability the following steps are generally applicable.

Flat Slab Gels: Separating Gel Only

11. Ensure glass plates are cleaned as in 1 above, and mark height of gel required.

12. Two, square glass plates may be sandwiched together in various ways using thin plastic (Lucite, plexiglass), rubber, or glass strips (spacers) around three edges. The spacers should be of perfectly matching thickness to ensure good contact and prevent leaking of the gels. Before forming the sandwich the outer edge of the plates can be lined with a very thin layer of vacuum grease (not silicone)[(2)], or the edges of the formed sandwich can be sealed with molten 1·5% agar[(1)], or the edges may be greased and sealed with a special adhesive tape (Uniscil/gradipore). The plates are clamped together and held vertically. They are self supporting if sufficient are prepared together or they may be supported individually[(2)].

13. Prepare required gel mixture as detailed in step 3.

14. Pour the gel mixture into the space between the two plates until it is filled to required height. In order to form slots in the upper surface of the gel in which to apply the samples, immediately insert a plastic (Lucite) "comb" of identical thickness to the spacer strips into the top surface of the gel mixture. The "comb" is simply a plastic strip with vertical slots (about 2 mm. wide) spaced about 5 mm apart. The "teeth of the comb" should fit closely against the surfaces of the two glass plates so that no gel can penetrate.

15. Gels may be used immediately or stored as detailed in step 6. After polymerization, the comb is removed to leave sample wells, the same thickness as the gel, and extending transversely from one glass plate to the other.

Flat Slab Gels: Separating and Stacking Gels. These are used with the discontinuous buffer systems (Table 12.2).

Proceed with initial preparation of separating gels as outlined in steps 11–13. However, in this case sample wells are not made in the separating gel but are made later in the stacking gel, therefore, after step 13 proceed as follows:

16. Pour the gel mixture into the space between the two plates until it is filled to required height.

17. Immediately overlay the gel mixture with water or buffer as detailed in step 5. After polymerization (interface formed) promptly remove the liquid overlayer, rinse the gel surface with buffer and drain well. Insert a plastic "comb" (as detailed in step 14) into the free edge of the sandwich and position it 0·5 cm. above the surface of the separating gel.

18. Prepare stacking gel mixture as detailed in step 7.

19. Pour stacking gel mixture into the sandwich, taking care that no air bubbles remain trapped at the base of the teeth of the "comb".

20. After polymerization, the "comb" is removed to expose sample wells. It is best to use the gels straightaway.

Sample Preparation

General Comments. The following points apply for sample preparation for both continuous and discontinuous buffers. It is important that proteins are prepared in buffers more dilute than that into which they will migrate either in the stacking gel (discontinuous) or directly into the separating gel (continuous). This feature helps to concentrate and sharpen the protein band.

It is necessary to completely reduce any protein disulphide bonds so that in the presence of SDS the constituent polypeptide chains are not constrained from unfolding, as this affects their mobility(3). Reduction is usually with 2-mercaptoethanol. The proteins must be denatured and bind their maximal amount of SDS(4,5,6). This is most readily and uniformly achieved by heating the protein samples briefly at 100° in the presence of 1–2% SDS. In order to satisfactorily apply the sample to the gel its density must be higher than that of the electrode buffer (see "Electrophoresis" later). This is achieved by addition of sucrose or glycerol. One should recognize the possibility of proteolysis of protein samples caused either by sample impurities or inherent proteolytic activity of the protein sample. This aspect will not be considered here but, if it is suspected (e.g. loss of certain protein bands or appearance of "extra" low molecular weight bands) various alternative methods of preparation are available to identify and to eliminate the problem (see 1, 7).

Preparation of Samples: Continuous Phosphate Buffer

(i) Add sodium phosphate buffer (0·01 M, pH 7·0) to the protein sample (particulate, lyophilized). Although other buffers of various ionic strengths, and with pH values in the range 4–9, and even solutions containing ammonium sulphate 10% can apparently be used without difficulty[7], it is generally advisable to dialyse protein solutions containing such components, and especially those containing SDS precipitants e.g. guanidinium HCl, against 0·01 M sodium phosphate pH 7·0.

(ii) Add a convenient volume of this protein solution to a tube containing a solution of sodium phosphate pH 7·0, SDS, 2-mercaptoethanol and sucrose (or glycerol) to bring the sample to a final concentration of 0·01 M sodium phosphate pH 7·0, 2% SDS, 5% 2-mercaptoethanol and 10% sucrose (or glycerol). The final protein concentration may be from 0·05–1·0 mg./ml. but is most conveniently 1–2 mg./ml.

(iii) Heat the protein sample in a boiling water bath for 2 min. Cool the sample to room temperature and it may be used directly.

(iv) If a marker is required add bromophenol blue to a final concentration of 0·001%.

Preparation of Samples: Discontinuous Buffers

(v) Add Tris-HCl buffer (0·0625 M, pH 6·8) (Reagent 6 diluted 8 times) to the protein sample (observe analogous precautions to (i) if required).

(vi) Add a convenient volume of this protein solution to a tube containing a solution of Tris-HCl pH 6·8, SDS, 2 mercapto-ethanol and sucrose (or glycerol) to bring the sample to a final concentration of 0·0625 M Tris-HCl pH 6·8, 2% SDS, 5% 2-mercaptoethanol and 10% sucrose (or glycerol)[8]. Final protein concentration is as in (ii) above.

(vii) Repeat steps (iii) and (iv) above.

Molecular Weight Standards. To obtain meaningful and reliable molecular weight estimates by SDS gel electrophoresis it is necessary to employ protein markers that have well characterized polypeptide chain molecular weights. In Table 12.3 are listed a number of protein standards that are commercially available and of proven reliability in these systems. The molecular weight range 12,000 to 70,000 is well covered but few standards are available above this range. This problem is overcome by the use of series of oligomeric proteins that are chemically cross-linked by glutaraldehyde (Table 12.3)[9]. The theoretical implications of the use of these protein oligomers is considered below.

It is convenient to prepare a mixture of suitable marker proteins

TABLE 12.3

Molecular Weights of standard protein markers

Proteins	Molecular Weight of Polypeptide chain[a]
RNA polymerase (E. coli)	150,000 and 160,000
γ-Globulin	150,000
Bovine serum albumin (dimer)	136,000
β-Galactosidase (E. coli)[b]	130,000
Phosphorylase a (muscle)[b]	94,000
Bovine serum albumin[c]	68,000
L-Amino acid oxidase (snake venom)	63,000
Catalase (liver)[b]	60,000
Pyruvate kinase (muscle)[b]	57,000
Glutamate dehydrogenase (liver)[b]	53,000
Fumarase (muscle)[b]	49,000
Ovalbumin[c]	43,000
Alcohol dehydrogenase (liver)[b]	41,000
Aldolase (muscle)[b]	40,000
D-Amino acid oxidase (kidney)[b]	37,000
Lactate dehydrogenase (muscle)[b]	36,000
Glyceraldehyde-3-phosphate dehydrogenase (muscle)[b]	36,000
Pepsin[c][d]	35,000
Carboxypeptidase A	34,600
Carbonic anhydrase[c]	29,000
Chymotrypsinogen[c]	25,700
Trypsin[c][d]	23,300
Myoglobin[c]	17,200
Haemoglobin subunit[b][c]	15,500
Lysozyme (egg white)[c]	14,300
Cytochrome c[c]	12,400

[a] References for mol. wts. are given in (2, 7, 10, 11).
[b] Proteins occur as oligomers under native conditions.
[c] Proteins that have been successfully converted into oligomeric series by use of glutaraldehyde (9 and unpublished data of author).
[d] Samples containing these proteins may be susceptible to proteolysis.

(about 0·1 mg. of each per ml.) in the required sample solution (see steps (ii) and (vi) above) by the procedure detailed above for the unknown protein sample. This mixture may be stored in the deep freeze and warmed up prior to use. Alternatively, marker proteins may be included with the unknown protein throughout the steps of protein sample preparation.

Electrophoresis Procedures

Sample Loading. The procedure is essentially the same for both rods and flat slabs, used with either a continuous or discontinuous buffer system.

1. Add the required reservoir buffer to the lower compartment of the electrophoresis apparatus and mount the rods or flat slabs in the apparatus so that the bottoms of the gels are in the buffer. Air bubbles can be removed from the bottom of the rods or sandwich using a syringe with a bent needle.

2. Fill the sample wells (flat slabs), or the rod tubes with the required reservoir buffer. Add some reservoir buffer to the upper compartment to check for leaks, grease if necessary.

3. Add the prepared protein sample, underneath the reservoir buffer, onto the top of the gel from a syringe or micropipette, e.g. Eppendorf with disposable tips. A short piece of disposable polythene tubing can be attached to the syringe needle to obviate washing the syringe. The higher density of the protein sample solution allows a compact layer to be formed on the top of the gel.

Ideally sample volumes of 10–30 μl. should be applied; larger volumes (up to about 150 μl.) can be handled but may lead to band broadening. About 20 μg. of protein per sample (1–10 μg. for each polypeptide) gives optimal results, but this varies somewhat depending on the nature of the protein and the particular gel procedure adopted.

4. Carefully add required reservoir buffer to the upper compartment.

Electrophoresis. This stage is again similar in outline for all types of SDS-gel systems. Electrophoresis is performed at room temperature. The exact current, voltage, time of a run depends on the system employed e.g. buffer system, gel porosity etc. Some suggested conditions are given below but a trial run will establish optimal conditions in a particular case. Resolution does not seem very susceptible to electrophoretic conditions, and times may be adjusted (within reason) to the convenience of the experimenter. Maizel[1] considers that "resolution is often better with long time-low voltage procedures", in contrast Ames[2] (see 6 below) using a discontinuous slab-gel system recently reported that "higher current (up to 30 ma. . . . about 1·5 hr.) has given excellent results; no distortion of bands occurred and sharpness of smaller molecular weight bands was improved."

5. Connect the power supply to the electrophoresis apparatus, with the anode at the bottom (for the negatively charged SDS-proteins to migrate downwards).

6. Electrophoresis can be stopped when tracking dye (bromophenol blue) approaches the bottom of the gel. With standard cylindrical rods (7 cm. × 0·6 cm. diam.) and a continuous buffer system apply 8 ma per gel for about 3 hours for 5% gels, 5 hours for 10% gels, and 8 hours for 15% gels[7]. With similar size gels in discontinuous buffer, apply 3 ma per gel for about 5 hrs. with 10% gels. A simple extrapolation from these conditions may be applied to flat slabs of various sizes. For example, using a discontinuous buffer with a flat gel slab 15 × 15 × 0·08 cm. apply 12 ma. per gel for about 4 hr[2].

Removal of Gels. Different techniques are required for cylindrical and for flat-slab gels.

7. Cylindrical rods: using a syringe with a long, fine needle squirt water in between the gel surface and the tube wall, whilst slowly rotating the tube. This loosens the gel and lubricates its surface, and the gel will usually slide easily from the tube. If required, pressure may be exerted on the loosened gel by applying a pipette bulb to one end of the tube.

Gels with high (above 10%) acrylamide are very rigid and more difficult to remove. They may be removed by first carefully breaking the gel tube with a hammer and rinsing away the broken glass fragments[1].

8. Flat slabs: The gels are easily removed by prying the glass plates apart gently, and if necessary, holding the plates under running water to loosen the gel slab off the glass surface.

Detection of Proteins

Staining. Various staining solutions are available and several are detailed below. Coomassie Brilliant Blue R250 is used in preference to Amido Black as it is more sensitive. In earlier procedures it was common to first precipitate the proteins and leach out SDS by soaking the gels in 10% trichloroacetic acid (TCA) or 20% sulphosalicyclic acid[11]. This procedure is now carried out simultaneously with staining.

Staining Solutions. The solutions are stable at room temperature but any precipitate that may form should be filtered off before use.

1. Dissolve Coomassie Brilliant Blue R250 (0·2%) in 50% methanol in water to which 7% by volume of glacial acetic acid is added[1].

2. Dissolve 1·25 g. Coomassie Brilliant Blue R250 in 227 ml. of methanol, add 46 ml. glacial acetic acid and water to 500 ml.[7].

3. Coomassie Brilliant Blue R250 (0·05%) is dissolved in water-methanol-TCA (5:5:1, V/V/W)[9].

A cylindrical gel rod is put into a test tube filled with one of the solutions 1–3 and left for 2 hours or more at room temperature. A gel slab can be placed in a flat dish containing staining solution; this is slowly mixed with a magnetic stirrer, taking care not to damage the gel. Handling the slab is facilitated by laying it on a piece of nylon stocking stretched over a plastic frame. Several rods may be stained together in a dish, if supported similarly on a corrugated frame. Stains 1 and 2 have been widely used by many authors, however, it has been my experience that when proteins are stained (or treated) with solutions that contain acetic acid the complete removal of background dye during subsequent destaining is often difficult; stain 3 does not suffer from this disadvantage and yet stains very quickly and with high sensitivity.

Destaining. After staining is complete rinse the gel with water. Excess dye must then be removed from within the gels for the protein bands to be clearly seen. This may be achieved by "soaking" or by electrophoresis.

Soaking. Add the rinsed gel to a clean tube/dish containing a destaining solution (see below). Destaining is improved by gently stirring or shaking, and by adding some ion-exchange resin or woollen pellets (Uniscil) to adsorb dye from solution. It is usually carried out at room temperature, when a period of about 6–24 hr. is generally needed, but it may be speeded up by warming e.g. 40°.

Destaining Solutions. These are used in conjunction with the corresponding staining solutions (see above).

1. 7% acetic acid[1].

2. 50 ml. of methanol, 75 ml. of glacial acetic acid, and water to 1 litre[7].

3. Water-methanol-trichloroacetic acid (38:2:3, v/v/w)[9].

Electrophoretic Destaining. The anionic nature of Coomassie Brilliant

Blue permits its removal by electrophoresis and this may take place transversely or longitudinally. Various commercial apparatus are available for this procedure and come complete with instructions. A brief outline is given in [7]. Transverse electrophoresis is the most rapid way to remove excess dye (half an hour). Destaining solution 2 is commonly used as buffer in this procedure. The backgrounds are generally not as clear as are obtained by soaking.

Various experimental aspects of staining and destaining are considered by Maizel[1].

Alternative Methods of Detection.

(i) Direct measurement of peptide bond absorbance in the ultra violet (210 nm.) is effectively prevented by the enormous ultra violet absorbance of the acrylamide itself. However, protein bands of sufficiently high concentrations can be scanned directly at 280 nm. at which wavelength background absorbance is not too serious. The gel is first extracted with buffer to remove unwanted UV absorbing materials, mounted in an appropriate silica holder and scanned using e.g. a Gilford Spectophotometer equipped with linear transport attachment or a Joyce Loebl Chromoscan. The Gilford machine can scan at any wavelength which can therefore be selected for maximum sensitivity in any particular case. The relative positions of the protein bands can be determined from the densitometer tracing (see later).

(ii) Prior reaction of proteins with dansyl chloride[12,13,14,20] allows bands to be visualized directly by their fluorescence under ultra violet light. One should however be aware that extensive dansylation is likely to affect protein mobility. Fluorescamine, a similar but even more sensitive reagent has been used in a similar manner[15,58]. Proteins may also be prestained with Remazol Brilliant Blue R and their separation followed by eye[16].

(iii) If radioactively labelled proteins are employed their presence may be detected by autoradiography and/or quantitated by scintillation counting of sectioned gels. These techniques are considered below.

Storage of Gels and Preservation of Gel Patterns. Various methods are available for storing the gels, either wet (a) or dry (b), or for preserving their information photographically or by photometry or radioautography.

(a) Stained bands remain visible almost indefinitely when gels are stored in 7·5% acetic acid in sealed tubes. Ample testimony to this fact is afforded by the serried ranks of anonymous tubes to be found in many laboratories carrying out gel electrophoresis.

(b) It is frequently more convenient to dry the gels when they can be handled more easily, and stored in a notebook. The procedure is that of Fairbanks *et al.*[17] described in Maizel[1]. It can be used with either gel rods or slabs; the former must first be sliced longitudinally (an appropriate template for holding the gel for slicing is simply made in any lab workshop or can be purchased commercially) to a thickness of about 0·08 cm. Thick gel slabs can also be sliced, or preferably slabs of the above thickness can be prepared initially and used directly[2]. Simple drying of gels causes them to shrink and become irregular. This tendency is eliminated if they are dried on a filter paper backing. The simple vacuum drying system is shown in Fig. 12.3 and can be

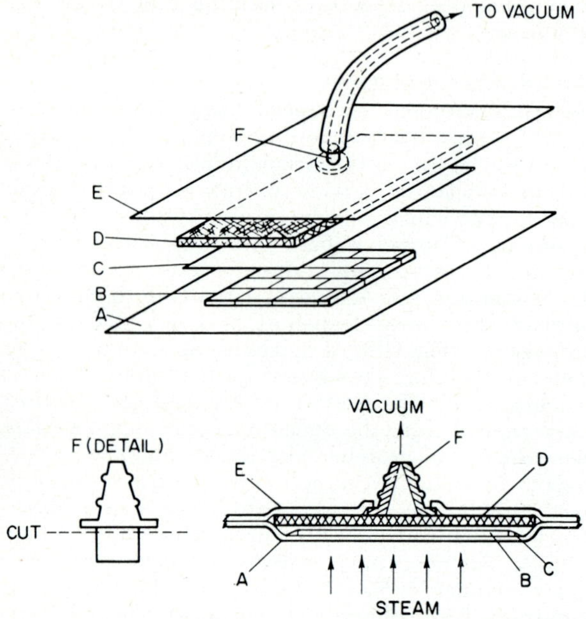

Fig. 12.3. Apparatus for drying polyacrylamide gels. It is simply made and its overall dimensions may be varied as required. In its original form it comprised (A) a sheet of silicone rubber (1/32 × 12 × 12 inches) onto which gel slices (1/32 inch thick) were laid (B). A piece of Whatman 3 mm. filter paper (5 × 6 inches) (C) was placed over the gels and a piece of porous polypropylene (5 × 6 inches) laid on top of that (D) with smooth side down. A further sheet of silicone rubber (E) was placed on top, with a 5/16 inch hole in the centre through which a shortened polypropylene tubing connector was inserted. A vacuum was applied as indicated and the apparatus was placed gel-side down on a steam bath. (Reproduced from Maizel[1] with permission.)

made quite easily. The slices of gel rods, or slab gels are placed on the bottom silicone rubber sheet and are covered first with a piece of filter paper, then with a porous plastic sheet, and finally the top silicone rubber sheet. A vacuum is applied while the apparatus is heated on a steam bath; drying is complete in about 15 min. The dry gel adheres to the paper backing and can be handled like a piece of paper.

(c) Gels can be readily photographed using transmitted light. With rods it is convenient to line them up in a shallow compartmentalized, plastic trough containing a little 7% acetic acid. Slabs also, are easier to handle and less liable to distortion if they are placed in a shallow liquid rather than on a dry surface. Gels may of course be photographed after drying as in (b). One such example is given in Fig. 12.4. Mobilities can be conveniently measured from photographs.

(d) In addition to densitometer scanning of unstained gels in the UV (see above) it is simple and convenient to scan stained gels in the visible region of the spectrum (570 nm.)[62]. A number of

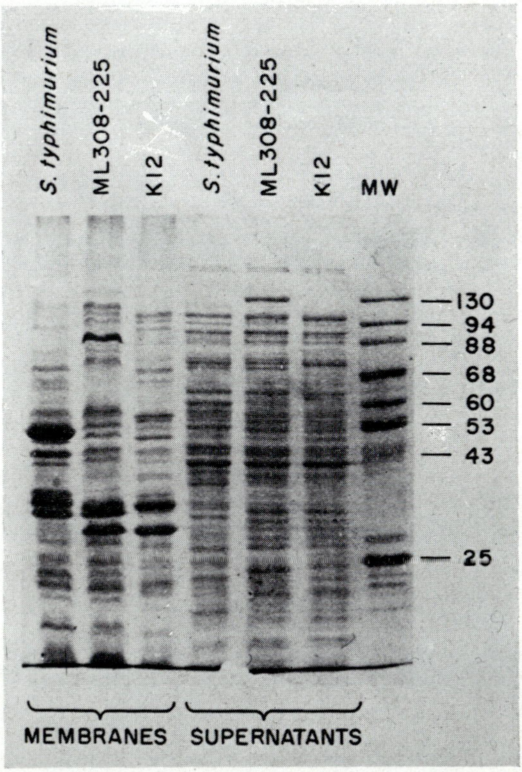

FIG. 12.4. Flat-slab, discontinuous buffer electrophoresis using a 9% acrylamide gel. Samples were prepared and electrophoresis was performed as described in the text. Patterns compare protein complement from membranes and supernatants of *Salmonella typhimurium* and *Escherichia coli* (ML308-225 and K12). Protein standards used (mol. wt. in parentheses) were β-galactosidase (130,000); phosphorylase a (94,000); ornithine decarboxylase (88,000); bovine serum albumin (68,000); catalase (60,000); glutamic dehydrogenase (53,000); ovalbumin (43,000); histidine-binding protein J (25,500). Reproduced from Ames[2] with permission.

specific instruments are available as well as attachments for spectrophotometers (see above). Again the ability to scan at any wavelength (as with the Gilford) permits separate densitometer tracings of e.g. proteins/glycoproteins/lipoproteins which may be differentially stained.

An example of the densitometer tracings of a series of rod gels is given in Fig. 12.5. This procedure yields a "magnified" representation of the gel pattern on graduated paper which facilitates accurate measurement of mobilities. However, this technique is often disappointing in its capacity to resolve closely spaced bands that are readily apparent by eye.

(e) Radioautographs are best prepared by exposing to x-ray film the dried paper-backed gels (see (b) above) in which radiolabelled proteins have been separated. Exposure time should be determined by trial and error but a band containing about 10^4 cpm. should give a reasonable exposure in one day. Differential radiolabelling of biological materials may be revealed by comparison of the stained gel pattern with its radioautograph. Densitometry of the radioautograph provides some measure of quantitation. More sensitive and accurate quantitation is obtained by sectioning the gel and determining radioactivity in each piece using a conventional liquid scintillation counter. Gels may be sectioned wet (preferably after being frozen at $-20°$) or after drying. Flat slabs are first cut longitudinally and then sectioned transversely as for cylindrical rods. A number of devices are available for sectioning the gels. For example, a self-made device in which a number of razor blades are drilled with two holes in their top edge, so that they can be slotted onto metal rods, washers are slotted on alternately to space the blades to the required section size. Automatic gel slicers complete with sample dispensers are commercially available (e.g. from Savant, Gilson)[1].

Drying the gels (b) obviates the need for such sophisticated equipment, for a dried stained gel can simply be cut with fine scissors.

A variety of scintillation mixtures may be used, including Brays (using dioxane), or ones based on toluene and a solubilizer (see [1] for discussion). One which gives good reproducibility and sensitivity is described by Ames[2] who used it with dried slabs. The mixture is made from toluene (3·73 litres); NCS-solubilizer (143 ml.): PPO (14 g): POPOP (0·21 g.). The excized dry strip is added to a scintillation vial containing 50 μl. water, the vial is tilted and the strip soaked for 15 min., 10 ml. of scintillation mixture is added, the vial is capped, well shaken and incubated at 37° overnight (or shorter times at higher temperature).

Interpretation of Results

Measurement of Mobilities

Gel dimensions change during staining and destaining so it is best to equilibrate the gels in 7·5% acetic acid before measuring mobilities.

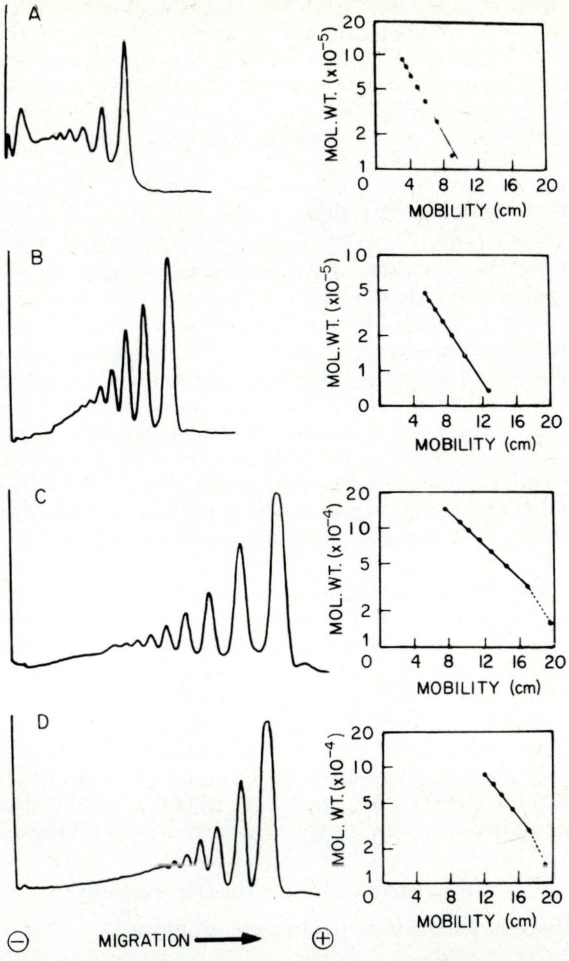

FIG. 12.5. Densitometer tracings of glutaraldehyde polymerised proteins run on cylindrical gels (5% acrylamide), in continuous buffer system as described in the text. Bands correspond to monomer, dimer, trimer etc. reading from right to left. (a) Asparaginase ("monomer" mol. wt. 135,000); (b) Bovine serum albumin (67,000); (c) Haemoglobin monomer (15,500); (d) lysozyme (14,300). Insets are plots of logarithm of molecular weight against mobility (cm.) obtained from the tracings. Mobilities of different samples are not directly comparable as electrophoretic conditions varied. See original paper for further details and procedure for protein polymerization. Reproduced from Payne[9] with permission.

Migration distances are measured from the top of the gel either directly from the stained or fluorescent gel or from photographs, densitometer tracings or radioautographs. It is usual to measure mobilities relative to internal marker proteins. Alternatively, distances may be measured relative to the marker dye, in which case it is necessary to mark its position (with india ink or a piece of copper wire) prior to protein staining.

Estimation of Molecular Weights

Shapiro, Vinuela and Maizel[17] first observed that in an SDS-phosphate buffer system the mobility of a protein was related to its molecular weight; specifically, they reported that log mol. wt. was inversely related to mobility. This observation has been confirmed and extended by many workers using continuous[7,9,10,11,18] and discontinuous[2,8,19] buffer systems. In practice therefore, the unknown molecular weight of a protein may be obtained from its mobility, using a plot of the mobilities of marker proteins against the logarithm of their known molecular weights. Some typical log mol. wt. versus mobility plots are shown for several oligomeric protein series in Fig. 12.5. Various authors have shown that in such a plot most of the protein markers shown in Table 12.3 lie on a straight line using gels of about 7–10% acrylamide[1,7,10,11].

For a given type of gel, resolution of the highest and lowest molecular weight proteins may be poor and it is therefore important that gels of optimum porosity (acrylamide concentration) are used for accurate determination of the unknown molecular weight. The influence of various factors on protein mobility are discussed below but in practice the following gel concentrations (acrylamide is varied while bis ratio is kept constant, see Table 12.1, 12.2) will usually be found suitable for the molecular weight ranges quoted: 15%, mol. wt. below 40,000; 10%, mol. wt. 10,000–70,000; 5%, mol. wt. 25,000–250,000. Resolution is invariably improved by use of the discontinuous buffer system.

Discussion of Empirical Observations

Effect of Various Parameters on Protein Mobility

Influence of the Degree of SDS Binding. Implicit in the empirical relationship between polypeptide chain molecular weight and electrophoretic mobility is the idea of a universal mechanism for SDS-protein interactions, and evidence for this is available. With a variety of different proteins it is found that at SDS monomer concentrations above 0·02% (8×10^{-4} M) that 1·4 g. of SDS is bound per gram of protein[4,5,21]. This corresponds to approximately one SDS molecule to two amino acid residues. It is therefore important that during sample preparation the weight ratio of SDS to protein is about 4:1.

The properties of protein/SDS complexes, as studied by hydrodynamic and optical techniques, are consistent with them behaving as rodlike particles in which the particle length varies with the molecular weight of the polypeptide chain[4,5,22]. It would seem to be a popular misconception that these are likely to be fully extended, "stiff" rods, in fact, the protein-SDS complexes contain a high degree of secondary structure (this feature seems essential to explain the adherence of chemically cross-linked proteins to the usual mobility—mol. wt. relationship, see below). It should be mentioned that the extent of SDS binding determined by use of equilibrium dialysis[4,5,21,22] is different from that found by other techniques[6,23], especially at high ionic strengths. It is a matter of debate as to which conditions are best related to protein samples that are prepared at 100° and electrophoresed at room temperature. Various electrophoretic species with different mobilities are observed with low ratios of SDS to protein[24]. Some proteins are relatively resistant to SDS binding, but others, that in the earlier literature were reported to be anomalous in this respect[23], are likely to be less resistant using the current sample preparation procedure at 100°. Glycoproteins show a decreased SDS binding ratio, that generally leads to decreased electrophoretic mobility and erroneous molecular weight determinations[21,25–29,36,37]. The presence of disulphide bonds limits unfolding of polypeptide chains and may lead to decreased SDS binding with consequent lower mobilities[3–5,21]. Finally, it is possible that the extent of SDS binding may be influenced by impurities in the SDS itself (e.g. longer chain alkyl groups). It has been reported that anomalous migration patterns are observed when using certain commercial sources of SDS, and that this may lead to erroneous molecular weight determinations, especially with the discontinuous buffer system[34,53].

In general SDS appears to bind tightly to proteins[59], sufficiently so that it has even been suggested that SDS need not be added to the buffer solutions during electrophoresis[50], however this does not appear to be a generally recommended procedure[18].

Effect of Protein Charge. Highly charged polypeptides e.g. polylysylglutamic acid[21] bind decreased amounts of SDS and consequently their electrophoretic mobilities are lower than expected. Proteins that have had their intrinsic charge modified greatly through chemical substitution e.g. succinylation, maleylation also show decreased electrophoretic mobility, but it is not clearly established whether this is caused by decreased SDS binding[21,30,31].

Usually, the high degree of SDS binding "swamps out" the intrinsic charge on the protein. However, proteins with high intrinsic charge may sufficiently affect the net charge of the SDS-complex as to cause deviations from the usual mobility—molecular weight relationship[20]. For example, positively charged histones bind SDS in the usual ratio[4,5] but show unusually low electrophoretic mobilities[32]. It is nevertheless possible to accommodate these deviations in a different type of mol. wt./mobility plot ([33], and see theory below).

Effect of Protein Size. Using a particular gel type, deviations from the usual mol. wt.-mobility relationship are found if protein size (mol. wt.) is significantly different from the optimal ranges given above. This merely confirms that extrapolated values are inaccurate, but appropriate changes in concentrations of acrylamide will improve accuracy.

However, low molecular weight polypeptide chains ($<$15,000) present a particular problem. They are generally very poorly resolved and may even migrate with the marker dye. This effect has been observed in gel filtration with SDS[22] and in polyacrylamide gel electrophoresis with[20,11] and without[35] SDS; it clearly leads to erroneous molecular weight determinations. An explanation based on the loss of asymmetry of small SDS-protein complexes, (in which the diameter of a "rodlike" particle approximates its length) does not seem entirely satisfactory[7,20,22,35]; the problem is best considered from a theoretical aspect[33,38,39]. In general, the shape and intrinsic charge of small polypeptides may be relatively more important in determining their mobility in SDS-gels than with large proteins[40]. Use of increased concentrations of bis and inclusion of 8 M urea in the gels seems to improve the results[40,61]. A promising alternative approach would be to form cross linked polypeptides, higher homologues of which should obey the usual mol. wt. mobility relation without need for special gel preparation; this has been shown to be true for lysozyme (mol. wt. 14,300) and haemoglobin (mol. wt. 15,500)[9].

Effects of Protein Cross-links. It is well documented that disulphide bonds limit SDS binding and cause decreased electrophoretic mobility and a consequent deviation from expected molecular weight[3]. It is to be expected therefore, that chemically cross linked proteins might also fail to bind SDS maximally and show abnormal mobilities. This is certainly true with extensively cross linked proteins[3], but a number of reports suggest that oligomeric proteins, prepared by using various cross-linking agents, can conform to the usual mol. wt.-mobility relationship[9,41–46]. In certain cases it is clear that the protein oligomers are linked together by very few crosslinks[9]. However, the chemical nature of the crosslink means that the proteins are not joined end to end but are linked through internal (side-chain) residues that are randomly positioned along the polypeptide chain. It is apparent that this mode of oligomer synthesis means that the maximum linear dimension of (say) a pentamer of mol. wt. (say) 5 $\times$ 20,000 cannot be equal to the overall length of a single polypeptide mol. wt. 100,000. The fact remains however, that in SDS gels, these oligomers and equivalent single-chain polypeptides migrate a similar distance determined by the molecular weight of their combined or individual chains respectively Fig. 12.6. These results are extremely difficult to reconcile with the conventional view that the "SDS complexes behave like rodlike particles in which the particle length varies uniquely with the molecular weight of the polypeptide chain" and that ". . . the hydrodynamic properties are a function only of the molecular length of the polypeptide chain"[4,5,7]. In contrast, these results imply that the predominately important feature governing electrophoretic mobility of SDS-protein complexes is their net charge. Thus, providing the polypeptide chains possess sufficient conformational flexibility such that they can bind their (normal) maximum complement of SDS molecules it appears to be of lesser importance whether the polypeptide chain is linear (as in a normal large protein) or even arranged in a "star-form" (as is likely with cross linked oligomers). Corroborative evidence for this view is provided by the anomalous migration of highly charged polypeptides (see above) in which the intrinsic protein charge is sufficient to significantly influence the net charge of the complex.

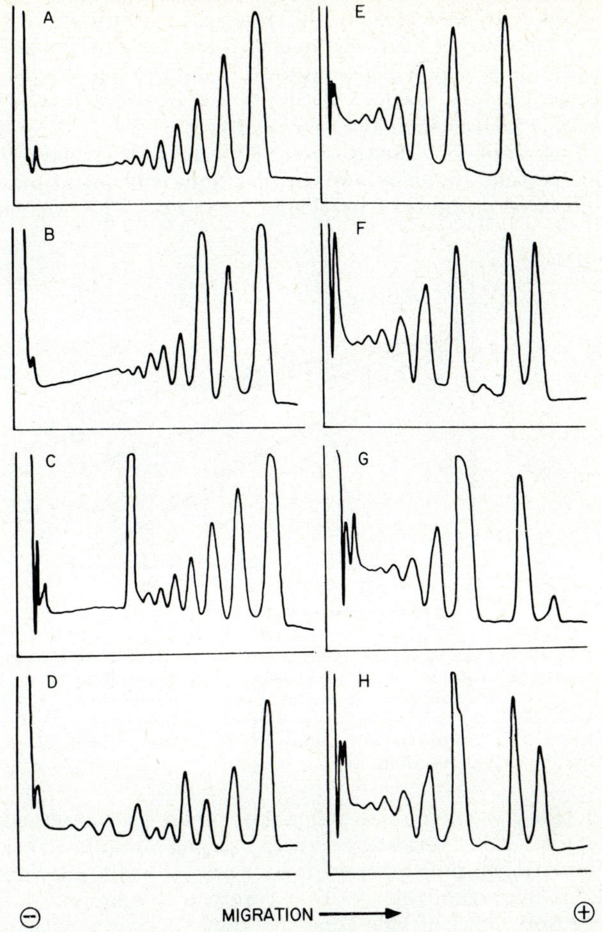

Fig. 12.6. Densitometer tracings of single proteins and protein polymers co-electrophoresed on cylindrical rods (5% acrylamide) in continuous buffer system at 5mA./gel for 3 hours (a–d) or 3–5 hours (e–h) as described in the text. Bands correspond to monomer, dimer, trimer etc. reading from right to left. (a) Haemoglobin polymers (monomer subunit mol. wt. 16,000); (b) Haemoglobin polymers plus ovalbumin (46,000); (c) Haemoglobin polymers plus asparaginase (135,000); (d) Haemoglobin polymers plus bovine serum albumin polymers (monomer mol. wt. 67,000); (e) Albumin polymers, as in (d); (f) Albumin polymers plus ovalbumin; (g) Albumin polymers plus asparaginase; (h) albumin polymers plus asparaginase and ovalbumin. See original paper for further details and procedure for protein polymerization. Reproduced from Payne[9] with permission.

An explanation for the observed mobilities of these crosslinked oligomers may be inherent in the theory of Rodbard and Chrambach[33,38,39] in which a relationship is shown to exist between the surface area of a protein and its electrophoretic mobility. The theory involves geometric probability factors involving *static*, steric factors and not hydrodynamic factors such as Stokes radius.

Effect of Acrylamide Concentration. The percent acrylamide in a gel affects polypeptide mobility and an optimum concentration exists for the resolution of proteins of particular mol. wts. This was considered above.

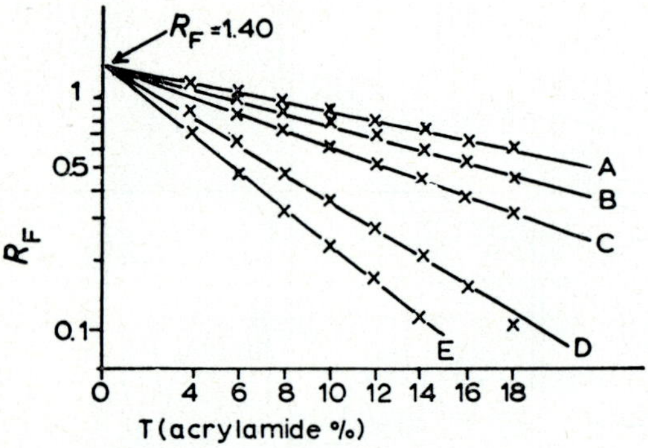

FIG. 12.7. A Ferguson plot of Rf (mobility relative to dye marker) of various proteins versus total acrylamide concentration (%) in SDS gel electrophoresis in a continuous buffer system. A, cytochrome c; B, myoglobin; C, chymotrypsinogen A; D, Ovalbumin; E, bovine serum albumin. Reproduced from Hayashi *et al*[33] with permission.

A good test for anomalous migration of an uncharacterized polypeptide is to run it in a series of gels with varying amounts of acrylamide. A linear relationship is observed between percent acrylamide and log Rf (Rf is migration relative to a common dye or protein marker). In a "Ferguson plot" of this type the slope increases with molecular weight (Fig. 12.7)[47]. When the lines are extrapolated to zero acrylamide concentration T=0; (it is more usual for T to equal the total acrylamide concentration i.e. acrylamide plus bis, rather than monomer alone) all marker proteins share a common R_{FO} value (R_{FO} is the relative mobility in the free solution of the same buffer system); anomalous proteins show up as having a different R_{FO} value[1,33].

The above linear relationship has been observed by a number of workers studying electrophoresis either with[1,7,11,19,33,36,38,39] or without SDS[35,38,39,41]. Various explanations have been offered[33] for the slight scatter sometimes seen in R_{FO} values of marker proteins, especially when using discontinuous buffer systems[19,36].

Gels with low acrylamide concentrations (about 3%), used for high molecular weight proteins, can be stiffened by addition of about 0·5% agarose.

Effect of Bis Concentration. It is usual to keep the ratio of monomeric acrylamide to cross-linker (bis) at 30:0·8. However, for a given total monomer concentration varying the proportion of bis alters the effective average pore size. Without bis no gel forms, as the bis concentration is increased a minimum pore size is reached when acrylamide:bis is about 20:1, addition of further bis apparently increases the effective pore size because polyacrylamide chains become randomly cross linked with large spaces between them[1,49]. Log molecular weight versus mobility plots for continuous gels containing constant acrylamide and varying bis concentrations suggest that decreasing the bis concentration from the standard amount may allow a wider range of molecular weights to be separated on a given percent acrylamide gel[10].

Effect of TEMED Concentration. Allison *et al.*[51] recently reported that increasing the TEMED concentration from that normally used in the continuous buffer system caused a generalized decrease in mobility with some proteins being affected more than others. This has important implications for molecular weight determinations. The cause of the effect was not established but appeared to occur during electrophoresis, (perhaps through TEMED binding to proteins and altering their charge) rather than at the polymerization stage.

Reproducibility

The electrophoretic mobility of a polypeptide relative to a marker can be obtained reproducibly with an accuracy of $\pm 2\%$ taking normal care in preparation and running. Absolute migration distances may vary more than this depending on batch to batch variation in gels, making it important to include markers in each run. When using a discontinuous buffer system it may be more accurate to measure mobility ($cm^2/V.s.$) rather than distance of migration (cm.)[19,33].

Sensitivity

This aspect has been considered by others[1,7]. In general, sensitivity is governed by band width, which in turn depends on sample volume, amount of protein, the efficiency of "stacking", the gel type and the conditions and duration of electrophoresis. For sharp bands, sample volume should be small (about 10 μl.), total sample protein should not be too high (this depends on actual gel size, 0·1–0·5 μg. of protein in a band are generally detected), "stacking" is improved by low electrolyte concentrations in the sample and especially by use of a discontinuous system.

Resolution

Resolution is dictated by the molecular weight difference between polypeptides. It is however strongly influenced by many of the parameters considered in the earlier discussion of empirical observations, and an optimum choice must be made in any instance based on a trial run. Fast moving (low mol. wt.) proteins are generally less well resolved than slower ones. Resolution is improved by increasing the length of gels, to 15–30 cm. as convenient.

Criteria for Selecting Experimental Procedure

The procedure adopted obviously depends on the nature of the problem under study. For example, if it is required to detect small differences in individual polypeptides within a spectrum of proteins from cytoplasmic or membrane fractions, then the discontinuous, gel-slab system is the method of choice. This is well illustrated by the work of Ames and her colleagues[2,48]. In initial runs it is convenient to use 7–10% gels, which may if required be modified later depending on the molecular weight range found to be of interest.

To determine the molecular weight of an essentially pure protein sample, a continuous, gel-rod system is most convenient. It is desirable to perform initial electrophoresis using a 7·5% gel, and a range of molecular weight standards from about 15,000–70,000 daltons. Once an approximate molecular weight has been determined for the unknown protein, then a gel of appropriate porosity (% acrylamide) coupled with related mol. wt. standards may be used. It is an important control that the protein under investigation should be run on a series of gels containing different acrylamide concentrations as a check for possible anomalous migration. Finally, it is always advisable to check the molecular weight by independent chemical and physical methods.

Protein studies between the two above cases must be judged on their merits. Discontinuous systems are somewhat more complicated to run but give better resolution, gel-slabs are especially useful for comparing large numbers of similar samples. Gel dimensions may be scaled up to permit application in a preparative form. A more detailed discussion and extension of some of the points raised here is given by Maizel[1].

Theoretical Considerations

A theoretical basis for gel electrophoresis has been proposed by Chrambach and Rodbard[38,39] and discussed by others[1,33]. In certain cases, it is important to have an appreciation of this as an aid to interpretation of results, and the reader is referred to the above papers.

One point only will be mentioned as an alternative way to determine molecular weights. According to Rodbard and Chrambach, the slope in the Ferguson plot (Fig. 12.7) Kr or retardation coefficient, is related to the surface area of the molecule. Under conditions of complete conformational flexibility, allowing maximum SDS binding, the surface area (and hence amount of SDS bound) is directly proportional to molecular weight of the polypeptide. It follows that Kr α mol. wt. Various other relationships between molecular weight and empirically determined parameters may be developed, that in certain instances may be more applicable than the usual log mol. wt. versus mobility plot.

Acknowledgements

I am indebted to the various authors quoted in the text for permission to reproduce figures from their published work. I am grateful to Dr. D. Wright for his helpful comments and criticism of the manuscript.

REFERENCES

1. Maizel, J. V. In: "Methods in Virology" (eds. K. Maramorosch and H. Korprowski) vol. **5,** Academic Press, New York, 1971, pp. 179–246.
2. Ames, G. F. L. *J. biol. Chem.*, 1974, **249,** 634.
3. Griffith, I. P. *Biochem. J.*, 1972, **126,** 553.
4. Reynolds, J. A., and Tanford, C. *J. Biol. Chem.*, 1970, **245,** 5161.
5. Reynolds, J. A., and Tanford, C. *Proc. Nat. Acad. Sci. U.S.*, 1970, **66,** 1002.
6. Allen, G. *Biochem. J.*, 1974, **137,** 575.
7. Weber, K., Pringle, J. R., and Osborn, M. In: "Methods in Enzymology" vol. **26C** (C. H. W. Hirs and S. N. Timasheff, editors), 1972, p. 3.
8. Laemmli, U. K. *Nature,* 1970, **227,** 680.
9. Payne, J. W. *Biochem. J.*, 1973, **135,** 867.
10. Weber, K., and Osborn, M. *J. biol. Chem.*, 1969, **244,** 4406.
11. Dunker, A. K., and Rueckert, R. R. *J. biol. Chem.*, 1969, **244,** 5074.
12. Inouye, M. *J. biol. Chem.*, 1971, **246,** 4834.
13. Shelton, K. R. *Biochem. Biophys. Res. Communs.*, 1971, **43,** 367.
14. Talbot, D. N., and Yphantis, D. A. *Analyt. Biochem.*, 1971, **44,** 246.
15. Pace, J. L., Kemper, D. L., and Ragland, W. L. *Biochem. Biophys. Res. Communs.*, 1974, **57,** 482.
16. Griffith, I. P. *Analyt. Biochem.*, 1972, **46,** 402.
17. Shapiro, A. L., Vinuela, E., and Maizel, J. V. *Biochem. Biophys. Res. Communs.* 1967, **28,** 815.
18. Shapiro, A. L., and Maizel, J. V. *Analyt. Biochem.*, 1969, **29,** 505.
19. Neville, D. M. *J. biol. Chem.*, 1971, **246,** 6328.
20. Williams, J. G., and Gratzer, W. B. *J. Chromatog.*, 1971, **57,** 121.
21. Pitt-Rivers, R., and Impiombato, F. S. A. *Biochem. J.*, 1968, **109,** 825.
22. Fish, W. W., Reynolds, J. A., and Tanford, C. *J. biol. Chem.*, 1970, **245,** 5166.
23. Nelson, C. A. *J. biol. Chem.*, 1971, **246,** 3895.
24. Katz, S., Shaw, M. E., Chillag, S., and Miller, J. E. *J. biol. Chem.*, 1972, **247,** 5228.
25. Segrest, J. P., Jackson, R. L., Andrews, E. P., and Marchesi, V. I. *Biochem. Biophys. Res. Comm.*, 1971, **44,** 390.
26. Bretscher, M. S. *Nature New Biology*, 1971, **21,** 229.
27. Schubert, D. *J. Mol. Biol.*, 1970, **50,** 287.
28. Glossmann, H., and Neville, D. M. *J. biol. Chem.*, 1971, **246,** 6339.
29. Voyles, B. A., and Moskovitz, M. *Biochim. Biophys. Acta*, 1974, **351,** 178.
30. Tung, J. S., and Knight, C. A. *Biochem. Biophys. Res. Communs.*, 1971, **42,** 1117.
31. Arndt, D. J., and Berg, P., *J. biol. Chem.*, 1970, **245,** 665.
32. Panyim, S., and Chalkley, R. *J. biol. Chem.*, 1971, **246,** 7557.
33. Hayashi, K., Matsutera, E., and Ohba, Y. *Biochim. Biophys. Acta*, 1974, **342,** 185.
34. Swaney, J. B., Woude, G. F. V., and Bachrach, H. L. *Analyt. Biochem.*, 1974, **58,** 337.
35. O'Sullivan, S. A., and Lewis, L. N. *J. Chromat.*, 1973, **87,** 109.
36. Banker, G. A., and Cotman, C. W. *J. biol. Chem.*, 1972, **247,** 5856.
37. Anderson, M., Cawston, T., and Cheeseman, G. C. *Biochem. J.*, 1974, **139,** 653.
38. Chrambach, A., and Rodbard, D. *Science*, 1971, **172,** 440.
39. Rodbard, D., and Chrambach, A. *Proc. Nat. Acad. Sci. U.S.*, 1970, **65,** 970.
40. Swank, R. T., and Munkres, K. D. *Analyt. Biochem.*, 1971, **39,** 462.
41. Hedrick, J. L., and Smith, A. J. *Arch. Biochem. Biophys.*, 1968, **126,** 155.
42. Handschumacher, R. E., and Gaumond, C. *Mol. Pharm.*, 1972, **8,** 59.
43. Carpenter, F. H., and Harrington, K. T. *J. biol. Chem.*, 1972, **247,** 5580.
44. Wolf, B., Lausarot, P. M., Lesnaw, J. A., and Reichmann, M. E. *Biochim. Biophys. Acta*, 1970, **200,** 180.
45. Wolf, B., Lesnaw, J. A., and Reichmann, M. E. *Eur. J. Biochem.*, 1970, **13,** 519.
46. Lehrer, S. S. *Biochem. Biophys. Res. Communs.*, 1972, **48,** 967.
47. Ferguson, K. A. *Metabolism*, 1964, **13,** 985.
48. Ames, G. F. L., Spudich, E. N., and Nikaido, H. *J. Bact.*, 1974, **117,** 406.
49. Gordon, A. H. In: "Laboratory Techniques in Biochemistry and Molecular

Biology" (eds. T. S. Work and E. Work), 1969, Vol. **1,** North Holland, Amsterdam, p. 1.
50. Stoklosa, J. T., and Latz, H. W. *Biochem. Biophys. Res. Communs.*, 1974, **58,** 74.
51. Allison, J. H., Agrawal, H. C., and Moore, B. W. *Analyt. Biochem.*, 1971, **58,** 592.
52. Davis, B. J. *Ann. N.Y. Acad. Sci.*, 1964, **121,** 404.
53. Studier, F. W. *J. mol. Biol.*, 1973, **79,** 233.
54. Sargent, J. R. "Methods in zone electrophoresis", 1969, published by B.D.H. Chemicals Ltd., Poole, U.K.
55. Adams, J. M., Jeppesen, P. G. N., Sanger, F., and Barrell, B. G. *Nature*, 1970, **227,** 570.
56. Lenard, J. *Biochem. Biophys. Res. Communs.*, 1971, **45,** 662.
57. Weber, K., and Kuter, D. J. *J. biol. Chem.*, 1971, **246,** 4504.
58. Eng, P. R., and Parkes, C. O. *Analyt. Biochem.*, 1974, **59,** 323.
59. Rosenberg, R. M., Crespi, H. L., and Katz, J. J. *Biochim. Biophys. Acta*, 1969, **175,** 31.
60. Loening, U. E. *Biochem. J.*, 1967, **107,** 251.
61. Wada, H., and Snell, E. E. *Analyt. Biochem.*, 1972, **46,** 548.
62. Fishbein, W. N. *Analyt. Biochem.*, 1972, **46,** 388.

CHAPTER 13

QUANTITATIVE IMMUNOELECTROPHORESIS

John Versey

Basic Principles.

In Chapter 6 the way in which antibodies combine with protein antigens to form specific precipitates was discussed. The quantitative immunoelectrophoretic methods to be presented in this chapter are all based upon the electrophoretic migration of antigens into antibody-containing gel. This results in the specific immunoprecipitation of the antigens by means of the corresponding precipitating antibodies. Each antibody/antigen system results in the formation of an individual precipitate, usually in the form of a peak. The area enclosed by this peak is proportional to the antigen/antibody ratio. Under suitable conditions the majority of the antibody molecules will not move in the gel, whilst antigen molecules with a different electrophoretic mobility from that of the antibody will move in the gel.

At the start of the electrophoretic run the antigens migrate into the gel in excess of the antibody molecules in the surrounding gel. Small soluble immune complexes are formed which migrate, binding increasing numbers of antibody molecules until sufficient antibody molecules have become bound for precipitation to occur. This point is often known as the "equivalence point". From this moment on the position of the precipitate will not alter even if electrophoresis is continued. Experiments have shown that the number of antibody molecules necessary to give rise to precipitation in immunoelectrophoresis is less than that calculated by Heidelberger and Kendal in the original experiments with precipitation in tubes.[1] During continued electrophoresis more antibody will combine with the precipitate to make it more distinct.

The area enclosed by the precipitate will depend upon both the con centrations of the antigen and the antibody in the system, and is given in the following formula:

$$\text{area} = k \times \frac{\text{concentration of the antigen}}{\text{concentration of the antibody}}$$

Antisera raised in most mammalian species have been found to give successful results in quantitative immunoelectrophoresis. The conditions for the electrophoresis should be such that the antibodies should not migrate during the run, so that the distribution of antibodies should remain constant around the origin. Such electrophoretic separations are usually carried out at pH 8·6, at which point all proteins are negatively charged and will migrate towards the anode. Gamma globulins will possess a weak negative charge but this will effectively be counter-balanced by electroendosmosis if carried out in a 1 per cent agarose gel. In this case the weak endosmotic migration of the buffer in agarose is sufficient. The use of agar is not to be recommended for these techniques, owing to the migration of antibodies, caused by the strong electro-endosmosis in the gel.

In general only those proteins which have a higher anodic electrophoretic migration rate can be quantitated by this technique. This will include those proteins from the albumin to the fast gamma regions of an electrophoretic separation. Anodal migration rates can be increased for many proteins by acetylation,[2] formylation,[3] or carbamylation.[4] These procedures reduce the *p*K value for all proteins, thereby increasing the net negative charge on the proteins at *p*H 8·6. With these procedures it has become possible to measure the immunoglubulins. A procedure for formylation is outlined later in this chapter.

In general there are in routine use today, two applications of quantitative immunoelectrophoresis. These are the Rocket electrophoresis method originally described by Laurell[5] and the two dimensional immunoelectrophoresis originally described by Ressler.[6] Since the introduction of these techniques, many other variations have been produced. They are in general research procedures and are mentioned at the end of this chapter.

Apparatus for Quantitative Immunoelectrophoresis

Electrophoretic Apparatus.

The apparatus consists of a flat cooling plate bridging two electrode buffer compartments as shown in Fig. 13.1. Cooling is required to maintain constant temperatures within the gel during electrophoresis in order that precipitation will not be disturbed. The required temperature will be maintained if cooling water is recirculated at a temperature of 4°C. A Grant low temperature bath and recirculator (down to +1°C) is suitable. Polyethylene glycol can be used as a coolant. If the cooling plate is made using epoxy resins, however, this coolant is unsuitable. In order that no evaporation should occur from the surface of the gel, the whole apparatus should be protected by means of a cover. If cooling is too intense, condensation may occur both on the plate and in the origin wells. Too little cooling may result in temperature gradients throughout the gel. Both will result in the formation of artefacts, which are discussed later. The apparatus should be horizontally level, again to prevent artefacts. Both the electrodes and the buffer vessels should be easily accessible for cleaning, since both tend to accumulate contaminating particles. The buffer vessels themselves should be of sufficient capacity to hold about one litre of buffer, which is adequate to prevent *p*H changes during electrophoresis. Current should be supplied by means of a power supply set on stabilized voltage. An output of 70–100 V will generally give rise to about 1–3 V/cm. in the gel (low voltage electrophoresis) while 250–300 V will give 8–10 V/cm. in the gel (high voltage electrophoresis). The voltages in the gel can conveniently be measured by means of a small voltmeter, placing the probes in the gel at the extremities.

Wicks.

Although wicks may be made of many materials, such as agarose or paper, the most suitable material is absorbent lint. Only lint which is

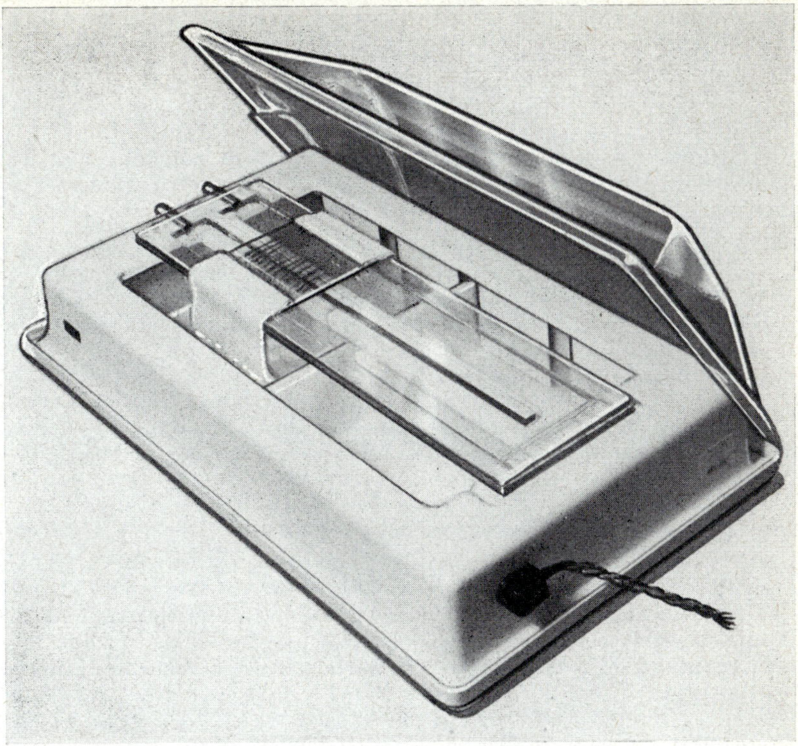

Fig. 13.1. Commercial electrophoresis apparatus assembled for rocket electrophoresis available from P.C.D. Ltd., Farnborough, Hampshire.

fairly thick is suitable for quantitative immunoelectrophoresis e.g. Johnson and Johnson, Slough, England. Wicks should be thoroughly washed before use and are cut out so that they are the width of the plate to be electrophoresed and long enough to extend from the plate to below the surface of the buffer in the buffer vessel. The lint is used folded double thickness, and after use the wicks may be washed and used several times. Specific wick patterns will be mentioned where necessary.

Gel Punch.

Stainless steel tube cut approximately five cm. long and sharpened on the inside surface of the tube, is suitable for gel punching. Several sizes are required according to the volume of sample to be applied:

1. Diameter 1·7 mm. . . . . 2 μl.
2. Diameter 2·5 mm. . . . . 5 μl.
3. Diameter 4 mm. . . . . 10 μl.

These volumes apply to a 1·5 mm. thick gel. When the wells have been cut the holes may be sucked out using a pasteur pipette connected to a water pump.

Template.

A template with the positions marked for well punching may be made from paper and stuck onto the underside of a piece of glass. The finished plate may then be cut with origin wells a uniform distance apart.

Glass Plates.

Glass plates cut from rolled glass are suitable. Several sizes are used:

5 × 5	Slide cover glasses from Kodak or Ilford, England
8 × 8 cm	transparency glasses from Ilford, England
10 × 10 cm	specifically cut to size
10 × 20 cm.	chromatography plates

Before use all slides must be cleaned with methanol.

Horizontal Table.

A level surface is required for pouring the plates for electrophoresis. A suitable surface can be made using a piece of plate glass to which has been attached adjustable legs. The surface must be checked using a spirit level.

Water Bath.

Agarose will set at temperatures less than 42°C. A practical temperature to maintain all glassware, into which agarose solutions are to be poured, and agarose is at 52°C, at which temperature only a little antiserum will be denatured.

Micropipette.

A variable micropipette 1–10 μl. for the application of samples is necessary. A suitable one may be purchased from Shandon Southern Instruments Ltd. or Hamilton, The Hague, Holland.

Reagent for Electrophoresis

Agarose.

Agarose can be obtained from many sources (L'Industrie Biologique Francaise, France; Miles-Serevac, Stoke Poges, Bucks, England) but only agarose which has a low endosmotic effect is suitable. Agarose is made up to 15 g/l. distilled water and is aliquoted into glass containers holding approximately 30 ml. This is then allowed to solidify and stired at 4°C. A bottle is merely brought out and melted in a boiling water bath when required; it is kept liquid in the 52° water bath.

Buffer.

Barbitone buffer *p*H 8·6, 0·12 M stock buffer.

206 g. sodium barbitone
40 g. barbitone

made up to 10 l. in distilled water. 10 g. of sodium azide is added as a preservative (alternatively 10 ml. of a 50 g./l. solution of thymol in isopropanol may be added).

Buffer for use in electrophoresis is made by diluting the stock buffer 1 in 6 to produce a solution *p*H 8·6, 0·02 M. This working buffer has a reduced buffering capacity, but minimizes heat production during electrophoresis. For this reason large amounts (1 l.) of buffer must generally be used in the buffer vessels.

Buffer for use in plate preparation is prepared by diluting the stock buffer 1 in 3 to produce a solution *p*H 8·6, 0·04 M. This buffer is mixed with the 15 g./l. agarose solution in equal volumes, to produce gels of 7·5 g./l. agarose in 0·02 M buffer.

Staining Solution.

Either:

(i) Coomassie brilliant blue . . . . . . 5 g./l.

or

(ii) Amido black 10B . . . . 10 g./l.

dissolved in a solution containing:

methanol	45 parts
water	45 parts
glacial acetic acid	10 parts

This solution is left overnight and then filtered. Coomassie brilliant blue is about three times as sensitive as amido black. Slides may be stained in this solution for about fifteen minutes.

Destaining solutions.

methanol	45 parts
water	45 parts
glacial acetic acid	10 parts

After staining the slides are washed twice in this solution for fifteen minutes. A final rinse in methanol will clear any residual background. Care must be taken since all the stain may be washed out if the slides are left longer than 5 minutes.

Antisera.

Most commercially available antisera raised in sheep or rabbits are suitable for quantitative immunoelectrophoresis. Purified immunoglobulin preparations available from Dako, Denmark, give very clear backgrounds.

Formylation.

Formaldehyde will react with free sulphydryl or amino groups to form condensation products:

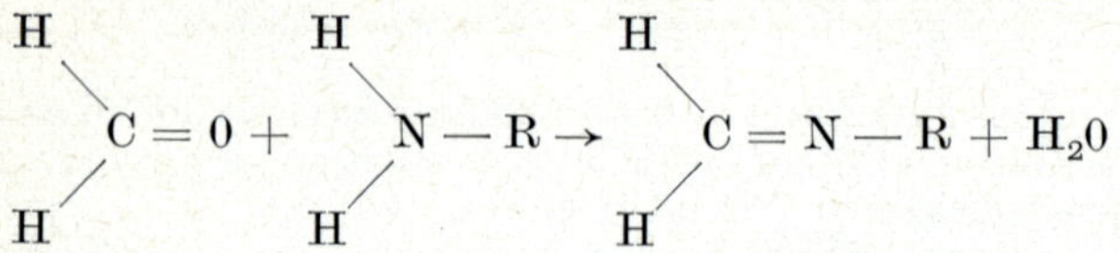

This reaction suppresses the basic groups and results in the formation of an acid, increasing the net negative charge on the protein. It is known as the Sorensen titration. Proteins with a *p*I near 8·0 will thus be increased in charge by this method, e.g. the immunoglobulins. Protein solutions are diluted to the required concentration for analysis in 0·36 per cent (w/v) formaldehyde solution, and then incubated for twenty minutes at room temperature. For the investigation of immunoglobulins a 1 in 10 dilution of serum was found to be most suitable, with an application volume of 2 μl. for IgG and 4 μl. for IgA and IgM, using suitable antisera.

Rocket Immunoelectrophoresis

"Rocket" immunoelectrophoresis is a simple, quick and reproducible method for the estimation of the concentration of single proteins in a protein mixture. First described by Laurell in 1966[(5)] the method involves the electrophoresis of proteins into agarose containing the required monospecific antiserum. Since the samples are electrophoresed in one direction only, from origin wells of constant size, the heights of the peaks formed will be directly proportional to the concentration of the specific protein under investigation. This relationship only applies if the peak heights are between 5 mm. and 60 mm. long. The precision of the system has been shown to be greater than 95 per cent. A complete plate is shown in Fig. 13.2.

Preparation of the plate.

Rocket electrophoresis can be carried out on many different sizes of glass plate. The most suitable sizes are 8 × 8 cm. and 20 × 10 cm. Electrophoresis is always run in gels with a thickness between 1 and 2 mm. with an optimum at 1·5 mm. thick. If the gel is too thick, temperature gradients will form, impeding precipitation. If the gel is too thin, small surface irregularities will become critical, impairing quantitation.

(*a*) *8 × 8 cm.* 5 ml. of 15 g./l. agarose solution are mixed with 5 ml. of of 0·04 M barbitone buffer containing the required volume of antiserum. This is usually between 50 and 100 μl. but will vary according to the titre of the antiserum. All antiserum must be titrated initially. Thus different volumes of antisera should be added to test plates and rocket electrophoresis carried out using constant antigen volumes. The concentration chosen for future studies is that which gives an optimum peak height of between 1 and 5 cm. for the standard preparations.

(*b*) *20 × 10 cm.* 15 ml. of 15 g./l. agarose solution are mixed with 15 ml. of 0·04 M barbitone buffer, again containing the required volume of antiserum.

The solution after mixing is carefully poured onto the glass plate which is resting on the horizontal table. To assist uniformity, whilst pouring the gel, and particularly for the larger plates, the plate may be prewarmed.

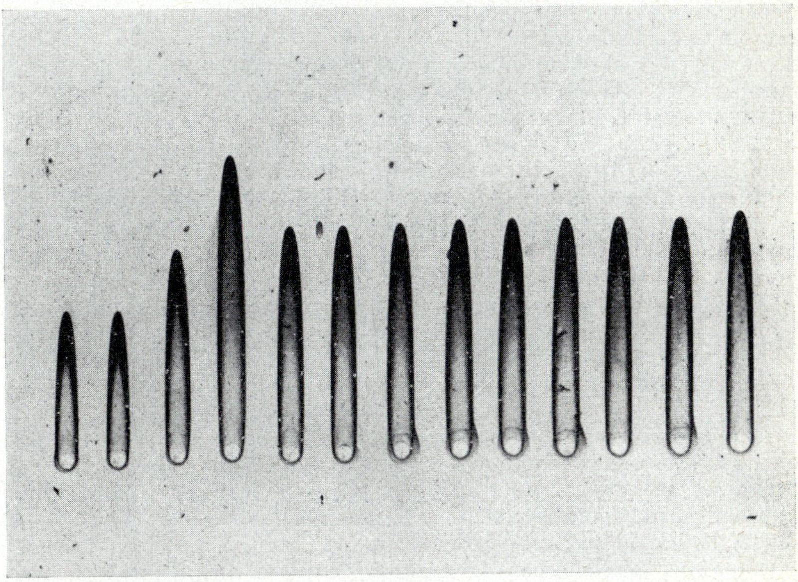

FIG. 13.2. A completed rocket electrophoresis slide produced using human albumin against sheep anti human serum albumin. Note the standard curve on the left.

Alternatively, a mould may be used as described by Laurell[8]. Care must be taken to avoid air bubbles, which can be burst if necessary by lightly flaming the upper surface with a bunsen flame. The gel takes about fifteen minutes to solidify. Subsequently origin wells may be punched out about 1 cm. from the cathode edge of the plate using the gel punches. A template should be used to ensure that the holes are cut in a straight line, about 5 mm. apart. Fourteen wells can be cut on an 8 × 8 cm. plate and thirty five on a 20 × 10 cm. plate. The wells are then evacuated by vacuum suction. The size of all punches used will be determined by the amount of sample to be applied. The larger the sample volume applied, the more accurate the results obtained.

Electrophoresis.

The plate is placed in the electrophoresis apparatus and the wicks placed to overlap the agarose surface by about five millimeters. Contact should be uniform across the whole width of the plate. Cooling is started and a low voltage of 1 V/cm. applied across the gel. The current should be such that the samples when applied will migrate towards the anode. This voltage is measured using the voltmeter placed in the agarose adjacent to the wicks. Samples and standards are applied with the current on, by micropipette.

Switching on the current before application avoids the production of diffusion rings around the origin wells. The exact volume applied will depend upon both the type of sample to be applied, and the particular protein to be investigated. Suitable applications are in general between 2 and 5 μl. and should be judged to give peak heights between 5 and 2 cm. Samples should therefore be diluted accordingly. As an example for the analysis of albumin in serum, using Dako antiserum (0·1 ml. in an 8×8 cm. plate), serum samples should be diluted 1 in 50 and a 2 μl. application used. Such an example is shown in Fig. 13.2. Double dilutions of a standard solution with additional dilutions within the normal range, should also be applied together with any necessary quality control samples. For a large plate, more than one standard curve should be applied.

After application of the samples, the electrophoresis is continued at a high voltage (8–10 V/cm.) or alternatively at a low voltage (1–3 V/cm.). These voltages should be measured using the voltmeter as before. The length of time to complete the electro-phoresis varies from protein to protein. Those proteins with a high electrophoretic migration rate at *p*H 8·6 such as albumin, will be at completion in approximately two hours on high voltage electrophoresis. Proteins with slower electrophoretic migration rates at *p*H 8·6 take longer to reach completion – up to five hours at the high voltage. All proteins should be at completion overnight (16–20 hours) on a slow electrophoretic run.

After this time the plate is removed, covered with a piece of filter paper, to give uniform contact, then cellulose paper layered on top about 1 cm. thick to absorb excess moisture, and a light weight applied for 15 minutes. The paper is then removed and the plate dried in a stream of warm air from a hairdryer. The plate can then be stained as outlined in the staining schedule.

Quantitation

Measurement of the peak height can be achieved using a ruler, measuring from the tip of the peak to the top of the origin well. To assist measurement the slide can be enlarged in a photographic enlarger. Calculation of the concentration of the protein in the unknown sample may be completed by comparing the height of the peak with a graph obtained from the peak heights of the peak with a graph obtained from the peak heights of the standard dilutions against their concentrations. If correctly applied, the peak heights will give a linear standard curve within the limits determined by the size of the plate and the origin wells used. Secondary standard preparations are available commercially for many proteins. The range of standards used will depend upon the concentration of the protein present in the unknown sample. Standardization as a whole is discussed later in this chapter.

The sensitivity of the method depends upon the relative concentrations of antibody and antigen in the system. By reducing the antibody concentration, the sensitivity can be increased, but less than 0.3 mg./l. of a protein is rarely detectable by protein staining.[9] Radiolabelling can increase sensitivity 60-fold.[10]

The precision of the method has been found for most proteins to be greater than 95 per cent. Errors and artefacts of the system are discussed later in this chapter.

Uses.

Most proteins can be estimated by this procedure, provided that a specific antiserum is readily available. Quantitation of the protein itself depends upon the availability of standards. Only when purified material is available, can a true value for the concentration be calculated. Concentrations can of course be expressed as a percentage of a mean normal pool.

In general proteins which have a net anodic migration rate at *p*H 8·6 can be estimated by this procedure, without alteration to the basic method. Proteins which have a cathodic or net zero mobility are unsuitable for this technique but, after formylation (see p. 000), they become negatively charged and can then be investigated by this procedure. Immunoglobulins so treated have been found to have a precision and reproducibility which is comparable with their estimation by the automated immunoprecipitin method developed by Technicon.[7] A comparison of the results is shown in Table 13.1. Carbamylation,[4] has been used extensively in a similar manner to increase the charge on many proteins. In addition to the immunoglobulins, transferrin,[11] carbonic anhydrases.[12] and fibrinogen[13] have all been investigated.

TABLE 13.1

A comparison of immunogolobulin measurement by formylated rocket electrophoresis, Automated immunoprecipitation, and single radial immunodiffusion, Slater 1974(7).

		Formylated Rockets g./l.	A.I.P. g./l.	S.R.D. g./l.
IgG	Mean	8·8	9·1	9·3
	S.D.	0·4%	0·39%	1·45
	coef. variation	5·8%	5·4%	14·7%
IgA	Mean	0·77	0·72	0·78
	S.D.	0.07	0·07	0·13
	coef. variation	8·8%	9·4%	17·1%
IgM	Mean	1·29	—	1·31
	S.D.	0·14	—	0·16
	coef. variation	11·0%	—	11·9%

However, carbamylation is a lengthy procedure unsuitable for routine use, so that formylation is now recommended.

Two dimensional immunoelectrophoresis

By combining electrophoretic separation in agarose with migration into an antibody-containing gel, a separation can be achieved which makes this technique far superior to classical immunoelectrophoresis. The proteins which can now be resolved number more than one hundred, many of which remain unidentified. In addition, the area enclosed by each precipitate is proportional to the antigen/antibody ratio of the system. Quantitation then can be achieved by planimetry. The reproducibility of the method is such that coefficients of variation have been achieved varying between 2–15 per cent, depending on the antigen. Two basic methods are in use: the macrotechnique, and the semi-automated method. A completed plate is shown in Fig. 13.3.

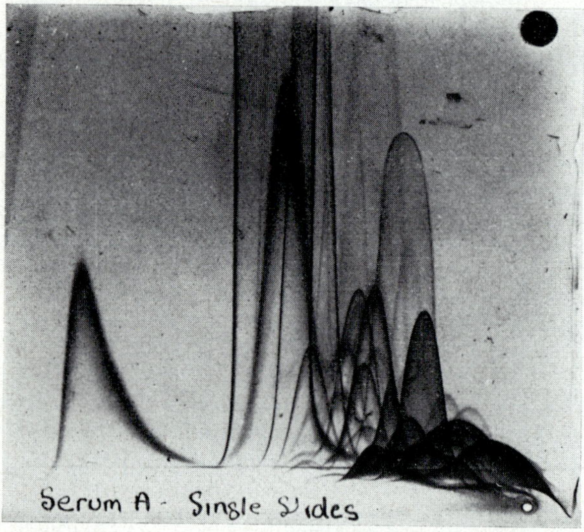

FIG. 13.3. Two dimensional electrophoretic pattern of normal human serum.

(a) Manual two-dimensional immunoelectrophoresis

Introduced initially by Laurell in 1965[14] and modified by Clarke and Freeman in 1966,[2] the technique is normally carried out on 10 cm. × 10 cm. glass plates. 7·5 ml. of 2 per cent agarose are mixed with 7·5 ml. of 0·04 M barbitone buffer, and poured onto a clear glass plate on the level table. After the plate has set, 4 wells (2·5 mm. diameter) are punched in the gel by means of the gel punches and template, and the holes are evacuated by vacuum suction. The volume of sample usually applied is between four and five microlitres from a micropipette, but, as with rocket electrophoresis, the volume must be adjusted to give suitable peak heights. After application the plate is placed on the electrophoresis apparatus and the wicks connected. Lint is cut as before. The sample is then electrophoresed towards the anode at 8–10 V/cm., using cooling. Electrophoresis usually takes about 1 hour, the use of a little aqueous bromophenol blue solution added to the origin wells will trace the migration of the albumin.

After electrophoresis the plate is removed and a strip of agarose containing the separated proteins removed by cutting with, for example, a disposable microtome knife or razor blade and template; the cut should be made as close to the origin well as possible, so that the proteins will migrate only a short distance before reaching the antiserum in the second dimension. The strip is placed parallel and about 0·5 cm. from the edge of a second plate on the level table. 6 ml. of 2 per cent agarose and 6 ml. of 0·04 M barbitone buffer containing the required amount of antiserum is mixed and then carefully poured onto the remainder of the glass plate. Care at this point must be taken to make sure the layer of agarose is even across the whole surface of the plate. After 10 minutes, the completed plate is returned to the electrophoreis is apparatus, and the lint wicks reconnected so that electrophoresis can be continued. The direction of electrophoresis is at right angles to the initial separation, so that the proteins are driven into the second dimension region, and is carried out at 1–3 V/cm. overnight. Cooling in this run is unnecessary. After the electrophoresis, the gel is pressed, dried and stained as before.

(b) Semi-automated two-dimensional immunoelectrophoresis

This procedure was introduced to simplify crossed immunoelectrophoresis.[15] The technique is basically the same as the manual method, but there are certain important differences. The size of the electrophoretic plates has been reduced, thereby saving materials, and the plate has been prepoured to contain both a region for the first electrophoretic separation, and an antibody containing region for the second stage of the electrophoresis. The sample is applied and electrophoresis carried out initially in one direction and then in the other without further interference. This changeover is automatic and occurs in a modified electrophoretic apparatus which is described below.

Preparing the plate.

2·25 ml. of barbitone buffer *p*H 8·6, 0·04 M, is mixed with 2·25 ml. of 15 g./l. agarose, and poured onto a 5 × 5 cm. glass plate resting on the level table. After 10 minutes the plate is placed on the template (Fig. 13.4) and gel slices 1·2 cm. wide are cut off using the microtome knife, and are transferred to the edge of a second plate. 1·5 ml. of 15 g./l. agarose and 1·5 ml. of barbitone buffer 0·04 M containing the required volume of antiserum (usually between 50–100 μl.) are mixed and then poured onto the remainder of the second plate. Care should be taken to avoid air bubbles. Origin wells are then punched in the first dimension gel on the finished plate. Usually the small gel punch is used and samples then applied by micropipette (approximately 2 μl.).

Caution. For this technique, origin wells must be cut in the bottom left and the bottom right corners of each successive plate, alternately, since the electrophoretic apparatus differs in the direction of the applied from plate to plate. This is explained in the next section.

Electrophoresis

The PCD tray equipment used in the rocket electrophoresis system is modified for this procedure, according to PCD instructions. The

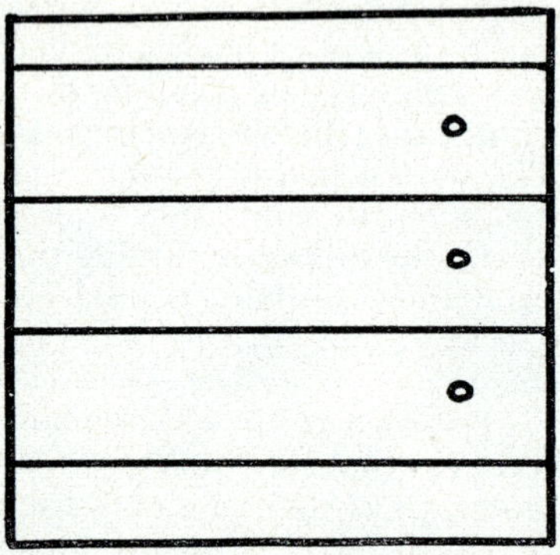

Fig. 13.4. Template for semi-automated two dimensional immuno-electrophoresis.

apparatus has been designed to supply current from either of two directions at right angles. By changing the input leads from one terminal to the other, the direction of the current can be effectively changed as shown in Fig. 13.5.

This allows the electrophoresis of the plate in one direction and a change automatically to a second dimension at right angles, to be developed. Special carriers provided with the apparatus are fitted in place on the equipment and wicks placed on these carriers, after having been moistened with buffer. The pattern of the wicks are cut according to Fig. 13.6. The wicks are put in place in the correct buffer vessels (the relevant vessel and wick can be determined from Figs. 13.5 and 13.6), and should be placed in position so that when the plate is placed on top of the wicks, the centres of the first dimension wick protrusions are in line with the origin well, and the centre line of the first dimension electrophoretic region. The plate itself should be placed inverted, the agarose in contact with the lint wicks.

The buffer vessels each contain approximately 150 ml. of buffer. Low buffer volumes are satisfactory since the wick length controls *p*H changes which occur in the vessels. The connections are made so that current flows through the 1st dimension region, and electrophoresis is carried out at 8–10 V/cm. for approximately 40 minutes. Using the PCD BES IV power supply (PCD Ltd., Farnborough, Hampshire), the voltages for both the 1st and 2nd dimensions can be set before starting the run. The timer is adjusted to the required timing and the supply will automatically change to the second dimension.

If an ordinary power supply is being used, the second dimension connections must be exchanged with those of the first dimension on the power supply after the necessary first dimension time has elapsed. The extent to which electrophoresis has occurred can be visualised by

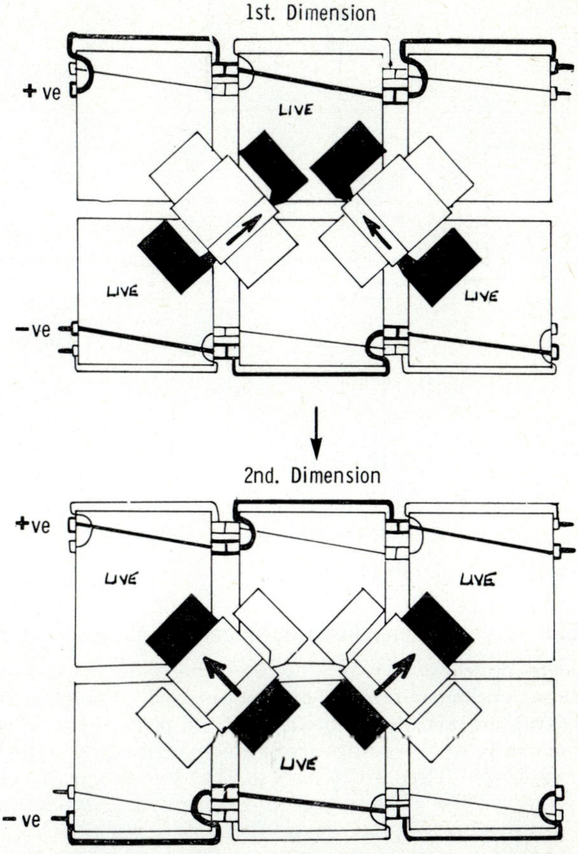

FIG. 13.5. Diagrammatic representation of the semi automated two-dimensional immunoelectrophoretic apparatus. illustrating the change from 1st to 2nd dimension. Note that the current is supplied through certain tanks and passes through others so that effectively the buffer vessels are alternately live and dead electrically. By changing the position of the input leads the direction of the current flow, when connections are made, can effectively be changed.

the addition of a small volume of aqueous bromophenol blue solution to the sample well. This will bind to albumin and electrophorese in conjunction. Once the second dimension is complete (in 16–20 hours at 1–3 V/cm.) the gel may be pressed, dried and stained.

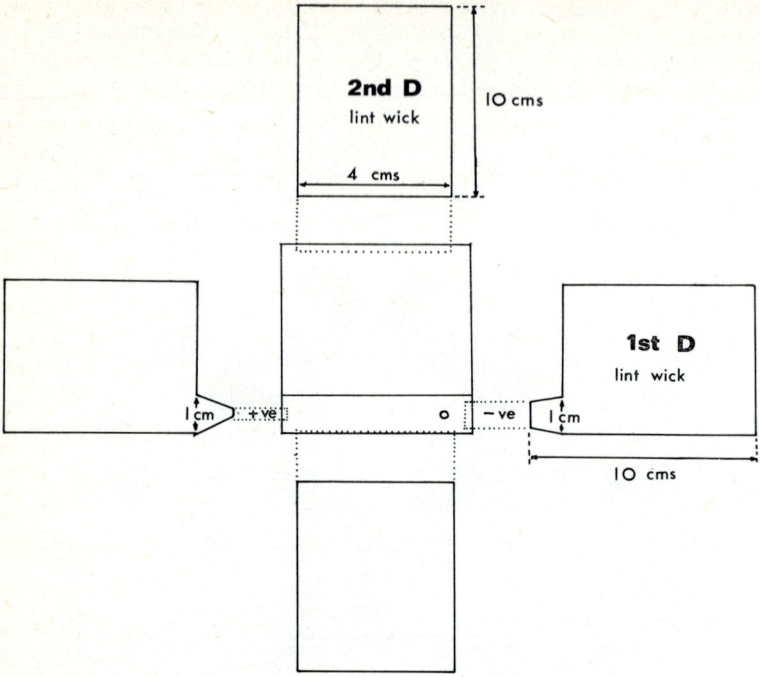

FIG. 13.6. Wick patterns for the semi-jautomated system.

Identification of two-dimensional immunoelectrophoresis patterns

Once electrophoresis is complete and the slide dried and stained, identification and measurement can take place. The procedure is the same for both methods. Identification of the peaks may be carried out immunochemically by the addition of known antibodies to the antiserum gel before running. This will result in the depression of the relevant precipitin peak in the electrophoretic pattern, in comparison with an untreated sample. Alternatively, addition of a purified protein to the sample will result in increased peak areas. The methods and variations to carry out these procedures have been well described by Axelsen.[16,17] Reference may also be made to a standard reference pattern obtained from pooled normal samples. Many purified human proteins are now commercially available, and more than twenty antisera to human serum proteins. In addition to immunochemical analysis, biochemical staining of specific proteins can be achieved. Grabar and Burtin[18] have described a wide number of staining procedures for caeruloplasmin, lipoproteins, pseudocholinesterase and glycoproteins.

Quantitation.

Quantitation of the individual antigen can be performed by measuring the area enclosed by the precipitate and comparing it to a standard peak. This standard can be material added in known concentration to the sample before running, such as carbamylated transferrin, or acetylated albumin, or a known serum protein can be used as a protein standard. In the latter method the concentration of that standard protein in the sample is calculated by a previous rocket electrophoresis. Transferrin has been found to be suitable for such a procedure.

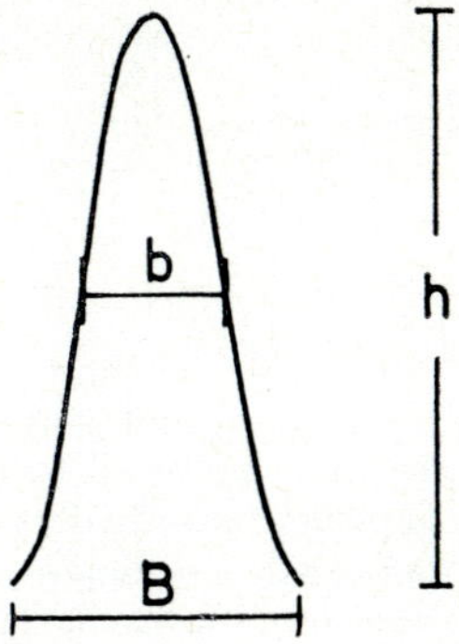

FIG. 13.7. Illustration of simple area measurement using height multiplied by the width.

Calculation of the areas of the peaks themselves may be attained by many methods but only two basic procedures will be considered.

1. *Planimetry*. Many forms of planimeter are available. These include complicated X, Y digitizers such as that produced by DMac, or a lower priced version by PCD. Simple planimeters which, although less accurate, give reasonably reproducible results, may be purchased from Gallenkamp Ltd. In all cases the area under the curve will be obtained by following the outside of the enlarged curve. Such measurements give a figure which can be compared directly with the standard peak and are generally within 1 per cent reproducible.

2. *Height × width*. Provided that the peaks produced are reasonably symmetrical the areas of the peaks may be estimated by simple triangular estimation. This method involves the measurement of the height of the peak, multiplied by the width of the baseline.[19] More accurately the width at half the height may be used (Fig. 13.7.). Since the baseline is easily discernible, measurement is easier than at the half height position. For this type of estimation, the height alone should not be used since this gives values with much higher coefficients of variation.

Area measurement is as consistent as measurement with the DMac and gives correlation coefficients which are in very good agreement. For peaks which are not symmetrical, peak areas by this method tend to be overestimated (Table 13.2). Interestingly area measurements by simple triangular measurement give coefficients of variation generally lower than those by planimetry largely because it is difficult to follow the edge of the peak.

TABLE 13.2

The estimation of peak areas by planimetry using a D.Mac digitizer, and simple area measurements. Versey and Slater, 1973 (19).

Peak	Mean Area mm^2		S.D. (coef. of v.)	
	D.Mac	$\frac{1}{2}$B × Ht	D.Mac	$\frac{1}{2}$B × Ht
C3	247	257	8·4 (3·1)	11·2 (4·4)
C4	100	111	9·0 (9·0)	12·4 (11·1)
H	285	299	8·1 (2·8)	14·2 (4·7)
α_2M	237	290	17·4 (7·3)	14·6 (5·0)

All area measurements by these methods give results with a high degree of precision. The actual precision obtained varies from protein to protein. If peak areas are large, and well-formed, coefficients of variation are generally small. As peak areas are decreased these coefficients of variation proportionally increase. Coefficients obtained by two dimensional immunoelectrophoresis are shown in Table 13.3. The precision

TABLE 13.3

The precision of area measurement by two dimensional electrophoresis.

Protein	Mean area mm^2	S.D.	coef. variation
Transferrin	302	3·2	1·0
C3	274	8·4	3·1
C4	100	9·0	9·0
GBG	48	5·0	10·4
haemopexin	285	8·1	2·8
α_2M	237	17·4	7·3
caeruloplasmin	266	19·0	7·1
Gc globulin	211	19·0	9·0
haploglobin	247	18·5	7·5
α_1 lipoprotein	122	12·0	9·8

also tends to be greater in general for smaller plates on the automated system, probably because peaks tend to be taller and sharper, since there is no diffusion time between first and second dimensions.

Sensitivity.

It has been reported that visible precipitates may be detected using Coomassie brilliant blue staining, with as little protein as 0·1 mg./l. Sample sizes are generally of the order 1–4 μl. but sensitivity can be increased by either increasing the sample application, or decreasing antiserum strength. The sensitivity when necessary can be increased 60 times, using radiolabelled antibodies.

Uses.

Two dimensional immunoelectrophoresis has many applications in a wide field. It is particularly adaptable to the study of protein heterogenicity, polymorphism and fragmentation of proteins, and abnormal fractions. In addition, an analysis of a wide range of proteins can be carried out simultaneously, giving a broad spectrum of the protein

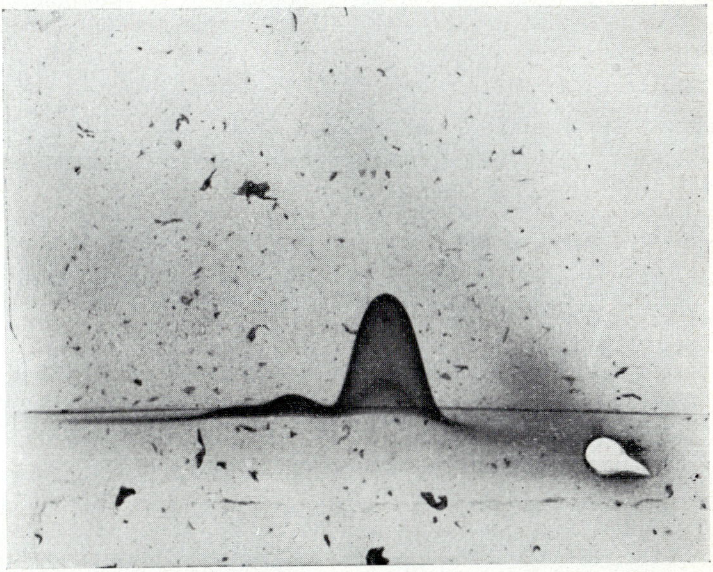

FIG. 13.8. Two dimensional immunoelectrophoretic slide of the analysis of the complement component C3 in a patient with systemic lupus erythenatosus.

pattern. Using such procedures, it has been possible to observe patterns in fluids such as serum, urine, cerebrospinal fluid, jejunal juice. Protein patterns have been investigated in a wide variety of disorders, including tuberculosis and sarcoidosis,[30] haemochromatosis,[21] chronic bronchitis[22] and following renal transplantation.[23] As with other procedures, immunoglobulin estimation is only possible after pretreatment of the sample with formaldehyde.

In this laboratory two-dimensional immunoelectrophoresis has mainly been applied to the study of complement.[24] The activation of many complement components results in the formation of a fragment which has a faster electrophoretic mobility. A suitable example of this is shown in Fig. 13.8. in which C3 has been activated to produce the two fragments C3b and C3c. Such activation has been investigated in a variety of disorders and has been taken to indicate complement activation *in vivo*.[25]

By virtue of its high resolving power, two-dimensional immunoelectrophoresis has been shown to be particularly suitable for the analysis of hereditary polymorphism and microheterogenicity. Such polymorphism has now been demonstrated for many plasma proteins including the Gc-globulins, α_1-antitrypsin, prothrombin, transferrin, haptoglobin and the complement components C3 and C4. Examples of microheterogenicity include the trailing effect of α_2-macroglobulin, and its complex formation with plasmin.

Abnormal proteins that can be demonstrated by two-dimensional electrophoresis include the well-known M components, which are clearly visible even at low concentration. An abnormal fraction of α_1-lipoprotein has been demonstrated in plasma from alcoholics.[26] Fragmentation of proteins, whilst not necessarily altering electrophoretic mobility, will commonly uncover altered antigenic sites. This will result in a different reaction with the antiserum and result in spur formation.[27]

Although two dimensional electrophoresis has mainly been applied in human biology and medicine so far, it is beginning to be used in forensic medicine and in the study of bacterial and fungal antigens.[28]

We have found that two-dimensional immunoelectrophoresis has two other important functions. Since it is a very sensitive procedure, the technique is particularly suitable for analysing both antigen and antibody preparations for purity. The presence of small amounts of contaminant are readily detected. This function is very important when highly purified protein solutions are required.

Problems and artefact produced by quantitative immunoelectrophoresis.

Many technical problems can occur during the course of an immunoelectrophoretic run and it is advisable to be fully aware of them before commencing any run. These problems have been fully described elsewhere[27,29] so that it is only necessary to describe them in brief:

Artefacts due to excessive current flow. When the voltage is too high, origin wells may empty of buffer too quickly so that the rate of migration of the protein may be retarded. This can result in the phenomenon of "doubling" or the "tunnel effect" described by Laurell[5] in which the migrating precipitation front becomes faster at the bottom surface of the gel than at the top. This effect can also occur if the gel is too thick in which case temperature gradients can be set up.

Excessive overheating coupled with vigorous cooling can result in water condensing on the gel surface. This will alter the precipitation point and result in irregular precipitation spurring and trailing.

Misshapen peaks. Sometimes the peaks produced both by rocket electrophoresis and by two dimensional immunoelectrophoresis, may be curved or irregular. This is generally caused by poor contact between the wick and the gel, or by unequal gel thickness. In both cases the electric field and consequently antigen migration will not be linear. Condensation on the gel surface may also serve to attach current by lowering the resistance at that point.

Faulty precipitation. No precipitation upon completion of electrophoresis may sometimes be seen. Apart from obvious causes such as no antisera in the gel, or the incorrect antisera, such failures are often due to faulty electrical connections. Such examples include anode and cathode being reversed or disconnected, an electric field which is too high or insufficient cooling. Precipitation may also be prevented if marked *p*H changes occur in the gel, if there is a poor reaction between the antibody and the antigen, or when there is an imbalance in the proportions of the antibody and antigen.

Extra precipitation peaks may occur if the antiserum contains extra antibodies against other components, or if there is partial immunochemical identity between the proteins. In addition, contamination of the slide by proteins or other substances migrating from the wick material, may sometimes cause precipitation. Such reactions may be prevented by placing a piece of dialysis membrane between the wick and the agarose gel, as described by Krøll.[30] Current interruptions may also cause spurious bands.

Reading before completion. If the electrophoretic run is stopped before all the antigen has been precipitated, misshapen rockets may be formed. Under such circumstances there is no constant relationship between the peak height and concentration of antigen.

These incomplete peaks will take many forms dependent upon the point at which the electrophoretic run was halted. When stopped very prematurely, the top of the peak will be completely missing, a blurred ill-defined precipitate taking its place. Nearer completion, the precipitate, although clearly visible, will be markedly rounded, no sharp point to the precipitate being present. This should not be confused with the rounded peaks produced by slow electrophoretic runs.

Other uses and variations.

Quantitative immunoelectrophoresis then has been applied to a wide range of investigations where protein concentrations are required. Rocket electrophoresis is a sensitive technique for single protein studies, which is highly accurate, and with the advent of formylation, all proteins can now be studied. Two dimensional immunoelectrophoresis, with its many applications, may now suitably be applied both to the analysis of protein patterns, particularly with a view to screening for particular types of disease, and to the analysis of individual or limited numbers of proteins.

In addition to these two applications of the basic procedure, a large number of variations have been suggested to give more scope. Such procedures are beyond the scope of this chapter, but full details of the methods have been described elsewhere.[31,32,33]

REFERENCES

1. Heidelberger, M., and Kendall, F. E. *J. Exper. Med.*, 1935, **62**, 467.
2. Clarke, H. G. M., and Freeman, T. *Prot. Biol. Fluids*, 1967, **14**, 503.
3. Nanto, Teppo. *Science Tools.* (In press.)
4. Weeke, B. *Scand. J. Clin. Lab. Invest.*, 1968, **21**, 351.
5. Laurell, C-B. *Analyt. Biochem.*, 1966, **15**, 45.

6. Ressler, N. *Clin. Chim. Acta.* 1960, **5**, 795.
7. Slater, L. *Annals Clin. Biochemistry.* (In press.)
8. Johansson, B. G. *Scand. J. Clin. Lab. Invest.* **29**, 1972, Supple. **124**, 7.
9. Johansson, B. G., and Malnquist, J. *Scand. J. Clin. Lab. Invest.* 1971, **27**, 255.
10. Kindmark, C-O., and Thorell, J. I. *Scand. J. Clin. Lab. Invest.* **29**, 1972, Supple. **124**, 49.
11. Weeke, B. *Scand. J. Clin. Lab. Invest.*, 1970, **25**, 269.
12. Norgaard-Pedersen, B. and Mondrup, M. *Scand. J. Clin. Lab. Invest.*, 1971, **27**, 169.
13. Nielsen, H. G., and Weeke, B. *Clin. Chim. Acta*, 1971, **35**, 281.
14. Laurell, C-B. *Analyt. Biocyem.*, 1965, **10**, 358.
15. Davies, D. R., Spurr, E., and Versey, J. M. B. *Clinical Science*, 1970, **40**, 411.
16. Axelsen, N. H., Bock, E., and Kroll, J. *Scand. J. Immuno.* **2**, 1973, Suppl. **1**, 91.
17. Bock, E., and Axelsen, N. H. *Scand. J. Immunol.* **2**, 1973, Supple. **1**, 95.
18. Grabar, P., and Burtin, P. *Immunoelectrophoretic Analysis*. Elsevier, Amsterdam, 1964.
19. Versey, J. M. B., and Slater, L. *Annals C. in. Biochem.*, 1973, **10**, 1.
20. Clarke, H. G. M., Freeman T., Hickman, R., Pryse-Phillips, W. E. M. *Thorax*, 1970, **25**, 423.
21. Amin, A. H., Clarke, H. G., Freeman, T., Murray Lyon, I. M., Smith, P. M., and Williams, R. *Clin. Science*, 1970, **38**, 613.
22. Ryley, H. C., and Brogan, T. D. *J. Clin. Path.*, 1973, **26**, 852.
23. Weeke, B., Weeke, E., and Bendixen, G. *Acta Med. Scand.*, 1971, **189**, 119.
24. Versey, J. M. B. *Annals Clin. Biochem.*, 1973. **10**, 100.
25. Versey, J. M. B., Hobbs, J. R., and Holt, P. J. L. *Annals Rheum. Dis.*, 1973, **32**, 557.
26. Johansson, B. G., and Laurell, C-B. *Scand. J. Clin. Lab. Invest.*, 1969, **23**, 231.
27. Clarke, H. G. M., and Freeman, T. *Clinical Science*, 1968, **35**, 403.
28. Svendsen, P. J., and Axelsen, N. H. *J. Immunol. Methods*, 1972, **1**, 169.
29. Laurell, C-B. *Scand. J. clin. Lab. Invest.* **29**, 1972, Suppl. **124**, 21.
30. Krøl, J. Unpublished.
31. *Scand. J. Immunology* **2**, 1973. Supple. **1**. Ed. Axelsen, N. H., Kroll, J., and Weeke, B.
32. Krøl, J. *Scand. J. Clin. Lab. Invest.*, 1968, **22**, 79.
33. Svendsen, P. J., and Rose, C. *Science Tools*, 1970, **17**, 13.

CHAPTER 14

PREPARATIVE BLOCK ELECTROPHORESIS

Jacqueline B. Weiss

Low Voltage Preparative Block Electrophoresis

IN the past, either starch[1] or ethanolized cellulose[2] have been used with considerable success as supporting media for preparative block electrophoresis. However, because neither of these media are totally insoluble, they are at a severe disadvantage if the materials to be separated contain a carbohydrate component. Furthermore, they both give rise to considerable electro-osmotic flow. Although such a back-flow may occasionally improve the electrophoretic separation of a mixture containing constituents with closely related mobilities (hæmoglobins), it will often do so only over relatively long distances and at the expense of increasing the zone spreading, which has been shown to depend mainly on the distance which the migrating material has moved through the supporting medium.[3]

Pevikon Block Electrophoresis

Polyvinyl chloride (Geon 426) was introduced in 1956[4] as a supporting medium for the separation of glycoproteins and acid glycosaminoglycans and it had the obvious advantage that it did not interfere in the analysis, of the eluted material, for carbohydrate. Nevertheless, despite this improvement, the problem of cathodal electro-osmotic flow was as great with this medium as with those previously in use. In 1960, a new polyvinyl resin, which was a copolymer of polyvinyl chloride and polyvinyl acetate,[5] was proposed as a supporting medium. This material, "Pevikon C-870" had been developed for the production of flooring, floor tiles and heavily filled products. Despite these rather modest beginnings it has proved itself to be a valuable support in preparative block electrophoresis. Perhaps the major advantage which Pevikon has over its predecessors is that it gives rise to a negligibly small electro-osmotic flow. In addition it is a completely non-swelling medium and it has been demonstrated that its microscopic structure is such as to considerably reduce zone spreading.

Physical and Chemical Characteristics of Pevikon C-870*.

Vinyl chloride-vinyl acetate suspension copolymer.

Vinyl acetate content	13 per cent
K-value, Fikentscher (1 per cent solution in cyclo-hexanone) . .	47
Relative viscosity at 20°C (1 per cent solution in cyclo-hexanone) .	1.60
Chlorine content	49.0 per cent

* Shandon Southern Instruments, Camberley, Surrey.

Moisture	max. 0.3 per cent
Sulphate ash	0
Solubility in ether	0·9 per cent
Solubility in boiling water	0·07 per cent
Bulk density	750 g/l.
Grain size distribution	
above 0·250 mm.	0 per cent
0·150–0·250 mm.	1 per cent
0·100–0·150 mm.	35 per cent
0·060–0·100 mm.	20 per cent
–0·060 mm.	45 per cent

The liquid content of a poured Pevikon block connected and equilibrated with the buffer tanks is approximately 50 per cent.

Preparation of Pevikon C-870 for Use as a Supporting Medium. Small quantities of metal ions, particularly Fe^{++}, are present as contaminants in the Pevikon supplied by the manufacturers. In order to remove these, the powder is shaken for 10 minutes with 10 volumes of 4N hydrochloric acid. The resultant suspension is poured into a sintered glass funnel and the majority of acid removed by suction. The powder is now washed on the funnel with a further 2 $\times$ 5 volumes of 4N hydrochloric acid and then copiously with distilled water. The damp powder is then shaken with 2N acetic acid (10 vol.) for about 15 minutes, replaced on the sinter and washed with distilled water until the washings are at a neutral *p*H. Finally, the washed powder is suspended in successive portions of distilled water (3 vol.) and decanted to remove any "fines" which may be present.

Apparatus. The apparatus consists of a plate glass support (60 cm. $\times$ 25 cm.), two plastic bars (60 cm. $\times$ 1·5 cm. $\times$ 1·5 cm.) each graduated along their length in 1 cm. divisions and two buffer tanks wide enough to accommodate the plate glass support (Fig. 14.1). Detachable plastic holders carrying platinum electrodes are used as such an arrangement has the advantage that the same pair of electrodes may serve in multiple electrophoretic systems. However, it will be obvious that this is not an essential requirement. The equipment is set up as in Fig. 14.1 and the supporting plate and plastic bars are covered with a thin polythene sheet. Packs of Whatman 3 MM filter paper 5 cm. wide and of such a length as to fit exactly between the two plastic bars, are placed on top of the polythene sheeting at either end of the glass supporting plate. The paper pads serve the purpose of completing the four sides into which the block will be poured and also of removing excess moisture from the poured block. A recent substitute for paper pads has been found in "Labshammies" obtainable from Jencon's of Hemel Hempstead, Hertfordshire. These sponges have a remarkable capacity for absorbing liquid and are far more efficient than paper. After rinsing they may be reused any number of times. They can be easily cut to fit the size of the block.

Preparation of Pevikon Block. A homogeneous suspension of the washed Pevikon powder in an appropriate buffer is prepared to a viscosity which is just sufficiently fluid to pour. The suspension is then poured into the supporting arrangement so as to fill it to the depth of the plastic bars. After a few minutes the paper packs or labshammies at either end may become saturated and provided the block is sufficiently set these should be quickly changed for fresh ones. The pads should now be left in position until the block is ready for connection to the electrode assembly. Failure to do this may result in "melting" of the block. The surface of the solidified block generally has a rather foamy appearance and this should be treated by *gently* rolling a hard rubber roller over the surface. The block is now ready for application of the test solution.

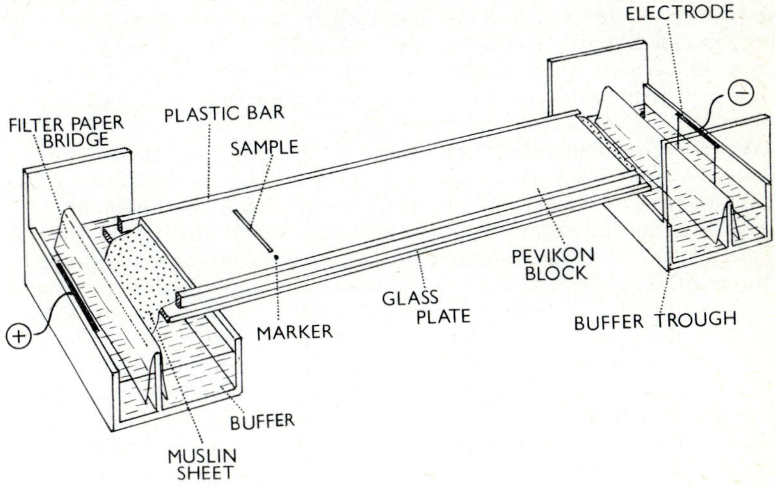

FIG. 14.1. Apparatus for Pevikon block electrophoresis; for description see text.

Application of Sample. A slit is cut in the block with a solid-backed safety razor blade. The distance of the slit from the cathodal or anodal end will vary with the type of material to be separated and with the buffer system used. It must however never extend to a width less than 2 cm. from each plastic side bar. In order to ensure that a satisfactorily narrow initial zone is obtained it is advisable to place two strips of Whatman 3 MM filter paper (2 cm. width), which have been moistened with buffer solution, about ½ cm. from either edge of the slit along its entire length. The slit is now widened slightly by inserting into it a strip of plastic (2 mm. thick) of its same length. The solution to be applied is made into a thin slurry with washed Pevikon powder, the plastic strip is removed, and the slurry is rapidly "poured" into the slit by means of a wide bore Pasteur pipette. The quantity of sample slurry should be sufficient to completely fill the slit to the level of the top surface after the liquid has infiltrated the block. In this way at least 10 ml. of sample can be applied to a block 25 cm. wide. The moistened paper strips are removed and the rubber roller is applied just over the slit to ensure that it is completely closed. The surface of the block is covered with a thin sheet of polythene, care being taken to ensure that no air bubbles are trapped between the polythene sheet and the surface of the block. The pads at either end of the block are now removed and the ends of the block are quickly trimmed with a razor blade before being connected to the buffer tanks by wicks made of two thicknesses of knitted nylon or orlon material. Washed gauze or lint can be used instead of the nylon material but this is not recommended if glycoproteins or acid glycosaminoglycans are being separated.

The electrophoretic run should always be carried out in a cold room as considerable heat is generated.

Voltage. The potential gradient and the time of the run will differ according to the nature of the materials to be separated and these will be described under the appropriate headings. However a potential gradient of more than 5 V/cm. usually gives rise to excessive heating effects, which cannot be adequately dealt with simply by carrying out the electrophoretic run in the cold room.

Isolation of Separated Fractions. After the run, the connecting material is removed from each end of the block and replaced by fresh filter paper pads which are left in position for a few moments. The polythene sheet is removed and the block is cut into 1 cm. wide segments, using the lines marked on the plastic side bars as guides. A sintered filter tube (10 cm. $\times$ 3 cm. above the sinter) is drawn out along its lower length and connected through a rubber bung to a test tube with a side arm. The drawn out end should be allowed to protrude about 1·5 cm. through the rubber bung. If a Buchner flask is used a test tube is placed inside the flask in such a manner as to be held in place by the protruding glass tube. The cut segments are transferred carefully to the filter tube and tamped down with a glass rod which has been flattened at one end. Suction is applied, the powder is again tamped down and washed with two successive portions of 1 ml. distilled water, suction and tamping being repeated after each wash. The same filter tube can be used repeatedly, the dried Pevikon being quickly blown out between each use by means of compressed air and the test tube inside the Buchner flask being removed with forceps and replaced. The protein recovery by this method is usually in the order of 90 per cent. A few very fine particles of Pevikon may pass through the sinter but these very quickly settle and a clear supernatant can be poured gently off these particles after the tubes have been left standing for about 2 hours.

A great advantage of Pevikon block electrophoresis is that biological materials do not adsorb to it and consequently the separated fractions can be eluted in a relatively small volume of liquid.

Zone Spreading. Hæmoglobin is a rather useful protein for the visual observation of band widths. After a run of 16 hours and a migration distance of 20 cm., a band of Hæmoglobin A can be shown to have spread to approximately twice its original width.

Use of "Marker" Dyes. It is realised that some workers may be disinclined to use preparative block methods because of the relative tediousness involved in the elution of so many samples. In the case of a visible protein such as hæmoglobin it will be obvious that only a small fraction of the block need be eluted. However it is also possible to "visualise" other materials by the use of markers. A selection of cationic or anionic dyes (whichever group is applicable) should be run on paper electrophoresis under the same conditions that will apply for the block separation. Under such conditions many dyes tend to streak but this should be ignored as streaking will not occur on Pevikon blocks. The furthest point travelled by the dye streak will therefore be equivalent to its position as a discrete spot during block electrophoresis. Once a dye has been found which has similar electrophoretic properties to the material which it is desired to elute, it can be applied as a spot in the space between sample and edge of the Pevikon block. This is best done by making a hole in the block with a capillary tube and filling the hole with dye solution from a fine hypodermic needle.

We have used this method for the separation of a specific urinary glycopeptide and found it to be most satisfactory. In our case Acridine orange was the dye of choice.

Applications

Separation of Glycopeptides and Glycoproteins. The choice of buffer is influenced by the same factors as described in Chapter 1. In the case of preparative electrophoresis there is an obvious advantage in using a volatile buffer. The pattern obtained when a mixture of urinary glycopeptides were subjected to electrophoresis in a pyridine-acetate buffer at *p*H 6·5 potential difference 5 V/cm. length, is shown in Fig. 14.2. In this case the sample was applied 10 cm. from the cathodal end and the running time was 12 hours.

For glycoproteins a rather longer running time is usually required. This can be roughly assessed by reference to the electrophoretic characteristics of the material on paper. For example, if a potential gradient of 7·5 V/cm. length had given a separation span of 10 cm. in 5 hours using paper electrophoresis, then a 25 hour running time and a potential gradient of 5 V/cm. length could be expected to give a separation scan of the order of 33 cm. on a Pevikon block.

It is advisable to check such calculations by experiment and in order to avoid wastage of valuable material, it is useful to pour a block only 2 cm. wide and to apply the material along an 0·5 cm. slit. The elution profile of such a block will not be very clear, due to surface tension effects, but it will, nevertheless, be adequate to determine the length of the separation scan.

Pevikon particles have not been found to interfere to any extent with the colorimetric methods for estimation of carbohydrates or proteins.[5] In the uronic acid estimation,[6] however, a grey colour develops during the stage which involves boiling the test solution with concentrated sulphuric acid and it is advisable to include a blank, eluted from a segment containing no applied material.

Acid glycoproteins from serum separate very well at *p*H 8·6 in an Oxoid barbitone/Acetate buffer ($\mu = 0{\cdot}1$). A running time of 36 hours with a potential gradient of 4·5 V/cm. reveals three acid glycoprotein peaks.[5]

Separation of Acid Glycosaminoglycans. Pevikon block electrophoresis is ideal for the separation of this group of compounds. In a phosphate buffer at neutral *p*H ($\mu = 0{\cdot}1$), separation of hyaluronic acid from chondroitin sulphate can be achieved in six hours. The sulphated glycosaminoglycans separate better at pH 2·5 in a formic acid buffer the running time required is 24 hours.

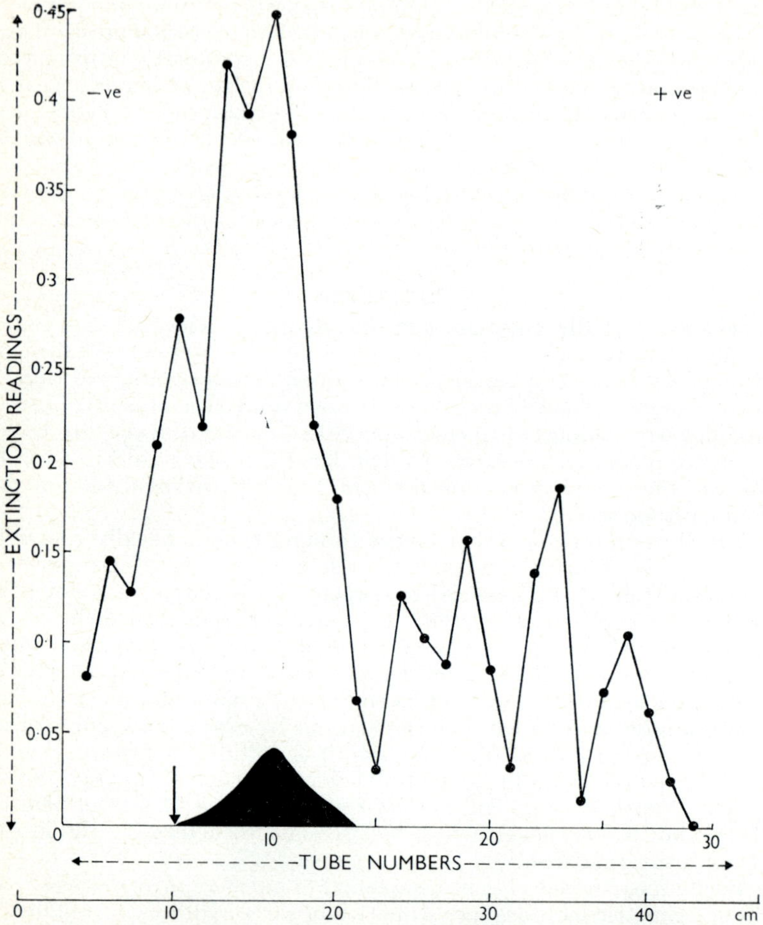

FIG. 14.2. Separation of urinary glycopeptides on Pevikon C 870. Extinction readings refer to the colour produced using the orcinol-sulphuric acid reagent. The shaded area indicates urinary pigments.

Separation of Proteins. Proteins, such as serum proteins, can usually be separated well on Pevikon using a barbitone buffer made by adjusting 0·05M sodium barbitone (10·3 g/l.) to pH 8·0–8·6 with HCl. Under these conditions, blocks 20 cm. wide × 1 cm. deep and running at 5 V/cm. usually give around 90 mA. The progress of the separation can be followed by taking prints from the block as follows. A strip of paper, the length of the block, is taken and held edge on to the block so that buffer soaks up into it; this is preferable to lying the strip flat on the block as powder would adhere to the strip thus spoiling both the paper and the block surface. The paper strip is dried and then stained for protein in the usual way. When eluting proteins from

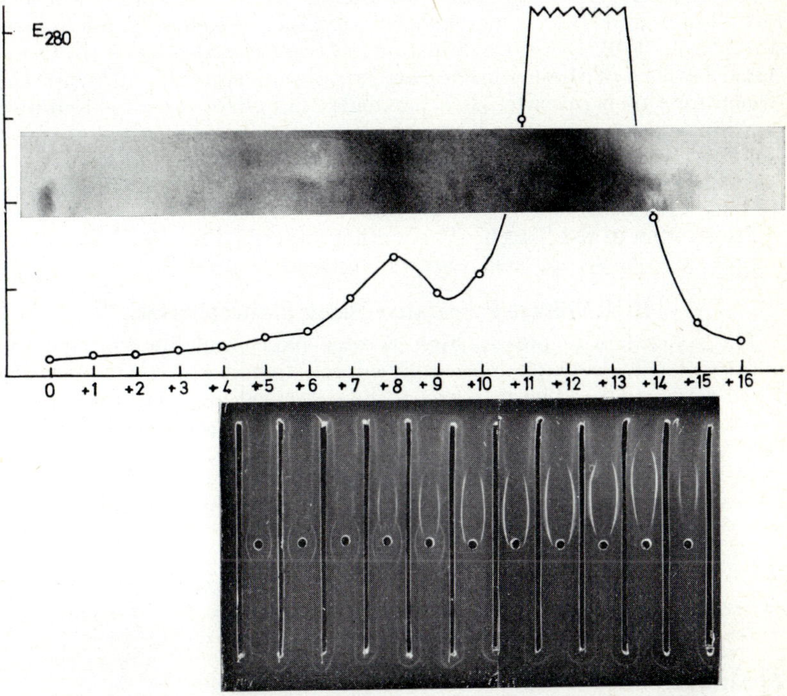

FIG. 14.3. Use of Pevikon block electrophoresis as the final stage of purification of calliphorin from an extract of *Calliphora erythrocephala* (blowfly). The stained paper test strip above indicated a major and a minor protein band, and accordingly Pevikon segments were cut. The extinctions of these segment eluates match the test strip. The corresponding immuno-electrophoretic patterns below show that the material isolated from the major band, segments 10–15, was pure calliphorin, and the minor band was due to a contaminant.

Pevikon sections, particular care must be taken to avoid frothing as this may lead to surface denaturation. The optical density of eluted fractions can be read at 280 μ. and the barbitone buffer salts, which also absorb at this wavelength, will yield the base line.

Starch Block Electrophoresis

Starch block electrophoresis still has a considerable vogue and except in the specific instances mentioned on p. 367 it continues to be a useful preparative separation technique. Nevertheless, the advantages of Pevikon are considerable and it seems likely that this material will eventually become the more commonly used supporting medium.

Starch. Potato starch varies greatly in its calcium and phosphorus content and the proportion of these ions appears to affect the electrophoretic behaviour of some proteins, particularly the hæmoglobins. Potato starch (superior grade), obtainable from K. J. Mulder & Co. Ltd., London, E.C.3, is particularly suitable. Reagent grade potato starch from J. T. Baker Chemical Co., U.S.A., can also be used. Occasional batches of these starches are not always suitable, although the reason for this is obscure. If a particular sample of starch is found to be unsuitable, a repeat order to the firm concerned should state that a sample from a different batch is required.

Apparatus. The apparatus described for Pevikon block can be used without alteration.

Preparation of the Starch. The method described for Pevikon is used except that the strong acid wash is omitted.

High Voltage Preparative Block Electrophoresis

The general principles of high voltage paper and low voltage preparative block electrophoresis apply equally to high voltage preparative block separations. The methods to be described have been used with Pevikon blocks only but they should be equally applicable with starch as a supporting medium.

In the past undue temperature rise during the electrophoretic run with consequent evaporation from the surface of the block has limited the application of high voltage to preparative block techniques. However, a machine has recently been designed (Chandos Products Ltd., New Mills, Cheshire, U.K.) that successfully overcomes problems of excessive heat generation and enables preparative block separations to be carried out at potential gradients of 150 V/cm. (Fig. 14.4.)

Apparatus.

The apparatus consists of a base plate, 33 cm. × 16 cm., of aluminium machined to a tolerance of 0·025 mm. (0·001 in.) which is evenly refrigerated by means of a high-capacity refrigeration unit with direct expansion into coils cast inside the aluminium block. The power supply incorporates a thyristor stabilising device and either voltage or current may be stabilised with optional limitation of the other parameter. A safety trip is incorporated in the electronic circuitry which is coupled to micro switches. These disconnect the high voltage power supply if the lid is opened during the run. Such a safety device is extremely important especially when the apparatus is being used in an open laboratory.

In the high voltage system it is important to use plastic sheeting to cover the aluminium base plate and this sheeting must be of sufficient thickness to act as an adequate insulator between the block and the metal plate. However too thick a plastic sheet will negate the cooling properties of the block and invalidate its use for high voltage separations. We have found that a sheet of polythene film 0·38 mm. (0.15 in.) thick as an initial covering for the aluminium base plate and a second sheet of the same thickness overlaying the plastic bars between which the block is poured provides adequate electrical protection without too greatly interfering with heat transfer. The use of two sheets also minimises the risk of "pinholing" since the likelihood of a pinhole fault in both sheets occurring in an identical position is small. The first plastic sheeting is run continuously across the base plate and beneath the electrode vessels. A thin coating of vaseline holds the bars in contact with this plastic sheet and prevents them slipping when they are themselves covered with the second sheet of plastic. When this is in position the block is poured.

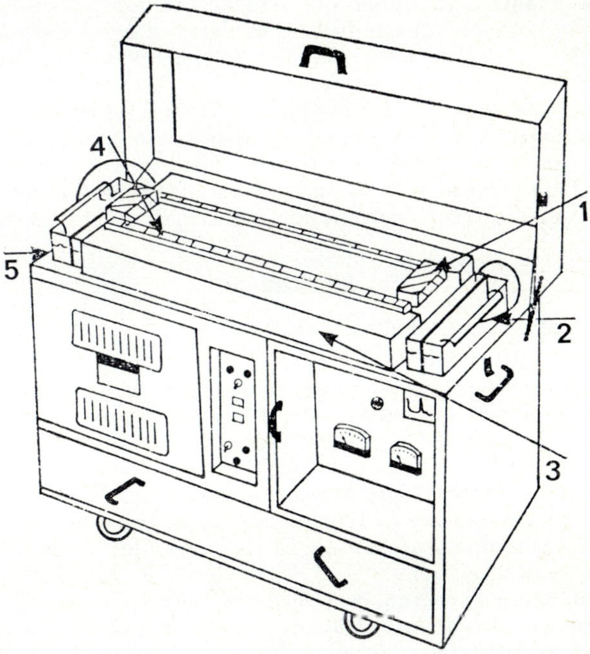

FIG. 14.4

Apparatus for high voltage preparative block electrophoresis.

The apparatus is shown with pads and plastic bars in position ready for the block to be poured but the plastic sheet covering the perspex bars and base-plate is not shown.

1. Labshammy or paper pad
2. Electrodes
3. Cooled Base-plate
4. Perspex bars
5. Electrode bath

Application of sample

It is important not to pierce the polythene covering when cutting a slit in the block. A safety razor is not therefore satisfactory. In general a blunt kitchen knife is used and care is taken not to cut to the extreme base of the block. This rule applies equally to trimming of the ends of the block before connecting it by wicks to the buffer tanks (see p. 370). To prevent evaporation from the wicks during the run they are covered with the same plastic sheeting which covers the completed block. The block is refrigerated to approximately 2° before the current is switched on.

Buffer Solutions.

For peptide mixtures volatile buffers are the most satisfactory, but where immunoglobulins and other proteins are to be separated salt buffers are needed to keep the proteins in solution. A simple and speedy method of evaluating a buffer for use in high voltage work is by means of a conductivity meter. A satisfactory model used in the author's laboratory is a direct reacting linear conductivity meter supplied by Chandos Intercontinental, Chandos Works, New Mills, Stockport, Cheshire.

The pyridine acetate buffer used in the method for peptide separation had a conductivity of 1·0 mmho. and this was the same conductivity as the dilute barbitone buffer used for the separation of serum globulins. For very high voltage work a conductivity of 0·5 mmho. can be recommended. Below this level current flow is too small to obtain a satisfactory result.

Applications

Separation of Serum Globulins.

The choice of buffer in this system is influenced by the problems involved in keeping the globulins in solution. Too low an ionic strength causes the precipitating out of some immunoglobulins and too high an ionic strength results in an unsatisfactory heat production. A barbitone buffer *p*H 8·8 and I 0·02 (conductivity 1·0 mmho) was found to be most suitable. For a block 33 cm × 1 cm.. a potential gradient of 70 V/cm. length is applied and the current remains constant at 6 mA/cm. width. No significant temperature rise occurs during a 6 hr. run with a starting temperature of 2°.

An 8 ml. volume of serum is applied to the cathodic side and after the run a print is taken by the method described in the section on separation of proteins p. 371. The portion of the block containing the protein is eluted in 1 cm. sections and the protein content of each fraction determined by its extinction at 280 nm. Immunoglobulins can be determined by an automated immunoprecipitin assay (Larson *et al.*, 1970) or by immunoelectrophoretic methods. A typical separation can be seen in Fig. 14.5. (Weiss & Brenchley, 1973.)

Separation of urinary peptides.

Small peptides from normal human urine are prepared by the method of Lote and Weiss, 1971. Peptides in a series of molecular weight ranges are eluted from a Sephadex G15 column and each range may be further separated by high voltage preparative block electrophoresis. A pyridine acetate buffer, *p*H 5·3 conductivity 1·0 mmho (100 ml. of pyridine and 32 ml. of acetic acid in 10 litres of distilled water) is suitable. Peptide material is generally applied to the centre of a block and a potential gradient of 110 V/cm. length is suitable. Typically at least

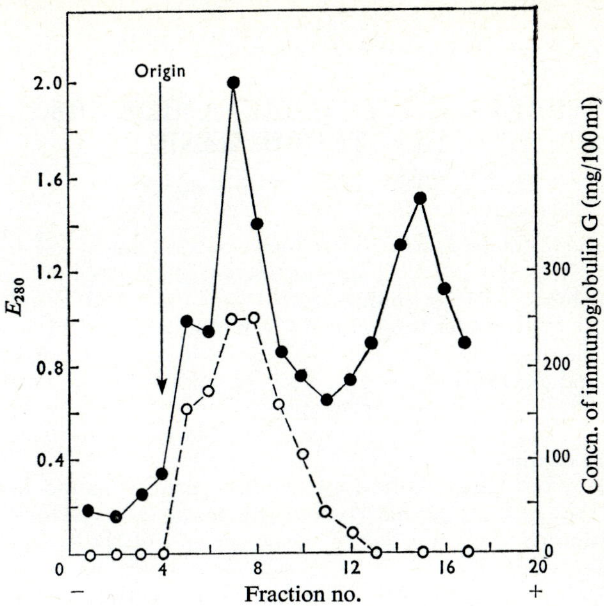

FIG. 14.5

Elution pattern of serum globulins separated by high voltage preparative block electrophoresis. (●, E_{280}; ○, concn of IgG)

IgG was determined by an automated immunoprecipitin assay. The shoulder in the immunoglobulin peak almost certainly represents denatured lipoprotein material remaining at the origin. Immunoelectrophoresis of each fraction indicated a slow IgG component in fraction 5 and two faster IgG components in fractions 7 and 11 respectively. Transferrin was the sole component of fraction 15. (Immunoelectrophoresis was carried out in 1 per cent agarose in barbitone buffer *p*H 8·6 and I 0·02 at 2·5 V/cm. for 100 min. Sheep anti-(whole human serum) serum (Wellcome Reagents Ltd., London N.W.1., U.K.) was used.

80 peptides may be separated in the MW range 600–1,500. The block is best eluted in 0·6 cm. sections for good resolution.

REFERENCES

1. Kunkel, H. G. and Trautman, R. "Electrophoresis", ed. M. Bier, Academic Press, 1959.
2. Flodin, P., and Kupke, D. W. *Biochim. Biophys. Acta*, 1956, **21,** 368.
3. Porath, J. *Biochim. Biophys. Acta*, 1956, **22,** 151.
4. Muller-Eberhard, H. J., and Kunkel, H. G. *J. Exper. Med.*, 1956, **104,** 253.
5. Muller-Eberhard, H. J. *J. Clin. Lab. Invest.*, 1960, **12,** 33.
6. Dische, Z. *J. biol. Chem.*, 1950, **183,** 489.
7. Munn, E. A., Feinstein, A., and Greville, G. D. (unpub.).
8. Larson, C., Orenstein, P. and Ritchie, R.F. 1970, in Technicon International Congress Vol. 1: Clinical, pp 91-94, Technicon Instruments Co., Tarrytown, N.Y.
9. Weiss, J. B., and Brenchley, P. 1973. Biochemical Society Transactions **1,** 571-574
10. Lote, C. J., and Weiss, J.B. 1971, FEBS Lett. **16,** 81-84

CHAPTER 15

PREPARATIVE ACRYLAMIDE DISC ELECTROPHORESIS

SECTION 1: *Ivor Smith*

THE excellence of separation of protein bands on analytical acrylamide gels, both disc and flat-bed, was an incentive to attempt to scale up the procedures, and techniques have been described based on both methods.[2–9] However the majority of methods described in the literature and the commercially available apparatus are based on a modified disc method as described below. Thus preparative disc electrophoresis (PDE) is based on the same principles as analytical DE and the major differences in apparatus design are those due to and derived from the scaling up of dimensions.

The main problems involved are the dissipation of ohmic heat in and through the enlarged gel matrix and the design of a method to collect the fractions as they run out of the lower end of the gel. Apparatus and methods described so far have been used to deal with quantities of 50–100 mg. of total protein although, as in section 2, considerably larger quantities can be treated without much enlargement of the gel but in a more versatile apparatus.

In the apparatus described here, the gel is polymerized in the annular space between two cooled surfaces; 20–40 ml. of gel are used, this corresponding to about 20–40 analytical gels. Protein bands emerge from the bottom of the gel into a Perspex, ring-shaped elution chamber through which buffer is passed slowly but continuously to a fraction collector, usually via some form of monitor such as a UV analyser.

Preparation of Buffers, Gels, etc. As this is a scaled-up version of the disc analytical procedure, all the buffers, gels, etc., can be used and prepared as previously described. Frequently no changes whatsoever are necessary other than the preparation of a somewhat larger volume of mixture in order to fill the larger piece of apparatus. The most frequent change, however, is the inclusion of urea into the gel matrix and this means inclusion of urea into all the components of the gel mixture; urea is not required in either the electrode or flush buffer. One effect of the urea is to alter the optimum monomer concentration for the separation such that a less concentrated gel is found suitable. To take one example, as used for the separation of bovine serum albumen shown in Figs. 15.3 and 15.4:

(*a*) Buffer 3 (see p. 219).

HCl, 10 N	4·8 ml.
Tris	36·3 g.
DMAP (or Temed) . . .	0·4 ml. (0·46 ml.)
Urea	36 g.
Water to	100 ml. (heat)

(*b*)	Monomer mixture:	
	Acrylamide	30 g.
	Bis	0·8 g.
	Urea	36 g.
	Water to	100 ml.
(*c*)	Urea	36 g.
	Water to	100 ml.
(*d*)	Ammonium persulphate	140 mg.
	Urea	36 g.
	Water to	100 ml.

Mix 100 ml. *a*, 20 ml. *b*, 10 ml. *c* and 40 ml. *d*; evacuate to deaerate and use immediately.

Electrode Buffer. Tris, 12 g., plus glycine, 57·6 g., up to 2 l. with water.

Flush Buffer. Tris, 52 g., acetic acid, 14 ml., up to 1 l.

Apparatus* For Preparative Electrophoresis

The apparatus consists of anode and cathode buffer chambers, an annular glass column in which the gel is polymerized between cooled surfaces, a Perspex (Lucite) elution chamber and a number of glass joints. All of the parts are clamped together to form a single, free-standing apparatus. All joints are smeared with silicone grease before use.

Electrode Chambers. Both the upper and lower chambers are rectangular Perspex (Lucite) boxes made to contain 1·0 and 1·5 l. respectively. Each includes a circular platinum electrode suspended within the chamber on a Perspex support. Connections from both chambers to the electrophoresis unit are via standard glass joints. Buffers are best removed from these chambers by suction via a water pump and any rinsing water should be removed similarly. However, in the apparatus described, the whole electrophoresis unit can be removed and inverted over a sink to pour out all the upper buffer. A buffer recirculation facility is included. It is essential to empty the upper chamber before removing it from the electrophoresis unit.

Electrophoresis Column. Electrophoresis is carried out in an all-glass unit of overall height approximately 25 cm. and built in the form of a vertical double-surface condenser. Thus the gel sits within an annular space such that it is cooled both externally and through its middle. The thickness of the gel is 6 mm. and the column height may be varied up to a maximum of about 12 cm. In order to retain the liquid gel within the column prior to setting, a porous foam-plastic ring, which just fits the annular space, is placed in position at the lower end of the column. The separated ions migrate through this ring unaffected. In order to ensure complete cooling of the total gel length, the inner cooling tube projects below the level of the plastics ring and outer flange into the elution chamber. Thus the electrophoresis chamber can stand vertically on a flat base and consequently, as an aid to filling the unit, a special glass plate is provided to ensure that the unit will stand vertically and so allow a true horizontal gel surface to form.

* This apparatus is currently being manufactured by Shandon Southern Instruments, Camberley, Sy., and Sewickley, Pa.

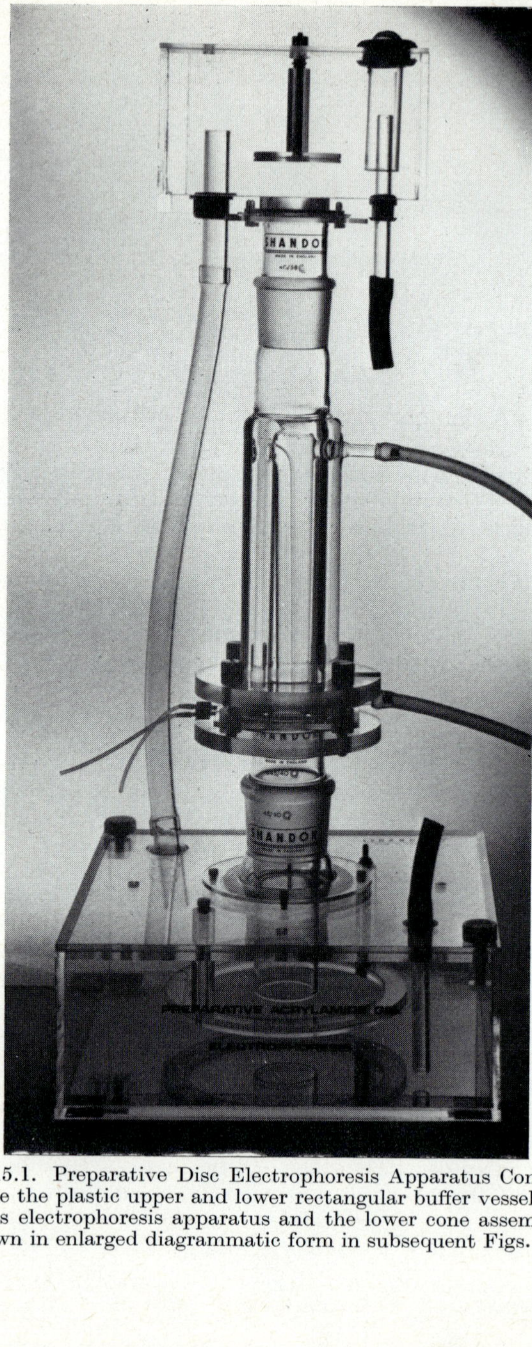

FIG. 15.1. Preparative Disc Electrophoresis Apparatus Complete. Note the plastic upper and lower rectangular buffer vessels; the glass electrophoresis apparatus and the lower cone assembly as shown in enlarged diagrammatic form in subsequent Figs.

The Annular Elution Chamber. The annular elution chamber is a Perspex disc 3 mm. thick and with a central hole, somewhat larger than the external diameter of the gel, drilled out. Two diametrically opposite tubes are inserted through the edge to provide an inlet and outlet for the flowing or flush buffer which carries away the separated components

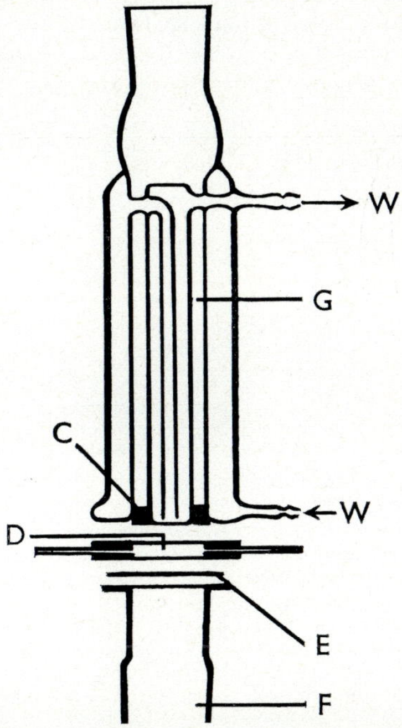

FIG. 15.2. Diagrammatic version of the glass electrophoresis unit and membrane area. W, water inlet and outlet for cooling; G, annular space in which gel is polymerized; C, the annular plastic ring; D, circular plastic unit through which the flush buffer flows. Note that the lower end of the glass unit fills most of the space in this unit and that the buffer volume is thus very small; E, dialysis or cellophane membrane. This flanged glass cone projects right down through the glass socket and into the buffer and is filled with gel to ensure contact between the glass chamber containing the electrophoresis gel and the buffer.

as they emerge from the gel column. The chamber rests on a sheet of dialysis tubing or cellophane which, itself, sits on a flanged cone. Thus a small cell is formed between the electrophoresis gel and the membrane which prevents any macromolecule passing through to the lower electrode vessel and ensures that all such large molecules will be swept out via the buffer. Buffer is continuously pumped through the chamber by means of a Delta or peristaltic pump at a rate which may vary between 20 and 1 ml. per hour as described subsequently.

The Lower Cone. The function of this lower cone is to ensure electrical continuity between the electrophoresis gel-elution chamber unit and the buffer in the lower electrode vessel. As a liquid contact would fall back into the lower vessel, the cone is filled with large (2–3 per cent) or small (7 per cent) pore gel polymerized as usual. The dialysis membrane sits squarely on the gel surface to ensure good contact.

This lower cone sits in and projects through an upward projecting socket. The length of the lower cone is some 16 cm., its diameter 3·5 cm. and its volume approximately 180 ml. Such a dead volume of gel seems unsatisfactory but can be reduced, if desired, in two ways. First, using the same cone, a flat-stoppered test tube can be placed

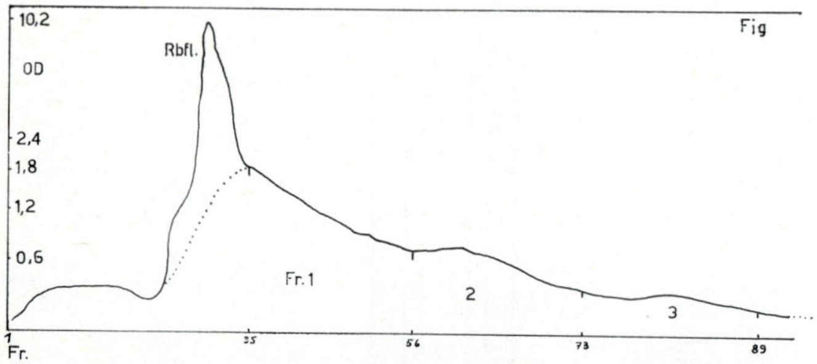

FIG. 15.3. Separation of oligomeric forms of 30 mg. bovine serum albumen. Height of gel 8 cm.; flush buffer 12 ml./hour unchanged; pH 8·5 using urea buffer described in the text. In this experiment the sample was applied above a small spacer gel at 2·5 per cent concentration and containing the Tris-phosphate buffer described earlier, the spacer gel was polymerized with riboflavin. Riboflavin being a small fluorescent molecule always travels out with the front and forms an effective marker if the albumen is either not stained or absent from the sample. All acrylamide contains fluorescent impurities which can be removed by recrystallization as described in Chapter 12 or, with continuous buffer systems, by a preliminary run before sample application.

vertically within the cone before the addition of gel monomer mixture such that the gel will polymerize to form an annular ring around the tube in a manner similar to that formed by the electrophoresis gel around the internal cooling tube. Alternatively, the long lower cone can be replaced by a similarly flanged but short cone without a lower joint and, after removing the upward projecting socket attached to the lower electrode vessel, inserted directly into the buffer. In this case, however, the separation chamber will need its own stand to support it.

Clamping Rings. The electrophoresis column, elution chamber and lower cone are clamped together with a pair of rings to ensure maximum apparatus stability and freedom from leaks. Freedom from leaks is also ensured by adequate silicone greasing of all joints as mentioned elsewhere.

Setting Up the Apparatus.

1. The lower cone (original or modified version). Silicone grease the flange of the cone and stand it, flange down, on a glass plate. Prepare sufficient volume of gel mixture and, after ensuring that no bubbles are trapped by quickly degassing, under vacuum if necessary, completely

fill the cone. Cover the upper end with a piece of polythene film to ensure polymerization of the surface; polymerization should be complete in about 15 minutes. This unit may be used for a number of identical experiments especially if continuous buffer methods are used. Between experiments, the gel must be prevented from drying out and this is best done by keeping it in the cold in a plastic bag containing water or buffer.

2. Circular dialysis membranes of diameter equal to that of the cone flange should be cut and soaked in water in a closed vessel until required.

3. Filling the electrophoresis chamber with gel. Soak the plastic ring in water or diluted buffer (or gel mixture); this should be done in a vacuum flask to ensure removal of all bubbles which might have been present in the ring. Place the ring on a counter-bored Perspex plate so that it is concentric with the depression, using forceps to manipulate the ring. Grease the ground base of the chamber and stand it on the plate so that the protruding centre cooling tube fits into the depression and the ring just fits into the annulus. Pipette 20–40 ml. gel mixture gently into the annulus above the ring which will give a column of 5–10 cm. height. Carefully but quickly overlay the liquid gel surface with water as follows: fit a length of flexible plastic tubing to a 2 ml. pipette and, having filled it, lower the end of the plastic tubing so that it rests just above the liquid surface. Allow the water to flow out as gently and evenly as possible, raising the pipette to keep the plastic tube level just above the water surface. A sharp refractile boundary should form at the interface if the operation has been correctly performed. Polymerization should be completed in about 15 minutes. The water layer is then removed by suction, using the original pipette. The gel surface can be rinsed with upper electrode buffer if desired but this is not generally necessary.

4. Setting up the flow cell. Carefully slide the lower cone off its glass plate and invert it; regrease the flange if necessary. Remove and blot a dialysis membrane and place it in position on the gel surface, smoothing out any air bubbles or creases to ensure good contact between the membrane and gel. Grease both sides of the annular elution chamber and place it down onto the membrane on the flange. Remove the electrophoresis chamber with its attached plastic ring from its base plate and place on the annular chamber. Clamp all the parts together with Perspex clamp rings. Depending on whether the original or the modified cone is used, set up the clamped unit in its final position in relation to the lower electrode vessel.

5. Start the cooling water flowing as fast as possible through the electrophoresis cell; ordinary tap or refrigerated water can be used.

6. Connecting the pump. Connect one side of the elution chamber to a syringe containing flush buffer and force buffer into the cell, tilting the apparatus if necessary, to ensure the elimination of the last air bubbles from the cell. Remove the syringe and connect to the flush buffer reservoir via the peristaltic pump ensuring that no bubbles are introduced at this stage; an air bubble trap can be inserted between the pump and the cell. Connect the exit tube of the cell to a monitoring device such as a UV analyser which leads to a fraction collector;

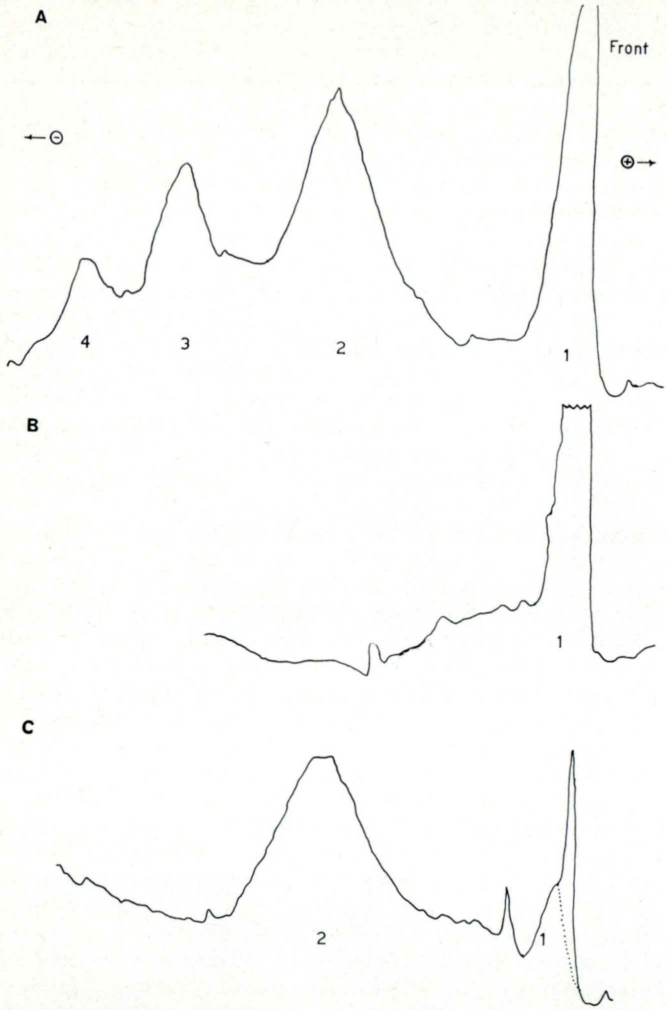

FIG. 15.4. Densitometric analytical disc electrophoresis tracings of A, original BSA sample, B, recovered fraction 1, and C recovered fraction 2; fractions 3 and 4 fall as expected and the curves are not included (see Fig. 15.3). The vertical axis represents optical density obtained by means of a Joyce-Loebl double beam photometer; the gels are placed between two microscope slides and clamped together such that they are slightly flattened and present a better surface to the scanning unit. Quantities of sample applied were: A, 0·25 mg. BSA; B, 0·35 mg. from fractions 35–55; C, 0·35 mg. from fractions 56–72 (Fig. 15.3). Samples recovered from the centres of peaks show no cross contamination.

effluent buffer can be run directly to the collector if no monitor is available or necessary.

7. Application of the sample. A volume of sample containing not more than 60 mg. protein can now be applied to the gel surface using one of two methods. A liquid sample, such as serum, should be treated with solid urea, sucrose or sephadex plus a few drops of 1 per cent bromophenol blue such that its density is greater than that of the upper buffer; it is then applied to the gel surface by overlaying as described in 3 above, in the polymerization stage. Alternatively, the sample solution is mixed with an equal volume of gel monomer solution (containing 5 per cent acrylamide and 1·25 per cent BIS in buffer-urea) and polymerized on top of the running gel; this latter procedure is rarely necessary.

8. Addition of upper buffer and electrode vessel. Upper buffer is taken up in a 25 ml. pipette, the tip is placed on the top surface of the inner cooling tube and the buffer allowed to flow out slowly. Because of the lower density of the buffer (no urea) it floats easily on the surface of even a liquid sample. Satisfactory addition can be observed particularly well when the sample contains bromophenol blue—minor imperfections are of no importance as the greater mobility of proteins in free solution more than compensates for this. Further quantities of buffer are applied until the level is above that of the top of the cooling tube. The upper electrode chamber is placed in position and further buffer added until about 1 l. is in position in the chamber.

9. Current. Switch on the current, anode in the lower chamber, and apply approximately 30 mA for about an hour or until the total sample has passed into the gel surface. Increase the current to 60–80 mA which will require about 500 V in the original apparatus described here.

10. Fraction collection. As the BPB reaches the annular cell, commence flush buffer flow of 12 ml./hour and, for serum, run for up to 20 hours. The buffer flow can be diminished stepwise by, say, 2 ml./hour every 3–4 hours.

11. Subsequently fractions may be concentrated and desalted by one of the methods previously described or by the procedure referred to later. Resulting samples can be prepared free of salts and urea for final freeze-drying. The separation is, of course, controlled by the analytical DE.

12. Dismantling and cleaning. Buffer should be sucked out from the upper electrode chamber, via the water pump, which is then lifted off the electrophoresis chamber. Likewise the annular chamber can be sucked free of flush buffer after disconnecting from the pump supply. Separate the electrophoresis chamber from the lower electrode vessel. Unclamp and separate the cone, cell and chamber. Store the lower gel, if desired, in a moist plastic bag in the cold.

The gel is removed by very carefully rimming and rotating as with the analytical tubes and using a long thin probe and lubricant. When the gel is freed from the outer wall, a jet of water will help to remove it from the apparatus. Any siliconing should be restricted to the lower 5–10 cm. so that flow of buffer down onto the sample surface (see 8 above) should not be adversely affected.

General Discussion. Preparative procedures may have the aim of isolating a single component amongst a few, as in the case of an abnormal hæmoglobin, or amongst many as in the case of an enzyme present in whole serum. The approach to isolation of a component will therefore vary and a number of the factors which affect separation and band width will be considered. Those factors which have been discussed under the analytical procedure have a similar effect here and so the discussion will consider mainly the additional factors present in the preparative set up.

With whole serum, some 16 hours may be required for the emergence of all those proteins which can penetrate into the gel matrix although the albumen will have come through rapidly. During this time, the slower migrating proteins will have undergone some diffusion in the gel and, because they also emerge slowly, will appear spread over many fractions. In order to overcome this, the rate of flow of flush buffer is usually reduced in a stepwise manner from the original of, say, 10–20 ml./hour down to 1–2 ml./hour in two or three steps. Alternatively the gel concentration or height can be reduced but this may cause a bunching of the first fractions.

In the separation of bovine serum albumen shown here, the well separated fractions correspond to monomer, dimer, trimer, etc., and there is consequently plenty of scope for alteration of flush buffer flow, gel concentration and height of column. In a like manner, Jovin *et al.*[2] obtained excellent separations of 40 mg. of hæmoglobins A and S (which differ by two units of charge per molecule) through 50 ml. of gel and Maizel separated 20 mg. quantities of ribonuclease, trypsin and chymotrypsin when 60 mg. of mixture was applied to a 10 ml. gel in 8 M urea. With straightforward separations of this type, the effluent can be continuously monitored with a UV analyser such as the Uvicord (LKB) and, subsequently, individual fractions can be bulked for investigation.

The approach to the isolation of one component in a complex mixture is rather different and, to some extent, depends on the concentration of the substance sought. Thus, in an attempt to isolate a low concentration enzyme from serum, the gel could be overloaded considerably in the sense that individual protein bands might form a continuous overlap when investigated by the Uvicord but the enzyme could be found in the sequence via its activity. It would not be necessary to investigate the whole series of fractions as the compounds emerge in the same sequence as found in the analytical separation and this would serve as a good guide; the existance of isozymes would, of course, complicate this procedure and it might then be simpler to investigate the whole sequence of fractions. Subsequently the bulked enzyme fraction could be re-run in a more concentrated gel which would have the effect of further separating the enzyme from those proteins which previously ran similarly. Furthermore, the application of a somewhat smaller total quantity of protein would allow a greater sharpening of bands which could be heightened by the use of a faster flush buffer flow rate. If the components of interest appear in the early part of the separation then a somewhat longer or more concentrated gel column can be used whereas if they are in the rear a quite short or less concentrated gel column would be satisfactory and, in this latter case, would result in a much quicker and sharper separation. In all cases, the degree of purification and separation can be determined by the analytical technique.

Urea is used to depolymerize macromolecules and, whenever it is incorporated into the gel, it should also be incorporated at the same or somewhat greater concentration in the sample by the addition of solid urea. Further, when the sample is applied as a solution directly to the gel surface it must be as a more dense liquid than the electrode buffer; samples containing urea are necessarily more dense than the buffer as the latter need not contain urea but, if other reasons exist, the density of sample can be increased by the addition of sucrose, ficoll, sephadex, etc., as described for the analytical procedures.

Concentration of Separated Fractions. The separated fractions may be present in quite small or quite large fractions. Small fractions may be freeze-dried but larger fractions may often require a preliminary concentration. The larger fractions may also contain considerable quantities of buffer salts which it is usually desired to remove. In Chapter 1, a number of concentration procedures are described and methods using acrylamide gel rods prepared with water or micro-dialysis are particularly suitable. A further method described by Jovin *et al.*[(2)] is also worthy of mention. All of these methods also result in removal of most of the buffer salts.

A Pyrex tube is closed with a rubber stopper which penetrates 3–4 mm. into the tube. A suitable volume of 3 per cent gel containing buffer is polymerized in the tube and, when set, the stopper is removed. A dialysis membrane or Viscap is placed over the gel end of the tube and held in position with a Tygon sleeve or rubber band, thus forming a small, lower chamber. The bulked fractions are placed above the gel, in the tube which dips into buffer, and covered with a layer of fresh buffer; if necessary urea or sucrose is added to the bulk. Electrophoresis is carried out and the protein collects in the small volume between the gel and the membrane. After protein migration, which will be rapid in the dilute gel, the tube is removed and inverted. The membrane is removed and the sample withdrawn. The dimensions of the tube can be varied within wide limits depending on the size of the bulked fractions and the quantity of material present. Cooling can be used, if necessary, in an apparatus such as a small condenser. Dialysis can be interposed before removal of the membrane. Finally the sample is freeze-dried.

PREPARATIVE ACRYLAMIDE GEL ELECTROPHORESIS

SECTION 2

A. D. Brownstone

THE use of polyacrylamide gel as a support medium for the electrophoretic separation of mixtures of charged macromolecules has become widespread because of the excellent resolution it affords and the ease with which it is made and handled. Since its introduction[1] numerous designs of apparatus for use of the gel on a preparative scale have appeared[2–9]. With a few exceptions[8,9] the gel is used in the form of a vertical cylinder contained in a tube fitted with a cooling jacket. Some designs also employ a vertical cooling finger to assist in the removal of the ohmic heat produced. It is an unfortunate corollary of this type of cooling that a radial temperature gradient is set up across the gel and this causes band distortion, – material migrating faster in the warmer region. In designing an apparatus for preparative use a major problem is the recovery of the separated zones without affecting the resolution obtained in the gel and without excessively diluting the recovered-material. Most published designs allow the separated zones to move out of the lower surface of the gel cylinder into a continuously flowing stream buffer which transports the material to a fraction collector. In some designs a dialysis type of membrane is used to prevent migration of the material out of the collecting chanber whilst in others contact of the material with a membrane is avoided.[3,4,8] Since this type of membrane has a slight negative charge, *p*H and buffer ion concentration changes take place in its vicinity (Bethe-Toropoff effect) which, under some conditions, have been reported to risk denaturing certain proteins.[8] However, this effect may be minimised by increasing the concentration of the eluting buffer and/or treating the membrane to remove the charge.

The apparatus described below employs an intermittent collection system and the largest of three sizes of the apparatus in regular use is capable of handling up to about 1,500 mg. of protein. Since the original description was published[7] some modifications have been made which improve the resolution and increase its capacity.

General Description.

The gel is cast in a mould as a cylindrical block. Three sizes of mould have been used – namely, 6 cm, 9 cm and 14 cm in diameter. Each of the moulds may be adjusted to cast gels 4 cm., 5 cm. or 6 cm. long. For use, the block of gel is transferred to a tubular gel holder about 17 cm. long, the lower end of which is closed by a dialysis membrane (Fig. 15.5). The diameter of the tube is 5 per cent less than that of the corresponding mould so that the gel is slightly compressed as it is fitted into the holder. The gel is initially supported in contact with the membrane. Small diameter plastic tubes passing vertically through the gel serve as inlet and outlet tubes for the collecting buffer. The lower part of the gel holder is surrounded by a transparent heat insulating jacket which considerably reduces heat losses from the walls of the gel and so cuts down the formation of radial temperature gradients. Since cooling occurs mainly from the upper and lower surfaces of the gel, the longitudinal temperature gradient so formed does not give rise to band distortion, but it does impose a practical limit of about 6 cm. to the length of the gel.

The gel-holding tube is supported in a tank of buffer, the temperature of which is thermostatically controlled. Buffer is pumped from the tank into the space above the gel and returns to the tank through a hole in the wall of the tube about 1 cm. in diameter and 4 cm. from the upper end of the tube. The level of buffer in the the tank corresponds to the hole so that there is no hydrostatic load on the gel. Electrodes are positioned about 3 cm. from the upper and lower surfaces of the gel. The design of the electrodes ensures that the electric field between them is as uniform as possible over the whole area of the gel upper and lower surfaces.

Intermittent collection system.

A predetermined volume of the collecting buffer is pumped via the inlet tube to between the lower surface of the gel and the membrane, where it spreads out into a thin layer. After remaining there for a preset time, which may be varied from 7 to 50 minutes, it is removed via the outlet tube and passed through a pressure sensitive switch, pump and UV absorptiometer to a fraction collector. When all the fluid has been pumped out, the gel and the membrane are in contact and the drop in pressure in the system opens the presure sensitive switch which in turn stops the outlet pump and starts the inlet pump. The latter runs for a preset time pumping in fresh collecting buffer.

The efficiency of the system is such that all the protein which has emerged from the gel in one electrophoresis period is removed from the system in three pumping cycles. The fraction volumes are approximately 1 ml. for the 6 cm. diameter gel, 2·5 ml. for the 9 cm. diameter gel and 6 ml. for the 14 cm. diameter gel. In order to prevent protein being held against the membrane by the electric field, the power supply is first of all reversed for 10 seconds and then turned off whilst the outlet pump is working.

It can be seen from the above description that whilst it would be possible to operate the system manually, automatic timing and control of the pumps, power supply and fraction collection is highly desirable. A circuit diagram for a suitable control unit will be found at the end of this chapter.

Gel Temperature.

Since most of the heat developed in the gel is removed from the upper and lower surfaces there will be a temperature gradient rising to a maximum half way between them. The actual temperature will depend on a number of factors: (1) Temperature of the tank buffer (2) wattage being dissipated between the electrodes (3) length of gel (4) time interval between collecting cycles.

For example, the maximum temperature in a 5 cm. long gel was 14° when the tank temperature was 2°, the wattage was 40, (200 V; 200 mA), and the time interval was 15 minutes. The buffer in both the tank and gel was 0·025 M TRIS/Glycine at *p*H 8·7. The *radial* temperature difference was 4°. (With no insulating jacket the radial temperature difference was 12°).

Constructional Detail

Gel support tube (Fig. 15.5a).

This may be of glass or a suitable transparent plastic material such as Perspex. The wall thickness should be 5–6 mm. if plastic or 3–4 mm if

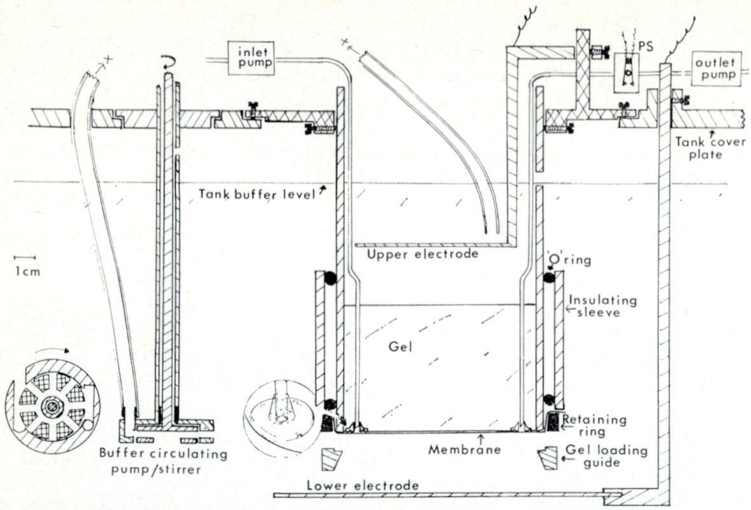

FIG. 15.5*a*. Showing how the insulated gel-holding tube, electrode and buffer recirculating pump, are mounted on the tank cover plate.

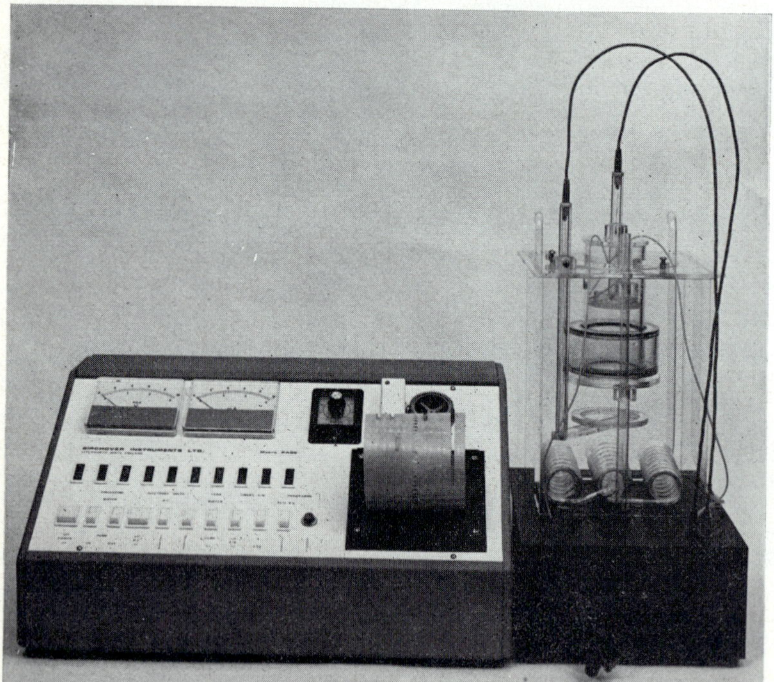

FIG. 15.5*b*. Commercially-available apparatus from *Birchover Instruments Ltd., The Spirella Building, Letchworth, Herts, SG6 4ET, England.*

glass. The lower end of the tube should be machined to be at right angles to the axis of the tube and the edges slightly bevelled to avoid cutting the dialysis membrane which closes this end of the tube. The length of the tube is 17–18 cm. and a 1 cm. diameter buffer overflow hole is bored through the wall 4–5 cm. from the upper end. The tube is supported in the tank by a Perspex ring fixed near the top of the tube by three nylon screws. The outer edge of the ring is stepped to fit in a suitably sized hole cut in the Perspex tank top. Three vertical levelling screws are fitted to the ring so that the surface of the gel may be adjusted to be horizontal.

Heat insulating jacket.

This is plastic tubing, similar to that used for the support tube, the internal diameter should be approximately 5 mm. greater than the external diameter of the support tube. The length of the jacket is about 8 cm. and it is sealed to the support by two suitably sized "O" rings, one at the top and the other at the bottom of the jacket. It is essential that these seals are water-tight so that the air space between the tubes is kept dry. In order to accommodate the membrane clamping ring, the lower end of the jacket is 1·3–1·5 cm. above the bottom of the support tube. This exposed part of the support tube, if Perspex, should be machined so that the outside is perfectly round and the wall thickness is reduced to 3 mm.

Membrane clamping ring.

This may be machined from 1 cm. thick Perspex or polypropylene. The internal wall is machined to have a 2° taper and the larger diameter end should just fit over the support tube with a wet membrane in place. As the ring is pushed on further, the membrane is gripped and stretched tightly over the end of the tube. For the smallest diameter of gel (6 cm.) the ring may be obtained by cutting a section from a suitably sized disposable polypropylene beaker. Fabricated rings should be 10 mm. × 3–5 mm. in section.

Membrane.

This is a single layer of dialysis membrane cut from $2\frac{3}{4}$ inch "Visking" dialysis tubing. Before fitting to the apparatus it should be thoroughly washed and soaked in water or the tank buffer.

Owing to the slight residual negative charge on this type of membrane it will be found that when using buffer systems at *p*H 7 and above, and with the anode below the membrane, there will be an electroosmotic flow of fluid through the membrane into the collecting chamber. The rate of flow will be greatest at high *p*H and with a high current passing. It will also be found that the *p*H of the collecting buffer may drop slightly and its conductivity increase.

When using an acid buffer, the reverse takes place, i.e. fluid is removed from the collecting chamber and the conductivity drops. This is a much more troublesome situation because the membrane may stick to the gel and prevent the proper functioning of the collection system, also the resistance of the collecting buffer layer will increase and may cause a cessation of the electrophoresis process if the time interval between pumping cycles is too long. Fortunately it is possible to treat the membranes chemically to overcome the above effects. Whilst this treatment is not essential when working above *p*H 7, the membranes must be treated for use below *p*H 7.

By treatment of the membrane with diethyaminoethyl chloride-hydrochloride, fixed positively-charged groups are introduced and it is possible to produce a membrane having a net positive charge. The quantity of the reagent required will have to be found empirically as it is dependent on the membrane in use and the exact condition of the treatment. A typical procedure is as follows: A suitably sized disc of the membrane is cut and weighed. It is then washed thoroughly in—four changes of distilled water for a total period of 30 minutes. Excess water is then removed from the surface using a filter paper or tissue, and the membrane is then immersed in *1N* NaOH (10 ml. per gm. of membrane) contained in a petri dish the diameter of which is only slightly larger than that of the membrane. The membrane is soaked in the NaOH solution for ten minutes, being turned over once or twice during this period.

Diethylaminoethyl chloride. Hydrochloride (DEAE, Cl, HCl) is dissolved in water to make a solution containing 10 mg./ml. To make an approximately neutral membrane, the amount of DEAE, Cl. HCl required is approximately 0·75–1·5 per cent of the weight of the membrane. The volume of the above solution containing this amount is added in ten equal aliquots at five-minute intervals to the NaOH solution. After each addition the lid of the dish is quickly replaced and the dish is gently rocked to ensure an even distribution of the reagent. Before each further addition the membrane is inverted and after the final addition the dish is rocked for a further 10 minutes. The membrane should then be washed in water until the washings are neutral.

In order to check the charge stage of the membrane, it is partially immersed for five minutes in an 0·01 per cent aqueous solution of Bromophenol Blue in water at about *p*H 7·5 (just enough TRIS being added to attain this *p*H). Another part of the membrane is then immersed for five minutes in an 0·02 per cent aqueous solution of Pyronine Y. The membrane is then washed in water until it is found that one or other of the dyes has washed out. A slight residual blue colour on the membrane indicates that it is positively charged whilst a slight pink colour indicates a negative charge. For use at *p*H 7 and above the membrane must retain a slight negative charge, whilst below 7 it must have a slight positive charge.

Acetylation of the membrane: Under conditions of electrophoresis the membrane becomes permeable to molecules having molecular weights below 17,000–20,000. In order to avoid such losses the membranes may be acetylated.[10]

A suitably sized piece of dialysis membrane is immersed in a mixture of 60 pts (by volume) of redistilled pyridine and 40 pts of acetic anhydride. The mixture may be heated at 60° for 6 hours to provide a very "tight" membrane, or left at room temperature overnight giving a slightly less tight membrane. The membrane is removed from the reaction mixture and rinsed in water for a short time to remove reagents on the surface.

Whilst still flexible it is fixed to the gel-support tube in the normal way and soaked in 0·1 N acetic acid until there is no further smell of pyridine. The membrane will now be hard and tough. The retaining ring is removed and whilst still on the tube the membrane is further washed with acetic acid to remove any pyridine trapped by the retaining ring. The membrane may now be removed from the tube and stored in water until required. When refitted to the tube a thin smear of vacuum grease on the end of the tube will prevent any small leaks which might otherwise occur.

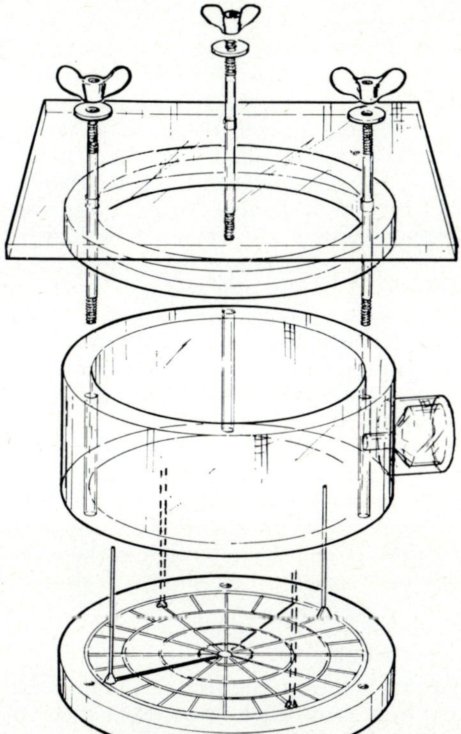

FIG. 15.6. Gel mould. The patterned base plate, cylindrical section, optional extension ring, and plain top plate are held together by 3 screwed rods. A light smear of grease on the mating surfaces prevents leakage. Note: In the case of the 14 cm. diameter mould, four fixing screws are used and four vertical channels are cast in the gel. (Reproduced by permission of Academic Press Inc., New York. Publishers of *Analytical Biochemistry*).

Gel mould (Fig. 15.6).

The mould is constructed mainly of Perspex or similar transparent material to which polyacrylamide gel does not adhere. It consists of three main parts – a plain upper plate, the cylindrical section and a base plate which is embossed so that a series of shallow channels are cast in the lower surface of the gel. These parts are clamped together to form a leak-free unit.

Base plate.

In order to make the patterned section, a negative is first made – aluminium sheet about 3 mm. thick being a suitable material. The channels are cut as shown; all channels are 0·5 mm. deep and wide except one which is 1 mm. deep (solid black in diagram). The central "well" is 3–5 mm. in diameter and 1 mm. deep. The diameter of the outer wall of the outer channel is the same as the internal diameter of the cylindrical section of the mould. Care must be taken to see that all the channels are properly interconnecting. The negative is carefully coated with a release agent such as a thin layer of silicone grease, and an outer wall of modelling clay (e.g. Plasticine) is made to retain the cold setting resin which is used to make a positive plate 8–10 mm. thick (a suitable resin being Holts Cataloy). The diameter of the plate should be about 2 cm. larger than that of the outer channel. Three holes are drilled into the resin plate and tapped to take the screwed rods which hold the mould together.

To cast vertical channels through the gel for the inlet and outlet tubes, stainless steel wires about 1 mm. diameter are held in Perspex mounting pieces which are themselves cemented into holes drilled through the plate. The mounting piece is formed from 5 mm. diameter Perspex rod one end of which is cone shaped. A hole is drilled along the axis of the rod into which is pushed a piece of silicone-rubber tubing 1 mm. bore, 1 mm. wall thickness. The stainless steel wire is pushed into the bore of the tubing. The outlet tube ends in the 1 mm. deep radial channel about 3–4 mm. in from the outer circular channel; the inlet tube ends in one of the short radial channels almost diametrically opposite to the outlet tube. In the case of the 14 cm. diameter mould there are two diametrically opposite inlet and outlet tubes, shown in dotted outline on the drawing (15.6).

Cylindrical section of mould.

This may be constructed from suitably sized Perspex tubing or from a number of rings cut from 1 cm. thick sheet. Four such rings are cemented together to form the basic mould and are turned to size in a lathe. Extra rings may be used to increase the length of the mould. The internal diameter of the mould should be 5 per cent greater than that of the corresponding gel-support tube. A filling hole, about 1 cm. in diameter is bored through the cylindrical wall and a small external reservoir, 10–15 ml. capacity, is fitted.

Top Plate.

This is a plain, 1 cm. thick Perspex sheet, having three equally spaced holes drilled through it to take the three screwed rods and nuts which hold the three sections of the mould together. Leakage is prevented by a slight smear of grease on the mating surfaces. The top plate is square, the side dimension being the same as the diameter of the bottom plate; it is positioned so that the mould can be stood with the end plate vertical and reservoir uppermost for filling.

End Pieces. (Fig. 15.5).

These are made from 8 mm. diameter rod. The flat face has three diametric channels cut with a fine saw. The rear part is cone shaped having identical dimension to the conical wire supports in the base plate. The axial hole is such that the end pieces are a push fit onto the plastic inlet and outlet tubes (for translucent vinyl No. 2, tubing from Portex Plastic, a No. 48 drill is suitable). A small section of nylon bolting cloth (120 mesh) may be cemented to these end pieces to prevent gel being sucked into the outlet tube.

GEL SUPPORT

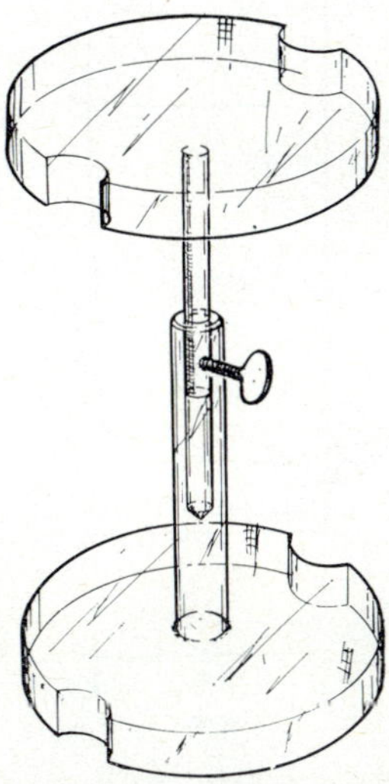

FIG. 15.7. Adjustable gel support (see text). (Reproduced by permission of Academic Press Inc., New York. Publishers of *Analytical Biochemistry*.)

Adjustable gel support (Fig. 15.7).

This is used to support the gel in the correct position whilst the dialysis membrane is being fitted. It is constructed from Perspex and the diameter of the plate is slightly less than that of the gel support tube into which it fits. The two cut-out sections are to accommodate the inlet and outlet tubes.

Gel loading guide.

This aids the loading of the gel into the support tube. It is a ring cut from 1 cm. thick Perspex sheet. The internal bore of the ring is coned so that the smaller diameter is the same as that of the tube whilst the larger diameter is slightly larger than that of the gel. The smaller diameter side of the ring is stepped to give easy location on the end of the tube. In use it is simply held onto the lower end of the support tube while it is pushed down over the gel – the inlet and outlet tubes having been threaded through the ring and the tube.

Pressure-sensitive switch.

This is used in conjunction with an automatic control system which is described in outline below (Fig. 15.8). It is mounted between the

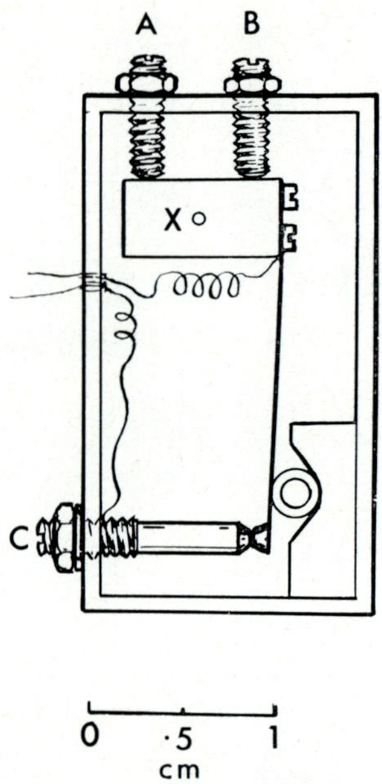

Fig. 15.8. Pressure-sensitive switch. (see text.) The spring strip contact is mounted on a small block which can pivot about an axis at X. The pressure it exerts on the tube is adjusted by screws A and B. The contact points should be gold-plated. (Reproduced by permission of Academic Press Inc., New York. Publishers of *Analytical Biochemistry*).

outlet pump and the gel assembly and consists of a short length of silicone rubber tubing which holds two contacts in the closed position. One of the contacts is mounted on a spring strip which presses on the tube. When all the fluid in the system has been removed the drop in pressure causes the tube to collapse slightly and allows the contacts to open. The silicone rubber tubing is 1·5 mm. internal diameter and 0·5 mm. wall thickness. The sensitivity may be improved if the section of tubing under the spring has been thinned by gently holding against a revolving grindstone. The spring strip should be adjusted until it is exerting a slight pressure on the tube. The screw mounted contact is then adjusted until the contacts just touch.

Inlet and Outlet Pump.

Those used with the original apparatus were finger pumps having an adjustable flow rate of from 0·5 to 3·5 ml./min. However simple peristaltic pumps having a similar flow rate have been used with success.

Tank buffer recirculating pump. (Fig. 15.5).

This is a simple centrifugal pump immersed in the tank. It is constructed of "Perspex" and serves two purposes, namely, to prevent the accumulation of gas bubbles under the membrane and to recirculate the buffer around the electrodes and gel as described above. When driven at about 1,000 r.p.m. the output from the recirculating side is 1 l./min.

Tank.

The size and capacity of the buffer tank will depend on the largest diameter gel which is to be used. Allowance must also be made to accommodate the cooling arrangements and recirculating pump. A tank suitable for all three gel sizes described may easily be constructed from 5 mm. thick Perspex sheet. Approximate dimensions are 19 cm. deep $\times$ 30 $\times$ 20 cm. A cover plate made from 1 cm. thick Perspex has holes cut out to take the recirculating pump, gel support tube and cooling arrangements. The lower electrode is also supported on a Perspex rod mounted through the tank top.

Electrodes.

These consist of 20 s.w.g. platinum wire wound toroidally round a Perspex ring 10 mm. wide cut from 3 mm. thick sheet. In addition, four equally spaced wires run diametrically across the ring. The diameter of the lower electrode should be the same as the largest gel to be used, whilst the upper electrode should fit the gel support tube with just enough clearance for the inlet and outlet tubes to pass.

Operation

Buffer system.

Buffer systems for use on the preparative scale are, with one or two reservations, similar to those used on the analytical scale. It should be remembered that since the buffer solution is continuously recirculated any electrolytic decomposition products which may be formed will pass through the gel. For this reason alkaline buffers containing chloride should not be used because of the electrolytic producton of hypochlorite.

Buffer having a high absorbtion in the 280 nm. regions, such as veronal, cannot be used if it is intended to detect the presence of protein by a UV absorptiometer. If possible, buffer systems having low conductivity should be chosen, in order to keep the time of the run to twenty-four hours or less. However, it is possible to use more highly conducting systems, and runs taking up to three days, if such a system is considered essential.

Most preparative scale experiments will be found to give satisfactory results when a continuous buffer system is used i.e. the same buffer composition is used in the gel, tank, and as collecting buffer. However, a discontinuous system may be used if required; in this case, removal of ionic impurities from the gel by pre-electrophoresis before the protein is

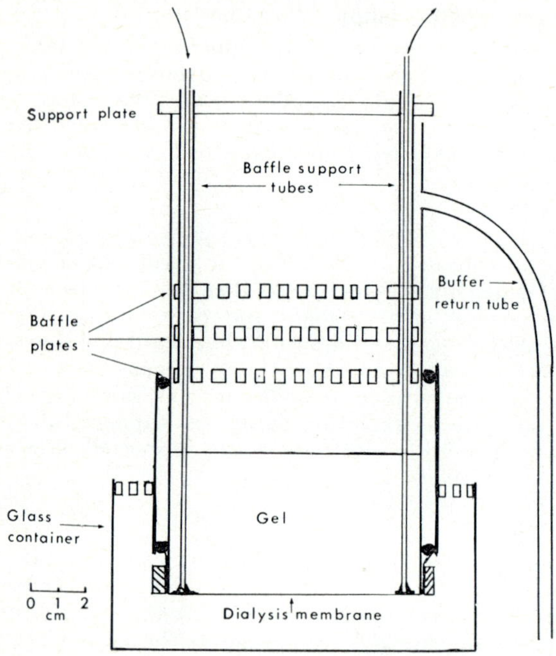

FIG. 15.9. Adaption of the apparatus for use with gels containin urea. The flow rate from both outlets of the circulating pump may have to be reduced in order to avoid mixing urea with the tank buffer. (Reproduced by permission of Academic Press Inc., New York. Publishers of *Analytical Biochemistry*.)

loaded, is carried out with the same buffer in the tank as is used in the gel, and the tank buffer is then changed. Use of sodium dodecyl sulphate presents no difficulties but the temperature of the buffer solution should *not* be *below* 10–12°C.

Urea-containing buffer may also be used, but the urea is not allowed to escape from the immediate vicinity of the gel. This is accomplished by inserting baffle plates in the tube between the gel and the upper electrode and surrounding the lower part of the gel support tube with a container to trap any urea which dialyses through the membrane (Fig. 15.9). A layer of urea-containing buffer about 1 cm. deep is placed over the upper surface of the gel. The collecting buffer also contains urea, but not the tank buffer. Should urea escape into the tank, electrolytic decomposition at the anode will give rise to a continuously increasing UV absorption.

When using buffers having a high conductivity in dilute solution, only a relatively low voltage (20–40 V) can be used to avoid excessive heating. This means that the mixture to be separated may take one or more hours to move into the gel from the solution in which it is loaded. By this time these will have considerable change in the ion concentration above the gel, accompanied by a possibly drastic *p*H change. This may be avoided by using a baffle system between the gel and upper electrode, as described above. In this way buffer recirculation round the electrodes may be maintained; the baffle may be removed when all the material has moved into the gel.

Typical buffer systems which have been used with success are listed with their approximate *p*H range. TRIS/glycine (*p*H 8·5–9·2); TRIS/boric acid (*p*H 8·5–9·2); TRIS/acetic acid (*p*H 7·5–8·5); TRIS/citric acid (*p*H 7·5–8·0); TRIS/ortho-phosphoric acid (*p*H 7·0–8·5). The most usual concentrations are 0·02–0·05 M with respect to TRIS, [Tris(hydroxymethyl) aminomethane mol. wt. 121·1]. NaOH/H_3BO_3 (boric acid) *p*H 8·5–9·0; NaOH/H_3PO_4 *p*H 6·5–7·5; β alanine/acetic acid *p*H 4–5. concentration ranges 0·01–0·05 M. Discontinuous buffer systems may be formed by using combination of a pair of the above systems, e.g. a gel made in TRIS/H_3PO_4 is run with a TRIS/glycine tank buffer. In all cases (above *p*H 7) the anion in the gel must have a higher mobility than (a) that of the tank buffer anion and (b) the fastest moving protein at the *p*H of the experiment.

Casting the gel. Caution.

Care must be taken to avoid contact with acrylamide and NN′ methylene bisacrylamide owing to their toxic nature. Only high grade material should be used and it should be kept in a refrigerator.

The monomer concentration required to give successful resolution of a protein mixture should be determined by experiment on the analytical scale, bearing in mind that any component to be collected will have to pass through the whole length of the preparative gel. The approximate total monomer concentration (95 per cent acrylamide + 5 per cent NN′ methylene BIS acrylamide) for resolving globular molecules having molecular weights from 10^4 to 10^6 Daltons is shown below:

Mol. Wt.	% Total Monomer concentration
$10^4 \rightarrow 5 \times 10^4$	$12 \rightarrow 8$
$5 \times 10^4 \rightarrow 10^5$	$8 \rightarrow 6{\cdot}5$
$10^5 \rightarrow 5 \times 10^5$	$6{\cdot}5 \rightarrow 5$
$5 \times 10^5 \rightarrow 10^6$	$5 \rightarrow 3{\cdot}5$

Gels having total monomer concentration of less than 3 per cent may be manipulated more easily if they are stiffened slightly by the inclusion of up to 0·3 per cent agarose. The required amount of solid monomer is dissolved in sufficient buffer so that the final solution will fill the mould and reservoir. Ammonium persulphate 0·08 per cent w/v is added and dissolved, then finally tetramethylethylene diamine (TEMED) 0·08 per cent v/v is added with stirring. The mould is placed on the bench with the end-plates vertical and the filling hole uppermost. In order to avoid trapping air bubbles the monomer solution is poured down a glass rod, inserted into the mould through the filling hole, until the reservoir is full. The glass rod is then removed and any air bubbles sticking to the wall of the mould should be dislodged by gentle tapping. As soon as the gel has set, 10–15 min. at room temperature, the mould is turned into the horizontal position and the top plate removed – this ensures that the gel is not deformed by shrinkage (which normally takes place on setting) away from the cylindical wall of the mould. The exposed gel is covered with a layer of buffer and allowed to stand for 20–30 minutes. The top plate may then be carefully replaced and the gel kept until required. Gels should be left at room temperature for about 2 hours before use.

Preparation of Agarose containing gels.

Agarose (0·3 per cent w/v) is first of all dissolved by boiling, in the required volume of buffer. The solution is then cooled to 40° and the other components added as described above. (In order to prevent premature polymerisation, the amounts of ammonium persulphate and TEMED should be reduced to 0·06 per cent w & v/v). After filling, the mould should be placed in a refrigerator for some hours so that the agarose gel will form.

Transfer of gel to the gel support tube. – Exposed surfaces of the gel should be kept wet.

The filling stalk is cut by carefully running a fine spatula round between the gel and mould. The mould is placed with the patterned plate uppermost, which is then removed, care being taken to ensure that the holes cast for the inlet and outlet tube are not damaged. A plain sheet of Perspex is then placed over the exposed gel and held in place whilst the mould is inverted. The upper plate and cylindrical section are next removed, leaving the gel exposed. The inlet and outlet tubes (each cut obliquely) are then gently pushed down through their respective holes in the gel. The gel is then inserted into the support tube with the aid of the guide ring as already described. The adjustable gel support is placed in the tube above the gel and the whole assembly inverted exposing the channelled end of the gel which should be level with the end of the tube. The inlet and outlet tubes are gently pulled through with forceps until enough is exposed to enable the end piece to be fitted. The tubes are then pulled back until the end pieces are flush with the gel surface.

Fitting the membrane.

The membrane should be well soaked in buffer, and then laid over the gel surface, the clamping ring placed over it and evenly pushed down over the tube. When the ring is about half-way down the membrane will be gripped and stretched slightly as the ring is pushed on further. There must be no creases in the membrane where it fits over the edge of the tube. The assembly is again inverted and after the adjustable support has been removed, the tube and gel are positioned in the buffer tank. The buffer level should coincide with the overflow hole. The upper electrode is now placed in position about 3 cm. from the gel surface. Similarly the lower electrode should be about 3 cm. below the membrane.

Clearing the gel.

Before the protein mixture is loaded onto the gel it is desirable that ionic impurities in the gel should be removed by electrophoresis. A thin (1–2 mm.) layer of 0·01 per cent Bromophenol blue, in the tank-buffer containing 5 per cent sucrose, should be place on the gel and run in before the buffer recirculation is started. When the blue band has emerged from the lower surface of the gel the tank buffer is replaced and the protein mixture can be loaded.

Loading the protein.

All solutions should be free from insoluble material. Before attempting a preparative scale experiment, a series of analytical runs with different loads should be carried out. In this way the maximum load per unit area of gel, for separation between adjacent bands, may be determined. It is important to check that little or no material is being trapped in the surface of the gel as this will lead to gel shrinkage and band distortion. It is desirable that proteins should enter the gel in as sharp a zone as possible. However if the mixture contains many slow moving components, the gel may become blocked and the remainder of the protein will be unable to move and will form a dense layer on or, just under the surface. If this should occur and no action is taken the upper part of the gel will shrink and any bands which are in the gel will become badly distorted. In sometimes happens that only some parts of the surface become blocked and this will lead to band distortion and "streaming", i.e. material may be slowly released from the surface and form a continuous stream through the gel. Surface blocking may sometimes be overcome by repeatedly reversing the polarity of the power supply every 3–4 minutes, and finally gently scraping the gel surface to remove the offending material before the run is continued. To avoid blocking the gel the protein may be applied as a less concentrated solution or in more concentrated *buffer* solution or in a combination of both. The maximum depth of the layer of protein solution should not exceed 1 cm. and the applied wattage should be restricted to 60 per cent of the normal maxium for the gel in use (6 cm. diameter gel 20 watts, 9 cm. diameter gel 40 watts, 13 cm. diameter gel 80 watts). The protein solution should contain 5 per cent sucrose and should be applied to the gel surface under the tank buffer. A little bromophenol blue added to the solution acts as a useful reference marker band. Buffer recirculation should be re-started when all the protein has entered the gel, and this can be checked if necessary by removing a sample of the layer for optical density measurement.

Performance of the system; (see Fig. 15.10).

This was checked using crystalline bovine serum albumin (Armour) as a test protein. The gel dimensions were: 9 cm. diameter $\times$ 5 cm. long;

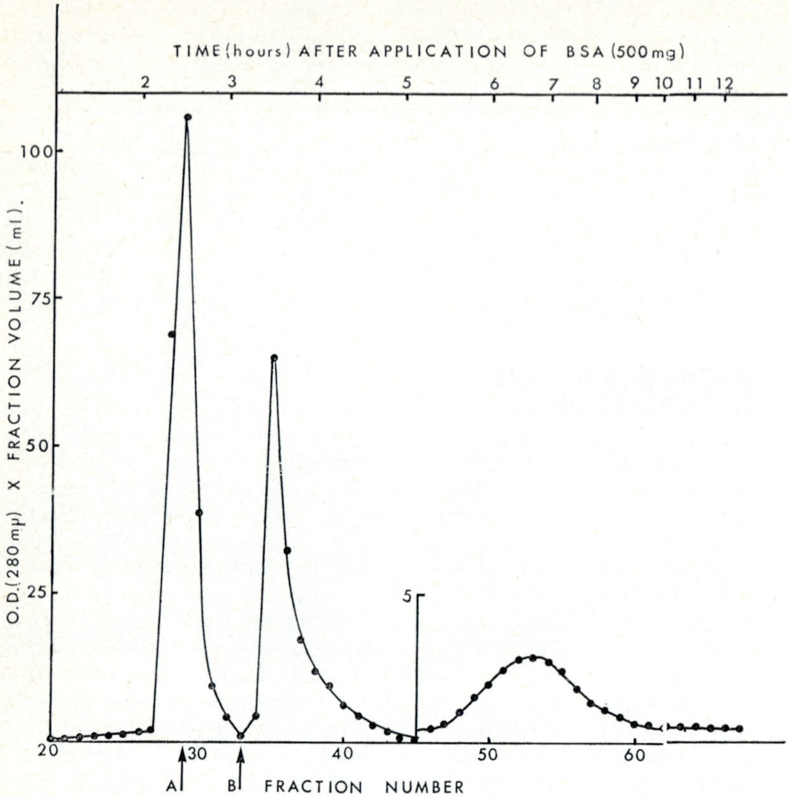

FIG. 15.10. Separation elution diagram using BSA as a test material; for explanation see text.

it contained 8 per cent total monomers (95 per cent Acrylamide, 5 per cent BIS) and was made in TRIS 0·03 M – Glycine 0·06 M. Pre-electrophoresis was carried out for 2·5 hrs. and the tank buffer changed before the albumin, 500 mg. in 15 ml. buffer + 5 per cent sucrose, was layered on the gel. The run was carried out at 160 v and 250 mA., until the ultraviolet absorptiometer tracing showed a peak for the albumin monomer. At point A the power supply was turned off but fractions were collected as usual; it can be seen that background optical density was reached after three cycles. Power was reapplied at point B and the remainder of the monomer band was eluted (2nd peak). The third peak, read on a different density scale, represents the dimer. No trimer was obtained and an analytical gel showed only the faintest trace of such a band. Recovery was always in the order of 95–98 per cent of the applied protein.

Preparation of Thyroglobulin 19S (see Fig. 15.11).

Trimmed pig thyroid gland was macerated and extracted in buffer (see below) using 2 ml. per gm. of thyroid. The extract, after clarification

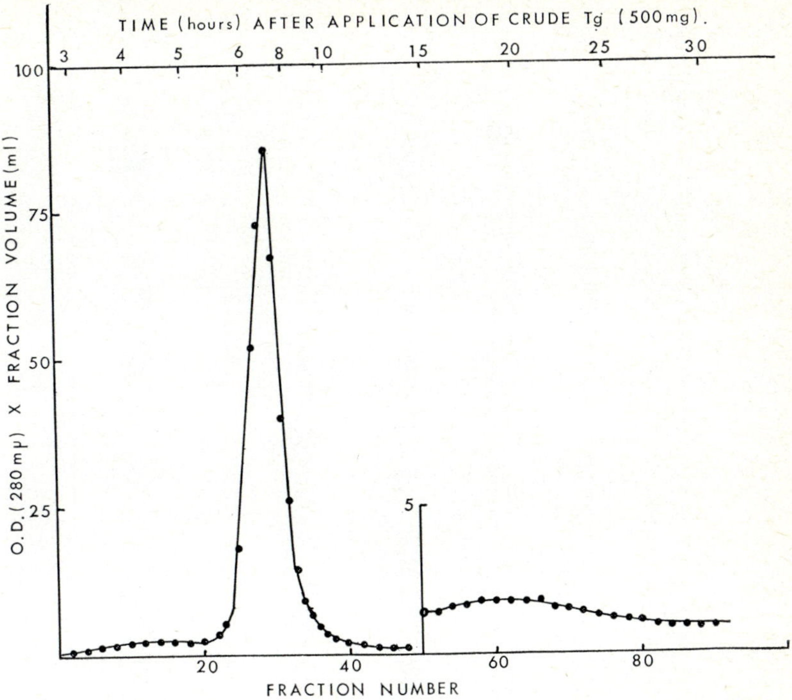

FIG. 15.11. Separation of 500 mg. crude thyroglobulin, for explanation see text.

by centrifugation, was partially purified by passage through a Sephadex G200 column which removes most of the protein smaller than the 19S fraction, but leaves small amounts of 27S and 31S components still present. A solution containing 500 mg. of this mixture in 40 ml. of *p*H 8·7 TRIS (21·6 g./l/)-Glycine (30·0 g./l.), buffer was applied to a 4·5 per cent total monomer gel, 9 cm. diameter by 4·5 cm. long, made in 1:10 dilution of the above buffer. The current was limited to 100 mA. for $2\frac{1}{2}$–3 hrs. by which time all the protein had moved into the gel. 380 mg. of 19S thyroglobulin was recovered.

Re-cycling experiments.

Since it is impracticable to use gels exceeding 6 cm. in length, an automatic re-cycling system has been developed. This may be of use when two adjacent bands are imperfectly separated after one passage through the gel. The fraction which it is desired to purify is re-loaded onto a fresh gel and re-run. On emerging, each band is diverted to a reservoir and then re-loaded onto the gel. This is carried out completely automatically and results in an increasing separation of the required band from the impurity band with each passage through the gel. The collecting buffer in a re-cycling experiment differs from the gel and

TABLE 15.1

Summary of conditions used in the purification of some other proteins. In all cases the monomer mixture contained 95 per cent acrylamide + 5 per cent NN′ methylene BIS acrylamide

Protein	% Total Monomer	Gel Length (cm)	Approx load of starting mixture per cm^2 gel surface	Pre-treatment of protein mixture	Gel buffer	Tank buffer	Collecting buffer	Reference
Rat α_1 (acute phase)	8	5	2·5 mg.	From serum via Sephadex G150 column	44·8 g. TRIS + 1·4 ml. ortho-phosphoric acid (Sp.gr. 1·75)/1., *p*H 8·8	1·08 g. NaOH + 3·09 g. boric acid/1., *p*H 9·2	as tank buffer	Based on 5
Human complement subcomponent Cl_q (reduced and alkylated)	5	6	0·5 mg.	Serum precipitation and DEAE-cellulose column	NaOH 4 g. + 3·2 ml. ortho-phosphoric acid (Sp.gr. 1·75) + 1 g. sodium dodecyl sulphate/1., *p*H 7·2	as gel buffer	as gel	11
Human β_1A globulin	4·5 rerun on 8%	4·5	7·5 mg.	Serum precipitation and Sephadex G150 column	TRIS, 2·0 g./l. + glycine 3·75 g./l. *p*H 8·7	as gel buffer	as gel buffer	12
Human Transferrin	8	4·5	10 mg.	From serum via Sephadex G150 column	,,	as gel buffer	as gel buffer	13
Mouse serum IgM	3	5·0	2·8 mg.	From serum via ammonium sulphate precipitation	,,	as gel buffer	as gel buffer	14

Fibrin fragments D & E	7	6	2·5 mg.		,,	as gel buffer	as gel buffer	unpublished
Vitellogenin	4·5	5	1·25 mg.	TEAE column	,,	as gel buffer	as gel buffer	unpublished
Human placental lactogen	10	5	1·45 mg.	commercial sample	,,	*Acetylated membrane used* as gel buffer	as gel buffer	unpublished
Clotting Factor 7	7·5	5	4·4 mg.		,,	as gel buffer	as gel buffer	unpublished
Adeno V Virus Penton	5·3	5	0·8 mg.	Fluorocarbon extraction and density graded	TRIS, 2·0 g./l. + glycine 3·75 g./l. or TRIS 3 g./l. + boric acid *p*H 8·7	as gel buffer	as gel buffer	15
Mouse J-chain	5·25	5	3·0 mg.		TRIS, 2·0 g. + glycine 3·75 g. urea 480 g./l. *p*H 8·7	TRIS 2·0 g. + glycine 3·75 g./l. *p*H 8·7	as gel buffer	unpublished
Cytochrome C	5	4.5	0·5 mg.	commercial sample	β-Alanine 9 g./l. + acetic acid to *p*H 4·4	as gel buffer	as gel buffer	7

The tank buffer temperature was +2° in all cases except when urea and/or S.D.S. were used, in which case it was 12°.

tank buffer and contains an anion which will form a zone sharpening "front" with the tank/gel anion, (e.g. a TRIS/Glycine gel/tank buffer is used with a TRIS/phosphate collecting buffer having the same *p*H and conductivity). Also, a "neutralized" membrane is used so that the conductivity of the collecting buffer remains reasonably constant. By adopting these procedures it will be found that a band is re-concentrated each time it is re-loaded and thus the diffusion of a band after a number of passages through the gel need be no greater than after one passage.

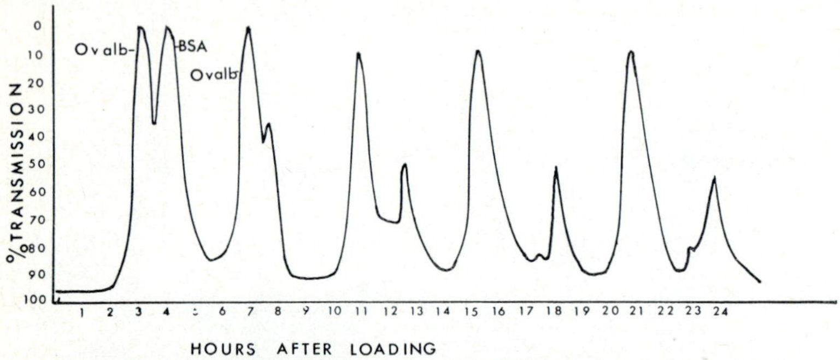

FIG. 15.12. Recycling experiment. 35 mg. each of bovine serum albumin monomer and ovalbumin (single band) were run on a 7·5 per cent gel, 6·5 cm. diameter, 5 cm. long, in TRIS, 1·5 g./l., boric acid 1·8 g./l., *p*H 8·7. The collecting buffer was TRIS, 1·2 g./l. + ortho-phosphoric acid (sp. gr. 1·75) 0·04 ml. *p*H 8·7. After one passage through the gel the BSA peak was discarded and the contaminated ovalbumin re-loaded. After each subsequent passage through the gel the BSA impurity separates further from the ovalbumin peak. The reloading was carried out automatically.

Fig. 15.12 shows the results of re-cycling a mixture of a B.S.A. monomer and ovalbumin, which is not completely resolved after only one passage through the gel.

Control Unit (Fig. 15.13)

Timer.

The sequence of events in the pumping cycle is initiated when the contacts (CSI) of the timing unit close. This unit should preferably have a variable speed control so that the time interval between pumping cycles may be varied from seven minutes to 50 minutes. The actual time for which the contacts remain closed should not exceed a maximum of approximately two minutes.

Operation sequence.

This is controlled by the microswitches operated by the cams driven by motor M. This is a small synchronous motor geared down to give a shaft speed of one revolution in four minutes. The sequence is as follows: CSI closes and turns on motor M. Almost immediately CS2 closes and keeps the motor on until the cams have completed one revolution (by which time CS1 will be open).

Motor M may be started manually by the push-button switch PB1. The button is held down until CS2 closes. As soon as motor M moves, CS3 closes for a short period and causes relay A to be energised. Contacts A1 close and the relay is locked on via the pressure-sensitive switch PS. Contacts A4 close, energising relays B and C. Contacts B1 close turning on the *outlet pump*. (Switches S1 S2 S3 and S4 enable either of the pumps to be turned on or off manually). The outlet pump continues to operate until all the collecting buffer has been removed and the pressure switch, PS, opens. The pumping speed of the outlet pump should be adjusted so that the buffer is removed in about 2 minutes when there is no electrophoresis taking place.

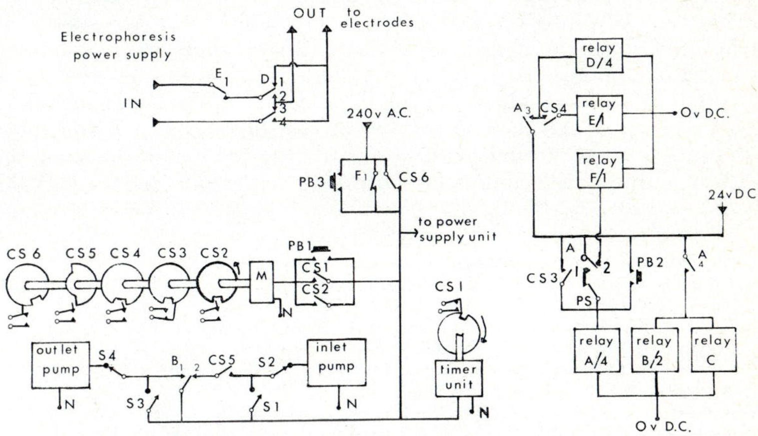

FIG. 15.13. Control Unit, circuit diagram.
Manually operated switches S1 S2 S3 S4.
Manually operated push button switches PB1, PB2, PB3.
Cam-operated switches CS1, CS2, CS3, CS4, CS5.
Pressure-sensitive switch PS.
—●N. Neutral side of 240 V AC supply.
Relay C may be used to actuate a fraction collector and recorder event —marker.
The electrophoresis power supply should be capable of delivering up to 1 amp at 250 volts.

When the pressure switch opens, relay A is unlocked and turns off relays B and C. Contacts B1 open and B2 close, turning off the outlet pump and turning on the inlet pump, provided CS5 is closed. This pump operates for the remainder of the four-minute period. Cam 5 is shaped so that the inlet pump cannot operate for more than two minutes during which time it should pump the required volume of fresh collecting buffer between the gel and membrane. If the volume of the collecting buffer increases due to electro-osmosis the outlet pump will take longer than two minutes to remove it and so the inlet pump will operate for a shorter period and proportionately less new buffer will be pumped in.

Relay C is used to actuate a fraction collector and a recorder event marker if required.

Electrophoresis power supply control.

When relay A is energised contact A3 closes and relay D is energised; this reverses the polarity of the electrodes. After a further ten to fifteen seconds CS4 contacts close and relay E is energised, opening contacts E1 and cutting off the power to the electrodes. The power is restored, with normal polarity, when the pressure switch opens and relay A is turned off.

Safety circuit.

If for any reason, the pressure switch fails to open, the outlet pump would run continously. To avoid this happening the 240 V AC supply to the apparatus is turned off if the outlet pump runs for more than approximately 3 minutes 45 seconds. Relay F is energised continuously, via contacts A2 whilst relay A is turned off. When relay A is energised, at the start of the pumping cycle, relay F is turned off and contacts F1 open and remain open so long as the outlet pump is working. The 240 V AC supply then passes through CS6. This switch opens for about ten to fifteen seconds at the the end of each revolution of motor M and if the outlet pump is still running at this time, contacts F1 will be open and the 240 V AC supply will be interupted until it is restored by operating PB3.

REFERENCES

1. Raymond, S., and Weintraub, L. *Science*, 1959, **130**. 711.
2. Jovin, T., Chrambach, A., and Naughton, M. A. *Anal. Biochem.*, 1964, **9**, 351.
3. Duesberg, P. H., and Ruekert, R. R. *Anal. Biochem.*, 1965, **11**. 342.
4. Hjerten, S., Jerstedt, S., and Tiselius, A. *Anal. Biochem.*, 1965, **11**, 211.
5. Gordon, A. H., and Louis, L. N. *Anal. Biochem.*, 1967, **21**, 190.
6. Schenkein, I., Levy, M., and Weis, P. *Anal. Biochem.*, 1968, **25**, 387.
7. Brownstone, A. D. *Anal. Biochem.*, 1969, **27**, 25.
8. Hjerten, S., Jerstedt, S., and Tiselius, A. *Anal. Biochem.*, 1969, **27**, 108.
9. Hodson, A. W., and Latner, A. L. *Anal. Biochem.*, 1971, **41**, 522.
10. Wink, H. *J. Polymer Sci.*, 1966, **4**, (Pt A–II), 830.
11. Reid, K. B. M., Lowe, D. M., and Porter, R. R., *Biochem. J.*, 1972, **130**, 749.
12. Charlwood, P. A. *Biochem. J.*, 1969, **115**, 897.
13. Charlwood, P. A. *Biochem. J.*., 1971, **125**, 1019.
14. Parkhouse, R. M. E., and Askonas, B. A. *Biochem. J.*, 1969, **115**, 163.
15. Winters, W. D., Brownstone, A. D., and Pereira, H. G. *J. gen. Virol.*, 1970, **9**, 105.

CHAPTER 16

MODELS FOR MACROMOLECULES, NUCLEIC ACIDS AND PROTEINS*

Ivor Smith, Margaret J. Smith and C. F. Doré

SEQUENCE analysis of proteins and nucleic acids has become a routine procedure in many laboratories, and the primary structures of a whole range of proteins and a number of nucleic acids, mainly *t*-RNA's, are now known. Within the last few years X-ray crystallography has been used to elucidate tertiary structures of proteins. The various techniques now serve to confirm the main body of data and to complement each other. From these results has emerged the general concept that many exchanges of, say, one aminoacid for another occur naturally, that these changes are permissible provided that they exert little or no effect on the tertiary structure but that they result in a chemical or biological defect when this structure is appreciably changed. Hence it has become desirable to have models which highlight the tertiary structure even at the expense of loss of information of side chains. For a complete appreciation of these primary, and tertiary structures a variety of types of model may be necessary and we describe two different types below.

All teachers will be aware of the problems involved in attempting to inculcate an understanding of three-dimensional or spacial chemistry. Hence, in designing such models, we have attempted to deal with teaching as well as research problems. Thus we have included this chapter on practical model building as a corollary to other chapters in the hope that it will be of value to all those interested in macromolecular structure.

Until recently all commercially available molecular construction kits have been based on the same type of unit namely the atom or atomic centre. Thus whether one has wanted to build a simple model representing a nucleotide or aminoacid on the one hand or, alternatively, a large, complex macromolecule representing a nucleic acid or protein, the basic building blocks involved have been the same atoms. Such models were originally designed to illustrate small molecules and their interactions, involving only primary structure; they have served and still do serve valuable purposes. However, when used to construct models of polymers or macromolecules involving secondary and tertiary structure (for which they were not designed) many disadvantages become evident and, following the discussion of these disadvantages together with the problems involved in the construction of

* This chapter is available as a separate booklet from Capital Biotechnic Developments Ltd., 66A Churchfield Rd., London, W.3., and Ealing Corp. Cambridge, Mass. It is supplied with all Bio-Bits and Atomunit Teaching Kits.

such macromolecular models, two new types of model will be discussed and illustrated.

Skeletal (ball and spring, rod) models were originally designed for the small molecules of organic chemistry. Hence an aminoacid or nucleotide built from such pieces occupies a length of some 6–12 inches, the construction of a trimer becomes a hazardous endeavour and it is, for most people, an impractical proposition to attempt to build a macromolecule.

To overcome some of the problems, accurately machined, well-fitting, space-filling models were produced. These have, indeed, played a valuable role as an aid to chemical and biological thinking but, again, they do not lend themselves to general and widespread use. The reasons for this may be listed as follows: the cost is generally prohibitive on all but the most advanced research levels, the time and skill required for construction means that, in general, they are not available for student use, their use value requires a degree of understanding not always possessed by the student but yet is only obtained by frequent use of the models. Current interest in biopolymers is now great at all levels of teaching and research, and the value of simple cheap models which can be used to demonstrate both the three dimensional structure of known macromolecules as well as the techniques by which these structures are determined would appear to be of value at the present time.

The design of models for the above purposes involves a number of assumptions, only some of which are scientific, and are listed as follows:

1. All models involve the acceptance of a code. Thus a carbon atom may be represented as a black sphere with four holes to accept valency "rods" at properly spaced angles or as a tetrahedral wire unit with a black spot at the centre. That the atom does not look like any of the representations is accepted automatically. It follows that molecules constructed from such models are no more than a coded representation also. The representation may be extremely accurate for such properties as bond angles, space occupation, etc., but, to press the point, the real molecule is not composed of coloured bits.

2. It follows from (1) that alternative codes are equally possible and may have other advantages or greater value in certain circumstances. Thus all aminoacids may, in principle, be represented by two "bits", namely a backbone bit common to all and an R group bit specific for each. Likewise ribo- and deoxyribo-nucleotides may each be represented by two different backbone bits, for the two main species, together with different bits for each of the different bases involved. Further, other units such as DNP, hydrogen bonds, etc., may be similarly represented by other specific bits.

3. The bits should be clearly and obviously different, small and yet sufficiently mechanically stable to enable macromolecules to be constructed.

4. The bits should be capable of allowing the demonstration of:

(*a*) Three dimensional structure of macromolecules such as the DNA double helix, tertiary structure of proteins, etc.

(*b*) The interaction of molecules such as hæm → globin, DNA →

RNA transcription, DNA replication, enzyme → substrate contact, etc.

(*c*) Techniques used in sequence determination, e.g. enzymic and chemical digestion, dinitrophenylation, etc.

(*d*) Techniques used in biological or chemical synthesis such as solid state synthesis of proteins and biosynthesis on templates.

(*e*) All bits, units and components should be sufficiently cheap such that the macromolecules for which they are designed may be constructed at a cost sufficiently small not to deter any potential used down to the elementary or lowest grade teaching level.

It will be readily recognized that the coding of, say, adenine by a thin diamond bit has involved the loss of a great deal of chemical information in the model. However, this is more than compensated for by the fact that, for example, the new double-stranded DNA model to be described shows over two complete turns of the helix, fully hydrogen-bonded, weighs under 200 g., is less than 40 cm. high and can be constructed for less than $20. Such a model illustrates base pairing, hydrogen bonding, replication and transcription by actual building rather than by simple observation and is sufficiently stable to be readily handled for direct observation and illustration of specific chemistry. By a simple modification of the backbone design, it will also allow illustration of splitting at both 3′ and 5′ positions as well as synthesis via nucleotide triphosphates. Similarly, models can be constructed which illustrate primary, secondary and tertiary structure as well as synthesis and analysis of proteins.

It is therefore evident that for complete appreciation of primary, secondary and tertiary structures two methods of model construction are required, each complementing the other. One type should be capable of giving information at the atomic level. The second should continue the information to the next stage, i.e. secondary and tertiary structures, where atomic groups may be considered as the basic unit.

Many versions of skeletal and space-filled atomic models are available. The new design of atomic model discussed below is capable of being used for both purposes and yet is still sufficiently cheap for general use. The open molecule is first constructed and then, if desired, plug-in mushroom units turn these into space-filled models. Atoms and bonds for these models are, of course, constructed to accurate dimensions.

The following discussion is intended as a contribution to the design and use of two sets of models; it should not be seen as an introduction to molecular biology for which a number of admirable texts exist and these are listed at the end of this chapter. The *Atlas of Protein Sequence and Structure*[1] has been used extensively as a source for most primary sequences and has been invaluable during the model building.

BIO-BITS*—MACROMOLECULAR MODELS

Models of macromolecules are now required for a number of separate and distinct purposes, some of which are complimentary whereas

* Bio-Bits[R] designed by Ivor Smith and Margaret J. Smith are available from Capital Biotechnic Developments Ltd., 66A Churchfield Rd., London, W.3, and Ealing Corp., Cambridge, Mass. Patent No. 1277731.

others are contradictory. It is desirable now to illustrate the following points:— (1) basic tertiary structure unencumbered by a vast array of constructional wire such that the molecule is lost behind its component atoms; (2) exact atomic distances and spacial distributions within the molecule; (3) molecular interactions; (4) the principles and techniques of sequence analysis and macromolecular synthesis—both chemical and biological. Bio-Bits have been designed with points 1, 3 and 4 in mind and Atomunits (next section) with points 2 and 3 in view.

The Bio-Bits basic building units represent atomic groupings such as the sugar, phosphate and aminoacid residues present in the backbones of nucleic acids and proteins, and the bases and R groups connected to these (see Figs. 16.1 and 16.6). The backbone colour code is dual so that the sugar-phosphate repeating units and the amino- and carboxyl-ends of the aminoacids arc obvious.

Bio-Bits may be used to build permanent, complete macromolecules both as three dimensional models and as primary chain structures. For permanent models the backbone is represented by a continuous length of colour-coded tubing through which runs a stabilising plastic-coated wire so that, once moulded into the correct three-dimensional shape, this shape is retained often without the need of further scaffolding. To illustrate chemical and biological techniques of analysis and synthesis, interconnecting monomeric backbone Bits are used.

Each monomer backbone, or monomeric equivalent in the permanent chain, has punched holes into which will fit the nucleic acid base Bits or the aminoacid R group Bits. These latter Bits each possess a small button-ended projection at one end and a hole at the far end into which additional Bits such as hydrogen bonds, DNP groups, etc. may be inserted. A further advantage of this Bit-addition procedure is that many additional Bits, for example DNP's, may be inserted into a large molecule which will still retain its obvious and original identity in a manner quite impossible when using atomic models. Monomers may be connected with the Bits off-set (out of plane) so that a greater impression of helical structure is obtained and, likewise, holes may be so punched in the continuous chains.

Nucleic Acid Bio-Bits

The backbones of DNA and RNA are composed of repeating 5′-phospho-deoxyribose and 5′-phosphoribose. They are constructed from green and yellow PVC tubing, respectively representing dexoyribose and ribose.

The monomer or monomeric equivalent is 2 cm. long and, at one end, holds a purple phosphate band 0·5 cm. long. The punched hole is close to this band (i.e. not central) and thus the complete unit represents the sugar-5′-phosphate biological building unit. For this reason the nylon connector projects from this end. Thus in chains constructed from monomer residues and also in permanent long chain backbones the chain direction is obvious as the base Bit is always close to the 5′-phosphate and far from the 3′-phosphate (see Fig. 16.1).

Complete molecular models are fabricated from a continuous length of backbone tubing of the correct colour. Here the phosphate groups or bands are arranged periodically along the backbone and holes are punched adjacent to them as described above. The dimensions are only approximately correct but, nevertheless, the double helical coil shows the correct ten base pairs per complete turn of the helix (see Figs. 16.2(*a*) and (*b*)). *t*-RNA similarly constructed will show the known base pairing within the molecule whilst simultaneously indicating those bases which do not pair because of spacial position or chemical modification as in the minor bases. Such a model also allows for the possibility of further pairing across the molecule, i.e. between unpaired bases of two opposite loops (see Fig. 16.4). Furthermore, in both the above cases, additional substituents may be inserted into the backbone by simply punching a further hole in it; one example would be a 2′-O-methyl found in some *t*-RNA's.

The purine and pyrimidine molecular residues are represented respectively by large or small, flattish, specifically-coloured Bits. Thus any given mononucleotide may be constructed from one such Bit inserted into the respective backbone monomer and, obviously, single-stranded polynucleotides are constructed similarly by inserting a number of these into a continuous chain or by joining a number of the monomers in sequence.

Hydrogen bonds are constructed from plastic coated wire which fit into the holes at the far end of the Bits. Red-coated wire represents a double hydrogen bond unit and blue-coated wire a triple hydrogen bond unit corresponding to the A═T and G≡C base pairs. With these units, double-stranded chains may be constructed and simple antiparallel DNA models result (see Fig. 16.2), which may also be used to illustrate replication; similarly double-stranded RNA models may be made. The DNA model is also suitable for illustrating transcription by adding ribonucleotides as shown in Fig. 16.2. Greater reality may be given to this by the addition of a pyrophosphate unit for connection to the mononucleotide so that the triphosphate may be shown approaching a chain and the pyrophosphate is removed prior to the joining on of the mononucleotide to the growing end.

A series of smaller, differently-shaped and colour coded Bits are used to indicate substituents into the base rings to form some of the minor bases. These can also be inserted into the holes in the major base Bits to indicate, for example, the conversion of cytosine into 5-methyl cytosine or into the backbone, as previously mentioned, to form 2′-O-methyl sugars. With these additional Bits, the clover-leaf forms of the *t*-RNA's may be constructed with or without the corresponding amino-acid in position (see Figs. 16.3 and 16.4). As the tertiary structures of these RNA's become known, so may they be modelled by means of additional cross-linking with hydrogen bonds and one such hypothetical structure is shown in Fig. 16.4*b*. Numerous other applications in biochemistry and genetics can be envisaged, and just a few examples will be given. Other residues, such as glucose residues, may be added in order to illustrate some of the phage structures. It is possible to construct transfer RNA's both for known structures and whilst the process of sequence determination is in progress as soon as small sequences or the total number of bases become known. Because the bases are separable from the backbone, known lengths of unknown sequence may be placed in position for subsequent completion as the sequence becomes known and as is now done with brackets on the written page; brackets may also be indicated on the model with a turn of wire or a pipecleaner. Further, the chain may be broken at any of the 5′ and 3′ positions to indicate enzymic hydrolysis and the chemical interconversion of bases may be indicated by, say, the removal of cytosine and its replacement by uracil in the chain sequence. Further transformations may be shown by the insertion of a "flag" into the hole in the base Bit. The open-ended design of these Bio-Bits enables as yet unknown bases or modifications to be inserted into the basic structures as these are found. Using chains composed of monomer units it is possible to demonstrate the process of crossing over and, consequently, the determination of the percentage linkage, complementation, alleles, transitions and transversions. Complete chains may be used to discuss genes, chromosomes, mutations, etc.

Protein Bio-Bits

The protein backbone is composed of a single residue or unit, namely: —NH.CH.CO—. This is designed as a blue tube (with a free blue, left-hand end to represent the amino end) together with a nylon tube interconnector and a red band at the right-hand end (to represent the acid end). Chains are built up from these monomer Bits. Each of the R groups is also specifically colour and size coded to identify it absolutely. Every common R group coded for on the template mechanism is represented by a separate Bit except those for Gln and Asn which use the same Bits as Glu and Asp together with a second Bit to represent the amide residue; this is so that the chemical procedures which remove this group may be illustrated. Hyp is made by the addition of a hydroxyl Bit to Pro, sulphate and phosphate are represented by additional Bits, similarly as are acetyl, formyl and other residues now known to occur in proteins. Sulphur bridges and hydrogen bonds are represented by coated wire connectors. Further Bits are available to illustrate those residues which are used during chemical manipulations as an aid to sequence studies, e.g. DNP, dansyl, aminoethyl (for pseudolysine) etc., (see Fig. 16.6, 16.7 and 16.8).

Complete chains are of continuous blue tubing with periodic red bands to indicate the carboxyl end of each monomer residue. Close to each band is punched its corresponding R group hole to indicate to which residue it belongs. This also serves as a direction marker so that even with an extended and coiled sequence the amino end and chain direction is immediately obvious (see Fig. 16.9).

The construction of permanent primary sequences of such molecules as vasopressin (together with a range of analogues), MSH, insulin, α and β globins, chymotrypsin, and even the immune globulins is a simple procedure requiring little time; the construction of even one long sequence is usually sufficient to imprint the total code in the mind.

Where the X-ray structure is known well enough to allow the construction of a three-dimensional molecule then the R groups can be offset around the main tube to heighten the helical effect though this may be simpler to perform after the backbone itself has been modelled (see Figs. 16.8, 16.9 and 16.12).

Analytical and synthetic procedures are readily demonstrable with monomer Bits (Figs. 16.10(*a*) and (*b*)). With the aid of extra Bits, to represent DNP and dansyl residues both end group and sequence studies may be illustrated; amino groups may be protected with acyl and *t*-BOC Bits; α and ω carboxyls may be covered with amide and *p*-nitrophenyl Bits; other Bits may represent phosphate, sulphate, sulphonate, monosaccharide units, etc. Enzymic attack can, of course, be represented by simply splitting the connection between susceptible monomers previously joined in sequence.

Protein Bio-Bits are of obvious value for a number of purposes. So far there has been no simple way of representing primary sequences (except on the written page) during the course of work on analytical sequencing. However, it is now a simple matter to plan a backbone as soon as the total number of residues is known. As small sequences are identified these may be separately constructed and then readily manipulated for overlaps found from multiple enzymic hydrolyses or chemical manipulations until the final structure enables the model to be completed. Corrections due to faulty original placements and residue inversions take mere moments to perform. Complete models representing the same molecule obtained from different species may easily be compared for homologies, transpositions and other changes.

Construction of Permanent Bio-Bit Models

A scale of 2 cm. per residue is used for all nucleic acid and protein models. Colour-coded tube is punched with either two or four holes around each circumference and colour-coded sleeves are placed in position, one close to each set of holes; such prepared tubing is supplied in 100 cm. lengths. Determine the number of residues required for the model, e.g. 21 for the A chain of insulin, and starting at a red carboxyl end cut off 21 residues so that the tube ends with a blue end just short of the twenty-second red sleeve. Measure and cut off a plastic coated wire about 1 cm. longer and insert into the tubing. Turn back 0·5 cm. of wire at each end and force back into the tube end to hold the wire firmly in place. Separate tubes may be joined either by overlapping wires in each direction, best done after the R groups are inserted, by means of staples from a paper-stapling machine or with a nylon tube which surrounds the coated wire but fits into the PVC tube.

β **MSH (pig).** Cut a length sufficient for 18 residues and lie with the blue at one end and the red sleeve at the right-hand end. Insert the residues as show in Fig. 16.10(*a*) and (*b*). To make enzymic fragments follow the figure.

Insulin. Cut two lengths of tube for the A and B chains, 21 and 30 residues long, and insert wire. Insert Bits as shown in Figure 16.9 until the two separate chains are completed. Using further pieces of wire form the three sulphur bridges, the intra-chain bridge first and the two inter-chain bridges to form the complete model of the primary structure. This model is suitable for the demonstration of techniques of sequence analysis.

Frame and Model Construction for 3D. The frame is of a collapsible packaway design. The molecular backbone is supported on spring-loaded wire; the springs fitting easily over $\frac{1}{16}$ in. mesh and the wires being practically invisible. Small ferrules slide along the wires and the backbone is held by a further piece of wire which loops into the ferrule (see Fig. 16.1). The three thicknesses of wire make a fairly tight fit in the ferrule so that, at this stage, no slipping occurs. When the whole is judged to be correctly placed along the vertical coordinate, the ferrule is pinched with a special tool fixing it permanently in position. The scale is 3 Å/2 cm and three such scales are used to fix the *X Y Z* coordinates. The vertical wire is first positioned in XY axes at top and bottom of the frame and the Z coordinate is then defined by the ferrule position. The frame mesh is currently $\frac{3}{8}$ in. (0.96 cm); but if the squares are considered as 1 cm this aids construction without appreciable loss in accuracy. Where helixes run approximately vertical, the suspension wires may need to pass through one or more turns of the same or another helix in order to preserve an accurate structure. This is easily accomplished by unhooking a spring and threading through the helix as required.

Two sets of the X-ray coordinates are required to position any aminoacid namely one for the α-carbon and a second from some point on the R group which serves to indicate the relative direction in which the whole group points in relation to the chain. The number of suspension wires required is between twelve and twenty for molecules such as myoglobin and lysozyme and obviously depends on the exact shape of the molecule. In general, for a molecule with a high helical content, the wires are best placed just outside the ends of each helix, but, for long helixes, a support in the middle helps with the final alignment. For non-helical molecules the choice is determined by convenience of building. When commencing building, four to six verticals are placed in position and a fifty-R-group length, 100 cm, of backbone is built into the system. Subsequently the R groups are inserted into one of the four holes available around the circumference of each residue. Building is continued in this manner until the model is complete.

Helixes can be constructed simply and quickly in the following way. One end of the helix tubing is held by the thumb on a piece of glass tubing about 2 cm in diameter and the blue backbone tubing is wound around the glass in a clockwise manner working away from the thumb. When the correct number of residues has been wound, remove the glass tube, adjust the backbone so that between 3.5 to 4 residues make one complete turn, stretch the helix lengthwise so that the two ends now correspond to the correct coordinate dimensions and assemble on the frame. Final adjustments can be made for each α-carbon and the R groups added to complete that part of the model.

The models illustrate many of the points of protein chemistry. First, all the helixes are immediately visible as are the non-helical portions which occur at the 'bends' in the chain where the molecule folds back on itself. Second, the cleft or active site is readily observed. Third, almost all of the polar side chains are seen to be on the outside of the molecule thus confering on it its extreme solubility whilst almost all the nonpolar side chains are buried within the structure. This property can be highlighted in the following way. All the white nonpolar chains fluoresce white and if the polar chains are painted with a dot of red fluorescent paint and the model viewed in ultraviolet light an external shell of red dots is seen to enclose the white bits completely; this makes an exceedingly effective and unforgettable demonstration of this general property of proteins. Similary active sites can be painted in other fluorescent colours.

Insulin Model. The A chain is made from a 21-residue blue backbone and the B chain from a 30-residue yellow backbone. Helical runs, antiparallel sheets, etc. are readily visible. For the construction of hypothetical proinsulin green backbone may be used for the C-peptide.

Myoglobin Model

XYZ coordinates of each α-carbon and of one position within its attached R group have been tabulated.[2,3] The peptide backbone contains 153 residues in a single chain, about 30 of which are in non-helical conformation and the remainder exist in eight separate helixes named A–H. Stretches of non-helix occur at each end of the molecule and serve also to separate helical segments where these make sharp changes in direction. The molecule contains no sulfur (cystine) bridges.

For the heme group, a thin, transparent, red sheet of Lucite, 6 cm square is generally found to be sufficiently adequate. It is suspended between the proximal and distal histidines (residues 64 and 93) and pushed well in towards the non-polar R groups. However, a shaped piece of Lucite can be constructed on the same scale as the main model in order to illustrate the side chains on the four pyrrole units and their hydrogen bond contacts with amino acid R groups.

Lysozyme Model. The tertiary structure of hens egg lysozyme was deduced by Phillips. The molecule contains 129 amino acids in a single chain which is crosslinked by four pairs of sulphur bridges. Phillips conceives the molecule in a number of parts which can be usefully taken as a guide to the model building (Fig. 16.3). The first forty residues (from the N-terminal end) form a compact structure containing most of the helical part of the molecule, and this can be built first. The next fourteen residues form an anti-parallel pleated sheet which points away from the first section forming a sort of wing and making the first part of the cleft; for the model, however, it is convenient to build a unit of twenty residues into the structure. Residues 61 to 87 form a separate third segment and residues 88 to 129 complete the structure. If desired, these four segments can be constructed using four colours of backbone tubing instead of the single blue backbone normally used. At this stage the four pairs of cysteines can be cross-linked as they all lie sufficiently close to the surface to enable this to be done without deranging the model.

(*Continued on page* 447.)

Parent Compound					Anti-metabolites
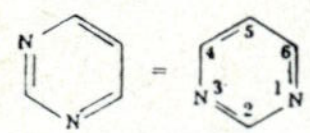 Pyrimidine Pyrimidine-Sugar $N_3 \rightarrow C_1'$	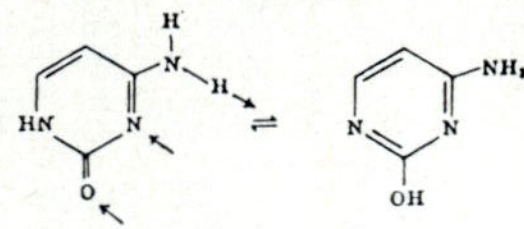 Cytosine, dark blue Cytidine = C	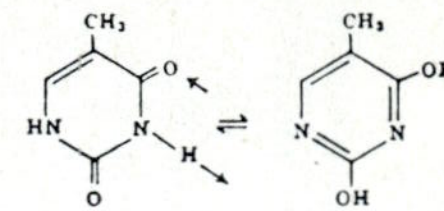 Thymine, dark red Thymidine = T	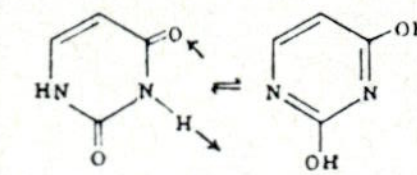 Uracil, white Uridine = U	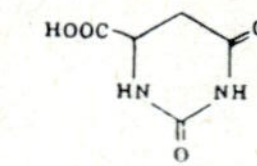 Orotic Acid Orotidine = O	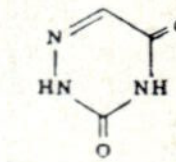 Azauracil yellow
Minor Bases Occur in small amounts usually in t-RNA's	5MC = 5 Methyl mauve ● NAC = N(6)Acetyl white 5HMC = 5 Hydroxymethyl gold ●	RT = Ribo Thymidine	D = 4:5 Dihydro U, light purple Q = sugar joined to C_5 not N_3, green bright OMU = 2'.O-Methyl S = thiouridin, green pale mottled		Also 5F-U, Br-U, Br-C } green ●
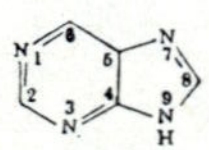 Purine Purine-Sugar $N_9 \rightarrow C_1'$	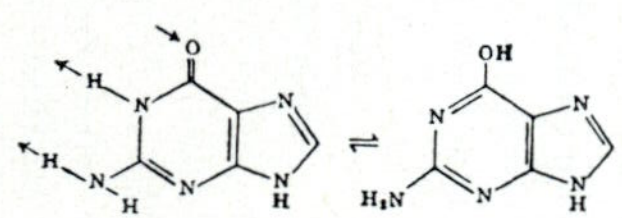 Guanine, light blue Guanosine = G	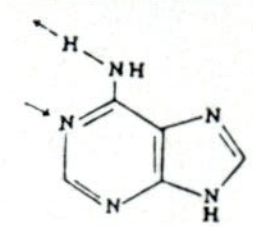 Adenine, light red Adenosine = A	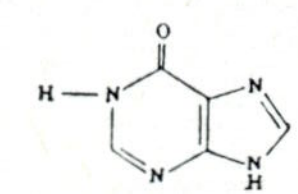 Hypoxanthine, black Inosine = I	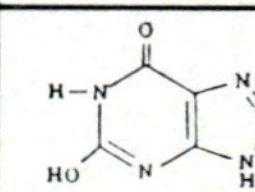 Xanthine Xanthosine = X large green mottled	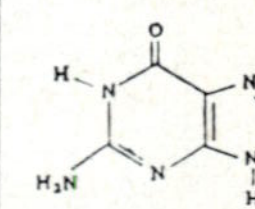 8 Azaguanine, yellow
Minor Bases Occur in small amounts usually in t-RNA's	DMG = N(2)dimethyl black △ 1MG = 1-methyl blue ● NMG = N(2)-methyl blue △ OMG = 2',O-methyl pink 7MG = brown	DMA = N(6)dimethyl red MA = 1-methyl blue ● NPA = N(6)isopentenyl 3MA = pearl ●	MI = 1-methyl blue ●		6-SH purine

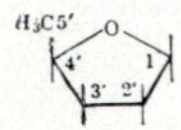	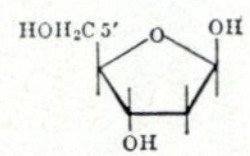	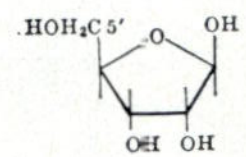	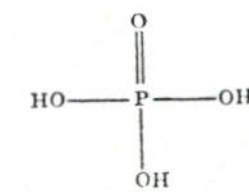
5-Methyl Tetrahydrofuran	Deoxyribose (D)	Ribose (R)	Phosphate (P) 5′, 3′, 2′ or 2′, 3′ cyclic

TABLE 16.1

Structure and Nomenclature of the Major Purine and Pyrimidine Bases, Nucleosides and other compounds of importance.

The bases exhibit keto-enol tautomerism but it is the former type of isomer which exhibits hydrogen bonding between base pairs. The one and three letter abbreviations refer to the nucleosides but the colours refer to the bases on the Bio-Bit design. Base pairing occurs between C and G, C≡G, and A and T, A=T, or A and U, A=U.

Further colour coding (see Fig. 16.1):

small sphere (part M_1) ●		tetrahedron (part M_2) Δ		multifacetted thick	
					mauve
AIG 1-methyl	blue	N(2)-methyl	blue G	N6PMS	dk blue
CUWG 2′,O-methyl . . .	pink	N(2)-dimethyl . . .	black G	N6P	lt blue
C 5-methyl	mauve	N(6)-dimethyl . . .	red A	TCP	pink
C 5-hydroxymethyl . . .	gold	N(6)-acetyl	white C		orange
U 5-F, 5-Cl or 5-Br . . .	green		gold	unknown subst. . . .	white
	white				Ivory (beige)
7Ml	brown (cubronze)				
3Ml	pearl lilac				

Note: When this chapter is available as a separate booklet the middle section, which contains most of the figures, may be pulled out. The figures can then be viewed with the relevant text in order to aid the model building.

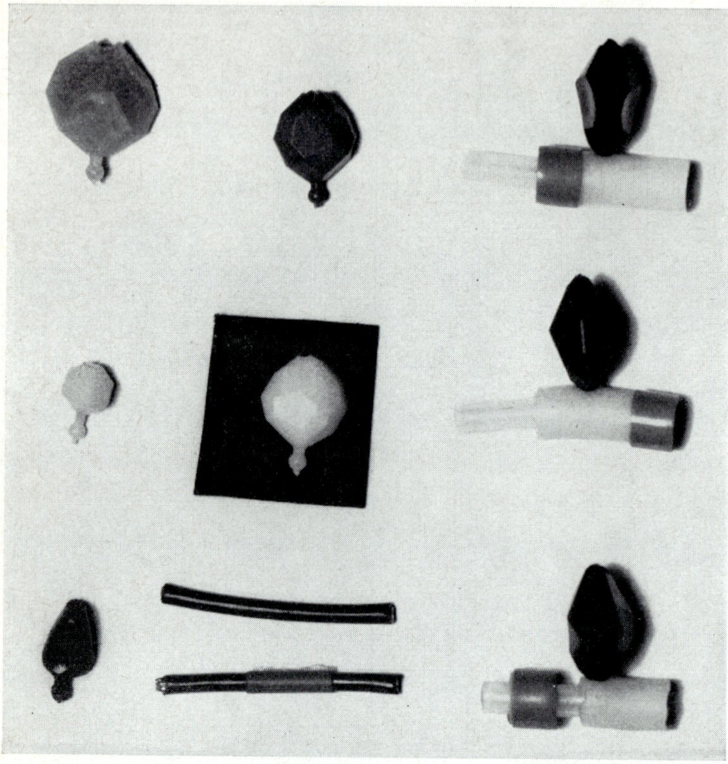

FIG. 16.1. Nucleic Acid Bio-Bits. A large, flattish, many-sided Bit is used to denote all purines and a small, but similar, Bit denotes pyrimidines. The individual bases are colour-coded as described in Table 16.1. Hydrogen bonds are colour-coded e.g. the double bonded A═T is red and the triple bonded G≡C is blue. Mono-nucleosides and nucleotides are constructed as illustrated. Parts M_1 and M_2 are available in a range of colours and are used to indicate substituents, e.g. methyl, acetyl, etc., in major bases to form minor bases.

Key:	Purines	Pyrimidines	5′ monoribotide
	M_1	Pyrimidines	3′ monoribotide
	M_2	Hydrogen bonds	5′ monoriboside with separate phosphate group

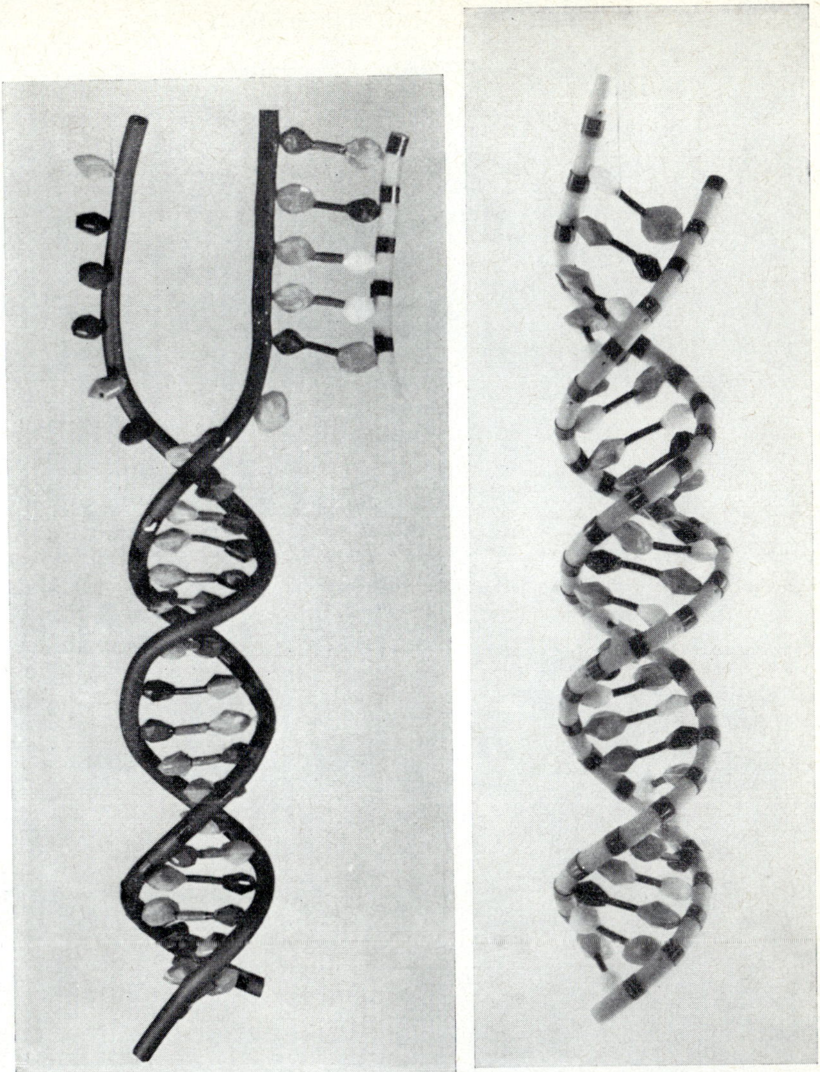

FIG. 16.2. RNA and DNA double helixes.

(*a*) DNA showing base pairing, ten base pairs per complete turn, the possibility of replication (top left) and transcription (top right), and the anti-parallel nature of the latter. Only one of the two green backbones is shown with phosphate bands in position in order to follow each backbone more easily.

(*b*) RNA double helix. Both yellow phosphoribose chains are shown complete and, again, running antiparallel. Note the more open helix of RNA compared to DNA. As definite sequences of these macromolecules are not known, the base pairs are added randomly along the DNA but the RNA can be made on a transcription basis from the particular DNA.

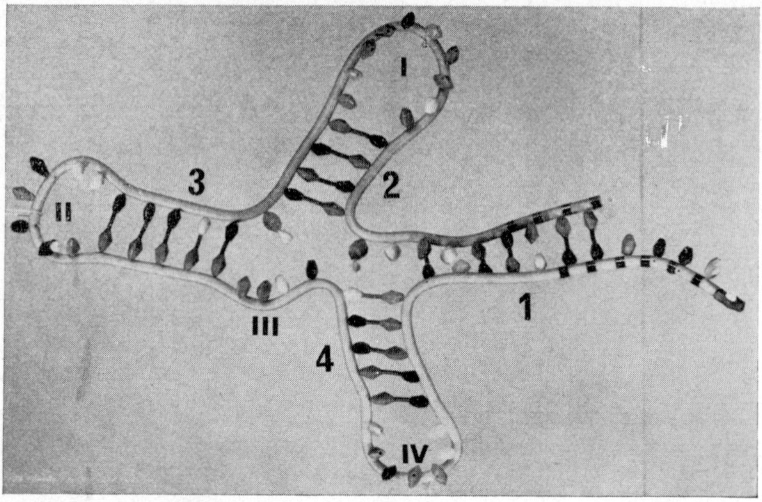

FIG. 16.3.

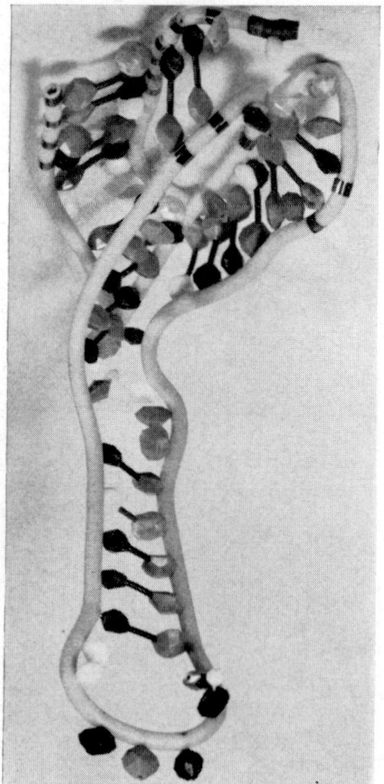

FIG. 16.4(*a*).

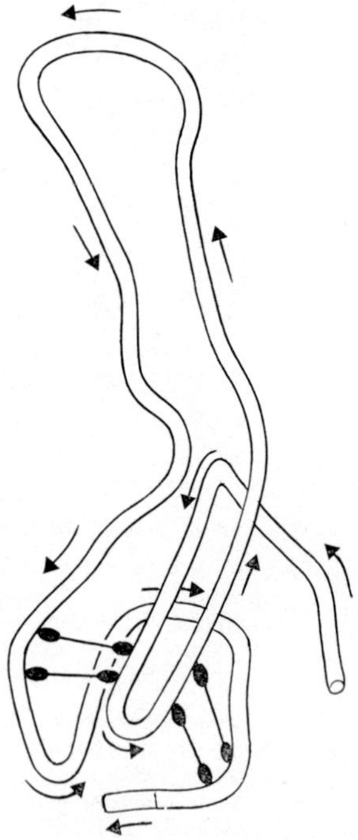

FIG. 16.4(*b*).

FIGS. 16.3, 16.4 and 16.5 t-ala-RNA^{ala} models and Sequence[1].

FIG. 16.3. Clover leaf form with four separate groups of base pairs, alanine at far right, and the three-base anticodon (nodoc) at far left. Phosphate bands are shown only at the two ends of the chain.

FIG. 16.4(a). Hypothetical tertiary structure with two extra sets of base pairs with their corresponding phosphate bands.

FIG.1 6.4(b). Drawing of backbone chain alone but indicating extra cross-linking of tertiary form.

```
1                          11                21
5′-pG G G C G U G   U MG   G C G C     G U A G D C G G D A
         1                   2                   I

                 31                    41
G C G C   DMG   C U C C C   U U I G C MI Q   G G G A G   A G U C
  2′              3              II             3′         III

 51                       61                     71
U C C G G   RT Q C G A U U   C C G G A   C U C G U C C   A C C A Ala
    4             IV             4′           1′
```

Minor bases, e.g. MG, are defined in Table 16.1.
Arabic numbers indicate a sequence which pairs, i.e. 1 with 1′.
Roman numbers indicate sequences in the loops which do not pair; these loops contain the minor bases and the anticodon or nodoc (IGC in II).
Further base pairing for Fig. 16.4. is between: 16 and 17 to 56 and 57, and 20 and 21 to 76 and 75.

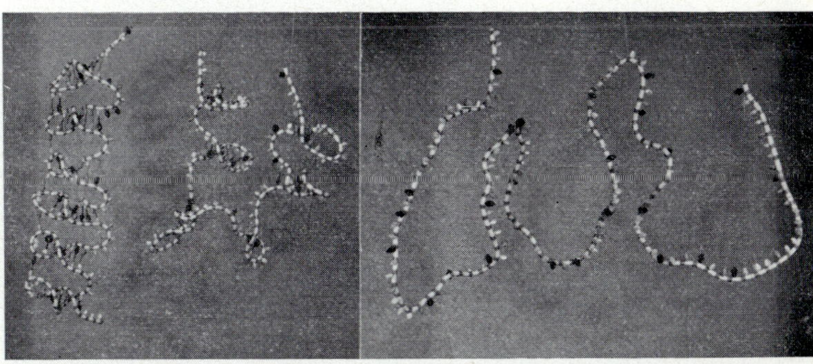

FIG. 16.5. Models representing possible conformations and configurational transitions of a high-polymer single-strand RNA macromolecule dependent on physicochemical changes in the environment. (Models based on a drawing by Spirin, A. S., in *Progr. Nuc. Acid Res.*, 1963, **1**, 335.) The hairpin loops are held together by further hydrogen-bonding.

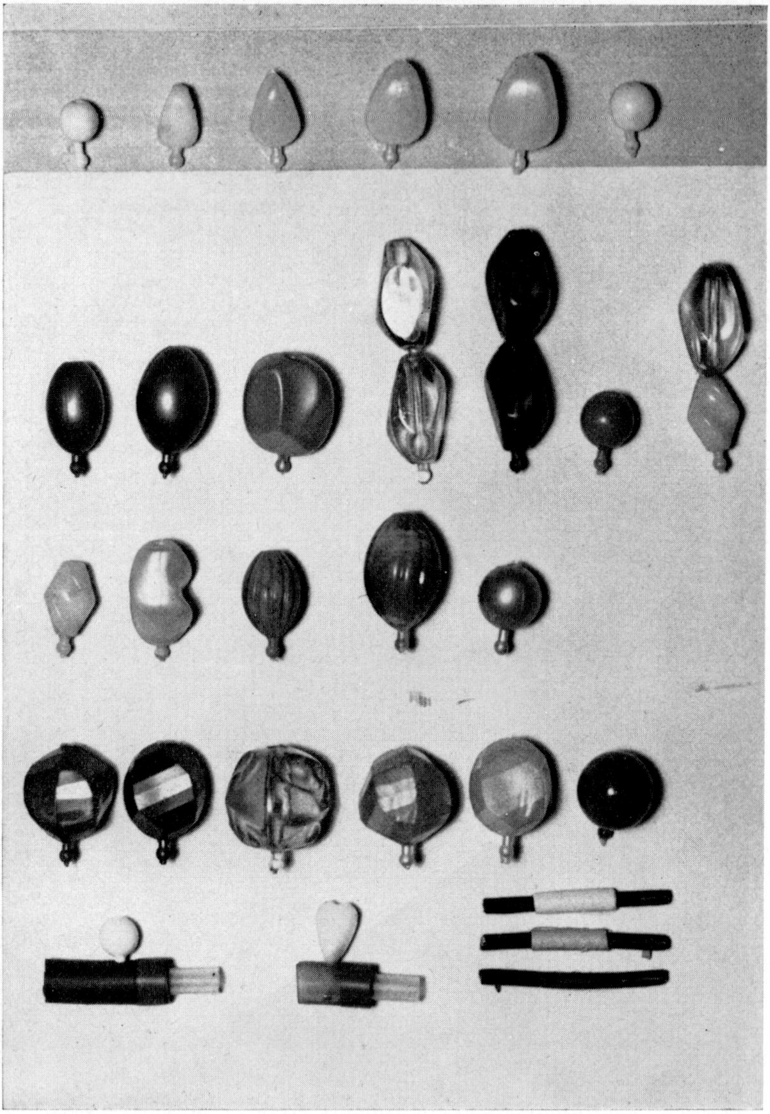

FIG. 16.6. Protein Bio-Bits, scale $\simeq \frac{7}{10}$. *Key:*

G	A	V	L	I	(Hy)	
D	E	H	K	R	(NH_2)	(ψLys)
C	M	S	T	P		
F	Y	W	(DNS)	(DNP)	(Ⓟ)	

Gly | Ac | various bonds

Note: all bonds are now full colour coded: s-s, yellow; salt bridges red; hydrogen, green; others, miscellaneous colours.

•H Glycine Gly[1] G[2] white	$\cdot CH_3$ Alanine Ala A white	CH_3 $\cdot \dot{C}H_2 \cdot CH_3$ Valine Val V white	CH_3 $\cdot CH_2 \cdot \dot{C}H \cdot CH_3$ Leucine Leu L white	CH_3 $\cdot \dot{C}H \cdot CH_2 \cdot CH_3$ Isoleucine Ile I white	•OH Hydroxy[8] Hy white
COOH $\cdot \dot{C}H_2$ Aspartic acid[4] Asp D red	CH_2COOH $\cdot \dot{C}H_2$ Glutamic acid[4] Glu E red	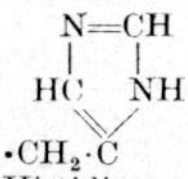 Histidine His H medium blue	$\cdot (CH_2)_4 \cdot NH_2$ Lysine Lys K light blue	$HN{:}\dot{C} \cdot NH_2$ $NH \cdot (CH_2)_3 \cdot$ Arginine Arg R dark blue	$\cdot NH_2$ Amino[4] medium blue
SH $\cdot \dot{C}H_2$ Cysteine Cys C yellow	$CH_3 \cdot S$ $\cdot CH_2 \cdot \dot{C}H_2$ Methionine Met M yellow	OH $\cdot \dot{C}H_2$ Serine Ser S light green	CH_3 $\cdot \dot{C}HOH$ Threonine Thr T dark green	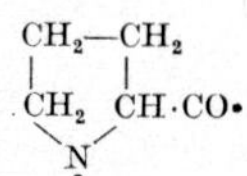 Proline Pro[5] P pink	$H_2N \cdot H_2C \cdot CH_2$ $\cdot CH_2 \cdot \dot{S}$ Pseudolysine (ψLys)[3] blue-yellow
 $\dot{C}H_2 \cdot$ Phenylalanine Phe F light purple	OH 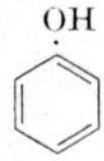 $\dot{C}H_2 \cdot$ Tyrosine Tyr Y dark purple	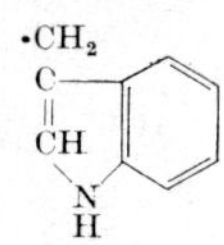 Tryptophan Trp W brown	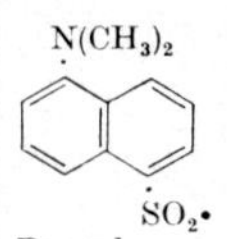 Dansyl (DNS)[6] green	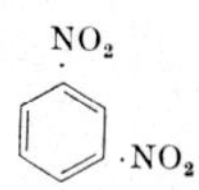 Dinitrophenyl (DNP)[6] yellow	$\cdot SO_3H$ $\cdot PO_3H$ Sulphate[9] or Phosphate red
Glycyl Gly	Acetyl Ac[7]	Various types of bond	Miscellaneous sulphur bridge hydrogen bond		

FIG. 16.7. Structure and nomenclature of the aminoacid R groups; colours refer to Bio-Bits design.

1. International three letter abbreviation.(1)
2. International one letter abbreviation.(1)
3. Term in general use. Bit and formula shown to illustrate formation of synthetic, non-natural, aminoacid useful in sequence studies.
4. Asparagine (Asn, N) formed by inserting amino Bit into aspartic acid. Glutamine (Gln, Q) formed similarly from amino and glutamic acid. Asx (B) used to indicate presence of Asp or Asn and Glx (Z) position of Glu or Gln or pGlu when exact identity not known; e.g. Asp found after mild acid hydrolysis which may have removed amide groups.
5. In proline the R group forms an integral part of the ring and cannot be shown separate from the N—C—CO which forms the peptide link or bond. Hydroxyproline (Hyp) is formed by inserting a hydroxy Bit into proline.
6. Used in sequence analysis.
7. Acetyl, formyl, etc. found in some peptides masking α-amino groups, e.g. α-MSH. White tube with red band.
8. Hy as in Hyp. hydroxyproline.
9. Sulphate cannot be represented as a single letter code (S = serine) and neither can phosphate in the aminoacid nomenclature although it is represented by *p* in the nucleic acid coding. This is frequently overcome using the symbols Ⓢ and Ⓟ unofficially.

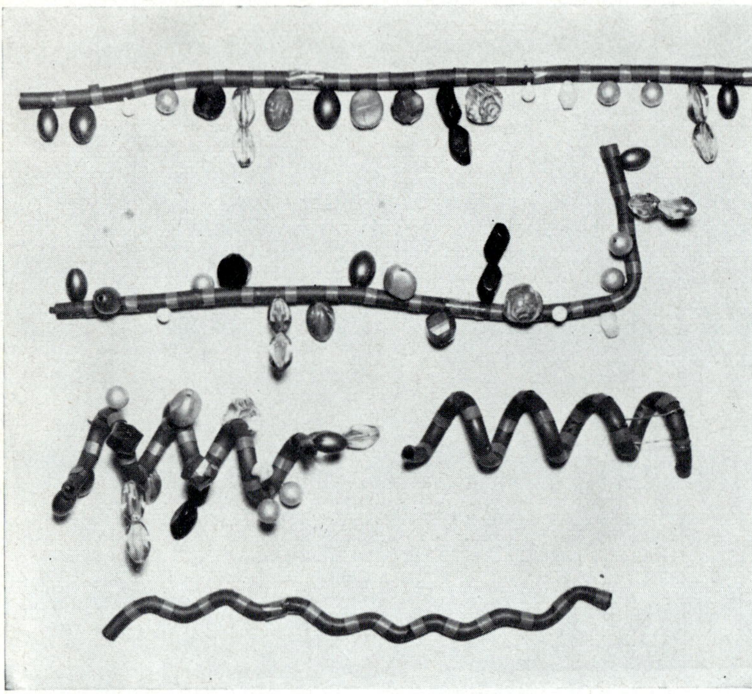

A
B
C
D
E

FIG. 16.8. Various usages of Protein Bio-Bits.

A. Straight chain model for illustration of primary structure, enzymic and chemical hydrolysis, homologies, etc. Note R group close to red carboxyl ring denoting chain direction.
B. R groups disposed around chain circumference with bend at proline position.
C. α helix backbone.
D. α helix with R groups—complete peptide.
E. Backbone for use in pleated chains with inter-chain hydrogen bonding.

Fig. 16.9. Insulin primary and tertiary structures.

(*a*) Primary structure for use in demonstrating the techniques of sequence analysis.
(*b*) Stereoview of dimer from 'outside'. A chain is dark and B chain is light. Note helical and anti-parallel structures with some hydrogen bonds shown.
(*c*) Stereoview to show the six disulphide bridges.

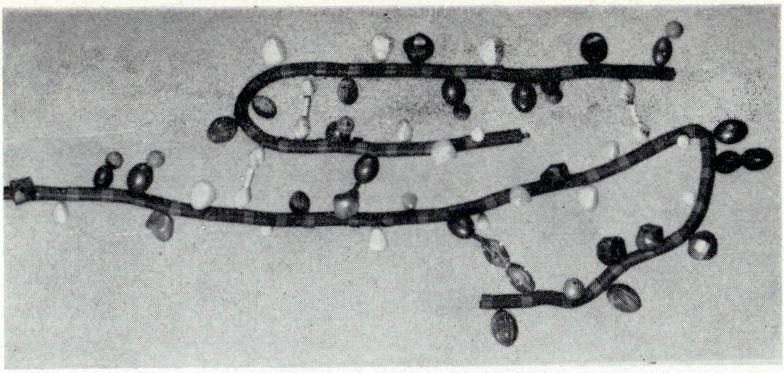

(a)

(b)

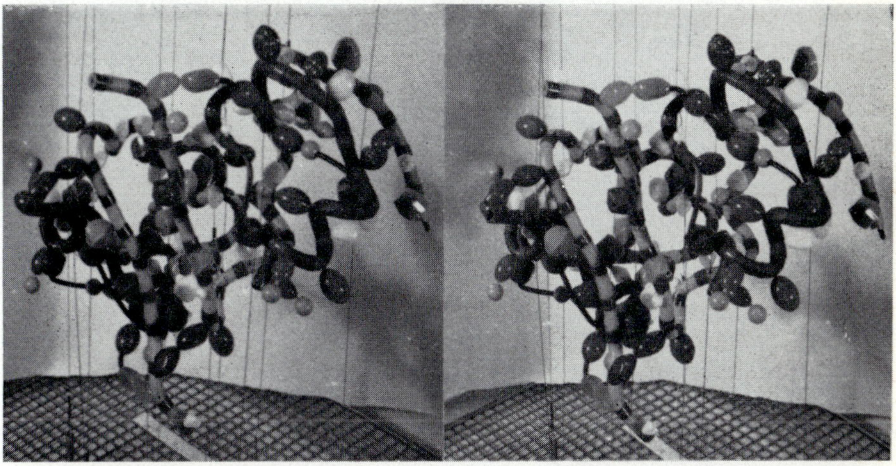

(c)

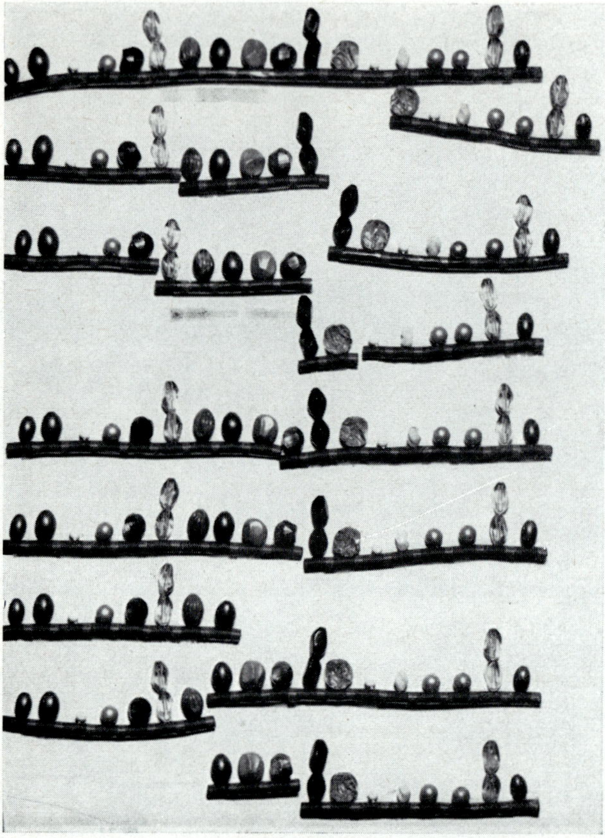

FIG. 16.10(*a*).

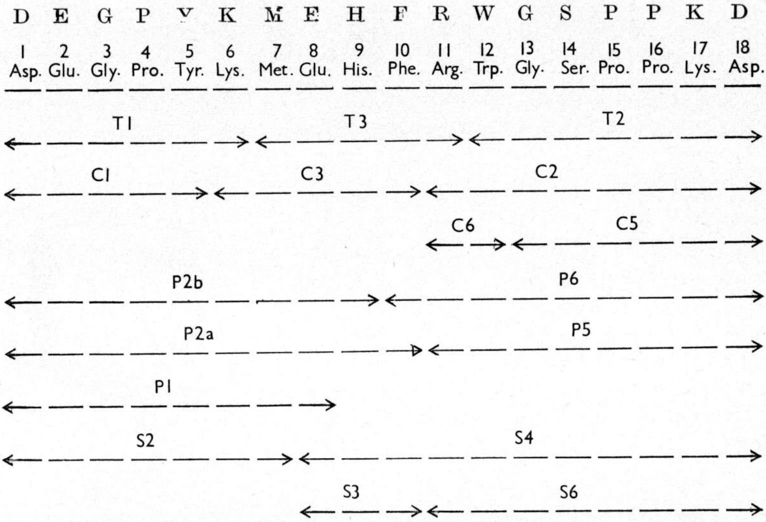

FIG. 16.10(*b*). Illustration of sequence analysis using different enzymes. MSH has been separately treated with trypsin (T), chymotrypsin (C), pepsin (P) and subtilisin (S). Each enzyme attacks specific bonds in the molecule to yield a set of smaller peptides.

(*a*) Using protein Bio-Bits.
(*b*) Explanation of (*a*) with chemical formulae.

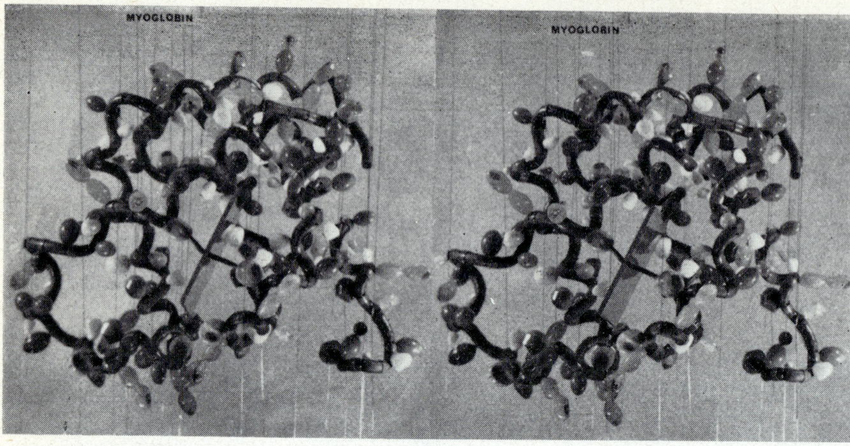

(*a*)

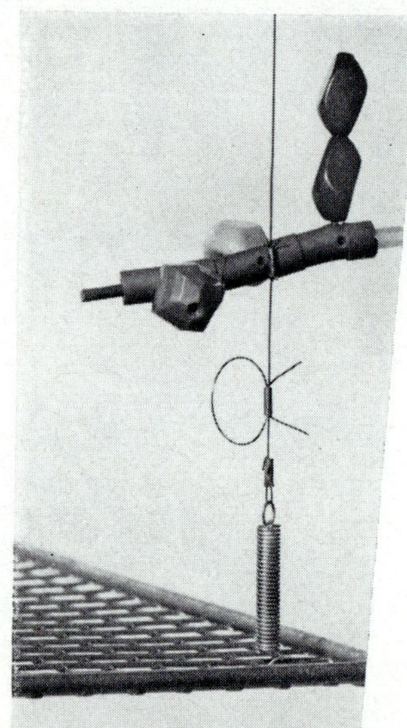

(*b*)

Fig. 16.11. Myoglobin and lysozyme.

(*a*) Stereoview of myoglobin showing helixes, the heme cleft, etc.

(*b*) Detail showing how spring loaded wire is fixed to the frame, the use of the ferrule and a tripeptide being supported in position.

(c)

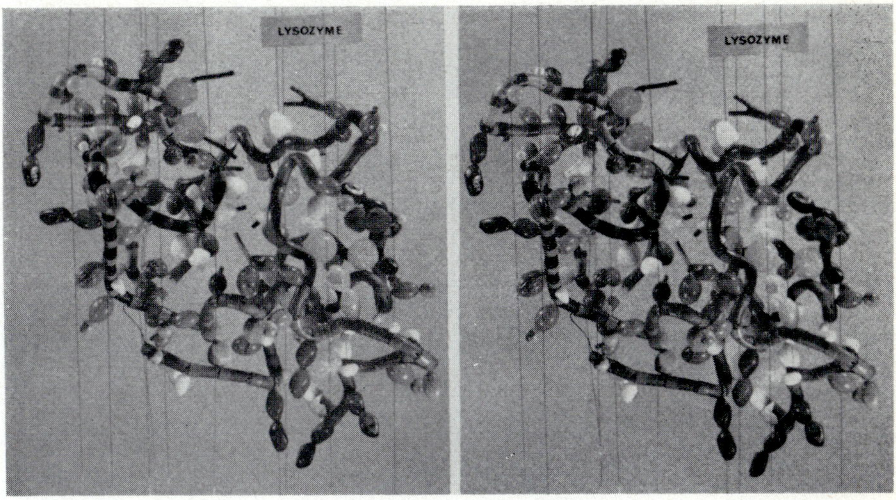

(d)

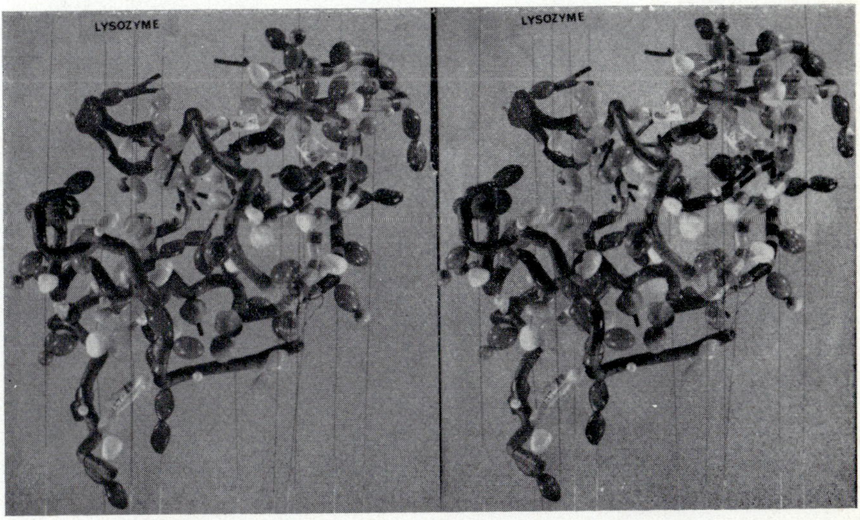

(c) Stereoview of lysozyme showing central cleft. Dark wires indicate hydrogen bonds for binding substrate and two lighter, central wires indicate the active site residues.

(d) Stereoview of lysozyme from the 'rear' showing some of the disulphide bridges.

(*a*)

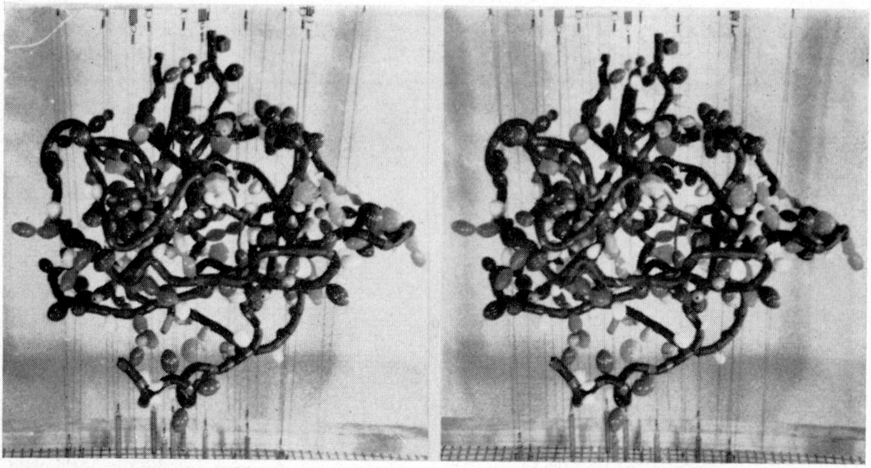

(*b*)

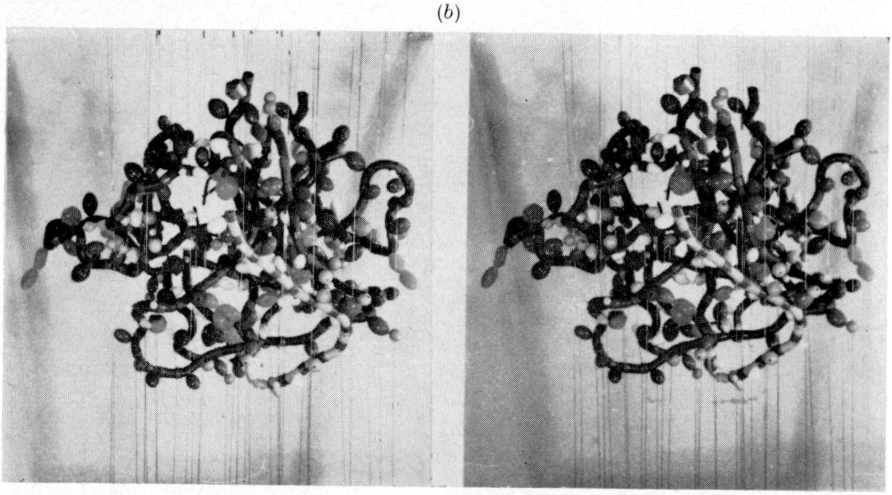

Fig. 16.12. Stereoviews of chymotrypsin.

(*a*) The active site 'hole'; bonds indicate Asp, His and Ser.

(*b*) View from the rear. Light chain at top is A chain joined by disulphide bridge to B chain. B16 terminus is hydrogen-bonded to Asp 194. C terminus of B chain and N terminus of C chain are also visible at bottom centre of the model.

(*a*)

(*b*)

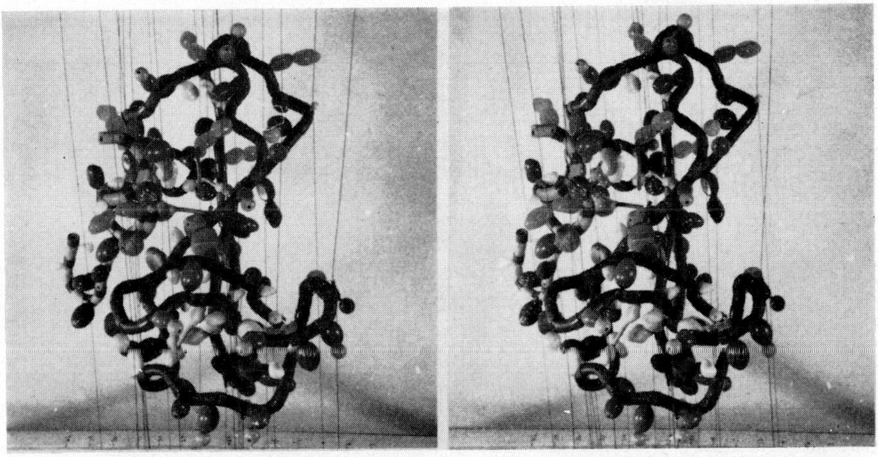

Fig. 16.13. Stereoviews of ribonuclease-S, RNAse-S.

(*a*) The central cleft with His, His and Lys marked. Note the number of Lys and Arg groups which lie on each side of the cleft. The S-peptide, light colored chain, is held in position by non-chemical bonding only.

(*b*) Side view showing some disulphide bridges, helical portion of S-peptide, etc.

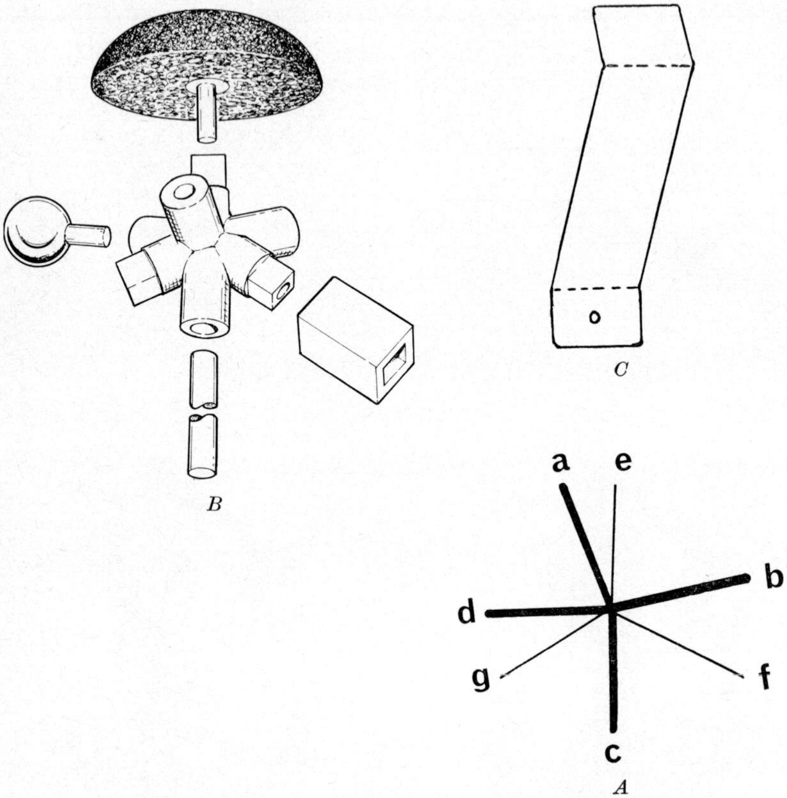

FIG. 16.14. The Atomunit.

A. Atomunit in diagrammatic form showing the disposition of the four tetrahedral (heavy) and three trigonal (light) arms. Note that arms e–c are at 180° and so can be used to align each unit. Tetrahedral arms B, C and D have an imprinted letter for easy identification.

B. Atomunit showing tetrahedral (round) and trigonal (square) arms, van der Waals' (mushroom) envelope, square bond lock, bond rod and round white hydrogen marker. Atomunit Colours: C, black; N, blue, O, red, P, purple; S, yellow; miscellaneous, white.

Note: the mushroom units have been so designed that they can be adapted for use with other skeletal models.

C. Simple tool which can be used to pull out bond rods from Atomunit.

TABLE 16.2

The Table indicates the Included Angle, and Symbol of Particular Arm Pairs of the Atomunit together with Examples of Compounds in which they are used

Symbol and number	Arms	Included angle (nominal)	Examples of use	
○	a–b b–c a–c b–d a–d c–d	109° 28′	Carbon: Phosphorus: Sulphur:	Paraffins, Pentoses Cyclohexanes, Peptide bonds Phosphates Sulphates
□	e–f e–g f–g	120°	Carbon and Nitrogen:	Nucleotides Peptide chains Benzene, Pyridine
D_1	c–e	180°	C—O—H···N:	Hydrogen bond
D_2	a–f	105°	C—S—S—C: H—O—H:	Di-sulphide bridge Water
U_3	a–g	94°	Oxygen: Carbon:	Sugars, Phosphates β-Lactam Benzyl-Penicillin

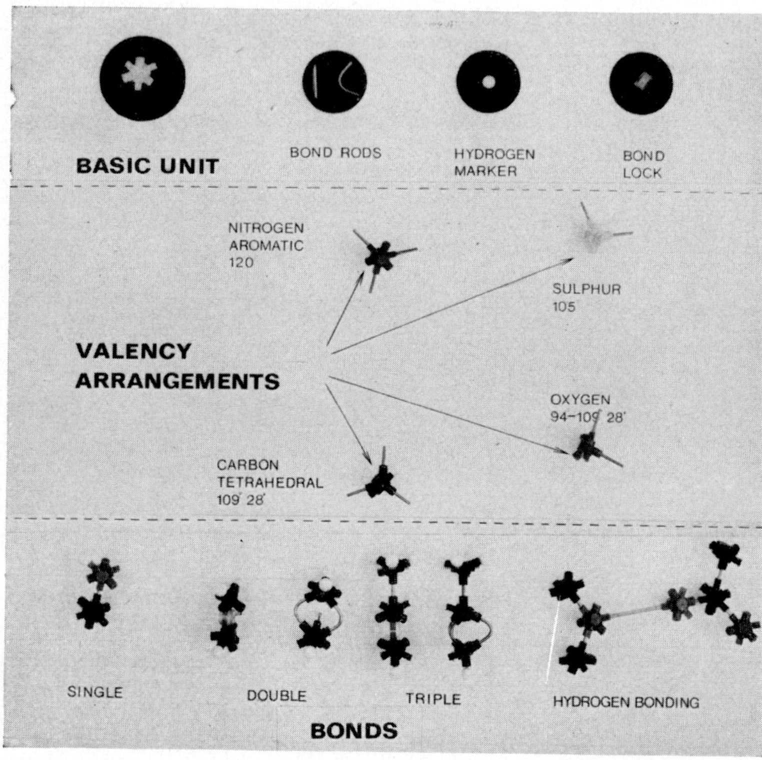

Fig. 16.15. Valency angle and bond arrangements using the Atomunit.

TABLE 16.3

Atomic and Hydrogen Bond Lengths of known occurrence in Peptides and Nucleic Acids(4)

These are average lengths as individual compounds frequently show minor differences due to their overall three dimensional structure. A scale of 2·0 cm./A has been used. However, from the calculated length of the bond, in millimetres, 4 mm. must be subtracted because the rod does not penetrate to the centre of the Atomunit but stops 2 mm. from this point. When increasing the model scale by increasing the rod length, the full length is first calculated and the 4 mm. then subtracted from this. Hydrogen markers have a hole running right through hence it is not necessary to reduce the bond length by 4 mm. This hole enables the same marker to be used with a hydrogen-bond rod. For the most accurate models, individual bonds should be cut appropriately to suit the X-ray data using a special cutter.

Bonding Atoms	Average Bond Length A	× 2·0 cm.	− K 4 mm.	Sample Structure	Atomunit Bond Rod Length cm.
N—C	1·41	2·82	2·42	Purines, Pyrimidines Peptide Bonds, etc.	2·4
C—C	1·47	2·94	2·54	Hexanes, Peptides, etc.	2·4
C═C	1·33	2·66	2·26	Purines, Benzyl Groups	2·4
C≡C	1·21	2·42	2·02	Acetylenes, etc.	2·0
C═O	1·20	2·40	2·00	Carboxyl Groups	2·0
C—S	1·79	3·58	3·18	Cysteine, Sulphoxides, Sulphones	3·2
C—Br	1·89	3·78	3·38	Bromobenzenes	3·2
O—P	1·65	3·30	2·90	Phosphates	2·9
C—H, N—H	1·04	2·08			2·0
O—H	0·96				
S—S	2·12	4·24	3·84		3·2
Hydrogen Bonds					
OH···O	2·8	5·6	5·2	Water	
OH···O	2·6	5·2	4·8	carboxylic acids	
NH···O	2·9	5·8	5·4	peptides, amides, nucleic acid base pairs	5·4
NH···N	3·0	6·0	5·6		
C≡N	1·13	2·26	1·86		2·0
C—O	1·40	2·80	2·40		2·4
C—Cl	1·74	3·48	3·08		3·2
C—P	1·8	3·6	3·2		3·2
S—H	1·3	2·6	2·2		2·4
P—H	1·4	2·8	2·4		2·4
P—Cl	2·0	4·0	3·6		3·2

FIG. 16.16. Atomunit models of an aminoacid and a dipeptide, the former being displayed as if it were to be the N-terminal end of the dipeptide. Both are shown in a generalized form where the whole R group is represented by a long white rod. The dipeptide is shown as if both carbonyl oxygens are hydrogen bonded to other aminoacids as they would be in an α-helix and with the free hydroxyl replaced by an NH as would occur in a tri-peptide although in black and white photographs this looks identical with an OH. Note the bond locks fixing each peptide bond in its plane and the two hydrogen bonds to NH groups of other aminoacids. Colour scheme N—C—C—O = black, blue, black, red.

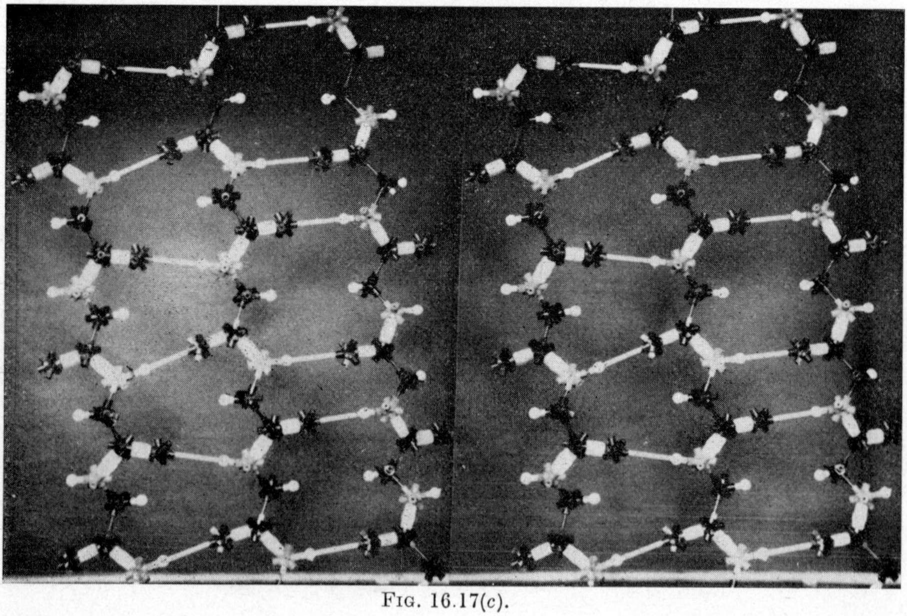

FIG. 16.17(*c*).

FIG. 16.17(*a*). Protein α-helix with R groups omitted to highlight the regular helical structure. Part of the helix is shown in skeletal form and part clad with van der Waals' envelopes to indicate space-filled form.

FIG. 16.17(*b*). C-terminal octapeptide of whale myoglobin, space-filled. (Courtesy of Dr. N. J. Crumpton.)

FIG. 16.17(*c*). On facing page. Stereopair of parallel (left) and anti-parallel (right) chains of peptide backbone.

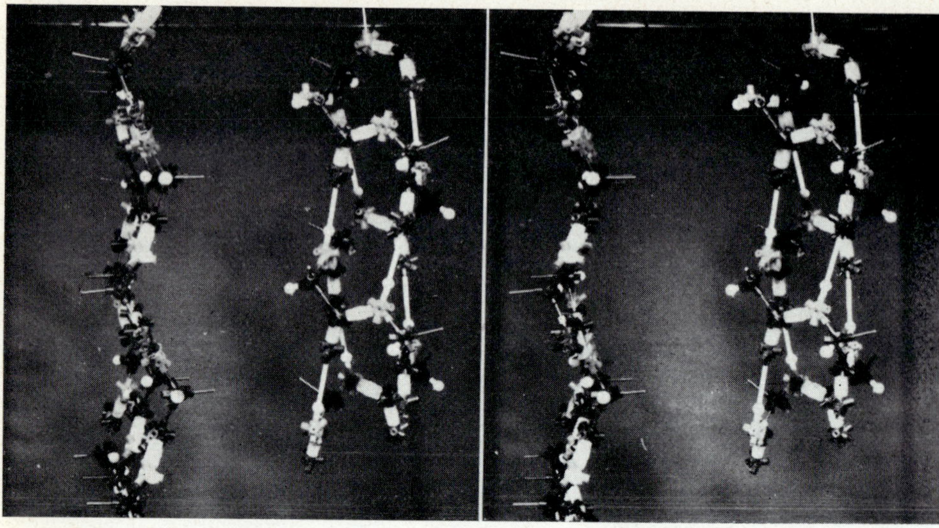

FIG. 16.17(*d*). Stereopair of model fig 16.17(*c*) taken edge on (left) and of α-helix backbone (right).

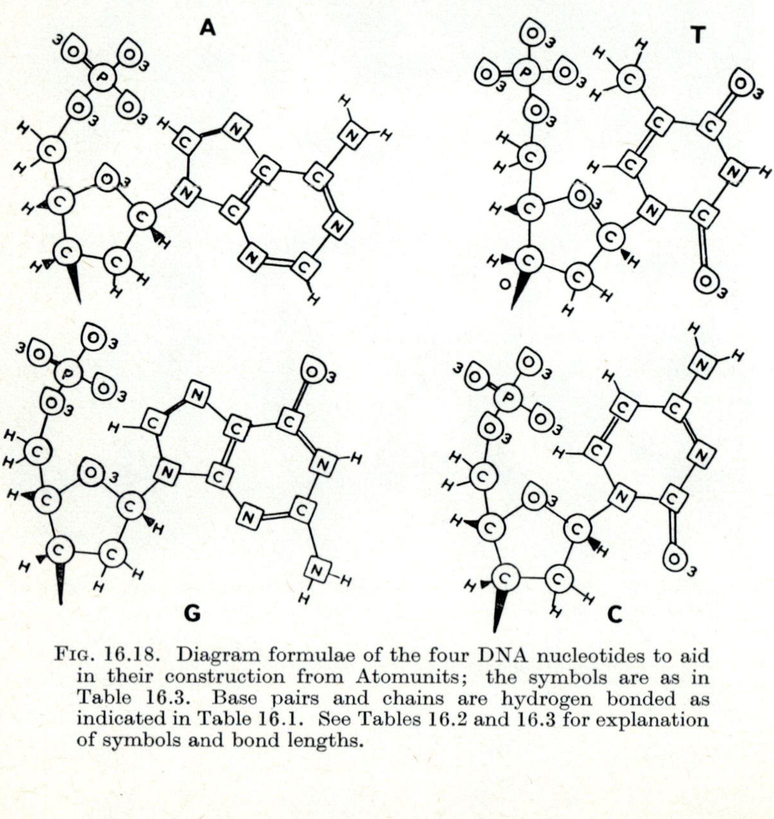

FIG. 16.18. Diagram formulae of the four DNA nucleotides to aid in their construction from Atomunits; the symbols are as in Table 16.3. Base pairs and chains are hydrogen bonded as indicated in Table 16.1. See Tables 16.2 and 16.3 for explanation of symbols and bond lengths.

FIG. 16.19. Part of an RNA double helix. At the top the helix curls round to form a hairpin loop but, lower down, a typical double helix is shown. Part of the model is skeletal and part space-filled. (Courtesy of Dr. R. A. Cox.)

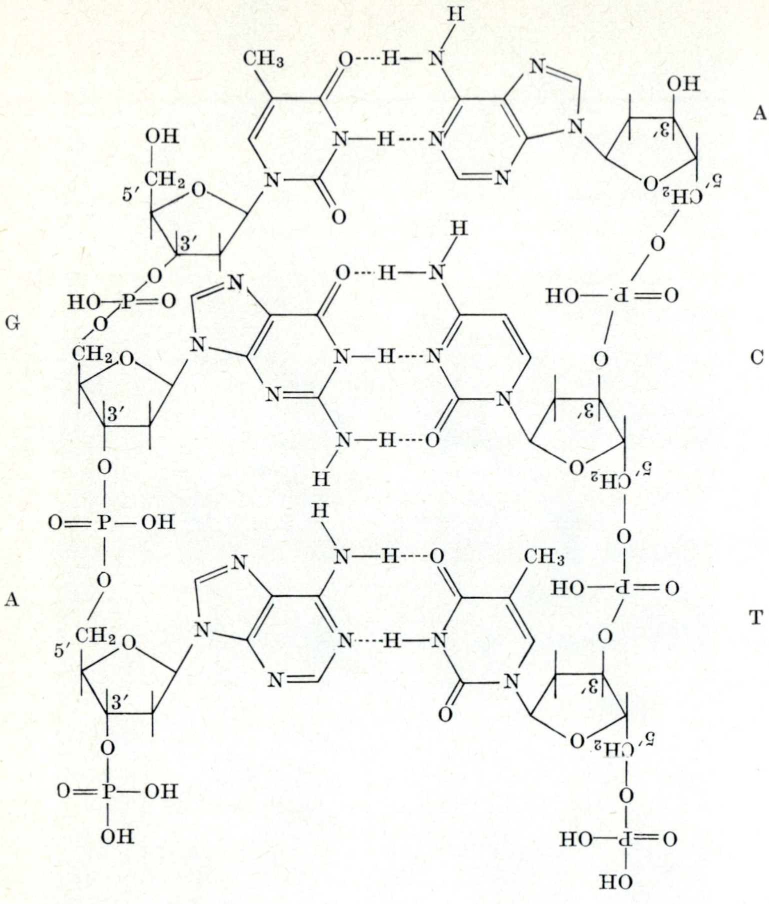

left hand chain ↓	*right hand chain* ↑
T-3′-p-5′-G-3′-p-5′-A-3′-p	d(p-5′-T-3′-p-5′-C-3′-p-5′-A)
TpGpAp	pTpCpA
TGAp	pTCA
TGA	TCA

FIG. 16.20. Illustration of base pairing and nomenclature of helix. The two chains run anti-parallel, i.e. they form in opposite directions. The sugar-phosphate chains run vertically in the plane of the paper whereas the base pairs lie horizontally at right angles to the plane of the paper. The nomenclature depends on the amount of information one wishes to convey. This formula of a triple base-pair may be used as a construction model.

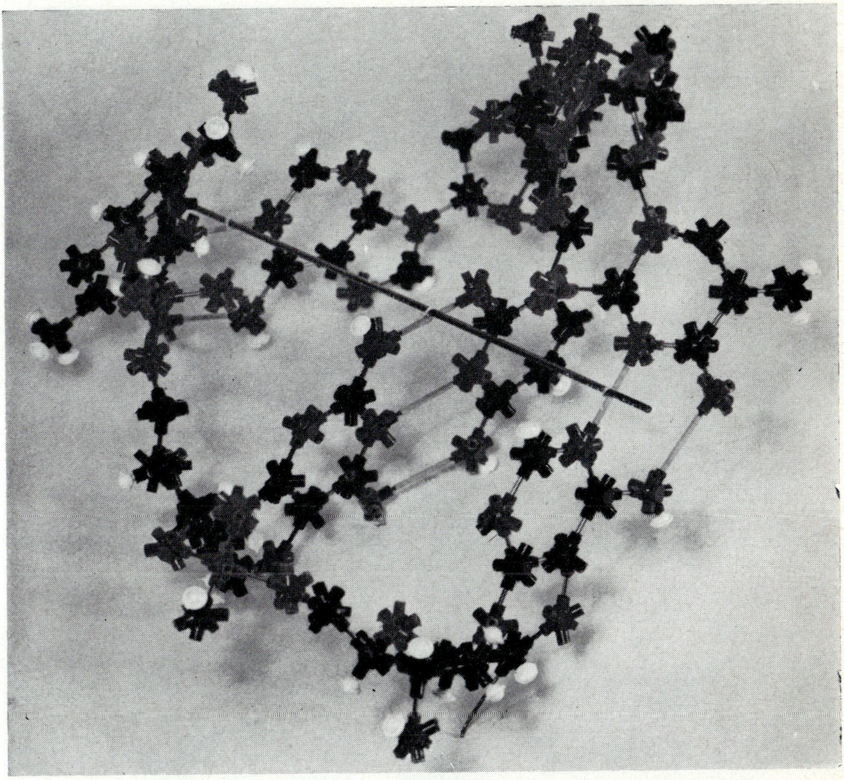

Fig. 16.21. Photograph of model corresponding to Fig. 16.20 (HM's have been omitted from the right hand sugar-phosphate chain). The rod through the middle of the figure represents the axis of the helix. All base pairs are flat and in parallel planes but the turn of the helix is evident

FIG. 16.22. Some small molecules. (*a*) chair form of cyclohexane, (*b*) phosphorous pentoxide P_4O_{10}, (*c*) octa-sulphur, (*d*) benzyl-penicillin.

Fig. 16.24. Chlorophyll-arachidic acid complex on frame. The model was built for the study of surface film interactions. Note the simple frame construction. The mesh is 10 cm spacing as standard (mesh in figure 7.5 cm for other reasons) and the frames are built as 2, 3 and 4 ft cubes. Vertical wire-rods are placed at any suitable position and held by press in clips. Cross wire rods are inserted into appropriate free holes in the atomunit and clipped to the verticals using suitable rod lengths. The frame system is extremely easy to build in as required, light weight and completely stable such that the completed model can be turned for viewing in any direction. If sufficient rod and clip structure is built into the interior structure it is possible to remove the model completely from the frame without deformation. [Figure by courtesy of Dr. J. Harris, Royal Institution, London]

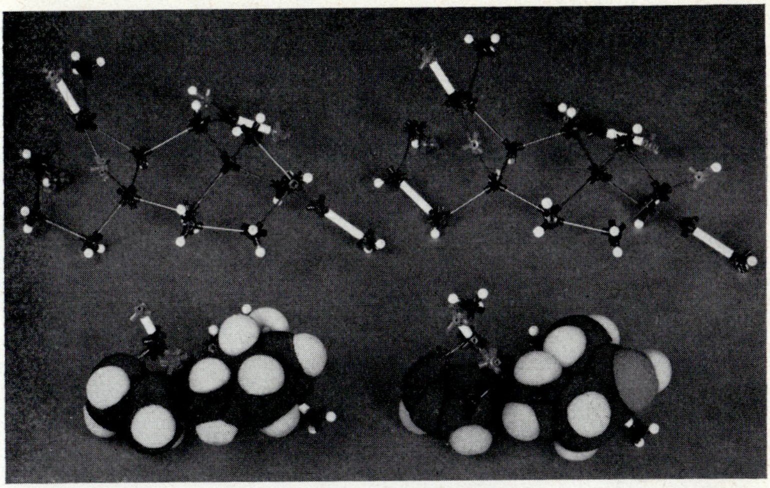

FIG. 16.23. The gibberellins. The upper figures show gibberellin and gibberellic acid built at double scale and the lower figures are spaced filled on the standard scale.

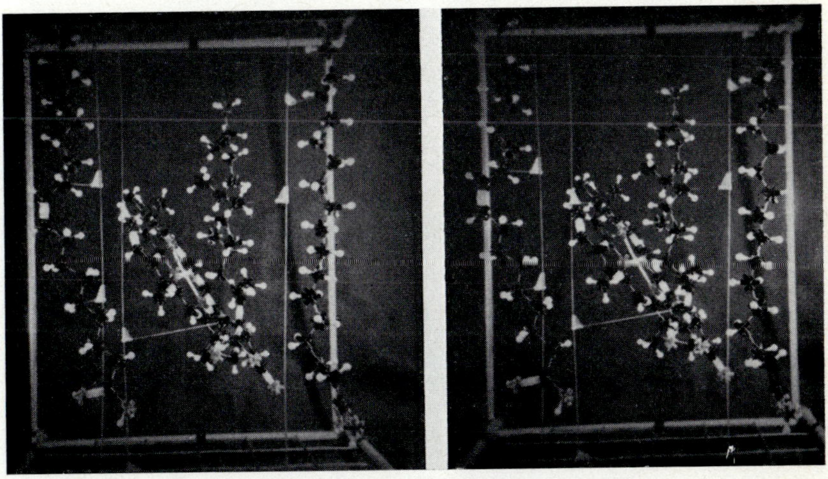

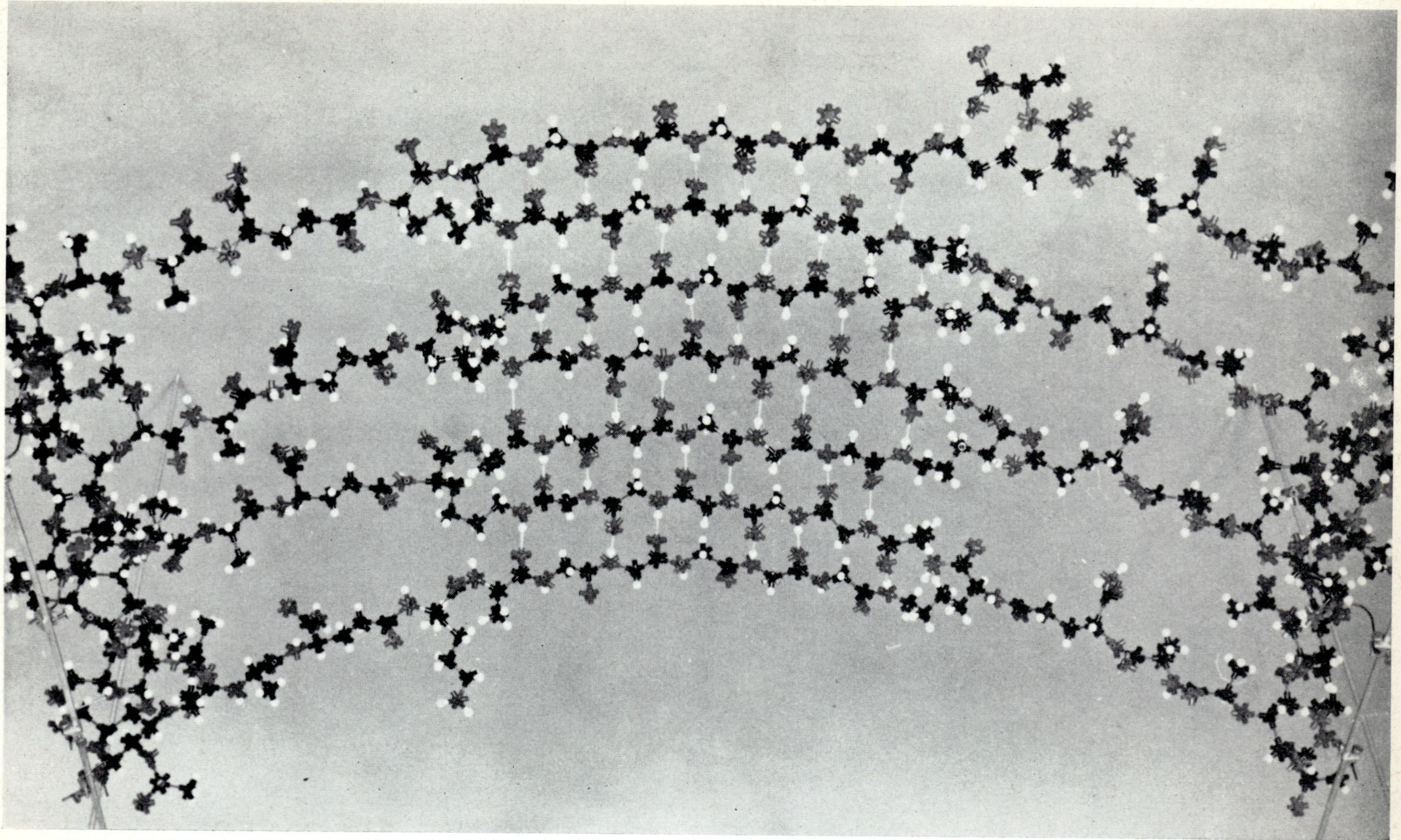

Fig. 16.25. Mucopeptide—hypothetical structure of mucopeptide of S. aureus (figure by courtesy of Drs. M. Kellerman and H. J. Rogers).

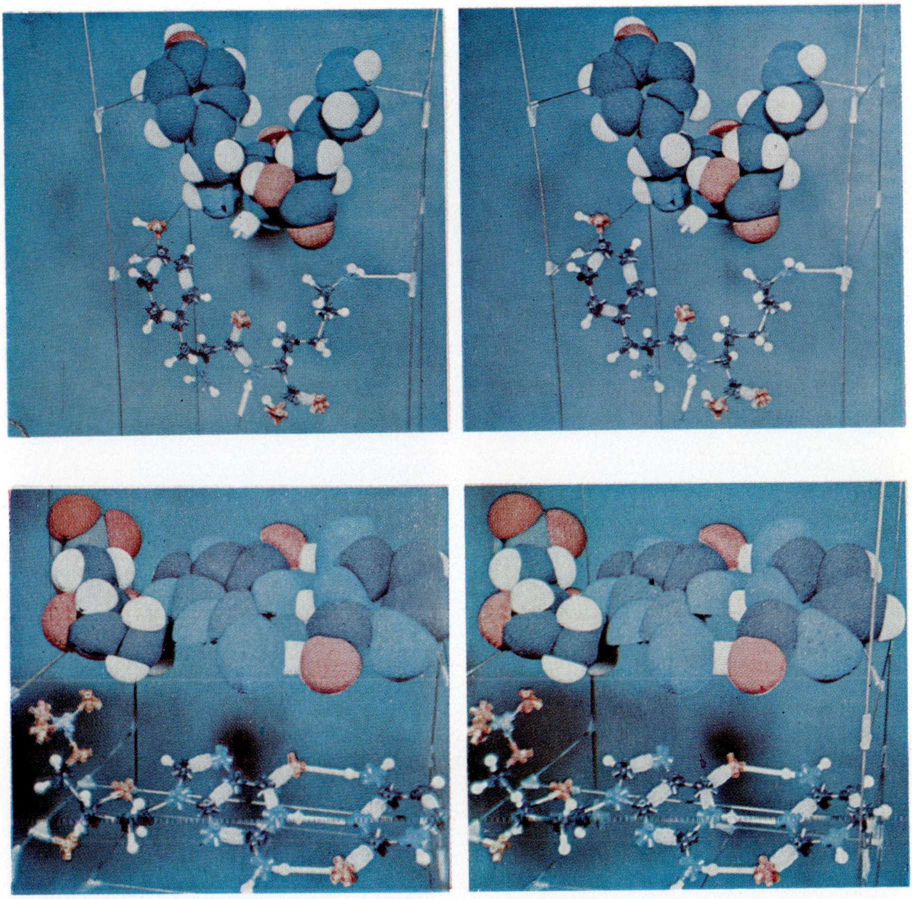

Tyrosyl-lysine (upper) p-5′ G ≡ C base pair (lower)

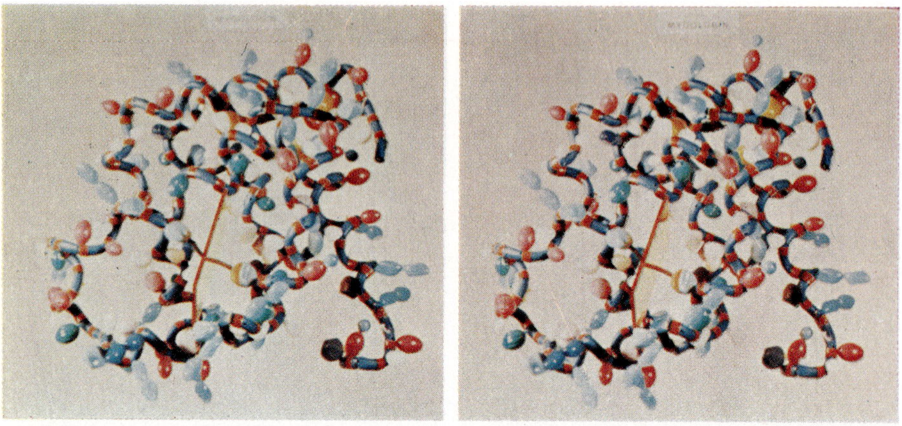

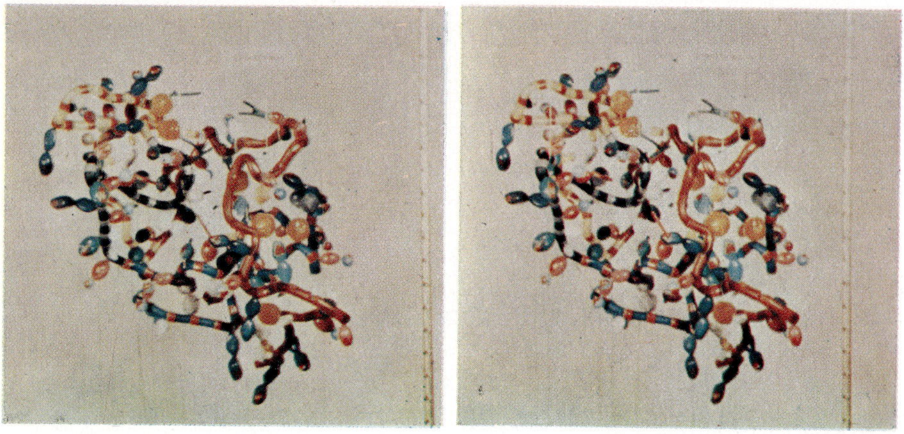

myoglobin (upper)

lysozyme (lower)

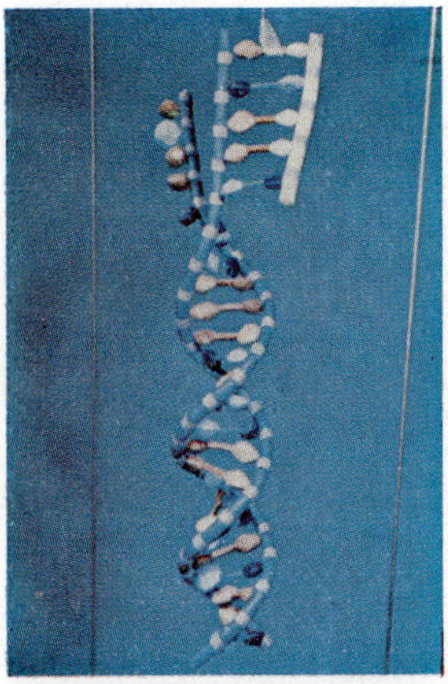

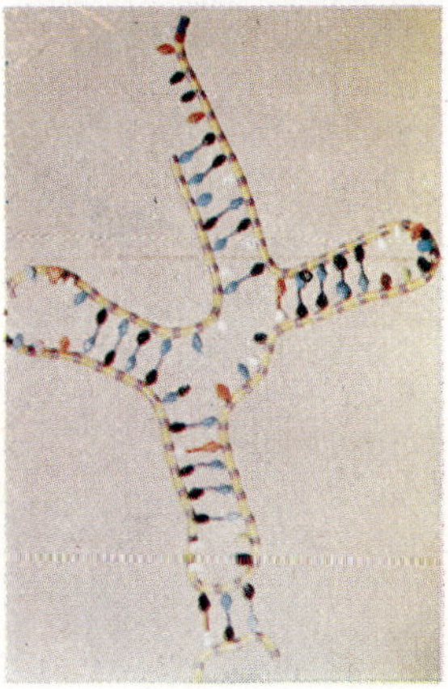

double helix; t-RNA

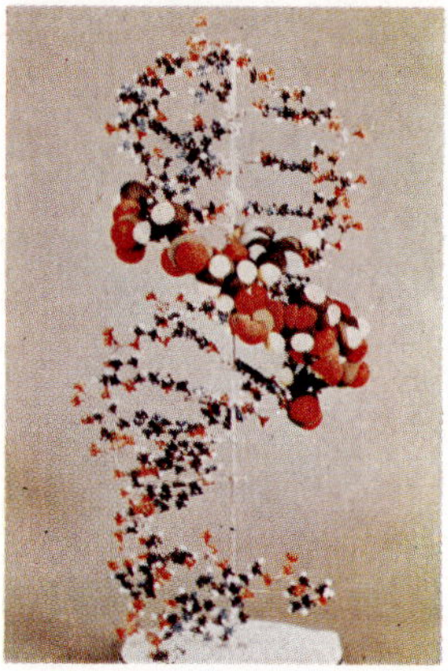

α-helix; RNA double helix

Seven main hydrogen bonds are known to bind the substrate and these may be indicated by green wires projecting from the R group bits. Two bonds, one from aspartic acid 52 and the other from glutamic acid 35, are known to take part in the enzymic attack on the substrate and these may be indicated by means of projecting red wires as can be seen in Figure 3. The cleft is again obvious and viewing in ultra-violet light after applying the red fluorescent paint as described above is equally instructive.

RN Ase, chymotrypsin, hemoglobin, cytochrome C, carboxypeptidase A, insulin, LDH and others. Data are also available.[3]

DNA. Double helix, replication and transcription (see Fig. 16.2).

1. Cut two lengths of tube for about 30 residues. Each tube should commence with a 5′ phosphate and terminate with a free sugar (green) end. The two tubes should be placed with the backbone running in opposite directions because of the anti-parallel nature of the helix.

2. Prepare 10–15 of each base pair by connecting A—C and G—T with the correspondingly coloured wire. Insert these base pairs randomly along the two chains as specific sequences are not known. The model will now look like a ladder.

3. Take a thin, sturdy rod such as eighth-inch dowel or a knitting needle and thread alternatively over and under the wires joining adjacent base pairs. Grasping the helix gently with the two hands, turn and compress it so that the left hand moves clockwise and the right hand anti-clockwise. A right-handed helix is so produced, i.e. looking through the model the helix appears to be turning clockwise into the distance. The degree of turning necessary is adjusted so that base-pairs 1, 11, 21, etc., lie vertically above each other in the same plane as there are ten base pairs per full turn. All the bases should be adjusted to lie flat, but in a plane of about 14° to the horizontal, with the chain vertical and further adjustment will enable the wide and narrow turns to be made obvious. Gently remove the dowels.

4. Replication and transcription are modelled as follows: Open up the base pairs at one end of the chain and, starting with a 3′ end, add monomer nucleotides across the hydrogen bonds and join with nylon links still maintaining the anti-parallel nature of addition.

RNA, Double Helix. Prepare as described under DNA 1–3 but using the appropriate bases and yellow ribose backbone. Remove the dowel rod and slightly expand the helix so that the vertical axis runs through the helical coil itself and not through the hydrogen bonds.

***t*-Ala-RNA^{ala}**. Alanine transfer RNA with the alanine in position. See Figs. 16.3 and 16.4.

1. First build the open chain with the bases correctly positioned and finishing with the aminoacid in place.

2. Add the hydrogen bonds to the correct base pairs starting from those around the anti-codon and gradually forming each of the four bonded segments.

3. Crosslink the base-pairs to form the final tertiary structure. Arrange this best so that the backbone of the whole model can be seen readily whilst standing without support.

An alternative procedure is to form the base-pairs first, inserting them into the correct holes which can be numbered to assist this. Then the non-paired residues can be added. Finally the tertiary structure is formed. A further refinement would be to convert the anticodon hydrogen-bonded portion to a helical shape.

ATOMUNIT MODELS*

As stated previously all molecular models kits available to-day (except Bio-Bits) use the atom as the basic reference unit. The difference between the two main types is how the atom is represented.

The space-filling type of atomic unit begins as a sphere representing, on a particular scale, the van der Waals' envelope. The sphere is then cut so that the flats are at defined angles and distances to allow other cut spheres to be attached in the correct configuration for the molecule it is to represent. The skeletal method relies on spokes radiating from a common centre, this being the atomic centre, and the spokes the covalent bonds. These units are connected by various means to form a spider-like network, each junction of two or more spokes being the centre of an atom. These two types of model are complementary, the space-filling units fail to show clearly ring formation, bond angles and internal structure which is more easily defined in the skeletal type but, on the other hand, the latter do not show steric hindrance and molecular packing as satisfactorily.

Atomunits were designed in an endeavour to combine the best features of both types and yet have a model system that was light in weight, easy to construct with the minimum of skill and technical knowledge and low in cost.

The construction unit which forms the basis for the model kit is a small plastic moulding, the plastic being chosen to give the correct degree of flexibility. The unit has been named "Atomunit" and these are colour-coded to represent individual multivalent elements such as carbon, oxygen, nitrogen, etc.

The units represent the atomic centres and in a model they will show the correct spatial relationship between each centre. The centres are joined by rods simulating the correct bond lengths, any stress set up in the model by the varying lengths of the bond rods will be reflected in the angles subtended by these rods. Double, triple, or hydrogen bonds may easily be simulated. Monovalent atoms (hydrogen, halogens) may be attached to the model where needed.

Using Atomunits, accurately scaled skeletal models may be produced of virtually any size simply by changing the lengths of the joining rods from one scale to another. To change a portion of a skeletal model to a space-filled model or to make a complete space-filled model only requires the attachment of separate mushroom-shaped flexible foam segments to that region or to the whole but these are suited to the smallest dimensional scale only. These segments have been named "Atomunit Envelopes" and represent the van der Waals' envelope.

* Atomunit models designed at the National Institute for Medical Research. Sole distributors Capital Biotechnic Developments, Ltd., 66A Churchfield Rd., London, W.3, England. Pat. No 1,125,840.

Description. The Atomunit comprises two groups of arms radiating from a common centre. The main group of four arms is tetrahedral, each arm being circular in section, the secondary group of three arms is in a trigonal configuration and each arm is square in section. These two groups are so disposed about each other that most angles commonly used are available, (Table 16.2). The unit allows for all the valency bond angle arrangements of any of the elements commonly used in organic chemistry (except some of the metals for which a further unit will be made available). In spite of the versatility of the Atomunit each weighs only 1·25 gms. In each of the seven arms is a socket, of identical depth, to accept a bond rod and grip it tightly. The centres of adjacent Atomunits are then accurately determined by the bond rod length. Each Atomunit possesses an 8 mm. diameter-punched face into which may be inserted bond rods carrying labels. With the more complex models, the main chain may be easily traced using coloured plastic-coated wire which can be threaded through 8BA solder tags whose ends are then pressed into the Atomunit arm sockets. The bond rods, of blackened mild steel, are available as pre-cut standard lengths or as stock length from which standard or non-standard lengths may be accurately cut using special cutter pliers fitted with a gauge stop. All bond rods are cut 4 mm. shorter than the calculated distance between atom centres to allow for non-penetration of the rod to the centre of the Atomunit which has a solid core. (See Figs. 16.14 and 16.15 and Tables 16.2 and 16.3.)

Where double or triple bonds occur in the model these are simulated by locking the adjacent Atomunits against rotation by the use of Bond Locks, which are square sectioned tubes that fit over the square ends of adjacent trigonal arms of any two Atomunits, the distance between the two units still being controlled by the bond rod. Hydrogen and the halogens are represented by small spherical units that may be plugged into the appropriate positions on the model. The hydrogen sphere is moulded with a continuous hole through one diameter. Hence it may be affixed with a small bond rod or, alternatively, a hydrogen bond will pass right through it to join two atomunits.

A skeletal model may be partly or completely space-filled by the addition of the Atomunit Envelopes. These are flexible, foamed-plastic, mushroom shapes coloured to the same code as the Atomunits, a spigot in the centre of each flat surface allowing it to be attached to any vacant position on any Atomunit. When a single Atomunit is clad with envelopes the simulated van der Waals' envelope is correct to a scale of 2·0 cms. per A. As Atomunit Envelopes are attached to a skeletal model they are deformed by adjacent envelopes to give a good representation of the complete van der Waals' envelope (Fig. 16.14).

Stages in the Construction of a Model

1. Decide the scale required for the model. If it is to be space-filled at certain parts then it must conform to a scale of: 2·0 cms. per A but, if not, any scale may be used that meets the requirements.

2. Estimate and select the number of bonds of each size.

3. In constructing a model having a repeat sequence it is preferably to commence with this part. Insert the correct rod and push it firmly into the hole of the Atomunit appropriate to the configuration: *a*; *b*; *c*; or *d* if it is tetrahedral; *e*, *f* or *g* if trigonal, for other angles consult Tables 16.2 and 16.3.

4. Holding the Atomunit with its rod, press the rod home into the hole of its adjacent unit. Repeat until the sequence is complete. Where rings are formed, gently ease the rod into the hole and, at the same time, strain the ring so that the bond arms are in line with the bond rod. At a double bond place a bond lock over one of the square ends before connecting the other unit with the bond rod.

5. Complete all sequences. Connect these with their connective units and complete the model with its side chains, hydrogen markers etc.

6. The model may reach such proportions that it is unable to support itself in the correct conformation and support rods will be required. For this the author has found the Biotechnics miniature scaffold system excellent; $\frac{1}{4}$ inch black aluminium tubes are used as main verticals and the stock bond rod material used to connect the model to the clamps.

7. Attach Atomunit Envelopes where required.

Construction of an L-aminoacid. $H_2N.CHR.COOH$. The carboxyl carbon is referred to as C_1 and the other (α carbon) as C_2; R group carbons are numbered as C_3 upwards when in the form of a chain (see Figs. 16.16 and 16.17). N is trigonal with two hydrogen markers (HM's) and arm g is attached to C_2. For C_2, HM is attached to arm d, the R group to arm c and arm a runs to C_1. C_1 is trigonal, arm g is attached to C_2, two oxygens are attached to arms e and f using oxygen arm e if hydrogen bonding is to be used as in a peptide helix or to any arm if not. An HM is attached to one oxygen (not the one taking part in the hydrogen bonding). D-aminoacids are constructed identically except that the HM and R group are interchanged on arms c and d of C_2.

R groups or side chains. These are constructed separately and then inserted into arm c of C_2. A selection of R groups is described as these will be used subsequently for the whale polypeptide described below.

Tyrosine. The benzene ring consists of six carbon units connected to each other by the bond rods, using two of the trigonal arms. As there are three double bonds in the ring each alternate connection will have a square bond restrainer straddling the adjacent square ends of the units. From one of the carbon units a bond rod will connect to ad oxygen in a trigonal form with a hydrogen marker occupying one of the other trigonal arms. From the ring carbon opposite the C—OH a bond rod will connect to a carbon in a tetrahedral form; two of the other arms will be completed with HM's, the remaining arm will join the chain through the spare arm of the C_2.

Lysine. Four carbon units are connected in a chain using two of the tetrahedral arms, the remaining two arms having hydrogen markers. The vacant arm of the end carbon is completed with a nitrogen attached by a trigonal arm with two HM's completing the three. The vacant arm of the first carbon connects to the chain as previously described.

Glutamic Acid. Two carbons connected tetrahedrally with HM's, the third carbon trigonal with oxygens on the remaining two arms, the oxygens being in trigonal form with HM on one of them. Connect to chain.

Leucine. Two carbons connected by tetrahedral arms. The first is completed with two HM's and a bond rod to the chain. Second carbon has one HM and two carbons attached also tetrahedrally. Finally completed with 3 HM's on each of the two carbons.

Glycine. Complete the C_2 with a hydrogen marker.

Glutamine. This will be constructed the same as glutamic acid except that the OH is changed for NH_2. Again, this will be in trigonal form.

Construction of an α-helix. To show the general properties of an α-helix, a minimum of 8–9 aminoacid residues should be joined. From the vast mass of data now available, a number of general rules have emerged.

1. The six atoms, C—CO—NH—C, surrounding the peptide link lie in a plane. This is accomplished by using bond locks across the C—N and C—O bonds (see Figs. 16.16 and 16.17).

2. The R groups on any two adjacent aminoacids lie in the trans position across the intermediate peptide bond.

3. The maximum amount of hydrogen bonding occurs. In each aminoacid, the hydrogen bonds from the NH and the CO point in opposite directions and are roughly parallel to the helix axis. This is accomplished by bond locks between units. The bond joins the O from one aminoacid to the N of the fourth aminoacid along the chain and a long hydrogen bond passes through a modified HM to fit into the N atomunit. To align the helix correctly, rotation is allowed around both tetrahedral bonds from each C_2 and all trigonal bonds are bond-locked (except those carrying HM's).

4. Proline, because of its unusual structure (see Fig. 16.7), only occurs in non-helical regions or in the first 1–3 residues of a helix where the imino-nitrogen need not form hydrogen bonds.

The helix is constructed as follows: Prepare a number of aminoacids as described above but with the following modifications:

(*a*) Do not place any R groups in position but leave a bond rod projecting from arm c of C_2. This will simplify the formation of the helix and enable the general structure to be seen more easily. R groups can be added when the helix is complete if desired.

(*b*) Place a bond lock between the C and O atoms of the carboxyl group which were previously connected trigonally; this forms the O part of the hydrogen bonding system.

(*c*) Remove the OH from arm e of C_1 as this carbon bond will join the NH, arm f, of the next aminoacid to form the first peptide bond; both these arms are trigonal and are connected with bond locks.

Connect up the aminoacid residues in sequence as just described leaving the amino group of the first aminoacid free. When the fifth aminoacid has been added, twist this part of the model clockwise so that a hydrogen bond can be inserted from the O of the first aminoacid into the N of the fifth. To accomplish this, a certain amount of rotation around bonds is necessary. However such rotation can occur only around the two tetrahedral bonds of carbon C_2 as described above. The hydrogen bond will be found to run roughly parallel to the long axis of the helix. Continue building through to the ninth aminoacid when the general a-helix properties will become evident. The carboxyl of this last aminoacid will be left as COOH.

Peptides and proteins. Construction of peptide and protein chains may start at either the N- or C-terminal ends; the α-helix described above started at the N-terminal end. The C-terminal octapeptide of whale myoglobin is shown in Fig. 16.17(*b*) and may be constructed on principles similar to those described above. Wherever helical portions occur, hydrogen bonds should be inserted to help support the structure.

Mono- and poly-saccharides and their derivatives. Pentoses may be constructed by reference to Fig. 16.18. Hexoses are constructed similarly but with the addition of one further carbon Atomunit in the ring. Di- and poly-saccharides are built up from monomer residues. Similarly, phospho derivatives may be formed as shown in the figure.

Purines and pyrimidines. These may be constructed as shown in Fig. 16.18. Correct base pairs can be hydrogen bonded (see Table 16.1) to form a completely planar model.

Mono- and poly-nucleotides. Mononucleotides can be constructed as shown in Figures 16.18 and 16.20. Practice in building AMP, ATP, UDPG, etc., is valuable as a preliminary to building such large molecules as DNA or RNA helices.

The double helix can be approached in a number of ways. Thus long backbone chains of sugar-phosphate repeating units may be built first and lined up in an anti-parallel fashion. Then base pairs are linked to their respective carbon Atomunits in the backbone with gradual turning of the helix. Because the whole molecule is so lightweight, the backbones need only be supported from above, thus leaving the structure free for working at any point. A complete double helix may be supported in this way although a more complex model such as that representing a tertiary RNA may need additional superstructure to aid in the intermediate building stages. When models are complete, part or all of the structure may be clad with van der Waals' envelopes using the mushroom units shown in Figs. 16.14 and 16.19.

Construction of small molecules. A number of simple molecules are illustrated in Figs. 16.22 and 16.23. For proteins and nucleic acids, relatively few double and no triple covalent bonds are required. Indeed, where double bonds are needed, these are best represented by means of bond-locks as the molecules concerned are planar and are held rigid in this way with the minimum of spatial requirement.

However, the small molecules require an alternative approach as the main interest often centres around the bond structure as, for example, in demonstrating geometric isomerism. The visual interpretation of double and triple bonds is difficult, but by using Atomunits with curved bond rods such bonds can be given a more realistic arrangement. Such curved bond rods are of different sizes for double and triple bonds.

Construction of models using the curved bond rod requires the use of the tetrahedral arms of the Atomunit, e.g. C=C, whereas in conventional construction the connections are through the square trigonal with a bond lock restricting rotation. With the curved rods the tetrahedral arms are used and two bond rods would curve from one Atomunit to the other. Similarly C≡C would have three rods joining. Such connections will still leave the unattached arm in the correct position. C=C has four arms planar to accept, for example, four hydrogen markers to form ethylene and C≡C has two arms at 180° to form linear acetylene.

REFERENCES

Introductory Reading

Clowes, R., "The Structure of Life," 1967 Pelican Book.
Rose, S., "The Chemistry of Life," 1967 Pelican Book.
"The Cell," *Life Science Library*.
"The Double Helix," Watson, J. D. Weidenfeld and Nicolson, 1968.

Scientific American

Allfrey, September, 1961, "How Cells make Molecules."
Nirenberg, March, 1963, "Genetic Code II."
Holley, February, 1966, "Transfer RNA."
Crick, October, 1966, "Genetic Code III."
Benzer, January, 1962, "Fine Structure of the Gene."
Jacob, June, 1961, "Virus and Genes."
Neurath, December 1964, "Protein digesting Enzymes."
Phillips, November, 1966, "3-D Structure of an Enzyme Molecule."
Hanawalt and Haynes, February, 1967, "Repair of DNA."
Yanofsky, May, 1967, "Gene Structure and Protein Structure."
Porter, October, 1967, "Structure of Antibodies."
Clark and Marcker, January, 1968, "How Proteins Start.'
Merrifield, March, 1968, "Automatic Synthesis of Proteins."
Bragg, July, 1969 "X-ray Crystallography."

Bio-organic Chemistry—Readings from *Scientific American*, W. H. Freeman and Co., San Francisco and London, 1968.

More Advanced Texts

Mahler, H. R., and Cordes, E. H. "Basic Biological Chemistry," Harper, 1968.
"Molecular Biology," *Brit. Med. Bulletin*, September 1965, **21,** (3).
Ingram, V. M., "The Biosynthesis of Macromolecules," Benjamin, 1966.
Watson, J. D., "Molecular Biology of the Gene," Benjamin, 1970.
Bernhard, S., "Enzymes: Structure and Function," Benjamin, 1965.
Dickerson, R. E. and Geis, I. "The Structure and Action of Proteins." Harper and Row 1969.

Practical Techniques for Structure Elucidation

Feinberg, J. G., and Smith, I., "Paper and Thin Layer Chromatography and Electrophoresis," Shandon Southern Instruments, Camberley, Surrey and Sewickley, Pa.; Longman, London.
Smith, I., Ed., "Chromatographic and Electrophoretic Techniques," 3rd edn.
Vol. 1, "Chromatography," (Chaps. 1, 2, 6, 7, and 39), 1969;
Vol. 2, "Zone Electrophoresis," (Chaps. 1, 5, 8), 1968,
Heinemann Medical Books, London and John Wiley, New York.

1. Dayhoff, M. O. and Eck, R. V., "Atlas of Protein Sequence and Structure," 5nd Edn., 1972, National Biomedical Research Foundation, Silver Spring, Maryland, 20901.
2. Smith, I., Smith, M. J. and Roberts, L., *J. Chem. Educ.*, 1970, **47,** 302.
3. X-ray coordinates are supplied with each special Biobit kit. For specific data see the following references:
   Chymotrypsin
   Biochem. Biophys. Res. Comm., 1969, **36,** 131.

Toward light source
KEEP
this line parallel to
the bottom of stereo picture.
Fold
HOW TO USE
1. Place viewer with open end toward light source. Both stereo pictures must be illuminated equally well, with no shadow on them from the viewer.
2. Make sure the bottom edge of the stereo picture is parallel to the line between the lenses. (Some defects of vision may necessitate a slight twist off the parallel.)
3. If you wear glasses, use them while viewing.
4. In some cases, focusing may be improved by raising or lowering the viewer.
5. Give your eyes fair amount of time to fuse the stereo pictures; only 8% of people cannot see stereo.
Fold
STEREO VIEWER
taylor-merchant
corporation
New York 36, N.Y.
VIEW FROM THIS SIDE
INSERT SLIDE HERE
APPLY PRESSURE HERE FOR FOCUS
Taylor
Merchant
stereo
Viewer

Phil. Trans. Roy. Soc. Lond., 1970. **B257,** 67.
Biochem. J., 1969, **112,** 261.
Ribonuclease-S
J. Biol. Chem., 1970, **24,** 308.
Myoglobin
Prog in Stereochem., 1969, Vol. 14, 299.
Hemoglobin
Harvey Lectures, 1967/8, **63**, 213.
Elastase
Phil. Trans. Roy. Soc. Lond., 1970, **B257**, 111
Insulin
Nature, 1969, **224**, 491
Lactic dehydrogenase, Malate dehydrogenase, etc.
J. Mol. Biol., 1973, **76,** 533.z
Carboxypeptidase A
Phil. Trans. Proc. Roy. Soc. London., 1970, **B157**, 177
Papain, *ibid*, 231 *Adv. Prot. Chem.*, 1971, **25,** 1.
Rubredoxin
J. Molec. Biol., 1970, **50**, 391

4. Tables of Interatomic Distances and Configurations in Molecules and Ions. Special Publication No. 11, Chemical Society, London.

Trypsin, *J. Mol. Biol.*, 1974, **83,** 185–208.
Carbonic Anhydrase C., *Nature New Biology*, 1972, **235,**
Cytochrome C., *J. Biol. Chem.*, 1971, **246,** 1511.
Cytochrome b_2. *Nature New Biol.*, 1971, **233,** 15.
Cytochrome C_5. *Biochem. Biophys. Res. Comm.*, 1973, **54,** 47.
Immune globulin Ig, *Proc. Nat. Acad. Sci.*
Immune globulin Fab, *Proc. Nat. Acad. Sci.*, 1973, **70,** 3,305.
Bence-Jones, *Protein, Biochem.*, 1973, **12,** 4,620
Phosphoglycerate mutase, *Nature*, 1974, **250,** 301.
Subtilisin, *Biochem. Biophys. Res. Comm.*, 1971, **45,** 337.
Staph Nuclease *J. Biol. Chem.*, 1971, **246,** 2,302.

Stereoviewers for printed page (upper) or slides (lower) are shown on page 454. Both are of lay-flat cardboard design with instructions for use printed on each viewer. (Available from Capital Biotechnic Developments).

Model Kits and Equipment Available for the Construction of Molecular Models

(A) Biobits kits

General Protein Kit
Special Protein Kit —Lysozyme
—Myoglobin
—Ribonuclease
—Chymotrypsin
—Insulin
—Hemoglobin

All special kits include backbone tube, R groups, support frame, X-ray coordinates and other data. Other special kits are, and will become, available for the construction of further specific protein structures as these are determined.

General DNA-RNA Kit

Special kits for construction of specific t-RNAs, 5S-RNA, etc., are available.

(B) Atomunits Kits

Atomunit Kit—General Atomic Model Kit
Atomunit Protein Permanent Demonstration Kit
Atomunit DNA Permanent Demonstration Kit
Atomunit 'Mushroom' Conversion Kit—skeletal to space-filled models

Permanent demonstration kits illustrate α-helixes, pleated sheets and double helix base pairs. They are provided for display during special teaching courses, and for general study purposes.

(C) Ancillary Equipment

(i) Special support frames for large models.
Biobit 3D support frame ($60 \times 60 \times 30$ cm).
Atomunit 3D support frame ($90 \times 90 \times 90$ cm).
Miniframe—a special easily constructed and rapidly demounted frame for smaller displays of models. Also very valuable as large-card supports for other demonstration purposes. Very light-weight, low-space, transportable kit.

(ii) Bulk packs of Biobit R groups, bases, colour-coded tubing numbered every tenth residue, atomunits and all other components of these kits. Various cutter tools, Mushrooms.

(iii) Copies of this manual as singles or packs

(iv) Stereoviewers for the printed page and for slides.

Full information on all the above kits and equipment, with detailed price lists, available on request.

CHAPTER 17

ELECTROPHORESIS OF RNA IN POLYACRYLAMIDE AND AGAROSE GELS

W. B. Gratzer and E. G. Richards

Factors affecting the mobility.

The mobility of RNA in free solution varies little with the molecular weight;[1] this is reckoned to be because the electrical charge and the frictional force originating from the viscosity of the solvent are roughly proportional to one another. In gels, agarose and polyacrylamide in particular, an additional factor is introduced, namely the molecular sieving effect of the gel. The result of this is that larger molecules are retarded relative to smaller ones, which in low concentration gels may travel with a mobility close to that in free solution. This retardation depends on the total concentration (T% w/w) of monomer in the gel, and to a lesser extent on the number of cross-links which in the case of polyacrylamide gels depends on the proportion of BIS to acrylamide in the monomer mixture (C% w/w).

Effect of Gel Concentration and Composition.

For polyacrylamide gels of constant proportion of BIS (constant C), both high and low-molecular weight RNA follow the Ferguson relation.[2,3] This is illustrated in Fig. 17.1 which shows linear plots of the

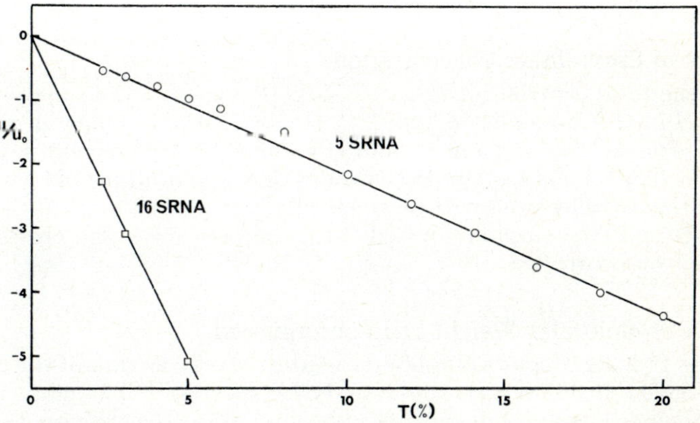

FIG. 17.1. Ferguson plots of natural logarithms of relative mobility (u/u_0) against total polyacrylamide gel concentration (T% w/v) for 5S and 16S RNA from *E. coli*.

natural logarithm of the mobility, u, (or distance moved in a unit time under unit potential gradient) against the gel concentration T, for 5S RNA and for 16S ribosomal RNA.

These plots imply a relation between u and T which may be expressed as:

$$\ln u = \ln u_0 - bT$$

where u_0 is the mobility extrapolated to zero gel concentration. This is found to be close to the value measured in free solution. The constant b is called the retardation coefficient, and is found to depend primarily on the molecular weight of the RNA, but also on its conformation as determined by its secondary and tertiary structure.[4] The Ferguson relation is also followed in the electrophoresis of RNA in agarose gels,[5] as shows in Fig. 17.2.

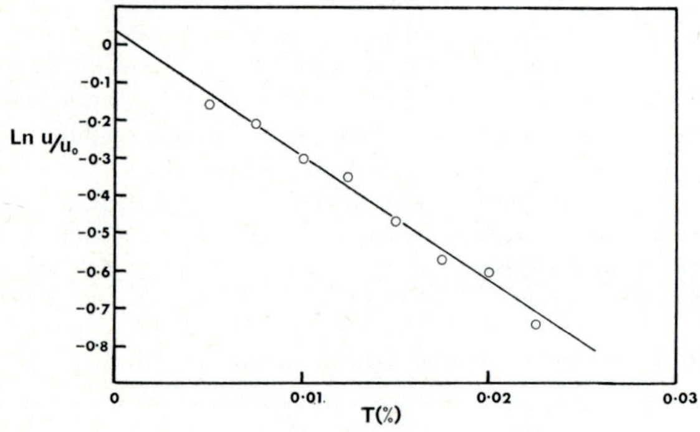

FIG. 17.2. Ferguson plots of natural logarithms of relative mobility (u/u_0) against total agar gel concentration (T% w/v). Data taken from.[5]

Effect of Cross-linker Concentration.

If the total gel concentration, T, is kept constant and the proportion of BIS in the monomer mixture, C, is varied, the mobility of an RNA species passes through a minimum at about $C = 5$–6 per cent depending on the value of T selected. It is usually best to maintain C at 5 per cent, since this produces gels with optimal physical properties,[6] the swelling and turbidity being minimised. In agarose gels, the cross-linking cannot be controlled.

Effect of Molecular Weight and Conformation.

Fig. 17.3 shows plots of mobility against the logarithm of the molecular weight, M, for several different RNA species.[7] The plots are seen to be roughly linear, but it cannot be emphasized too strongly that each plot is linear only over a limited range of M. If such plots are extrapolated beyond their range of validity, they lead to the absurd implications that RNA species of sufficiently high molecular weight would have negative mobilities, and that RNA species of sufficiently low molecular weight would have mobilities higher than that in free solution. Instead we would expect, and find in practice, that the plots deviate from linearity at both ends in the manner indicated by the dotted lines shown in Fig. 17.3.

If these limitations are borne in mind such plots provide useful calibration curves for determining in a semi-quantitative manner the molecular weight of new RNA species. It is convenient to load a gel with several RNA's of known molecular weight and to plot log M against the distance moved in a give time. As long as the molecular weights of the standards straddle that of the unknown, the latter may be safely estimated.

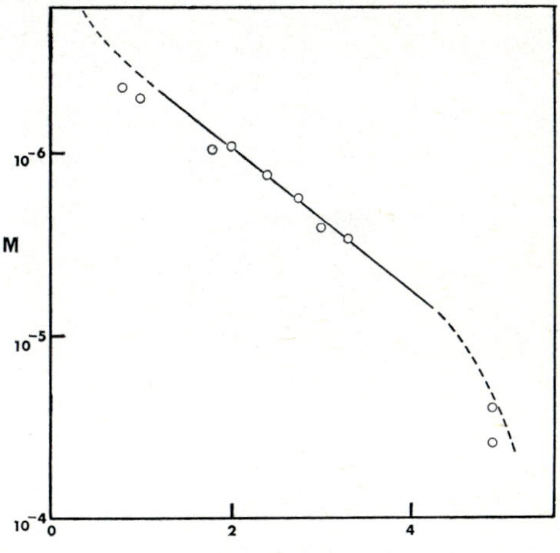

FIG. 17.3. Plot of logarithm of molecular weight of various RNA species against relative mobility in 2·25 per cent polyacrylamide gel. Data taken from.[7]

Close examination of plots of this kind however reveals that the points do not always lie on a smooth curve, and that deviations occur outside the limits of experimental error. The deviation is explained by the observation that the mobility depends on conformation as well as molecular weight. This is dramatically illustrated in Fig. 17.4 which shows a pattern obtained from a mixture of native and denatured 5S RNA in a 16 per cent gel. Two conformational isomers are clearly resolved. There is some evidence that these deviations can be reduced by plotting the Stokes radius (r) against the mobility.[13] The former quantity can be estimated from the molecular weight and sedimentation coefficient (s) according to the relation:

$$r = \frac{s\,(1 - \bar{v}\rho)}{M\eta}$$

where $\bar{v}$ is the partial specific volume and ρ and η the density and viscosity of the solvent which should be the same in the determination of the sedimentation coefficient, as in the electrophoresis. In so far as high-molecular weight RNA behaves as a random coil,[8] we would expect the Stokes radius to be roughly proportional to the square-root of the molecular weight. In practice this method of presenting molecular weight calibration curves is open to the objection that it is not based firmly on well-founded theoretical principles, and also that it does not overcome the conformation effect. A more useful stratagem is to destroy

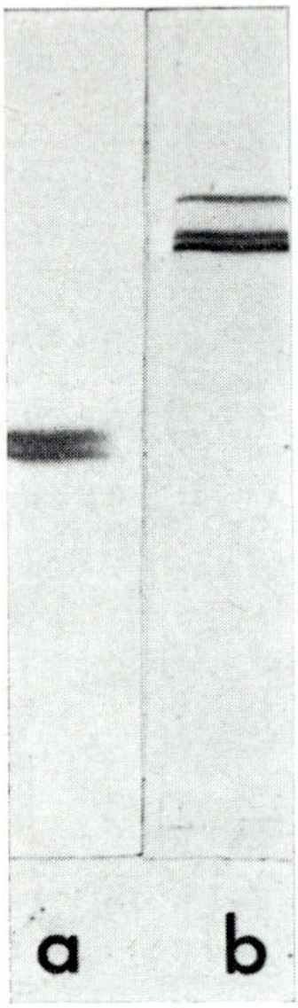

Fig. 17.4. Electrophoretic pattern obtained with partially denatured 5S RNA in *a*) 10% *b*) 16% polyacrylamide gels.

all secondary structure by running the gels in a denaturing solvent. The most satisfactory method is probably to work in formamide instead of water and Fig. 17.5. shows a plot of log M versus mobility in this solvent for several species of RNA and several fractionated synthetic polynucleotides. A smooth curve is obtained, though this also approximates to a straight line only over limited molecular weight range, for the same reasons as have been discussed above.

Technique for Aqueous Gels

Several options are open to the worker intending to examine RNA by gel electrophoresis.

Polyacrylamide or Agarose Gel Concentration.

Polyacrylamide gels cannot be prepared at a total monomer concentration (T) of less than about 2·2 per cent. Above this value the choice of gel concentration depends on the molecular weight range of

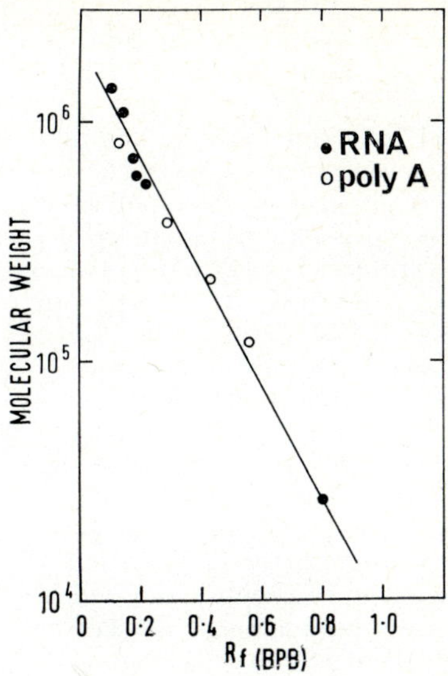

FIG. 17.5. Relation between electrophoretic mobility (relative to bromophenol blue) and molecular weight for RNA species and polyriboadenylic acid fractions in formamide-polyacrylamide gel.[16]

the RNA species it is desired to separate. A guide to this choice is provided in Table 17.1 which gives the appropriate gel concentration

TABLE 17.1

Approximate concentration of monomer ($T\%$ w/v) in polyacrylamide gels (with $C = 5\%$) required for 10 per cent mobility differences between RNA species differing in molecular weight by 10 per cent.

M	$T(\% w/v)$
10^4	28
3×10^4	12
10^5	6·6
3×10^4	4·2
10^6	2·9
3×10^6	2·2

to resolve RNA species differing in molecular weight by 10 per cent. This guide should be used with caution and it may be found necessary to perform preliminary experiments with several gel concentrations, to establish which one gives optimal resolution for the species in question. It is in general difficult to resolve both very high and very low-molecular weight species in the same gel. This can be overcome to some extent by the use of gradient gels.[9] For very high-molecular weight RNA (some viral species etc), 2·5 per cent acrylamide gels are too dense to permit the entry of the RNA. This difficulty may to some extent be overcome by reducing the BIS concentration from the normal 5 per cent of the total monomer to a lower value – 1 per cent say.

Even this method may fail for very high-molecular weight species and it is then best to use agar or agarose gels. The use of composite gels containing a mixture of agarose and polyacrylamide, has also been described.[10] Agarose gels can be made up at concentrations in the range ca. 0·3–4 per cent and behave like polyacrylamide as far as their effect on mobility is concerned. It is important to use good quality agarose (e.g. Seakem, British Drug House Ltd.) or agar, (Koch-Light Ionagar II) which possesses few ionic groups that would otherwise give rise to severe electroendosmotic flow effects.

Flat-bed or cylindrical gels.

The use of agarose is generally allied to the use of flat-bed gels and these are conveniently prepared and used in apparatuses of the type described by Akroyd (see chapter 10) and by DeWachter and Fiers.[11] The gel in this apparatus is about 1 mm. thick. A similar apparatus can be used for polyacrylamide gels. Thicker flat-bed gels can also be made by casting a block, of gel in a suitable mould or they can be obtained prepared commercially (e.g. Gradipore apparatus of Universal Scientific Co.; see chapters 10 and 12).

The advantage of flat-bed gels is that several samples can be run on the same gel to facilitate comparison of mobilities. Polyacrylamide gels can also be cast in cylindrical tubes. The material may be glass (or quartz) or perspex (plexiglass), and it will be found easier to remove the gel from the tube if this is made of precision bore tubing. Tubes of 5 mm. internal diameter are convenient for most purposes though other diameters may be used (see chapers 9–12). For microelectrophoresis, capillaries are used, and for details of such techniques the book by Neuhoff[12] should be consulted. The advantage of cylindrical gels lies in the simplicity of the apparatus required and in fact that the RNA is contained within the bounds of the tube. This is important in quantitative work, for in flat-bed gels the RNA concentration at the centre of a zone is to some extent diminished by sideways diffusion and other effects. Cylindrical gels have as a disadvantage the inconvenience of removing the gels from the tubes, particularly when the acrylamide concentration is high, but there are means of mitigating this difficulty.

Choice of Buffer.

In theory any buffer at any *p*H and at any concentration can be used for polyacrylamide gel electrophoresis of RNA. In practice the ionic strength of the buffer should not be too low, and 0·05 ionic strength is a practical lower limit. At very high buffer concentration the time taken for the separation becomes inconveniently long and resolution is impaired. For optimal resolution, 0·05 ionic strength buffer should be used.

We give in Table 17.2 a recipe for a gel at *p*H 7·8, suitable for RNA. The buffer concentration is the same in the gels as in the reservoir.

TABLE 17.2

Recipes for gel and reservoir solution for use with RNA at pH 7·5. These are given for 10 per cent gels, but the proportion of Cyanogum may be varied to give any desired gel concentration. To prepare gels, gel stock solution and ammonium persulphate solution should be mixed rapidly in the volume proportion 20:1 and quickly poured into the mould. The setting time should be 10–20 minutes. If it is too rapid, decrease the concentration of ammonium persulphate, and if too slow, increase it.

Gel stock solution (This will keep in the cold for at least a fortnight).

Tris	0·92 g.
0·1 M HCl	50 ml.
TEMED	0·25 ml.
Cyanogum 41	10·5 g.
Water	to 100 ml.

Ammonium persulphate (This should be made fresh each day)

Ammonium persulphate	0·35 g.
Water	to 25 ml.

Reservoir buffer

Tris	18·4 g.
M HCl	100 ml.
Water	to 2,000 ml.

Recipes for buffers of low *p*H found elsewhere in this book may also be used. Recipes for formamide containing gels are discussed below.

Polymerisation.

The most reproducible polymerisation catalysts are TEMED (tetramethylethylenediamine) or DMAPN (dimethylaminopropionitrile) with ammonium persulphate (APS) as initiator. TEMED is incorporated in the recipes listed in table 17.2. To initiate polymerisation the stock gel solution is mixed with a small volume of 10 per cent ammonium persulphate so that the final concentration is about 0·7 mg./ml. The optimum ammonium persulphate in the final mixture must be found by trial and error, and should result in a setting time of 10–15 minutes. Gels which set faster are difficult to manipulate and gels that set slower have irreproducible properties, and may be inhomogeneous. The optimum concentration of APS depends also on the gel concentration, T.

During the polymerisation process, air, which acts as an inhibitor must be rigorously excluded. This is accomplished in flatbed gels by covering the mould with a glass plate and in cylindrical gels by carefully layering water on top of the denser gel mixture. A simple device for facilitating this has been described.[13]

Some authors have recommended the recrystallisation of the acrylamide and BIS before use.[14] We have found this to make no practical difference and to be unnecessary. (See below for effects in ultraviolet scanning of gels). Indeed for gels to contain 5 per cent BIS, the commercially available Cyanogum 41 is adequate for most purposes.

Application of sample.

The RNA sample, preferably dissolved in water or other low ionic strength solvent, is mixed with an equal volume of 10 per cent sucrose containing a little bromophenol blue as tracker dye. This is then layered on top of the gel under a layer of reservoir buffer. This is conveniently done with the aid of an Hamilton syringe, which allows the application of loads of closely controlled volumes with minimal disturbance and mixing of the heavy sucrose solution with the reservoir buffer. A convenient volume to load on a 5 mm. diameter cylindrical gel is 40 μl. though volumes as large as 0·2 ml. can be used with only slight impairment of the resolution. The quantity of RNA in the load should be such that there is not more than 1 μg. of the components it is wished to observe in a 5 mm. gel. Smaller quantities down to 0·01 μg. can be detected with moderate ease, but the use of larger quantities leads to a gross deterioration of the resolution.(13) 1 μg. in 40 μl. gives an RNA concentration that may conveniently be measured by optical absorbance methods.

The use of spacer gels and discontinuous buffer systems has been described(15) but other than in exceptional circumstances (e.g. very dilute samples), offers no practical advantages, but rather a considerable complication of the method.

The reservoir buffer.

The volume of the two reservoirs should be as large as conveniently possible, and it is advantageous for vertical gels, whether flat bed or cylindrical, to be totally immersed in the buffer. This helps to dissipate heat generated by the ohmic resistance of the current-carrying gels. For the same reason glass tubes or parallel plates to hold the gel are to be preferred to plastic, since the latter has a much smaller thermal conductivity. For very long runs it is desirable to arrange for continuous mixing of the buffers in the anodic and cathodic reservoirs to prevent *p*H changes that may otherwise occur in consequence of liberation of electrode products. This may be satisfactorily accomplished by the aid of a small non-metallic aquarium pump that, can be made to pump the lower reservoir solution into the upper container, the latter being equipped with an overflow to allow the liquid to drain back into the lower compartment. Alternatively external electrodes as described below may be used. Generally the nature of the buffer and its concentration should be the same in both reservoirs as in the gel itself.

Applying the potential.

A current of about 5 mA per 5 mm. cylindrical gel is required, at a potential of 50–200 V depending on the buffer used. This may be obtained from commercially available power supplies. A constant-current power supply is preferable but not essential, except for the most precise applications.

The potential is applied by way of electrodes dipping into the two reservoir solutions. These may be made of platinum wire or of carbon rods. Electrode processes taking place in such a system lead to the liberation of acid in the cathodic compartment and alkali in the anodic compartment. In prolonged runs or with weakly buffered reservoir solutions this may lead to undesirable *p*H changes in the gel itself (possibly resulting in the precipitation of the RNA) and buffer circulation (see above) or reversible electrodes (see below) must be used.

The use of currents in excess of 5 mA per tube may lead to undesirable distortion of the zones by dint of heating effects and should be avoided even though the separation is more rapid. An exception to this is in preparative runs using much higher electrolyte concentrations when much higher currents must be used.

External reversible elecrtodes.

We give here the construction of efficient reversible electrodes, since they are desirable when problems arise concerning adequate buffering of aqueous gels, or when it is necessary to run electrophoreses for very long periods. Fig. 17.6 shows a mercury-calomel reversible electrode.

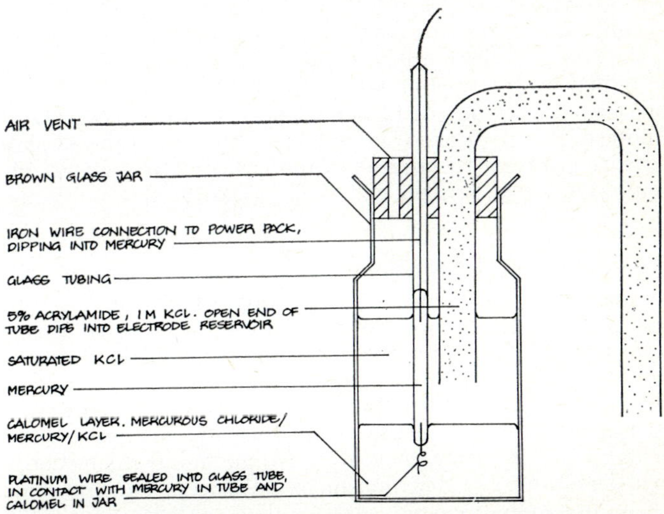

Fig. 17.6. Reversible calomel electrode. The vessel is a 250 ml. wide-mouthed reagent bottle. For description, see text. (From J. E. Pinder, Ph.D. thesis, University of London, 1974.).

The principle is that mercury is converted to calomel at one electrode, and calomel reduced to mercury at the other. After each run, the anode and cathode leads are reversed, and the electrodes will then maintain themselves for long periods.

The vessel is a 250 ml. wide-mouthed reagent bottle. The rubber stopper is drilled for a 1·5 cm. glass U-tube as a bridge to the reservoir, a narrower tube (0·6 cm. o.d.), reaching to the bottom of the bottle, and a vent for escape of electrode gases. A short length of fairly thick platinum wire is sealed into the bottom of the narrow tube, which should be made of soda glass. The procedure for this operation is first to heat a length of tubing in a flame and pull it out to make a thin-walled segment (like a Pasteur pipette). The platinum wire is threaded through about a 1 cm. length of this thin tube, which is then melted in the flame to make a bead on the wire. A suitable length of the soft-glass tubing is then prepared and the end melted with rotation in the flame and allowed nearly to close. The platinum wire held with tweezers is introduced through the small hole, still held in the flame, and the glass bead allowed to fuse into the tubing. After cooling, the end of the tube is best annealed in a very cool flame. The entire process is accomplished in a few moments. The tubes are pushed through the holes in the stopper with the aid of a little glycerol as lubricant.

To prepare the mercury-calomel paste, mercury and mercurous chloride are amalgamated by stirring, preferably in a mortar with a little saturated potassium chloride solution. The paste should be about 2 cm. deep, and it is important that the platinum wire penetrate below its surface. The bottle is then filled with saturated potassium chloride solution. Contact to the power lead is made by a wire (made of iron rather than copper), thrust well into the narrow tube, dipping into mercury in the bottom. The stopper and the top of the bottle should be well greased with Vaseline to eliminate "creep". The bridge is prepared by inverting the U-tube and setting in it a 5 per cent aqueous acrylamide gel in 1 M potassium chloride. When not in use the end of the gel should be covered with parafilm to prevent evaporation. After several runs the acrylamide gel in the bridges will begin to deteriorate, and has to be replaced.

Two-dimensional gel electrophoresis

Very spectacular fractionation of complex mixtures can often be obtained by the use of a two-dimensional technique, with a change in running conditions from the first to the perpendicular direction. One may for example, change from an aqueous neutral system to one at acid *p*H, at which some of the bases become protonated and the charge of the components in the mixture is differentially changed; or one may use a denaturing, or partly denaturing regime in one dimension, which again will generate a new distribution of mobilities. The first dimension may be run in a cylindrical gel, which is then applied transversely onto a flat gel, or both electrophoreses may be performed in the flat gel. We shall give here the details of a very satisfactory procedure, which makes use of a simple improvised apparatus, and follows closely the method evolved by DeWachter and Fiers.[11] It is desirable to operate at a higher acrylamide concentration in the second dimension to allow some sharpening to occur at the interface of the first and second gels.

The gel is cast between two glass plates, 2 mm, thick, 40 × 20 cm. for the first dimension, and 30 × 25 cm. for the second. The sides are sealed with tape around the perspex strip spacers (see Chapter 10 this vol.) and the cell is embedded in plasticine. A little of the acrylamide solution is poured in to a depth of about 1 cm., and allowed to set. This seals the bottom of the cell, and the gel layer may now be prepared by pouring in the desired acrylamide solution (e.g. 10 per cent for chain lengths of up to about 100 nucleotides in a mixture) in 0·025 M citric acid, 6 M urea. The solution is layered with water and allowed to set. The sample is then applied, either across the entire gel, or in separate slots, if slot-formers are used. The sample should be made up in 6 M urea, containing added sucrose and tracker dye. The separation in this system is best carried out in the cold. After electrophoresis, the strip of gel bearing the sample is cut out. This strip should be not more than about 5 mm. wide. The slicing should be done with a razor blade, using a ruler as a guide, since the edges must be straight. The strip is placed on a glass plate, which will form one of the walls of the electrophoresis cell for the second dimension, about 4 cm from the bottom. The spacers and the second plate are then added, the assembled cell placed upright, and the new gel, which may be 20 per cent acrylamide in a *p*H 8 buffer system (e.g. 0·04 M Tris-citrate, *p*H 8, as used by DeWachter and Fiers) is cast between the plates as before. The sample is run upwards, the urea remaining near the bottom of the gel. The change in the colour of the bromphenol blue from yellow to purple registers the transition in the buffer environment experienced by the RNA. After electrophoresis, staining or autoradiography ensue as before.

Formamide gels

The details of the formamide technique have been discussed by Pinder *et al.*[16] It has been shown[17] that single-stranded RNA species are completely denatured in formamide at room temperatures, both base-pairing and single-stranded stacking being sensibly eliminated. Double-stranded RNA is not melted unless the solution is heated. Melting ensues at about 40° (depending no doubt on base composition) and when the solution is cooled the RNA remains in a structureless state indefinitely. It has not been proved that very GC-rich double-stranded RNAs also conform with this behaviour, but all species so far examined do melt completely and apparently obey the same molecular weight-mobility law.

Choice of gel concentration.

Formamide gels are more porous than aqueous gels. A 4 per cent gel for example is appropriate for the molecular weight range of RNA for which a 2·5 per cent aqueous gel would normally be used. Thus 3–4 per cent gels are suitable for fractionation of high-molecular weight materials. For fractionation of tRNA fragments in formamide Staynov *et al.*[18] used 20 per cent acrylamide. At the same time the mechanical properties of the gels resemble those of aqueous gels of the same acrylamide concentration.

Pre-treatment of formamide.

As received from chemical suppliers, formamide contains a high concentration of ionic impurities, and gives an alkaline reaction. Such crude formamide will not satisfactorily support the polymerisation of acrylamide, and must first be subjected to a purifying treatment. Coloured batches of solvent are best discarded.

2 g. of coarse mesh mixed-bed ion-exchange resin, Amberlite Monobed, MB-1 (British Drug Houses, Ltd.,) is added to 50 ml. formamide and vigorously stirred for 2 hr. For reassurance it is best to follow the decrease in conductivity during this period (Fig. 17.7). It will drop from a typical starting value of 400 μmho to some 5 μmho. The accompanying drop in apparent *p*H may alternatively be followed. The resin is

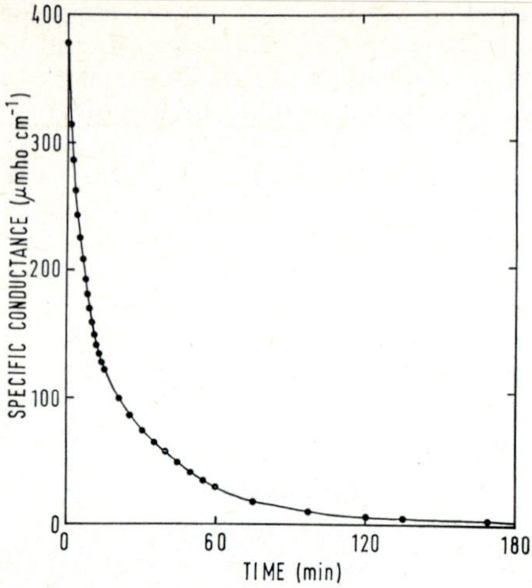

FIG. 17.7. Removal of ionic impurities from formamide. The graph shows the rate of diminution of the conductivity during treatment with a mixed-bed ion-exchange resin.(17)

then filtered off. One should not allow the solvent to stand for long periods over the resin, since impurities are progressively extracted. The formamide is best kept in the dark, and will keep without much deterioration for 3–4 days unbuffered, and is stable for much longer periods after buffering (see below).

Buffer systems and preparation of gels.

Early experiments were carried out in unbuffered sodium chloride solutions. (Staynov *et al.*, 1972), and this is quite satisfactory if external electrodes are used. However, on general grounds a buffered system is to be preferred. It must not be assumed that buffer ions will function in the same range in formamide as in water, and by no means all buffer systems are suitable. A 0·02 M barbital buffer (Staynov *et al.*, 1972) is satisfactory. For a standard eight-tube disc-gel apparatus (Chapter 9), the following quantities are satisfactory. For 4 per cent acrylamide gels, dissolve in about 20 ml. deionised formamide:

0·85 g. acrylamide
0·15 g. Bis
0·092 g. barbital (diethylbarbituric acid)
0·06 ml. TEMED

Adjust the apparent *p*H (with the standard glass and calomel electrode pair) to 9·0 with small volumes of 1N HCl, with vigorous magnetic stirring, to ensure rapid mixing and so prevent local decomposition of formamide. This is then made up to 25 ml. To this is added 0·2 ml. of fresh, 10 per cent (w/v) aqueous ammonium persulphate, and the solution is poured at once (without degassing) into the prepared glass tubes

and layered with 70 per cent (w/v) formamide-water from a hypodermic syringe. Setting usually takes about 30 mins.

The liquid is then poured off, and replaced with buffered formamide, prepared by dissolving 0·092 g. diethylbarbituric acid in deionised formamide, adjusting to apparent *p*H 9·0 with 1N sodium hydroxide with magnetic stirring as before, and making up to 25 ml.

Samples.

Ethanol-precipitates of RNA dissolve readily, lyophilised samples more slowly, in the buffered formamide; 5 per cent sucrose and a trace of bromphenol blue are incorporated, and the samples, routinely 10 μl. of 1 mg./ml. RNA, are layered on the gel surfaces.

Running conditions.

For economy, and because there is no detectable disadvantage, the reservoir solutions are 0·02 M aqueous sodium chloride. For reasons not so far understood, the use of aqueous barbital buffer in the reservoirs leads to distortion of zones. Contamination with electrode products, and large changes of *p*H in the gels is avoided by (i) ensuring that the runs are of sufficiently short duration; (ii) circulating buffer between the upper and lower reservoirs with a pump, as in some commercial apparatuses; or best of all (iii) using external elecrodes. The gels are run for about 1 hour at 5 mA/ tube (constant current), or until the tracker dye has nearly reached the bottom of the gels.

Staining and densitometry.

After electrophoresis, the formamide gels are removed from their tubes in the usual way, and stained overnight in any of the usual stains, e.g. 0·1 per cent pyronine Y in 0·5 per cent acetic acid. The gels are then destained in 10 per cent acetic acid. For direct denistometry in the ultraviolet, the gels are equilibrated with 1 per cent lanthanum acetate in 5 per cent acetic acid (overnight say, with one change of solvent thereafter). They may then be scanned at 260 nm. The formamide, which cuts off above 260 nm, is extracted by this process. Formamide-containing gels can be scanned at longer wavelengths, with loss of sensitivity. The lower working wavelength limit is probably about 280 nm.

Results of formamide gel electrophoresis: molecular weight-mobility relations.

Formamide gels give excellent results, with, we find, rather sharper zones and better resolution than aqueous. A typical gel is shown in Fig. 17.8. The results bear out the expectation that all RNA species in this destructuring medium are hydrodynamic homologues, and therefore adhere to a single molecular weight-mobility law, as indeed do the synthetic polymers, polyriboadenylic acid and polyribouridylic acid. This is shown in Fig. 17.3., the data of which refer to unbuffered gels. A disadvantage of formamide gels is that mobilities are not as reproducible from run to run as in aqueous systems, or even from tube to tube. For the most precise results therefore, it is best to run molecular weight markers in the same gel as the unknown, the mixture of markers being also run in another gel so that the zone(s) of unknown material can be identified. Naturally any nicks in the chain in regions in which there is normally base-pairing, cause the pieces to separate in formamide gels.

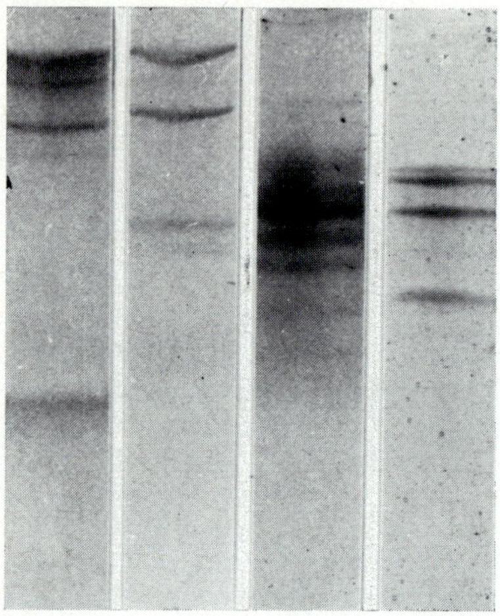

a b c d

FIG. 17.8. Typical electrophoresis patterns of RNA in formamide gels: (*a*) mixture of TMV, *E. coli* 23s, 16s and 5s rRNA; (*b*) rabbit reticulocyte RNA showing haemoglobin messenger bands; (*c*) slightly degraded rabbit reticulocyte rRNA, showing the appearance of "hidden" breaks; (*d*) bromegrass mosaic virus RNA.[16]

Detection and Quantitation of Gels.

The procedures given below were evolved for the processing of aqueous gel systems, but also apply, *mutatis mutandis*, to formamide gels.

The detection of the position of the zones of separated RNA and an estimate of the relative amount of RNA in each may be accomplished by direct scanning in the ultraviolet, or by staining the RNA. If on the other hand the RNA is radioactivitly labelled (best with ^{32}P), autoradiographic methods may be used.

Staining.

Several stains have been used to bind to the RNA in the zones, but Pyronine Y and Toluidine Blue are sensitive and convenient. Pyronine Y has the advantage for some applications that its absorbance maximum is close to 546 nm., the wavelength of the mercury green line.

The gels removed from the tubes, are immersed for 4–6 hours in a solution containing:

0·1% Pyronine Y
0·5% Acetic acid

Destaining should then be started without delay. If it is not there is a danger of progressive irreversible staining of the gel itself. The stained and washed gels are immersed in 0·5 per cent acetic acid for 24 hours, with periodic changes of solvent. An alternative procedure is to use one of the electrophoretic methods that have been described,[15] in a home-made or commercial apparatus. Gels that have been destained electrophoretically however are unsuitable for quantitative scanning, since the stain tends to be removed unevenly from different parts of the gel. Using the above methods, zones containing as little as 0·05 μg. can easily be detected by eye.

The binding of the dye to the RNA is a complex process that is not fully understood. For this reason one cannot be sure that the binding ratio is the same for all RNA species. Certainly it is less than one molecule of dye per nucleotide residue but the exact figure is hard to determine with precision.

Zones treated with Pyronine Y are effectively fixed as well as stained; in fact it is difficult if not impossible to remove the stain completely from the RNA or the stained RNA from the gels.

If an even smaller concentration of stain is used, so that the gel containing the dye is transparent, the stained zones of RNA are sometimes visible and may even be scanned without destaining. Prestaining is also possible, the coloured zones being detectable during the run if a little of the dye is added to the sample before the run; with some dyes the sensitivity of this process may be enhanced by making use of fluorescence e.g. viewing the gels under a "Minerolite" ultraviolet lamp (Woods' glass filter).

Scanning.

Gels that have been stained with pyronine Y and destained in this manner may be scanned at 546 nm., using a commercial gel-scanning attachment to a spectrophotometer. It is important that this should use a well collimated beam of monochromatic light, whose width should not exceed $\frac{1}{5}$ the width of the narrowest zone to be scanned. Usually this requires a beam width of 0·1 mm or less. Secondly, the height of the beam should not be greater than $\frac{1}{3}$ the diameter of the cylindrical gel. The latter should be placed in a rectangular cell (conveniently made of perspex), its width slightly less than the diameter of the gel. This ensures that the sides of the gel are slightly compressed to produce "flats".

If these precautions are taken, the area under the peak corresponding to a zone is proportional to the amount of RNA in that zone to an accuracy of about 5 per cent. Nevertheless the constant of proportionality may vary from one RNA species to another.

An alternative procedure is to scan the gels in the ultraviolet at 260–270 nm. without staining. This may be satisfactorily done even for 15 per cent gels. It is important to recognize however, that gels as ordinarily prepared contain ultraviolet-absorbing impurities. These may be removed by pre-running the gel for several hours before applying the sample. It will be found that an ultraviolet-absorbing material is removed by electrophoresis towards the cathode in the space of this time. Gels treated in this way are satisfactory transparent down to 255 mm. Re-crystallisation of the acrylamide is not therefore necessary.

Radioactive analyses.

There are two methods of dealing with radioactive zones in gels, namely autoradiography or extraction and counting. We will not discuss details here, since for the most part the procedures are the same as for proteins, and these are dealt with elsewhere in this book. One respect in which nucleic acids differ however is that they are commonly best labelled with ^{32}P, which is a high-energy β emitter, and thus very suitable for autoradiography. Flat gels are most convenient for the purpose: cylindrical gels must be sliced laterally to produce a flat surface. A device consisting of a frame into which the gel fits, and a slicer made up of three taut wires, has been described,(19) and gives good results. Whereas for autoradiography of ^{14}C-labelled samples the gel has to be dried down, the ^{32}P emission is able to penetrate the gel, and autoradiography can therefore be performed on the wet gel immediately after electrophoresis. The gel is covered with a layer of thin plastic film ("Saran" sandwich wrapping), and placed in contact with a sheet of X-ray film. For a strongly labelled sample an exposure of a few minutes to an hour should be sufficient. For application of autoradiography the location of RNA zones followed by extraction and sequencing, the standard sources on sequence techniques should be consulted.(20)

For counting, the gel must be sliced, and the slices transferred to planchettes for plaque counting, and extracted or chemically dissolved for scintillation counting (see Chapter 9). An apparatus which slices gels, ashes the slices and dispenses the combusted samples into scintillation vials for dissolving and counting is commercially available (Packard 306).

Preparative electrophoresis

Preparative acrylamide gel electrophoresis on a small scale may involve either extraction of the RNA from zones cut out of the gel, or continuous electrophoretic elution. The latter process can in principle be scaled up, but the problems inherent in the use of large masses of gel are familiar, and in general increased quantities can be bought only at the cost of diminished resolution.

Extraction of RNA from Zones

The first problem concerns the accurate location of zones. One method is to slice the gel into equal segments with one of the commercial or home-made models of slicer, and extract each one. This is time-consuming and better avoided, A second is to stain an identical gel, or preferably a strip from the same gel, and match this up (allowing for expansion in the stain) with the unstained gel. Autoradiography of ^{32}P–labelled RNA also gives rapidly and uniquely the positions of zones in the gel. Alternatively the RNA may be "prestained" with a fluorescent dye, such as acridine orange, and the zones picked out under

an ultraviolet lamp (Woods light). Extraction of stained zones is inefficient. A further approach is to scan the gel in the ultraviolet with a stainless steel wire laid across the gel, guessing at the correct location, until the position of the sharp deflection of the recorder pen indicates that the boundary of the required zone has been correctly marked. (Hamlyn and Martini, personal communication).

The zones, once isolated by slicing the gel with a razor blade can be extracted electrophoretically, or by diffusion. The latter method is the simplest. The slice is placed in 0·5 ml of 5 mM EDTA[21] and shaken overnight at 4°. Most of the RNA is extracted from a 2·4 per cent gel by this means, and a second extraction in the same volume gives a yield of up to 90 per cent. Heating at 100° makes the extraction much more rapid (say 1 hr.), but there is of course the likelihood of damage to the RNA. Some workers also include a denaturing agent, such as dimethylsulphoxide, in the extraction mixture. Another rapid method is to homogenise the slice with 1 ml. of buffer in a glass homogeniser. DeWachter and Fiers[11] use a 2 ml. plastic disposable syringe with a disc of filter paper (Whatman 52) in the bottom. The gel slice is placed on this filter paper and homogenised by vigorous rotation of the plunger in the bottom of the syringe. The plunger is then withdrawn carefully so as to leave the filter paper in place, and 0·5 ml. of 1 M sodium chloride is added. The syringe is agitated on a vortex mixer to bring the gel debris into suspension without disturbing the filter paper, and the solution is then gently expelled through the syringe nozzle. Sanger and Brownlee[20] have devised another strategy. The gel slice is homogenised in 10 ml. of 0·2 M sodium chloride, and the suspension is poured into a funnel containing a disc of Whatman DE81 ion-exchange paper. The gel is allowed to settle under gravity for some minutes, and the supernatant filtered under vacuum. The gel debris are rinsed off the filter paper with water. The paper is then dried on filter paper in a desiccator, and extracted with concentrated buffer (2 M triethylammonium carbonate, *p*H 10) in a centrifugal filter, made up of a conical tube into which the folded paper is placed, and which will fit, resting on the rim, into a glass centrifuge tube. Two or three centrifugal extractions with 0·1 ml. buffer extract most of the RNA. Alternatively the ion-exchange paper can be cut into a strip, sandwiched between two glass plates and the protruding end dipped in buffer. The buffer is drawn up by capillarity and into a capillary tube.

Electroelution techniques

Many apparatuses, including commercial designs are available. A simple arrangement that gives quite good results, though not of the highest resolution, is that of Popescu *et al.*[21] This is depicted in Fig. 17.9. The adaptor for a standard 0·6 mm. gel tube is the sawn-off end of a plastic 5 ml. disposable hypodermic syringe. Slots are cut in the plastic, and the glass tube is then readily pushed in. Plastic tubing, softened in a flame is fitted into or over the nozzle and buffer is flushed through it and and conical compartment below the glass tube with a peristaltic pump. The eluate is best monitored continuously with a micro-flow cell at 260 nm, or longer wavelengths if the RNA concentration is high.

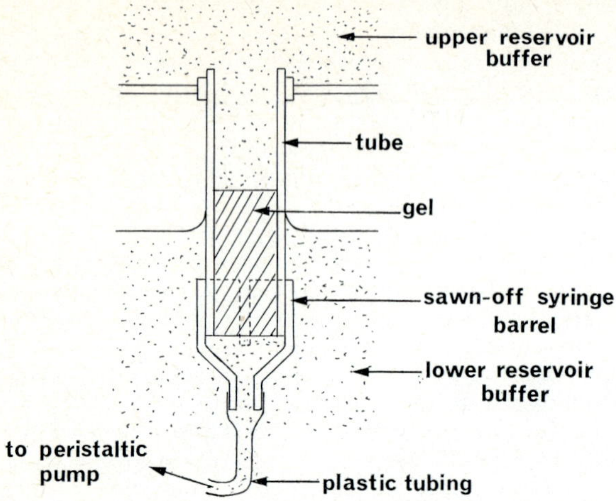

FIG. 17.9. Simple apparatus for small-scale preparative electrophoresis of RNA (after Popescu *et. al.*,).[22]

Electrophoresis is performed at 6 mA on a single tube at a pumping rate of 3 ml./h. Fractions of 0·5 ml. are collected, and several hundred μg. can be fractionated with good resolution. For a larger-scale fractionation in an instrument now commercially available, see Elson and Jovin,[22] or Chapter 15.

REFERENCES

1. Olivera, B. M., Baine, P., and Davidson, N. *Biopolymers*, 1964, **2**, 245.
2. Ferguson, K. A. *Metabolism*, 1964, **13**, 21.
3. Hedrick, J. L., and Smith, A. J. *Arch. Biochem.*, 1968, **126**, 155.
4. Richards, E. G., Lecanidou, R., and Geroch, M. E. *Eur. J. Biochem.*, 1973, **34**, 262.
5. Tsanev, R., Staynov, D., Kokileva, L., and Mladenova, I. *Anal. Biochem.*, 1969, **30**, 66.
6. Richards, E. G., and Temple, C. J. *Nature Phys. Sci.*, 1971, **230**, 92.
7. Bishop, D. H. L., Claybrook, N. R., and Spiegelman, S. *J. Mol. Biol.*, 1967, **26**, 372.
8. Boedtker, H. *J. Mol. Biol.*, 1960, **2**, 171.
9. Caton, J. R., and Goldstein, G. *Anal. Biochem.*, 1971, **42**, 14.
10. Peacock, A. C., and Dingman, C. W. *Biochemistry*, 1967, **6**, 1818; 1968, **7**, 668.
11. DeWachter, R., and Fiers, W. in *Methods in Enzymology*, 1971, **21**, 167; *Anal. Biochem.*, 1972, **49**, 184.
12. Neuhoff, V. "Micromethods in Molecular Biology", Springer-Verlag, Berlin, 1973.
13. Richards, E. G., and Lecanidou, R. *Anal. Biochem.*, 1971, **40**, 43.
14. Loening, U. E. *Biochem, J.*, 1967, **102**, 251.
15. Richards, E. G., Coll, J. A., and Gratzer, W. B. *Anal. Biochem.*, 1965, **12**, 452.
16. Pinder, J. C., Staynov, D. Z., and Gratzer, W. B. *Biochemistry*, 1974, **13**, 5367.
17. Pinder, J. C., Staynov, D. Z., and Gratzer, W. B. *Biochemistry*, 1974, **13**, 5373.
18. Staynov, D., Beltchev, B., Yaneva, M. ,and Gratzer, W. B. in "Electrophoresis and Isoelectric Focusing in Polyacrylamide Gel" (R. C. Allen and H. R. Maurer, Ed.)., de Gruyter, Berlin, 1974, p. 253.

19. Fairbanks, G., Levinthal, C., and Reeder, R. H. *Biochem. Biophys. Res. Commun.*, 1965, **20,** 343.
20. Brownlee, G. G. "Determination of Sequences in RNA", North-Holland, Amsterdam, 1972.
21. Popescu, L. M., Lazarus, L. H., and Goldblum, N. *Anal. Biochem.*, 1971, **40,** 247.
22. Elson, E., and Jovin, T. M. *Anal. Biochem.*, 1969. **27,** 193.

INDEX